T0296250

CAMBRIDGE LIBRARY COLLECTION

Books of enduring scholarly value

Physical Sciences

From ancient times, humans have tried to understand the workings of the world around them. The roots of modern physical science go back to the very earliest mechanical devices such as levers and rollers, the mixing of paints and dyes, and the importance of the heavenly bodies in early religious observance and navigation. The physical sciences as we know them today began to emerge as independent academic subjects during the early modern period, in the work of Newton and other 'natural philosophers', and numerous sub-disciplines developed during the centuries that followed. This part of the Cambridge Library Collection is devoted to landmark publications in this area which will be of interest to historians of science concerned with individual scientists, particular discoveries, and advances in scientific method, or with the establishment and development of scientific institutions around the world.

Reprint of Papers on Electrostatics and Magnetism

William Thomson, Baron Kelvin (1824–1907), born with a great talent for mathematics and physics, was educated at Glasgow and Cambridge. While only in his twenties, he was appointed to the University of Glasgow's Chair in Natural Philosophy, which he was to hold for over fifty years. He is best known for lending his name to the Kelvin unit of measurement for temperature, after his development of an absolute scale of temperature. This book is a corrected 1884 edition of Kelvin's 1872 collection of papers on electrostatics and magnetism. It includes all his work on these subjects previously published as articles in journals including the *Cambridge Mathematical Journal* and the *Transactions of the Royal Society*. Kelvin also wrote several new items to fill gaps in this collection, so that its coverage of the state of electromagnetic research in the late nineteenth century is comprehensive.

Cambridge University Press has long been a pioneer in the reissuing of out-of-print titles from its own backlist, producing digital reprints of books that are still sought after by scholars and students but could not be reprinted economically using traditional technology. The Cambridge Library Collection extends this activity to a wider range of books which are still of importance to researchers and professionals, either for the source material they contain, or as landmarks in the history of their academic discipline.

Drawing from the world-renowned collections in the Cambridge University Library, and guided by the advice of experts in each subject area, Cambridge University Press is using state-of-the-art scanning machines in its own Printing House to capture the content of each book selected for inclusion. The files are processed to give a consistently clear, crisp image, and the books finished to the high quality standard for which the Press is recognised around the world. The latest print-on-demand technology ensures that the books will remain available indefinitely, and that orders for single or multiple copies can quickly be supplied.

The Cambridge Library Collection will bring back to life books of enduring scholarly value (including out-of-copyright works originally issued by other publishers) across a wide range of disciplines in the humanities and social sciences and in science and technology.

Reprint of Papers on Electrostatics and Magnetism

LORD KELVIN

CAMBRIDGE UNIVERSITY PRESS

Cambridge, New York, Melbourne, Madrid, Cape Town,
Singapore, São Paolo, Delhi, Tokyo, Mexico City

Published in the United States of America by Cambridge University Press, New York

www.cambridge.org
Information on this title: www.cambridge.org/9781108029810

This edition first published 1884
This digitally printed version 2011

ISBN 978-1-108-02981-0 Paperback

REPRINT OF PAPERS

ON

ELECTROSTATICS

AND

MAGNETISM.

𝕮𝖆𝖒𝖇𝖗𝖎𝖉𝖌𝖊 :
PRINTED BY C. J. CLAY, M.A. & SON,
AT THE UNIVERSITY PRESS.

REPRINT OF PAPERS

ON

ELECTROSTATICS

AND

MAGNETISM

BY

SIR WILLIAM THOMSON, D.C.L., LL.D., F.R.S., F.R.S.E.,

FELLOW OF ST PETER'S COLLEGE, CAMBRIDGE, AND
PROFESSOR OF NATURAL PHILOSOPHY IN THE UNIVERSITY OF GLASGOW.

SECOND EDITION.

London:

MACMILLAN & CO.

1884

PREFACE TO THE FIRST EDITION.

This volume consists chiefly of reprinted articles, on Electrostatics and mathematically allied subjects, which originally appeared at different times during the last thirty years, in the *Cambridge Mathematical Journal*, the *Cambridge and Dublin Mathematical Journal*, Liouville's *Journal de Mathématiques*, the *Philosophical Magazine*, Nichol's *Cyclopædia*, the *Reports of the British Association*, the *Transactions* or *Proceedings of the Royal Societies of London and Edinburgh*, the *Royal Institution of Great Britain*, and the *Philosophical Societies of Manchester and Glasgow*. The remainder, constituting about a quarter of the whole, is now printed for the first time from manuscript, which, except a small part about twenty years old, entitled "Electromagnets," has been written for the present publication, to fill up roughly gaps in the collection. The original dates of the republished articles, the dates of all new matter appearing as insertions or notes in the course of those articles, and the dates of the fresh articles have all been carefully indicated.

The article on Atmospheric Electricity, extracted from Nichol's *Cyclopædia*, was originally written at the request of my late friend and colleague the Editor; and for the permission to reprint it I am indebted to his son, my colleague, Professor John Nichol, and to the Messrs. Griffin, the publishers of the *Cyclopædia*.

The present volume includes as nearly as may be all that I have hitherto written on electrostatics and magnetism. I have excluded from it electrical papers in which either thermo-dynamics or the kinetics of electricity is prominent. I intend that, as soon as possible, it shall be followed by a collected re-print of all my other papers hitherto published.

I take this opportunity of thanking Professors Clerk Max-well and Tait for much valuable assistance which they have given me in the course of this work.

WILLIAM THOMSON.

Yacht "Lalla Rookh,"
Lamlash, *Oct.* 12, 1872.

PREFACE TO THE SECOND EDITION.

This Second Edition is substantially a reprint of the First Edition; the only changes made being the correction of a few *errata* which had escaped detection in the volume as originally published.

W T.

The University,
Glasgow, *March* 12, 1884.

CONTENTS.

ARTICLE I.—ON THE UNIFORM MOTION OF HEAT IN HOMOGENEOUS SOLID BODIES, AND ITS CONNEXION WITH THE MATHEMATICAL THEORY OF ELECTRICITY.

SECTIONS

Temperature at any point within or without an Isothermal Surface . . . . 1—10
Uniform Motion of Heat in an Ellipsoid . . . . . 11—20
Attraction of a Homogeneous Ellipsoid on a point within or without it . . . . 21—24

II.—ON THE MATHEMATICAL THEORY OF ELECTRICITY IN EQUILIBRIUM.

DIVISION I.—ON THE ELEMENTARY LAWS OF STATICAL ELECTRICITY.

Investigations of Coulomb, Poisson, and Green 25
Examination of Harris's Experimental results . . . . 26—35
Faraday's researches on Electrostatical Induction . . . 36—50

III.—ON THE ELECTROSTATICAL CAPACITY OF A LEYDEN PHIAL AND OF A TELEGRAPH WIRE INSULATED IN THE AXIS OF A CYLINDRICAL CONDUCTING SHEATH.

Application of the Principles brought forward in the preceding Articles . . . . 51—56

IV.—ON THE MATHEMATICAL THEORY OF ELECTRICITY IN EQUILIBRIUM.

DIVISION II.—A STATEMENT OF THE PRINCIPLES ON WHICH THE MATHEMATICAL THEORY IS FOUNDED.

Object of the Article 57
The two kinds of Electricity . . . . . . 58—60
Electrical Quantity . . . . 61—62
Superposition of Electric Forces 63

SECTIONS

The Law of Force between Electrified bodies 64
Definition of the resultant Electric Force at a Point . . . 65
Electrical Equilibrium 66
Non-conductors of Electricity 67
Conductors of Electricity 68
Electrical Density at any Point of a charged Surface . . . 69
Exclusion of all Non-conductors except Air 70
Insulated Conductors 71
Recapitulation of the Fundamental Laws 72
Objects of the Mathematical Theory of Electricity . . . 73
Actual Progress in the Mathematical Theory of Electricity . . 74

V.—ON THE MATHEMATICAL THEORY OF ELECTRICITY IN EQUILIBRIUM.

DIVISION III.—GEOMETRICAL INVESTIGATION WITH REFERENCE TO THE DISTRIBUTION OF ELECTRICITY ON SPHERICAL CONDUCTORS.

Object of the Article 75
Insulated Conducting Sphere subject to no External Influence . 76
Determination of the Distribution 77
Verification of Law III. 78

Digression on the Division of Surfaces and Elements— ·

Object of the Digression 79
Explanation and Definition regarding Cones 80
The Solid Angle of a Cone, or a complete Conical Surface . . 81
Sum of all the Solid Angles round a Point $=4\pi$ 82
Sum of the Solid Angles of all the complete Conical Surfaces $=2\pi$ 83
Solid Angle subtended at a Point by a Terminated Surface . 84
Orthogonal and Oblique Sections of a Small Cone . . . 85
Area of the Segment cut from a Spherical Surface by a Small Cone 86
Theorem 87
Repulsion on an Element of the Electrified Surface . . . 88

INSULATED SPHERE SUBJECTED TO THE INFLUENCE OF AN ELECTRICAL POINT.

Object 89
Attraction of a Spherical Surface of which the Density varies inversely as the Cube of the Distance from a given point . 90—92
Application of the preceding Theorems to the Problem of Electrical Influence . . . . 93—95

EFFECTS OF ELECTRICAL INFLUENCE ON INTERNAL SPHERICAL AND ON PLANE CONDUCTING SURFACES

. . . . . 96—112

INSULATED SPHERE SUBJECT TO THE INFLUENCE OF A BODY OF ANY FORM ELECTRIFIED IN ANY GIVEN MANNER

. . . . 113—127

SECTIONS

VI.—ON THE MUTUAL ATTRACTION OR REPULSION
BETWEEN TWO ELECTRIFIED SPHERICAL CON-
DUCTORS . . . . 128—142

VII.—ON THE ATTRAQTIONS OF CONDUCTING AND NON-
CONDUCTING ELECTRIFIED BODIES . . . . 144—148

VIII.—DEMONSTRATION OF A FUNDAMENTAL PROPO-
SITION IN THE MECHANICAL THEORY OF ELECTRI-
CITY . . . . 149—155

IX.—NOTE ON INDUCED MAGNETISM IN A PLATE . . 156—162

X.—SUR UNE PROPRIÉTÉ DE LA COUCHE ÉLECTRIQUE
EN ÉQUILIBRE À LA SURFACE D'UN CORPS CON-
DUCTEUR. Par *M. J.* LIOUVILLE . . . . . . 163—164

Note on the preceding Paper 165

XI.—ON CERTAIN DEFINITE INTEGRALS SUGGESTED
BY PROBLEMS IN THE THEORY OF ELECTRICITY . 166—186

XII.—PROPOSITIONS IN THE THEORY OF ATTRACTION.

Part I. . . . . 187—198
Part II. . . . . 199—205

XIII.—THEOREMS WITH REFERENCE TO THE SOLUTION
OF CERTAIN PARTIAL DIFFERENTIAL EQUATIONS . 206

ADDITIONS TO A FRENCH TRANSLATION OF THE PRECEDING . . . 207

XIV.—ELECTRICAL IMAGES.

Extraite d'une lettre de *M.* William Thomson à *M.* Liouville . 208—210
Extraits de deux lettres addressées à *M.* Liouville. Par *M.*
William Thomson . . . . 211—220
Note au sujet de l'Article précedent. Par *M.* Liouville . . . 221—230

XV.—DETERMINATION OF THE DISTRIBUTION OF ELEC-
TRICITY ON A CIRCULAR SEGMENT OF PLANE OR
SPHERICAL CONDUCTING SURFACE, UNDER ANY
GIVEN INFLUENCE . . . . 231—248

XVI.—ATMOSPHERIC ELECTRICITY.

Preliminary Explanations . . . . 249—251
The whole Surface of the Earth electrified [generally negatively] . 252
The State of Electrification of the Air [unknown and cannot be
inferred with certainty from observations of the Electric density
of the Earth's Surface: observation from balloons wanted] . 253—261
Description of preliminary Experiments made to test aërial Electricity, and of the Instruments employed . . . . . 262—266

ROYAL INSTITUTION LECTURE.

SECTIONS

Earliest Observations of Atmospheric Electricity . . . . 267—268
Essential qualities of the Apparatus required for the observation
    of Atmospheric Electricity 269
Description of the Divided Ring Reflecting Electrometer . . 270—273
Description of the Common House Electrometer . . . . 274—276
Description of the Portable Electrometer 277
Burning Match and Water-dropping Collectors . . . . 278—279
Remarks on the origin, nature, and changes of Terrestrial Atmo-
    spheric Electricity . . . . 280—291
Kew Self-recording Atmospheric Electrometer, with specimen of
    the results . . . . 292—293

ON ELECTRICAL FREQUENCY 294

ON THE NECESSITY FOR INCESSANT RECORDING AND FOR SIMUL-
TANEOUS OBSERVATIONS IN DIFFERENT LOCALITIES TO INVESTI-
GATE ATMOSPHERIC ELECTRICITY 295
OBSERVATIONS ON ATMOSPHERIC ELECTRICITY . . . . . 296—300
ON SOME REMARKABLE EFFECTS OF LIGHTNING OBSERVED IN A
FARM HOUSE NEAR MONIMAIL 301

XVII.—SOUND PRODUCED BY THE DISCHARGE OF A
    CONDENSER . . . . 302—304

XVIII.—MEASUREMENT OF THE ELECTROSTATIC FORCE
    PRODUCED BY A DANIELL'S BATTERY.

Preliminary Explanation . . . . 305—306
Absolute Electrometer 307
Reduction to absolute measure, of the readings of torsional Electro-
    meters, by means of the Absolute Electrometer . . . . 308—313
General results of the Weighings . . . . . . 314—316
Postscript—corrected results . . . . 317—319

XIX.—MEASUREMENT OF THE ELECTROMOTIVE FORCE
    REQUIRED TO PRODUCE A SPARK IN AIR BETWEEN
    PARALLEL METAL PLATES AT DIFFERENT DIS-
    TANCES.

Description of the Experiments . . . . . . 320—321
Table I., Measurements by Absolute Electrometer . . . 322—323
Table II., Measurements by Portable Electrometer . . . 324
Table III., The two series compared—Additional Experiments,
    Tables IV. and V. 325
Table VI., Summary of results reduced to Absolute Measure . . 326

APPENDIX—Explanation of Terms. Measurement of quantities of
electricity. Electric density. Resultant electric force at any
point in an insulating fluid. Relation between Electric density

SECTIONS

on the surface of a conductor and electric force at points in the
air close to it. Electric pressure from the surface of a conductor
balanced by Air. Collected formulæ. Electric potential. Interpretation of measurement by Electrometer. Relation between
Electrostatic force and variation of Electric potential. Stratum
of Air between two parallel or nearly parallel plane or curved
metallic surfaces maintained at different potentials . . . 327—338
Absolute Electrometer 339
Additional Experiments 340

XX.—REPORT ON ELECTROMETERS AND ELECTROSTATIC
MEASUREMENTS.

Definition—Requisites for accurate Electrometry . . . . 341—342
Classification of Electrometers—
    I. Repulsion Electrometers
    II. Symmetrical Electrometers
    III. Attracted Disc Electrometer 343
Divided Ring Electrometer described, and adjustments explained . 344—357
Absolute Electrometer . . . . 358—363
New Absolute Electrometer . . . . 364—367
Portable Electrometer . . . . 368—378
Standard Electrometer . . . . 379—382
Long-range Electrometer . . . . 383—384
Idiostatic and Heterostatic Electrometers 385
Concluding remarks regarding Electrometers . . . . 386—390

XXI.—ATMOSPHERIC ELECTRICITY.

New Apparatus for observing Atmospheric Electricity . . . 391
Description and results of simultaneous observations made at two
    stations at different elevations in the University of Glasgow . 392
Effect of sudden changes of wind . . . . 393—395
Changes observed during a thunder-storm 396
Effects observed in the neighbourhood of an escape of high-pressure
    steam . . . . 397—399

XXII.—NEW PROOF OF CONTACT ELECTRICITY . . . 400

XXIII.—ELECTROPHORIC APPARATUS AND ILLUSTRATIONS
OF VOLTAIC THEORY.

On a Self-acting Apparatus for multiplying and maintaining Elec-
    tric Charges, with applications to illustrate the Voltaic Theory . 401—407
On a Uniform Electric Current Accumulator . . . . 408—411
On Volta-Convection by Flame . . . . 412—415

SECTIONS

On Electric Machines founded on Induction and Convection,—
Electric Replenisher, Potential-Equalizer, with applications . 416—426
On the Reciprocal Electrophorus . . . . 427—429

XXIV.—A MATHEMATICAL THEORY OF MAGNETISM.

Introduction . . . . 430—433

PART FIRST.—ON MAGNETS AND THE MUTUAL FORCE BETWEEN MAGNETS.

CHAPTER I.—PRELIMINARY DEFINITIONS AND EXPLANATIONS.

Definition of a Magnet . . . . 434—435
Action of the Earth on a Magnet sensibly a couple; Directive
tendency, Magnetic Axis, Dip, Polarity . . . . . 436—446
Distribution of Magnetism in a Magnet . . . . . 447—451

CHAPTER II.—ON THE LAWS OF MAGNETIC FORCE, AND ON THE DIS-
TRIBUTION OF MAGNETISM IN MAGNETIZED MATTER.

Mutual Action between two thin uniformly and longitudinally
Magnetized Bars . . . . 452—453
Strength of a Magnet, Unit Strength, Magnetic Moment, Inten-
sity and Direction of Magnetization . . . . . 454—462

CHAPTER III.—ON THE IMAGINARY MAGNETIC MATTER BY MEANS OF
WHICH THE POLARITY OF A MAGNETIZED BODY MAY BE REPRE-
SENTED . . . . 463—475

CHAPTER IV.—DETERMINATION OF THE MUTUAL ACTIONS BETWEEN
ANY GIVEN PORTIONS OF MAGNETIZED MATTER.

Explanations . . . . 476—478
"Resultant Magnetic Force at any Point" . . . . 479—480
The "Potential" . . . . 481—484
Potential at a point P due to a given Magnet . . . . 485—501
On the Expression of Mutual Action between two Magnets by
means of the Differential Coefficients of a Function of their
relative Positions . . . . 502—503

CHAPTER V.—ON SOLENOIDAL AND LAMELLAR DISTRIBUTIONS OF
MAGNETISM.

Explanations 504
Definitions and Explanations regarding Magnetic Solenoids . 505
Definitions and Explanations regarding Magnetic Shells . . 506
Solenoidal and Lamellar Distributions of Magnetisms . . 507
Complex Lamellar and Complex Solenoidal Distributions of
Magnetism . . . . 508—509
Action of a Magnetic Solenoid and of a Complex Solenoid . . 510—511
Potential at any point due to a Magnetic Shell—Action of
Magnetic Shells 512
Criterion of a Solenoidal Distribution of Magnetism . . . 513

SECTIONS

Criterion of a Lamellar Distribution of Magnetism . . . 514
Resultant Force, due to a lamellarly-magnetized Magnet, on any
    external or internal point . . . . 515—523

CHAPTER VI.—ON ELECTROMAGNETS.

Introductory Remarks 524
Investigation of the Action between two Galvanic Arcs, or be-
    tween a Galvanic Arc and a Magnetic Pole . . . . 525—530
Unit of strength for an Electric Current . . . . . 531—533
Hypothesis of Matter flowing 534
Division of Electromagnets into three Classes 535
Linear Electromagnets 536
Superficial Electromagnets 537
Solid Electromagnets 538
Analytical Investigation of the Conditions to which the Distri-
    bution of Galvanism in Solid and Superficial Electromagnets
    is subject . . . . 539—543
Applications 544
A similar Synthetic Solution indicated 545
Electromagnets and their respective equivalent Polar Magnets—
    Rules for Direction . . . . 546—550
Remarks and Additions . . . . 551—553
Original Investigation of § 517 referred to in § 518 . . . 554

XXV.—ON THE POTENTIAL OF A CLOSED GALVANIC
    CIRCUIT OF ANY FORM . . . . 555—560

XXVI.—CHAPTER VII.—ON THE MECHANICAL VALUES OF DISTRIBU-
    TIONS OF MATTER AND OF MAGNETS.
Mechanical Values of Distributions of Matter . . . . 561—563
Polar Magnets . . . . 564—568
Electromagnets . . . . 569—572

XXVII.—CHAPTER VIII.—HYDROKINETIC ANALOGY . . . . 573—583

XXVIII.—CHAPTER IX.—INVERSE PROBLEMS.
Definition—Divided into two Classes 584
Class I.—Force given for every point of space . . . . 585—588
Class II.—Force or component of force given through some
    portion of space . . . . 589—601

XXIX.—ON THE ELECTRIC CURRENTS BY WHICH THE
    PHENOMENA OF TERRESTRIAL MAGNETISM MAY BE
    PRODUCED . . . . 602—603
CHAPTER X.—MAGNETIC INDUCTION. .

XXX.—ON THE THEORY OF MAGNETIC INDUCTION IN
    CRYSTALLINE AND NON-CRYSTALLINE SUBSTANCES.

xiv Contents.

SECTIONS

Explanations and Definitions. Force at any point due to a Magnet.
Total magnetic force at a point. "A Field of magnetic force."
"A line of magnetic force." "A uniform field of magnetic force."
Resultant Distribution of Magnetism . . . . . . 604—605
Axioms of Magnetic Force 606
Laws of Magnetic Induction according to Poisson's Theory . . 607—609
Conclusions from these Laws . . . . 610—619
Appendix—Quotations from Poisson regarding Magne-Crystallic
action—Explanation. Demonstration . . . . . 620—624

XXXI.—MAGNETIC PERMEABILITY AND ANALOGUES IN
ELECTROSTATIC INDUCTION, CONDUCTION OF HEAT
AND FLUID MOTION . . . . 625—631

XXXII.—DIAGRAMS OF LINES OF FORCE; TO ILLUSTRATE
MAGNETIC PERMEABILITY . . . . . . 632—633

XXXIII.—ON THE FORCES EXPERIENCED BY SMALL
SPHERES UNDER MAGNETIC INFLUENCE; AND ON
SOME OF THE PHENOMENA PRESENTED BY DIA-
MAGNETIC SUBSTANCES.
Attraction of Ferromagnetics . . . . 634—642
Repulsion of Diamagnetics . . . . 643—646

XXXIV.—REMARKS ON THE FORCES EXPERIENCED BY
INDUCTIVELY MAGNETIZED FERROMAGNETIC OR
DIAMAGNETIC NON-CRYSTALLINE SUBSTANCES.
Faraday's Law of Attractions and Repulsions . . . . 647—653
Experimental illustrations of Faraday's Law . . . . 654—664
On the Stability of Small Inductively magnetized bodies in Posi-
tions of Equilibrium 665
On the relations of Ferromagnetic and Diamagnetic Magnetization
to the magnetizing force . . . . 666—668

XXXV.—ABSTRACT OF TWO COMMUNICATIONS—
On certain Magnetic Curves; with applications to Problems in the
Theories of Heat, Electricity, and Fluid Motion . . . 669
On the Equilibrium of elongated Masses of Ferromagnetic Sub-
stances in uniform and varied Fields of Force 669

XXXVI.—REMARQUES SUR LES OSCILLATIONS
d'aiguilles non cristallisées de faible pouvoir inductif paramagnétique ou diamagnétique, et sur d'autres phénomènes magnétiques produits par des corps cristallisés ou non cristallisés;
from the "Comptes Rendus" of the French Academy, 1854,
first half-year 670

XXXVII.—ELEMENTARY DEMONSTRATION OF PROPOSI-
TIONS IN THE THEORY OF MAGNETIC FORCE . . 671—673

SECTIONS

Examination of the Action experienced by an infinitely thin, uniformly and longitudinally Magnetized Bar, placed in a Non-uniform Field of Force, with its length direct along a line of Force . . . . 676—688

XXXVIII.—CORRESPONDENCE WITH PROFESSOR TYNDALL.
Letter to Professor Tyndall on the "Magnetic Medium," and on the effects of Compression . . . . 689—693
Letter from Professor Tyndall to Professor W. Thomson on Reciprocal Molecular Induction 694
Letter from Professor W. Thomson to Professor Tyndall, on the Reciprocal Action of Diamagnetic Particles . . . . 695—696

XXXIX.—INDUCTIVE SUSCEPTIBILITY OF A POLAR MAGNET . . . . 697—699

XL.—GENERAL PROBLEM OF MAGNETIC INDUCTION . 700—732

XLI.—HYDROKINETIC ANALOGY FOR THE MAGNETIC INFLUENCE OF AN IDEAL EXTREME DIAMAGNETIC.
On Forces experienced by Solids immersed in a Moving Liquid . 733—740
Extracts from two Letters to Professor Guthrie . . . . 741—743
Report of an Address on the Attractions and Repulsions due to Vibration, observed by Guthrie and Schellbach.—Hydrokinetic Analogy for Extreme Diamagnetic . . . . . . 744—750

XLII.—GENERAL HYDROKINETIC ANALOGY FOR INDUCED MAGNETISM.
Permeability in Hydrokinetic Analogy . . . . . . 751—756
Kinetic Energy a Minimum . . . . 757—758
Analogy of Force . . . . 759—763

I.—ON THE UNIFORM MOTION OF HEAT IN HOMOGENEOUS SOLID BODIES, AND ITS CONNEXION WITH THE MATHEMATICAL THEORY OF ELECTRICITY.*

(Art. III. of complete list in *Mathematical and Physical Papers*, Vol. I.)

[From *Cambridge Mathematical Journal*, Feb. 1842. Reprinted *Philosophical Magazine* (1854, first half-year).]

[Since the following article was written,† the writer finds that most of his ideas have been anticipated by M. Chasles in two Mémoires in the *Journal de Mathématiques*; the first, in vol. III., on the Determination of the Value of a certain Definite Integral, and the second, in vol. V., on a new Method of Determining the Attraction of an Ellipsoid on a Point without it. In the latter of these Mémoires, M. Chasles refers to a paper, by himself, in the twenty-fifth *Cahier* of the *Journal de l'Ecole Polytechnique*, in which it is probable there are still further anticipations, though the writer of the present article

---

* [*Note added June* 1854.]—This paper first appeared anonymously in the *Cambridge Mathematical Journal* in February 1842. The text is reprinted without alteration or addition. All the footnotes are of the present date (March 1854). The general conclusions established in it show that the laws of distribution of electric or magnetic force in any case whatever must be identical with the laws of distribution of the lines of motion of heat in certain perfectly defined circumstances. With developments and applications contained in a subsequent paper (II. below) on the Elementary Laws of Statical Electricity (*Cambridge and Dublin Mathematical Journal*, Nov. 1845), they constitute a full theory of the characteristics of lines of force, which have been so admirably investigated experimentally by Faraday, and complete the analogy with the theory of the conduction of heat, of which such terms as "conducting power for lines of force" (*Exp. Res.* §§ 2797—2802) involve the idea.

† [*Note added June* 1854.]—This preliminary notice was written some months later than the text which follows, and was communicated to the editor of the journal to be prefixed to the paper, which had been in his hands since the month of September 1841. The ideas in which the author had ascertained he had been anticipated by M. Chasles, were those by which he was led to the determination of the attraction of an ellipsoid given in the latter part of the paper. He found soon afterwards that he was anticipated by the same author in an enunciation of the general theorems regarding attraction; still later he found that both an enunciation and demonstration of the same general theorems had been given by Gauss, whose paper ap-

T. E.   1

has not had access to so late a volume of the latter journal.
Since, however, most of his methods are very different from
those of M. Chasles, which are nearly entirely geometrical, the
following article may be not uninteresting to some readers :—]

1. If an infinite homogeneous solid be submitted to the action
of certain constant sources of heat, the stationary temperature
at any point will vary according to its position; and through
every point there will be a surface, over the whole extent of
which the temperature is constant, which is therefore called an
isothermal surface. In this paper the case will be considered
in which these surfaces are finite, and consequently closed.

2. It is obvious that the temperature of any point without
a given isothermal surface, depends merely on the form and
temperature of the surface, being independent of the actual
sources of heat by which this temperature is produced, provided
there are no sources without the surface. The temperature
of an external point is consequently the same as if all the
sources were distributed over this surface in such a manner
as to produce the given constant temperature. Hence we may
consider the temperature of any point without the isothermal
surface, as the sum of the temperatures due to certain constant
sources of heat, distributed over that surface.

---

peared shortly after M. Chasles' enunciations; and after all, he found that
these theorems had been discovered and published in the most complete and
general manner, with rich applications to the theories of electricity and
magnetism, more than ten years previously, by Green! It was not until
early in 1845 that the author, after having inquired for it in vain for several
years, in consequence of an obscure allusion to it in one of Murphy's papers,
was fortunate enough to meet with a copy of the remarkable paper ("An
Essay on the Application of Mathematical Analysis to the Theories of
Electricity and Magnetism," by George Green, Nottingham, 1828) in which
this great advance in physical mathematics was first made. It is worth
remarking, that, referring to Green as the originator of the term, Murphy
gives a mistaken definition of "potential." It appears highly probable that
he may never have had access to Green's essay at all, and that this is the
explanation of the fact (of which any other explanation is scarcely conceivable), that in his Treatise on Electricity (Murphy's *Electricity*, Cambridge,
1833) he makes no allusion whatever to Green's discoveries, and gives a
theory in no respect pushed beyond what had been done by Poisson. All
the general theorems on attraction which Green and the other writers referred
to, demonstrated by various purely mathematical processes, are seen as
axiomatic truths in approaching the subject by the way laid down in the
paper which is now republished. The analogy with the conduction of heat
on which these views are founded, has not, so far as the author is aware,
been noticed by any other writer.

3. To find the temperature produced by a single source of heat, let r be the distance of any point from it, and let v be the temperature at that point. Then, since the temperature is the same for all points situated at the same distance from the source, it is readily shown that v is determined by the equation

$$-r^2 \frac{dv}{dr} = A.$$

Dividing both members by r^2, and integrating, we have

$$v = \frac{A}{r} + C.$$

Now let us suppose that the natural temperature of the solid, or the temperature at an infinite distance from the source, is zero: then we shall have $C = 0$, and consequently

$$v = \frac{A}{r} \quad\text{.................................. (1)}.$$

4. Hence that part of the temperature of a point without an isothermal surface which is due to the sources of heat situated on any element, $d\omega_1^2$, of the surface, is $\dfrac{\rho_1 d\omega_1^2}{r_1}$, where r_1 is the distance from the element to that point, and ρ_1 a quantity measuring the intensity of the sources of heat at different parts of the surface. Hence, the supposition being still made that there are no sources of heat without the surface, if v be the temperature at the external point, we have

$$v = \iint \frac{\rho_1 d\omega_1^2}{r_1} \quad\text{........................(2)},$$

the integrals being extended over the whole surface. The quantity ρ_1 must be determined by the condition

$$v = v_1 \quad\text{.............................. (3)},$$

for any point in the surface, v_1 being a given constant temperature.

5. Let us now consider what will be the temperature of a point within the surface, supposing all the sources of heat by which the surface is retained at the temperature v_1 to be distributed over it. Since there are no sources in the interior of the surface, it follows that as much heat must flow out from the interior across the surface as flows into the interior, from the sources of heat at the surface. Hence the total flux of heat from the original surface to an adjacent isothermal surface in

the interior is nothing. Hence also the flux of heat from this
latter surface to an adjacent isothermal surface in its interior
must be nothing; and so on through the whole of the body
within the original surface. Hence the temperature in the
interior is constant, and equal to v_1, and therefore, for points at
the surface, or within it, we have

$$\iint \frac{\rho_1 d\omega_1^2}{r_1} = v_1 \dots\dots\dots\dots\dots\dots (4).$$

Now, if we suppose the surface to be covered with an attractive medium, whose density at different points is proportional
to ρ_1, $-\dfrac{d}{dx}\iint \dfrac{\rho_1 d\omega_1^2}{r_1}$ will be the attraction, in the direction of
the axis of x, on a point whose rectangular co-ordinates are
x, y, z. Hence it follows that the attraction of this medium
on a point within the surface is nothing, and consequently ρ_1
is proportional to the intensity of electricity in a state of equilibrium on the surface, the attraction of electricity in a state
of equilibrium being nothing on an interior point. Since, at
the surface, the value of $\iint \dfrac{\rho_1 d\omega_1^2}{r_1}$ is constant, and since, on
that account, its value within the surface is constant also, it
follows, that if the attractive force on a point at the surface
is perpendicular to the surface, the attraction on a point within
the surface is nothing. Hence the sole condition of equilibrium of electricity, distributed over the surface of a body,
is, that it must be so distributed that the attraction on a point
at the surface, oppositely electrified, may be perpendicular to
the surface.

6. Since, at any of the isothermal surfaces, v is constant, it
follows that $-\dfrac{dv}{dn}$, where n is the length of a curve which cuts
all the surfaces perpendicularly, measured from a fixed point
to the point attracted, is the total attraction on the latter point;
and that this attraction is in a tangent to the curve n, or in
a normal to the isothermal surface passing through the point.
For the same reason also, if ρ_1 represent a flux of heat, and not
an electrical intensity, $-\dfrac{dv}{dn}$ will be the total flux of heat at the
variable extremity of n, and the direction of this flux will be

along n, or perpendicular to the isothermal surface. Hence, if a surface in an infinite solid be retained at a constant temperature, and if a conducting body, bounded by a similar surface, be electrified, the flux of heat, at any point, in the first case, will be proportional to the attraction on an electrical point, similarly situated, in the second; and the direction of the flux will correspond to that of the attraction.

7. Let $-\dfrac{dv_1}{dn_1}$ be the external value of $-\dfrac{dv}{dn}$ at the original surface, or the attraction on a point without it, and indefinitely near it. Now this attraction is composed of two parts; one the attraction of the adjacent element of the surface; and the other the attraction of all the rest of the surface. Hence, calling the former of these a, and the latter b, we have

$$-\frac{dv_1}{dn_1} = a + b.$$

Now, since the adjacent element of the surface may be taken as infinitely larger, in its linear dimensions, than the distance from it of the point attracted, its attraction will be the same as that of an infinite plane, of the density ρ_1. Hence a is independent of the distance of the point from the surface, and is equal to $2\pi\rho_1$. Hence

$$-\frac{dv_1}{dn_1} = 2\pi\rho_1 + b.$$

Now, for a point within the surface, the attraction of the adjacent element will be the same, but in a contrary direction, and the attraction of the rest of the surface will be the same, and in the same direction. Hence the attraction on a point within the surface, and indefinitely near it, is $-2\pi\rho_1 + b$; and consequently, since this is equal to nothing, we must have $b = 2\pi\rho_1$, and therefore $\qquad -\dfrac{dv_1}{dn_1} = 4\pi\rho_1$ ....(5).

Hence ρ_1 is equal to the total flux of heat, at any point of the surface, divided by 4π.

8. It also follows that if the attraction of matter spread over the surface be nothing on an interior point, the attraction on an exterior point, indefinitely near the surface, is perpendicular to the surface, and equal to the density of the matter at the part of the surface adjacent to that point, multiplied by 4π.

9. If v be the temperature at any isothermal surface, and ρ the intensity of the sources at any point of this surface, which would be necessary to sustain the temperature v, we have, by (5),

$$-\frac{dv}{dn} = 4\pi\rho,$$

which equation holds, whatever be the manner in which the actual sources of heat are arranged, whether over an isothermal surface or not; and the temperature produced in an external point by the former sources, is the same as that produced by the latter. Also, the total flux of heat across the isothermal surface, whose temperature is v, is equal to the total flux of heat from the actual sources. From this, and from what has been proved above, it follows that if a surface be described round a conducting or non-conducting electrified body, so that the attraction on points situated on this surface may be everywhere perpendicular to it, and if the electricity be removed from the original body, and distributed in equilibrium over this surface, its intensity at any point will be equal to the attraction of the original body on that point, divided by 4π, and its attraction on any point without it will be equal to the attraction of the original body on the same point.*

If we call E the total expenditure of heat, or the whole flux across any isothermal surface, we have, obviously,

$$E = -\iint\frac{dv_1}{dn_1} d\omega_1{}^2.$$

10. Now this quantity should be equal to the sum of the expenditures of heat from all the sources. To verify this, we must, in the first place, find the expenditure of a single source. Now the temperature produced by a single source is, by (1),

$v = \dfrac{A}{r}$, and hence the expenditure is obviously equal to

---

* [*Note added June* 1854.—After having established this remarkable theorem in the manner shown in the text, the author attempted to prove it by direct integration, but only succeeded in doing so upwards of a year later, when he obtained the demonstration published in a paper, "Propositions in the Theory of Attraction" (*Camb. Math. Jour.* Nov. 1842), which appeared almost contemporaneously with a paper by M. Sturm in Liouville's *Journal*, containing the same demonstration; exactly the same demonstration, as the author afterwards (in 1845) found, had been given fourteen years earlier by Green.]

$-\dfrac{dv}{dr} \times 4\pi r^2$, or to $4\pi A$. If $A = \rho_1 d\omega_1^2$, this becomes $4\pi\rho_1 d\omega_1^2$.

Hence the total expenditure is $\iint 4\pi\rho_1 d\omega_1^2$, or $-\iint\dfrac{dv_1}{dn_1}d\omega_1^2$, which agrees with the expression found above.

The following is an example of the application of these principles:—

Uniform Motion of Heat in an Ellipsoid.

11. The principles established above afford an easy method of determining the isothermal surfaces, and the corresponding temperatures, in the case in which the original isothermal surface is an ellipsoid.

The first step is to find ρ_1, which is proportional to the quantity of matter at any point in the surface of an ellipsoid, when the matter is so distributed that the attraction on a point within the ellipsoid is nothing. Now the attraction of a shell, bounded by two concentric similar ellipsoids, on a point within it, is nothing. If the shell be infinitely thin, its attraction will be the same as that of matter distributed over the surface of one of the ellipsoids in such a manner that the quantity on a given infinitely small area at any point is proportional to the thickness of the shell at the same point. Let a_1, b_1, c_1 be the semi-axes of one of the ellipsoids, $a_1 + \delta a_1$, $b_1 + \delta b_1$, $c_1 + \delta c_1$ those of the other. Let also p_1 be the perpendicular from the centre to the tangent plane at any point on the first ellipsoid, and $p_1 + \delta p_1$ the perpendicular from the centre to the tangent plane at a point similarly situated on the second. Then δp_1 is the thickness of the shell, since, the two ellipsoids being similar, the tangent planes at the points similarly situated on their surfaces are parallel. Also, on account of their similarity,

$\dfrac{\delta a_1}{a_1} = \dfrac{\delta b_1}{b_1} = \dfrac{\delta c_1}{c_1} = \dfrac{\delta p_1}{p_1}$, and consequently the thickness of the shell is proportional to p_1. Hence we have, by (5),

$$-\frac{1}{4\pi}\frac{dv_1}{dn_1} = \rho_1 = k_1 p_1 \dots\dots\dots(a),$$

where k_1 is a constant, to be determined by the condition $v = v_1$, at the surface of the ellipsoid.

12. To find the equation of the isothermal surface at which the temperature is $v_1 + dv_1$, let $-dv_1 = C$, in (a). Then we have

$k_1 p_1 dn_1 = \dfrac{C}{4\pi}$, or $p_1 dn_1 = \theta_1$, where θ_1 is an infinitely small constant quantity; and the required equation will be the equation of the surface traced by the extremity of the line dn_1, drawn externally perpendicular to the ellipsoid. Let x', y', z' be the co-ordinates of any point in that surface, and x, y, z those of the corresponding point in the ellipsoid. Then, calling $\alpha_1, \beta_1, \gamma_1$ the angles which a normal to the ellipsoid at the point whose co-ordinates are x, y, z makes with these co-ordinates, and supposing the axes of x, y, z to coincide with the axes of the ellipsoid, $2a_1$, $2b_1$, $2c_1$, respectively, we have

$$x' - x = dn_1 \cos \alpha_1 = \frac{\dfrac{x}{a_1^2} dn_1}{\sqrt{\left(\dfrac{x^2}{a_1^4} + \dfrac{y^2}{b_1^4} + \dfrac{z^2}{c_1^4} \right)}} = \frac{x}{a_1^2} p_1 dn_1 = \frac{x}{a_1^2} \theta_1,$$

or $x' - x = \dfrac{x'}{a_1^2} \theta_1$, since θ_1 is infinitely small, and therefore also $x' - x$; whence

$$x = x' \left(1 - \frac{\theta_1}{a_1^2} \right) = \frac{x'}{1 + \dfrac{\theta_1}{a_1^2}}.$$

In a similar manner we should find

$$y = \frac{y'}{1 + \dfrac{\theta_1}{b_1^2}}, \quad \text{and} \quad z = \frac{z'}{1 + \dfrac{\theta_1}{c_1^2}}.$$

But $\dfrac{x^2}{a_1^2} + \dfrac{y^2}{b_1^2} + \dfrac{z^2}{c_1^2} = 1$, and hence we have

$$\frac{x'^2}{a_1^2 \left(1 + \dfrac{\theta_1}{a_1^2} \right)^2} + \frac{y'^2}{b_1^2 \left(1 + \dfrac{\theta_1}{b_1^2} \right)^2} + \frac{z'^2}{c_1^2 \left(1 + \dfrac{\theta_1}{c_1^2} \right)^2} = 1,$$

or

$$\frac{x'^2}{a_1^2 + 2\theta_1} + \frac{y'^2}{b_1^2 + 2\theta_1} + \frac{z'^2}{c_1^2 + 2\theta_1} = 1,$$

for the equation to the isothermal surface whose temperature is $v_1 + dv_1$, and which is therefore an ellipsoid described from the same foci as the original isothermal ellipsoid. In exactly the same manner it might be shown that the isothermal surface whose temperature is $v_1 + dv_1 + dv_1'$, is an ellipsoid having the same foci as the ellipsoid whose temperature is $v_1 + dv_1$, and consequently, as the original ellipsoid also. By continuing this process it may be proved that all the isothermal surfaces are ellipsoids, having the same foci as the original one.

13. From the form of the equation found above for the isothermal ellipsoid whose temperature is $v_1 + dv_1$, it follows that θ_1 or $p_1 dn_1$ is $= a_1 da_1$, where da_1 is the increment of a_1, corresponding to the increment dn_1 of n_1. Hence, if a be one of the semi-axes of an ellipsoid, $a + da$ the corresponding semi-axis of another ellipsoid having the same foci, dn the thickness at any point of the shell bounded by the two ellipsoids, and p the perpendicular from the centre to the plane touching either ellipsoid at the same point, we have

$$\frac{dn}{da} = \frac{a}{p}.\dots\dots\dots\dots\dots\dots(b).$$

14. All that remains to be done is to find the temperature at the surface of any given ellipsoid, having the same foci as the original ellipsoid. For this purpose, let us first find the value of $-\dfrac{dv}{dn}$ at any point in the surface of the isothermal ellipsoid whose semi-axes are a, b, c. Now we have, from (a),

$$-\frac{dv}{dn} = 4\pi k p,$$

where k is constant for any point in the surface of the isothermal ellipsoid under consideration, and determined by the condition that the whole flux of heat across this surface must be equal to the whole flux across the surface of the original ellipsoid. Now the first of these quantities is equal to $4\pi k \iint p d\omega^2$ ($d\omega^2$ being an element of the surface), or to $4\pi \dfrac{ka}{\delta a} \iint \delta p d\omega^2$, since $\dfrac{\delta a}{a} = \dfrac{\delta p}{p}$. But $\iint \delta p d\omega^2$ is equal to the volume of a shell bounded by two similar ellipsoids, whose semi-axes are a, b, c, and $a + \delta a, b + \delta b, c + \delta c$, and is therefore readily shown to be equal to $4\pi \dfrac{\delta a}{a} abc$. Hence $4\pi \dfrac{ka}{\delta a} \iint \delta p d\omega^2$, or $4\pi k \iint p d\omega^2$, is equal to $4^2 \pi^2 k abc$. In a similar manner we have, for the flux of heat across the original isothermal surface, $4^2 \pi^2 k_1 a_1 b_1 c_1$, and therefore $4^2 \pi^2 k abc = 4^2 \pi^2 k_1 a_1 b_1 c_1$,

which gives $k = k_1 \dfrac{a_1 b_1 c_1}{abc}.$

Hence we have

$$-\frac{dv}{dn} = 4\pi k_1 \frac{a_1 b_1 c_1}{abc} p \quad\text{...................}(c).$$

15. The value of v may be found by integrating this equation. To effect this, since a, b, c are the semi-axes of an ellipsoid passing through the variable extremity of n, and having the same foci as the original ellipsoid, whose axes are a_1, b_1, c_1, we have $a^2 - a_1^2 = b^2 - b_1^2 = c^2 - c_1^2;$

which gives

$$\left.\begin{array}{c} b^2 = a^2 - f^2 \\ c^2 = a^2 - g^2 \end{array}\right\} \quad\text{...............}(d).$$

where $f^2 = a_1^2 - b_1^2,\ g^2 = a_1^2 - c_1^2$

Hence (c) becomes

$$-\frac{dv}{dn} = 4\pi k_1 \frac{a_1 b_1 c_1 p}{a\sqrt{(a^2 - f^2)}\ \sqrt{(a^2 - g^2)}}.$$

Now, by (b), $dn = \dfrac{a\,da}{p}$, and hence

$$dv = -4\pi k_1 \frac{a_1 b_1 c_1 da}{\sqrt{(a^2 - f^2)}\ \sqrt{(a^2 - g^2)}}.$$

Integrating this, we have

$$v = -4\pi k_1 a_1 b_1 c_1 \int \frac{da}{\sqrt{(a^2 - f^2)}\ \sqrt{(a^2 - g^2)}} + C \quad\text{.........}(e).$$

16. The two constants, k_1 and C, must be determined by the conditions $v = v_1$ when $a = a_1$, and $v = 0$ when $a = \infty$; the latter of which must be fulfilled, in order that the expression found for v may be equal to $\iint \frac{k_1 p_1 d\omega_1^2}{r_1}$.

17. To reduce the expression for v to an elliptic function, let us assume

$$\left.\begin{array}{c} a = f \operatorname{cosec} \phi \\ a_1 = f \operatorname{cosec} \phi_1 \end{array}\right\} \quad\text{...................}(f),$$

which we may do with propriety if f be the greater of the two quantities f and g, since a is always greater than either of them, as we see from (d). On this assumption, equation (e) becomes

$$v = \frac{4\pi k_1 a_1 b_1 c_1}{f} \int_0^\phi \frac{d\phi}{\sqrt{(1 - c'^2 \sin^2\phi)}} + C = \frac{4\pi k_1 a_1 b_1 c_1}{f} F_{c'}\phi + C$$

where

$$c' = \frac{g}{f} \quad\text{.........................}(g).$$

18. Determining from this the values of C and k_1 by the conditions mentioned above, we find $C = 0$, and

$$k_1 = \frac{fv_1}{4\pi a_1 b_1 c_1 F_{c'}\phi_1} \quad \dots\dots\dots\dots(h);$$

hence the expression for v becomes

$$v = v_1 \frac{F_{c'}\phi}{F_{c'}\phi_1} \quad \dots\dots\dots\dots\dots(k).$$

19. The results which have been obtained may be stated as follows:—

If, in an infinite solid, the surface of an ellipsoid be retained at a constant temperature, the temperature of any point in the solid will be the same as that of any other point in the surface of an ellipsoid described from the same foci, and passing through that point; and the flux of heat at any point in the surface of this ellipsoid will be proportional to the perpendicular from the centre to a plane touching it at the point, and inversely proportional to the volume of the ellipsoid.

20. This case of the uniform motion of heat was first solved by Lamé, in his Mémoire on Isothermal Surfaces, in Liouville's *Journal de Mathématiques*, vol. ii. p. 147, by showing that a series of isothermal surfaces of the second order will satisfy the equation $\dfrac{d^2v}{dx^2} + \dfrac{d^2v}{dy^2} + \dfrac{d^2v}{dz^2} = 0,$

provided they are all described from the same foci. The value which he finds for v agrees with (e), and he finds, for the flux of heat at any point, the expression

$$\frac{KA}{\sqrt{(\mu^2 - \nu^2)}\,\sqrt{(\mu^2 - \rho^2)}};$$

or, according to the notation which we have employed,

$$\frac{4\pi k_1 a_1 b_1 c_1}{\sqrt{(a^2 - \nu^2)}\,\sqrt{(a^2 - \rho^2)}},$$

where ν is the greater real semi-axis of the hyperboloid of one sheet, and ρ the real semi-axis of the hyperboloid of two sheets, described from the same foci as the original ellipsoid, and passing through the point considered. Hence a^2, ν^2, ρ^2 are the three roots of the equation

$$\frac{x^2}{u} + \frac{y^2}{u - f^2} + \frac{z^2}{u - g^2} = 1,$$

or
$$u^3-(f^2+g^2+x^2+y^2+z^2)u^2+\{f^2g^2+(f^2+g^2)x^2+g^2y^2+f^2z^2\}u-f^2g^2x^2=0.$$
Hence $\qquad a^2v^2\rho^2=f^2g^2x^2,$

and $\quad a^2v^2+a^2\rho^2+v^2\rho^2=f^2g^2+(f^2+g^2)x^2+g^2y^2+f^2z^2.$

Therefore,
$$(a^2-v^2)(a^2-\rho^2)=a^4-a^2v^2-a^2\rho^2-v^2\rho^2+\frac{2a^2v^2\rho^2}{a^2}$$

$$=a^4-\{f^2g^2+(f^2+g^2)x^2+g^2y^2+f^2z^2\}+2f^2g^2\frac{x^2}{a^2}$$

$$=a^4-(a^2-b^2)(a^2-c^2)-(2a^2-b^2-c^2)x^2-(a^2-c^2)y^2$$
$$\qquad -(a^2-b^2)z^2+2(a^2-b^2)(a^2-c^2)\frac{x^2}{a^2}$$

$$=a^4-(a^2-b^2)(a^2-c^2)-(b^2+c^2)x^2-(a^2-c^2)y^2$$
$$\qquad -(a^2-b^2)z^2+2b^2c^2\left(1-\frac{y^2}{b^2}-\frac{z^2}{c^2}\right)$$

$$=a^4-(a^2-b^2)(a^2-c^2)-(b^2+c^2)x^2-(a^2+c^2)y^2-(a^2+b^2)z^2+2b^2c^2$$
$$=a^2b^2+a^2c^2+b^2c^2-\{(b^2+c^2)x^2+(a^2+c^2)y^2+(a^2+b^2)z^2\};$$
which is readily shown, by substituting for $a^2b^2+a^2c^2+b^2c^2$
its equal $(a^2b^2+a^2c^2+b^2c^2)\left(\frac{x^2}{a^2}+\frac{y^2}{b^2}+\frac{z^2}{c^2}\right)$, to be equal to $\frac{a^2b^2c^2}{p^2}$.

Hence the expression for $-\dfrac{dv}{dn}$, given above, becomes
$$-\frac{dv}{dn}=4\pi k_1\frac{a_1b_1c_1}{abc}p,$$
which agrees with (c).

Attraction of a Homogeneous Ellipsoid on a Point within or without it.

21. If, in (c), we put $k_1=\dfrac{da_1}{a_1}$, the value of $-\dfrac{dv}{dn}$ at any point
will be the attraction on that point of a shell bounded by two
similar concentric ellipsoids, whose semi-axes are
$$a_1,\ a_1\sqrt{(1-e^2)},\ a_1\sqrt{(1-e'^2)},$$
and $\qquad a_1+da_1, (a_1+da_1)\sqrt{(1-e^2)}, (a_1+da_1)\sqrt{(1-e'^2)},$
where $\qquad a^2-b^2=a_1^2-b_1^2=a_1^2e^2$
and $\qquad a^2-c^2=a_1^2-c_1^2=a_1^2e'^2$ $\Big\}$ .... (1),
the density of the shell being unity. Now this attraction is in

a normal drawn through the point attracted to the surface of
the ellipsoid, whose semi-axes are a, b, c. If we call α, β, γ
the angles which this normal makes with the co-ordinates
x, y, z of the point attracted, we have

$$\cos \alpha = \frac{\dfrac{x}{a^2}}{\sqrt{\left(\dfrac{x^2}{a^4}+\dfrac{y^2}{b^4}+\dfrac{z^2}{c^4}\right)}}=\frac{px}{a^2},$$

and similarly, $\cos \beta = \dfrac{py}{b^2}$, $\cos \gamma = \dfrac{pz}{c^2}$.

Hence, calling dA, dB, dC the components of the attraction
parallel to the axes of co-ordinates, we have, from (c),

$$\left. \begin{aligned}
dA &= 4\pi x \frac{b_1 c_1}{a^3 bc}p^2 da_1 \\
dB &= 4\pi y \frac{b_1 c_1}{ab^3 c}p^2 da_1 \\
dC &= 4\pi z \frac{b_1 c_1}{abc^3}p^2 da_1
\end{aligned} \right\} \dots\dots\dots\dots(2).$$

22.   The integrals of these expressions, between the limits
$a_1 = 0$ and $a_1 = a_1$, are the components of the attraction of an
ellipsoid whose semi-axes are a_1', b_1', c_1', or a_1', $a_1'\sqrt{(1-e^2)}$,
$a_1'\sqrt{(1-e'^2)}$, on the point (x, y, z). Now, by (1), we may
express each of the quantities b, c, b_1, c_1, in terms of a and a_1,
and the equation

$$\frac{x^2}{a^2}+\frac{y^2}{b^2}+\frac{z^2}{c^2}=1, \text{ or } \frac{x^2}{a^2}+\frac{y^2}{a^2-e^2 a_1^2}+\frac{z^2}{a^2-e'^2 a_1^2}=1\dots(3),$$

enables us to express either of the quantities a, a_1 in terms of
the other. The simplest way, however, to integrate equations
(2), will be to express each in terms of a third quantity,

$$u = \frac{a_1}{a}\dots\dots\dots\dots\dots\dots(4).$$

Eliminating a from (3), by means of this quantity, we have

$$a_1^2 = u^2 x^2 + \frac{y^2}{u^{-2}-e^2}+\frac{z^2}{u^{-2}-e'^2}.$$

Hence $a_1 da_1 = \left\{ ux^2 + \dfrac{u^{-3}y^2}{(u^{-2}-e^2)^2}+\dfrac{u^{-3}z^2}{(u^{-2}-e'^2)^2}\right\} du$

$$= \left(\frac{x^2}{a^4}+\frac{y^2}{b^4}+\frac{z^2}{c^4}\right)a_1^4 u^{-3}du = a_1^4 p^{-2}u^{-3}du.$$

Also, from (4), we have $a = \dfrac{a_1}{u}$; from which we find, by (1),

$b = \dfrac{a_1}{u} \sqrt{(1 - e^2 u^2)}$, $c = \dfrac{a_1}{u} \sqrt{(1 - e'^2 u^2)}$.   By (1) also, $b_1 = a_1 \sqrt{(1 - e^2)}$, $c_1 = a_1 \sqrt{(1 - e'^2)}$.   Making these substitutions in (2), and integrating, we have, calling a' the value of a, when $a_1 = a_1'$,

$$
\left.
\begin{aligned}
A &= 4\pi x \sqrt{(1 - e^2)} \sqrt{(1 - e'^2)} \int_0^{\frac{a_1'}{a'}} \frac{u^2 du}{\sqrt{(1 - e^2 u^2)} \sqrt{(1 - e'^2 u^2)}} \\[2mm]
B &= 4\pi y \sqrt{(1 - e^2)} \sqrt{(1 - e'^2)} \int_0^{\frac{a_1'}{a'}} \frac{u^2 du}{(1 - e^2 u^2)^{\frac{3}{2}} (1 - e'^2 u^2)^{\frac{1}{2}}} \\[2mm]
C &= 4\pi z \sqrt{(1 - e^2)} \sqrt{(1 - e'^2)} \int_0^{\frac{a_1'}{a'}} \frac{u^2 du}{(1 - e^2 u^2)^{\frac{1}{2}} (1 - e'^2 u^2)^{\frac{3}{2}}}
\end{aligned}
\right\} \quad (5).
$$

23. If the point attracted be within the ellipsoid, the attraction of all the similar concentric shells without the point will be nothing; and hence the superior limit of u will be the value of $\dfrac{a_1}{a}$ at the surface of an ellipsoid, similar to the given one, and passing through the point attracted.

Now, in this case, $a_1 = a$, since a is one of the semi-axes of an ellipsoid passing through the point attracted, and having the same foci as another ellipsoid (passing through the same point) whose corresponding semi-axis is a_1. Hence, for an interior point, we have

$$
\left.
\begin{aligned}
A &= 4\pi x \sqrt{(1 - e^2)} \sqrt{(1 - e'^2)} \int_0^1 \frac{u^2 du}{\sqrt{(1 - e^2 u^2)} \sqrt{(1 - e'^2 u^2)}} \\[2mm]
B &= 4\pi y \sqrt{(1 - e^2)} \sqrt{(1 - e'^2)} \int_0^1 \frac{u^2 du}{(1 - e^2 u^2)^{\frac{3}{2}} (1 - e'^2 u^2)^{\frac{1}{2}}} \\[2mm]
C &= 4\pi z \sqrt{(1 - e^2)} \sqrt{(1 - e'^2)} \int_0^1 \frac{u^2 du}{(1 - e^2 u^2)^{\frac{1}{2}} (1 - e'^2 u^2)^{\frac{3}{2}}}
\end{aligned}
\right\} \quad \dots(6).
$$

24. These are the known expressions for the attraction of an ellipsoid on a point within it. Equations (5) agree with the expressions given in the Supplement to Liv. V. of Pontécoulant's *Théorie Analytique du Système du Monde*, where they are found by direct integration, by a method discovered by Poisson. They may also be readily deduced from equations (6) by Ivory's Theorem. Or, on the other hand, by a comparison of them, after reducing the limits of the integrals to 0 and 1, by substituting $\dfrac{a_1'}{a'} v$ for u, with equation (6), Ivory's Theorem may be readily demonstrated.

II.—ON THE MATHEMATICAL THEORY OF ELECTRICITY IN EQUILIBRIUM.

(Art. xviii. of complete list in *Mathematical and Physical Papers*, Vol. i.)

I.—ON THE ELEMENTARY LAWS OF STATICAL ELECTRICITY.*

[From *Cambridge and Dublin Mathematical Journal*, Nov. 1845. Reprinted *Philosophical Magazine*, 1854, second half-year, with additional Notes of date March 1854.]

25. The elementary laws which regulate the distribution of electricity on conducting bodies have been determined by means of direct experiments, by Coulomb, and in the form he has given them, which is independent of any hypothesis,† they have long been considered as rigorously established. The problem of the distribution of electricity in equilibrium on a conductor of any form was thus brought within the province of mathematical analysis; but the solution, even in the simplest cases, presented so much difficulty that Coulomb, after having investigated it experimentally for bodies of various forms, could only compare his measurements with the results of his theory by very rude processes of approximation. Without, however, giving rigorous solutions in particular cases, he examined the general problem with great care, and left nothing indefinite in the conditions to be satisfied, so that it was entirely by analytical difficulties that he was stopped. As an example of the

---

* This paper is a translation (with considerable additions) of one which appeared in Liouville's *Journal de Mathématiques*, 1845, p. 209.

† Coulomb has expressed his theory in such a manner that it can only be attacked in the way of proving his experimental results to be inaccurate. This is shown in the following remarkable passage in his sixth memoir, which follows a short discussion of some of the physical ideas then commonly held with reference to electricity. "*Je préviens pour mettre la théorie qui va suivre à l'abri de toute dispute systématique, que dans la supposition des deux fluides électriques, je n'ai d'autre intention que de présenter avec le moins d'élémens possible, les resultats de calcul et de l'expérience, et non d'indiquer les véritables causes de l'électricité. Je renverrai, à la fin de mon travail sur l'électricité, l'examen des principaux systèmes auxquels les phénomènes électriques ont donné naissance.*"—Histoire de l'Académie, 1788, p. 673.

success of his theoretical investigations, we may refer to the
well-known demonstration of the theorem (usually attributed
to Laplace) relative to the repulsion exercised by a charged
conductor on a point near its surface.*

The memoirs of Poisson, on the mathematical theory, contain the analytical determination of the distribution of electricity on two conducting spheres placed near one another,
the solution being worked out in numbers in the case of two
equal spheres in contact, which had been investigated experimentally by Coulomb (as well as in another case, not examined
by Coulomb, which is given as a specimen of the numerical
results that may be deduced from the formulæ). The calculated ratios of the intensities at different points of the surface
he is therefore enabled to compare with Coulomb's measurements, and he finds an agreement which is quite as close as
could be expected, when we consider the excessively difficult
and precarious nature of quantitative experiments in electricity:
but the most remarkable confirmation of the theory from these
researches is the entire agreement of the principal features,

---

* This theorem may be stated as follows:—Let *A* be a closed surface of
any form, and let matter, attracting inversely as the square of the distance, be
so distributed over it that the resultant attraction on an interior point is
nothing: the resultant attraction on an exterior point, indefinitely near any
part of the surface, will be perpendicular to the surface and equal to $4\pi\rho$,
if $\rho\omega$ be the quantity of matter on an element ω of the surface in the neighbourhood of the point. Coulomb's demonstration of this theorem may be
found in a preceding paper in the *Mathematical Journal*, Vol. iii. p. 74 (above,
I. 7). He gives it himself, in his sixth memoir on Electricity (*Histoire de
l'Académie*, 1788, p. 677), in connexion with an investigation of the theory
of the proof plane in which, by an error that is readily rectified, he arrives at
the result that a small insulated conducting disc, put in contact with an electrified conductor at any point, and then removed, carries with it as much electricity as lies on an element of the conductor at that point equal in area to the
two faces of the disc; the quantity actually removed being only half of this.
This result, however, does not at all affect the experimental use which he
makes of the proof plane, which is merely to find the ratios of the intensities
at different points of a charged conductor. As the complete theory of this
valuable instrument has not, so far as I am aware, been given in any English
work, I annex the following remarkably clear account of it, which is extracted from Pouillet's *Traité de Physique*:—"Quand le plan d'épreuve est
tangent à une surface, il se confond avec l'élément qu'il touche, il prend en
quelque sorte sa place relativement à l'électricité, ou plutôt il devient lui-
même l'élément sur lequel la fluide se répand; ainsi, quand on retire ce
plan, on fait la même chose que si l'on avait découpé sur la surface un
élément de même épaisseur et de même étendue que lui, et qu'on l'eût enlevé
pour le porter dans la balance sans qu'il perdît rien de l'électricité qui le

even in some very singular phenomena, of the experimental results with the theoretical deductions. For a complete account of the experiments we must refer to Coulomb's fifth memoir (*Histoire de l'Académie*, 1787), and for the mathematical investigations to the first and second memoirs of Poisson (*Mémoires de l'Institut*, 1811), or to the treatise on Electricity in the *Encyclopædia Metropolitana*, where the substance of Poisson's first memoir is given.

The mathematical theory received by far the most complete development which it has hitherto obtained, in Green's *Essay on the Application of Mathematical Analysis to the Theories of Electricity and Magnetism*,* in which a series of general theorems were demonstrated, and many interesting applications made to actual problems.†

Of late years some distinguished experimentalists have begun to doubt the truth of the laws established by Coulomb, and have made extensive researches with a view to discover the laws of certain phenomena which they considered incompatible with his theory. The most remarkable works of this kind

---

couvre; une fois séparé de la surface, cet élément n'aurait plus dans ses différents points qu'une épaisseur électrique moitié‘moindre, puisque la fluide devrait se répandre pour en couvrir les deux faces. Ce principe posé, l'expérience n'exige plus que de l'habitude et de la dextérité: après avoir touché un point de la surface avec le plan d'épreuve, on l'apporte dans la balance, où il partage son électricité avec le disque de l'aiguille qui lui est égale, et l'on observe la force de torsion à une distance connue. On répète la même experience en touchant un autre point, et le rapport des forces de torsion est le rapport des repulsions électriques; on en prend la racine carrée pour avoir le rapport des épaisseurs. Ainsi le génie de Coulomb a donné en même temps aux mathématiciens la loi fondamentale suivant laquelle la matière électrique s'attire et se repousse; et aux physiciens une balance nouvelle, et des principes d'expérience au moyen desquels ils peuvent en quelque sorte sonder l'épaisseur de l'électricité sur tous les corps, et determiner les pressions qu'elle exerce sur les obstacles qui l'arrêtent."

To this explanation it should be added that, when the proof plane is still very near the body to which it has been applied, the effect of mutual influence is such as to make the intensity be insensible at every point of the disc on the side next the conductor, and at each point of the conductor which is *under* the disc. It is only when the disc is removed to a considerable distance that the electricity spreads itself symmetrically on its two faces, and that the intensity at the point of the conductor to which it was applied, recovers its original value. It was the omission of this consideration that caused Coulomb to fall into the error alluded to above.

* Nottingham, 1828.

† This memoir of Green's has been unfortunately very little known, either in this country or on the Continent. Some of the principal theorems in it

have been undertaken independently by Mr Snow Harris and
Mr Faraday, and in their memoirs, published in the *Philosophical Transactions*, we find detailed accounts of their researches. All the experiments, however, which they have
made, having direct reference to the distribution of electricity
in equilibrium, are, I think, in full accordance with the laws
of Coulomb, and must therefore, instead of objections to his
theory, be considered as confirming it. As, however, many
have believed Coulomb's theory to be overturned by these
investigations, and as others have at least been led to entertain
doubts as to its certainty or accuracy, the following attempt
to explain the apparent difficulties is made the subject of the
first of a series of papers in which various parts of the mathematical theory of electricity, and corresponding problems in
the theories of magnetism and heat, will be considered.

26. We may commence by examining some experimental
results published in Mr Harris's first memoir *On the Elementary Laws of Electricity.** After describing the instruments
employed in his researches, Mr Harris gives the details of
some experiments with reference to the attraction exercised
by an insulated electrified body on an uninsulated conductor
placed in its neighbourhood. The first result which he an-

---

have been re-discovered within the last few years, and published in the following
works :—

    Comptes Rendus for Feb. 11th, 1839, where part of the series of theorems is
announced without demonstration, by Chasles.

    Gauss's memoir on "General Theorems relating to Attractive and Re-
pulsive Forces, varying inversely as the square of the distance," in the
Resultate aus den Beobachtungen des magnetischen Vereins im Jahre 1839,
Leipsic, -1840. (Translations of this paper have been published in Taylor's
Scientific Memoirs for April 1842, and in the Numbers of Liouville's *Journal*
for July and August 1842.)

    Mathematical Journal, vol. iii., Feb. 1842, in a paper "On the Uniform
Motion of Heat, etc." (I. above).

    Additions to the Connaissance des Tems for 1845 (published June 1842),
where Chasles supplies demonstrations of the theorems which he had previously
announced.

    I should add that it was not till the beginning of the present year (1845)
that I succeeded in meeting with Green's Essay. The allusion made to his
name with reference to the word "potential" (*Mathematical Journal*, vol. iii.
p. 190), was taken from a memoir of Murphy's, "On definite Integrals with
Physical Applications," in the *Cambridge Transactions*, where a mistaken
definition of that term, as used by Green, is given.

    * *Philosophical Transactions*, 1834.

nounces is that, when other circumstances remain the same, the attraction varies as the square of the quantity of electricity with which the insulated body is charged. It is readily seen, as was first remarked by Dr Whewell in his *Report on the Theories of Electricity, etc.*,* that this is a rigorous deduction from the mathematical theory, following from the fact that the quantity of electricity induced upon the uninsulated body is proportional to the charge on the electrified body by which it is attracted.

27. The remaining results have reference to the force of attraction at different distances, and with bodies of different forms opposed. As these are generally very irregular (such as "plane circular areas backed by small cones"), we should not, according to Coulomb's theory, expect any very simple laws, such as Mr Harris discovers, to be rigorously true. Accordingly, though they are announced by him without restriction, we must examine whether the experiments from which they have been deduced are of a sufficiently comprehensive character to lead to any general conclusions with respect to electrical action. Now, in the first place, we find that in all of them the attraction is "independent of the form of the unopposed parts" of the bodies, which will be the case only when the intensity of the induced electricity on the unopposed parts of the uninsulated body is insensible. According to the mathematical theory, and according to Mr Faraday's researches "on induction in curved lines," which will be referred to below, the intensity never absolutely vanishes at any point of the uninsulated body: but it is readily seen that in the case of Mr Harris's experiments, it will be so slight on the unopposed portions that it could not be perceived without experiments of a very refined nature, such as might be made by the proof plane of Coulomb, which is in fact, with a slight modification, the instrument employed by Mr Faraday in the investigation. Now to the degree of approximation to which the intensity on the unopposed parts may be neglected, the laws observed by Mr Harris when the opposed surfaces are plane may be readily deduced from the mathematical theory. Thus let v be the

* *British Association Report* for 1837.

potential in the interior of the charged body, A; a quantity which will depend solely on the state of the interior coating of the battery with which in Mr Harris's experiments A is connected, and will therefore be sensibly constant for different positions of A relative to the uninsulated opposed body, B. Let a be the distance between the plane opposed faces of A and B, and let S be the area of the opposed parts of these faces, which will in general be the area of the smaller, if they be unequal. When the distance a is so small that we may entirely neglect the intensity on all the unopposed parts of the bodies, it is readily shown from the mathematical theory that (since the difference of the potentials at the surfaces of A and B is v) the intensity of the electricity produced by induction at any point of the portion of the surface of B which is opposed to A, is $\dfrac{v}{4\pi a}$ Hence the attraction on any small element ω, of the portion S of the surface of B, will be in a direction perpendicular to the plane and equal to $2\pi\left(\dfrac{v}{4\pi a}\right)^2.*$ Hence the whole attraction on B is

$$\frac{v^2 S}{8\pi a^2}.$$

This formula expresses all the laws stated by Mr Harris as results of his experiments in the case when the opposed surfaces are plane.

28. When the opposed surfaces are curved, for instance when A and B are equal spheres, we can make no approximation analogous to that which has led us to so simple an expression in the case of opposed planes; and we find accordingly that no such simple law for the attraction in this case has been announced by Mr Harris. He has, however, found that it is expressed with tolerable accuracy by the formula

$$F = \frac{k}{c(c-2a)},$$

where c is the distance between the centres of the spheres, a the radius of each, k a constant, which will depend on a and on the charge of the battery with which A is in communica-

---

* See VII. below.

tion. Though, however, this formula may give results which do not differ very much from observation within a limited range of distances, it cannot, according to any theory, be considered as expressing the physical law of the phenomenon. For, according to it, when the balls are very distant, F ultimately varies as $\frac{1}{c^2}$. Now it is clear that the law of force must ultimately become the inverse cube of the distance, since the quantity of electricity induced upon B will be ultimately in the inverse ratio of the distance, and the attraction between the balls as the product of the quantities of electricity directly, and as the square of the distance inversely, and hence the formula given by Mr Harris cannot express the law of force when the balls are very distant. In the experiments by which his formula is tested, the force of attraction is measured by means of an ordinary balance and weights: the only comparison of results which he publishes is transcribed in the following table :—

Dist. of Centres.	Measured Force in Grains.	Values of $\frac{15c_1(c_1-2)}{c(c-2)}$.
$c_1 = 2\cdot3$	15	15
$c_2 = 2\cdot5$	$8\cdot25+$	$8\cdot28$
$c_3 = 2\cdot8$	$4\cdot6+$	$4\cdot62$
$c_4 = 3\cdot0$	$3\cdot5-$	$3\cdot45$

29. From this table we see that the formula is verified in three cases to the extent of accuracy of the experiments. Comparisons extended to a much wider range of distances would be required to establish it, and it would be necessary to take precautions to prevent the experimental results from being influenced by disturbing causes. In the experiments made by Mr Harris, we find that no precautions have been taken to avoid the disturbing influence of extraneous conductors, which, according to the descriptions and drawings he gives of his instruments, seem to exist very abundantly in the neighbourhood of the bodies operated upon, being partly metal in connexion with the insulated system with which the body A communicates, and partly uninsulated metal, in the fixed parts of the electro-

meter, and in the movable parts by which B is supported.
The general effect produced by the presence of such bodies
in disturbing the observed law of force, must be to make it
diminish less rapidly with the distance when A and B are
separated by a considerable interval: and it is probably owing,
at least in part, to such disturbing causes that Mr Harris's
results nearly agree, as far as they go, with a formula which
would ultimately give for the law of force the inverse square of
the distance between A and B, instead of the inverse cube.

30. The determination by the mathematical theory of the
attraction or repulsion between two electrified conducting
spheres has not hitherto, so far as I am aware, been attempted,
and would present considerable difficulty by means of the
formulæ ordinarily given for such problems. It may, however,
very readily be effected by means of a general theorem on the
attraction between electrified conductors, which will be given
in a subsequent paper.* Thus, if $F(c)$ be the force of attraction,
corresponding to the distance c between the centres, in the
particular case when the two spheres are equal (the radius of
each being unity), and the potential in the interior of one of
them is nothing (as will be the case when the body is uninsulated), the potential in the interior of the other being v,
I have found the following formulæ, which express $F(c)$ by a
converging series :—

$$F(c) = v^2 c \left(\frac{P_1}{Q_1^2} + \frac{P_2}{Q_2^2} + \frac{P_3}{Q_3^2} + \text{etc.} \right) \dots\dots\dots\dots(A),$$

where
$$\left. \begin{aligned} Q_1 &= c^2 - 1 \\ Q_2 &= (c^2 - 2) Q_1 - 1 \\ Q_{n+2} &= (c^2 - 2) Q_{n+1} - Q_n \end{aligned} \right\} \dots\dots\dots\dots(B),$$

$$\left. \begin{aligned} P_1 &= 1 \\ P_2 &= 2c^2 - 3 \\ P_{n+2} &= (c^2 - 2) P_{n+1} + (Q_{n+1} - P_n) \end{aligned} \right\} \dots\dots(C).$$

* [*Note added March* 1854.—The enunciation of the "general theorem"
alluded to, the investigation founded on it, by which the author first arrived
at the conclusion made use of here, and another demonstration of the same
conclusion, founded on the method of electrical images, and strictly synthetical in its character, are published, with comprehensive numerical results,
in the *Philosophical Magazine* for April 1853.]

31. These formulæ enable us to calculate Q_1, Q_2, Q_3, Q_4, etc., and then P_1, P_2, P_3, P_4, etc., successively, by a simple and uniform arithmetical process, for any particular value of c, I have thus calculated the values of $\dfrac{F(c)}{v^2}$ in five cases, the first four of which are those examined by Mr Harris, and have obtained the following results, each of which is true to five places of decimals :—

c.	$v^{-2}F(c)$.
2·3	0·32926
2·5	0·17423
2·8	0·09168
3·0	0·06592
4·0	0·02075

32. To compare these with Mr Harris's measurements, we may calculate the value of the potential in his battery, during the observations, by means of his first result, and thence find the attraction for the other three cases by means of the calculated values of $v^{-2}F(c)$. Thus we have $v^{-2} \times 15 = \cdot3293$, which gives
$$v^2 = 45\cdot56,$$
and hence
$$F(2\cdot5) = 7\cdot94,$$
$$F(2\cdot8) = 4\cdot18,$$
$$F(3) = 3\cdot00.$$

These numbers differ considerably from Mr Harris's results, but in the direction indicated by the considerations mentioned above.

33. The most important part of the researches of Mr Harris is that in which he investigates the insulating power of air of different densities. The result at which he arrives is, that the intensity necessary to produce a spark depends solely on the density of the air, and not otherwise on the pressure or temperature. He thus shows that the conducting power of flame, of heated bodies, and of a vacuum, are due solely to the rarefaction of the air in each case. He also shows that the intensities necessary to produce a spark are in the simple ratios of the densities of the air.

34. In a subsequent memoir, by the same author,* we find additional experiments on the elementary principles of the theory of electricity. The first series which is described, was made for the purpose of testing the truth of Coulomb's law, that the repulsion of two similarly charged points is inversely as the square of the distance, and directly as the product of the masses. In experiments of this kind in which accurate quantitative results are aimed at, many precautions are necessary. Thus all conducting bodies, except those operated upon, must be placed beyond the reach of influence, and the distance between the repelling bodies must be considerable with reference to their linear dimensions, so that the distribution of electricity on each may be uninfluenced by the presence of the other. Also the bodies should be spheres, so that the attraction may be the same as if the whole electricity of each were collected at its centre; and the distance to be measured will then be the distance between the centres. These conditions have been expressly mentioned by Coulomb, and they have been fulfilled, as far as possible, in his researches, as we see by the descriptions of the experiments made, which we find in his memoirs. He has thus arrived by direct measurement at the law, which we know by a mathematical demonstration,† founded upon independent experiments, to be the rigorous law of nature, for electrical action. None of these precautions, however, have been taken in the experiments described in Mr Harris's

---

* *Philosophical Transactions*, 1836.
† See Murphy's *Electricity*, p. 41, or Pratt's *Mechanics*, Art. 154.
[*Note added March* 1854.—Cavendish demonstrates mathematically that if the law of force be any other than the inverse square of the distance, electricity could not rest in equilibrium on the surface of a conductor. But experiment has shown that electricity does rest at the surface of a conductor. Hence the law of force must be the inverse square of the distance. Cavendish considered the second proposition as highly probable, but had not experimental evidence to support this opinion, in his published work (An attempt to explain the phænomena of Electricity by means of an Elastic Fluid). Since his time, the most perfect experimental evidence has been obtained that electricity resides at the surface of a conductor; in such facts, for instance, as the perfect equivalence in all electro-statical relations of a hollow metallic conductor of ever so thin substance, or of a gilt non-conductor (possessing a conducting film of not more than $\frac{1}{200000}$ of an inch thick) and a solid conductor of the same external form and dimensions; the minor premise of his syllogism is thus demonstrated, and the conclusion is therefore established.]

memoir, and the results are accordingly unavailable for the accurate *quantitative* verification of any law, on account of the numerous unknown disturbing circumstances by which they are affected. The phenomena which he observes, however, afford *qualitative* illustrations of the mathematical theory of a very interesting nature, as may be seen from the following examples of his results :—

(*a*) When the distance between the bodies is great with reference to their linear dimensions, the repulsion is inversely as the square of the distance, and directly as the product of the masses.

(*b*) When the distance is small, the action becomes apparently irregular. Thus if the quantities of electricity on the two bodies be equal, the force, which is always of repulsion, does not increase so rapidly when the bodies approach, as if it followed the law of the inverse square of the distance.

(*c*) If the charges be unequal, the repulsion ceases at a certain distance, and at all smaller distances there is attraction between the bodies.

35. These results are, with all their peculiarities, in full accordance with the theory of Coulomb, which indicates that, if the quantities of electricity be equal, and the bodies equal and similar, there will be repulsion in every position : but if there be any difference, however small, between the charges, the repulsion will necessarily cease, and attraction commence, before contact takes place, when one body is made to approach the other. Unless, however, the difference of the charges be sufficiently considerable, a spark may pass between the bodies, and render the charges equal, before attraction commences. In Mr Harris's experiments, in which the bodies seem to have been nearly oblate spheroids, the attraction is generally sensible before the distance is small enough to allow a spark to pass, if the charge on one be double of that on the other.

Mr Harris next proceeds to investigate the theory of the proof plane, and to examine whether it can be considered as indicating with certainty the intensity of electricity at any part of a charged body, and, principally from an experiment made on a charged non-conductor (a hollow sphere of glass), comes to a negative conclusion. It should be remembered,

however, that, the proof plane having never been applied to
determine the intensity at points of the surface of a charged
non-conductor, such conclusions in no way interfere with
adopted ideas. Since there can be no manner of doubt as to
the theory of this valuable instrument, as we find it explained
by M. Pouillet,* nor as to the experimental use of it made by
Coulomb, it is unnecessary to enter more at length on the
subject here.

36. Mr Faraday's researches on electrostatical induction,
which are published in a memoir forming the eleventh series
of his Experimental Researches in Electricity, were undertaken with a view to test an idea which he had long possessed,
that the forces of attraction and repulsion exercised by free
electricity, are not the resultant of actions exercised at a distance, but are propagated by means of molecular action among
the contiguous particles of the insulating medium surrounding
the electrified bodies, which he therefore calls the *dielectric.*
By this idea he has been led to some very remarkable views
upon induction, or, in fact, upon electrical action in general.
As it is impossible that the phenomena observed by Faraday
can be incompatible with the results of experiment which
constitute Coulomb's theory, it is to be expected that the
difference of his ideas from those of Coulomb must arise solely
from a different method of stating, and interpreting physically,
the same laws: and farther, it may, I think, be shown that
either method of viewing the subject, when carried sufficiently
far, may be made the foundation of a mathematical theory
which would lead to the elementary principles of the other as
consequences. This theory would accordingly be the expression of the ultimate law of the phenomena, independently of
any physical hypothesis we might, from other circumstances,
be led to adopt. That there are necessarily two distinct
elementary ways of viewing the theory of electricity, may
be seen from the following considerations, founded on the
principles developed in a previous paper in this Journal.†

---

* See foot-note on § 25.
† On the Uniform Motion of Heat, and its Connexion with the Mathematical Theory of Electricity (I. above).

37. Corresponding to every problem relative to the distribution of electricity on conductors, or to forces of attraction and repulsion exercised by electrified bodies, there is a problem in the uniform motion of heat which presents the same analytical conditions, and which, therefore, considered mathematically, is the same problem. Thus, let a conductor A, charged with a given quantity of electricity, be insulated in a hollow conducting shell, B, which we may suppose to be uninsulated. According to the mathematical theory, an equal quantity of electricity of the contrary kind will be attracted to the interior surface of B (or the surface of B, as we may call it to avoid circumlocution), and the distribution of this charge, and of the charge on A, will take place so that the resultant attraction at any point of each surface may be in the direction of the normal. This condition being satisfied, it will follow that there is no attraction on any point within A, or without the surface of B, that is, on any point within either of the conducting bodies. The most convenient mathematical expression for the condition of equilibrium, is that the potential at any point P* must have a constant value when P is on the surface of A, and the value nothing when P is on the surface of B; and it will follow from this that the potential will have the same constant value for any point within A, and will be equal to nothing for any point without the surface of B.

If A be subject to the influence of any uninsulated conductors, we must consider such bodies as belonging to the shell in which A is contained, and their surfaces as forming part of the surface of B: in such cases this surface will generally be the interior surface of the walls of the room in which A is contained, and of all uninsulated conductors in the room. If, however, we have to consider the case in which A is subject to no external influence, we must suppose every part of the surface of B to be very far from A. The most general problem we can contemplate in electricity (exclusively of the case in which the insulating medium is heterogeneous, and exercises a special action, which will be alluded to below), is to determine

---

* The term used by Green for the sum of the quotients obtained by dividing the product of each element of the surfaces of A and B, and its electrical intensity, by its distance from P.

the potential at any point when *A*, instead of being a single
conductor, is a group of separate insulated conductors charged
to different degrees, and when there are non-conductors electrified in a given manner, placed in the insulating medium, in
the neighbourhood. The conditions of equilibrium will still be
that the potential at each surface due to all the free electricity
must be constant, and the theorems stated above will still be
true: thus the attraction will be nothing in the interior of
each portion of *A*, and without the surface of *B*; and the
whole quantity of induced electricity on the latter surface will
be the algebraic sum of the charges of all the interior bodies
with its sign changed. When the potential due to such a
system is determined for every point, the component of the
resultant force at any point *P*, in any direction *PL*, may be
found by differentiation, being the limit of the difference
between the values of the potential at *P*, and at a point *Q*, in
PL, divided by *PQ*, when *Q* moves up towards and ultimately
coincides with *P*, and the direction of the force, on a *negative*
particle, being that in which the potential increases. By
Coulomb's theorem, the intensity at any point in one of the
conducting surfaces is equal to the attraction (on a negative
unit) at that point, divided by 4π.

38.   Now if we wish to consider the corresponding problem
in the theory of heat, we must suppose the space between *A*
and *B*, instead of being filled with a dielectric medium (that is
a non-conductor for electricity), to be occupied by any homogeneous solid body, and sources of heat or cold to be so distributed over the terminating surfaces, or the interior surface
of *B* and the surface of *A*, that the permanent temperature
at the first surface may be zero, and at the second shall have a
certain constant value, the same as that of the *potential* in the
case of electricity. If *A* consist of different isolated portions,
the temperature at the surface of each will have a constant
value, which is not necessarily the same for the different portions. The problem of *distributing sources of heat, according to
these conditions*, is mathematically identical with the problem
of distributing *electricity in equilibrium* on the surfaces of *A*
and *B*. In the case of heat, the *permanent temperature* at any
point replaces the *potential* at the corresponding point in the

electrical system, and consequently the *resultant flux of heat* replaces the *resultant attraction* of the electrified bodies, in direction and magnitude. The problem in each case is determinate, and we may therefore employ the elementary principles of one theory, as theorems, relative to the other. Thus, in the paper in which these considerations are developed, Coulomb's fundamental theorem relative to electricity is applied to the theory of heat ; and self-evident propositions in the latter theory are made the foundation of Green's theorems in electricity.* Now the laws of motion for heat which Fourier lays down in his *Théorie Analytique de la Chaleur*, are of that simple elementary kind which constitute a mathematical theory properly so called ; and therefore, when we find corresponding laws to be true for the phenomena presented by electrified bodies, we may make them the foundation of the mathematical theory of electricity: and this may be done if we consider them merely as actual truths, without adopting any physical hypothesis, although the idea they naturally suggest is that of the propagation of some effect by means of the mutual action of contiguous particles; just as Coulomb, although his laws naturally suggest the idea of material particles attracting or repelling one another at a distance, most carefully avoids making this a *physical hypothesis*, and confines himself to the consideration of the mechanical effects which he observes and their necessary consequences.†

39. All the views which Faraday has brought forward, and illustrated or demonstrated by experiment, lead to this method of establishing the mathematical theory, and, as far as the analysis is concerned, it would, in most *general* propositions, be even more simple, if possible, than that of Coulomb. (Of course the analysis of *particular* problems would be identical in the two methods.) It is thus that Faraday arrives at a knowledge of some of the most important of the general

---

* It was not until some time after that paper was published, that I was able to add the direct analytical demonstrations of the theorems, which are given in the papers on "General Propositions in the Theory of Attraction," *Camb. Math. Jour.*, vol. iii. pp. 189, 201 (XII. below), and which I have since found are the same as those originally given by Green.

† See first foot note on § 25.

theorems, which, from their nature, seemed destined never to be perceived except as mathematical truths. Thus, in his theory, the following proposition is an elementary principle:— Let any portion α of the surface of A be projected on B, by means of lines (which will be in general curved) possessing the property that the resultant electrical force at any point of each of them is in the direction of the tangent: the quantity of electricity produced by induction on this projection is equal to the quantity of the opposite kind of electricity on α.* The lines thus defined are what Faraday calls the "curved lines of inductive action." For a detailed account of the experiments by which these phenomena are investigated, reference must be made to Mr Faraday's own memoirs, published in the *Philosophical Transactions*, and in a separate form in his *Experimental Researches*.

40. The hypothesis adopted by Faraday, of the *propagation* of inductive action, naturally led him to the idea that its effects may be in some degree dependent upon the nature of the insulating medium or dielectric, by which, according to this view, it is transmitted. In the second part of his memoir he describes a series of researches instituted to put this to the test of experiment, and arrives at the following conclusions:—

---

* This theorem may be proved as follows:—

Let S be any closed surface, containing no part of the electrified bodies within it, which we may conceive to be described between A and B; let P be the component in the direction of the normal, of the resultant force at any point of the surface S, and let ds be an element of the surface at the same point. Then it may be easily proved (see *Camb. Math. Jour.*, vol. iii. p. 204) that $\iint P ds = 0$ (a), the integrations being extended over the entire surface. Now let S be supposed to consist of three parts; the portion α, of the surface of A; its projection β, on the interior surface of B; and the surface generated by the curved lines of projection. The value of P at each point of the latter portion of S will be nothing, since the tangent at any point of a line of projection is the direction of the force. Hence, if $[\iint P ds]$ and $(\iint P ds)$ denote the values of $\iint P ds$, for the portions α and β of S, the equation (a) becomes

$$[\iint P ds] + (\iint P ds) = 0.$$

But if ρ be the intensity of the distribution on the surface A or B, at any point, we have, by Coulomb's theorem,

$$\rho = \frac{P}{4\pi}.$$

Hence $[\iint \rho ds] + (\iint \rho ds) = 0,$

which is the theorem quoted in the text.

41. If the dielectric be air, the inductive action is quite independent of its density or temperature (which, as Mr Faraday remarks, agrees perfectly with previous results obtained by Mr Harris); and in general, if the dielectric be any gas or vapour capable of insulating a charge, the inductive action is invariable. Hence he concludes that "*all gases have the same power of*, or *capacity for*, sustaining induction through them (which might have been expected when it was found that no variation of density or pressure produced any effect)."

When the dielectric is solid, the induction is greater than through air, and varies according to the nature of the substance. Numbers which measure the "specific inductive capacities" of the dielectrics employed (sulphur, shell lac, glass, etc.) are deduced from the experiments.

42. To express these results in the language of the mathematical theory, let us recur to the supposition of a body, A, charged with a given quantity of electricity, and insulated in the interior of a closed conducting shell, B. The potential of the system at the interior surface of B, and at every point without this surface, will be nothing ; at the surface and in the interior of A it will have a constant value, which will depend on the form, magnitude, and relative position of the surfaces A and B, on the quantity of electricity on A, and, according to Faraday's discovery, on the *dielectric power* of the insulating medium which fills the space between A and B. If this be gaseous, neither its nature nor its state as to temperature, pressure, or density will affect the value of the potential in A ; but if it be a solid substance, such as sulphur or shell lac, the value of the potential will be less than when the space is occupied by air, and will vary with the nature of the insulating solid.

43. The result in the case of a gaseous dielectric is what would follow from Coulomb's theory, if we consider gases to be quite impermeable to electricity, and to be entirely unaffected by electrical influence. The phenomena observed with solid dielectrics, which agree with the circumstance observed by Nicholson, that the *dissimulating power* of a Leyden phial depends on the nature of the glass of which it is made, as well as on its thickness, have been by some attributed to a slight degree of conducting power, or of penetrability, pos-

sessed by solid insulators. This explanation, however, seems
to be very insufficient; and besides, Faraday has estimated the
nature of the effects of imperfect insulation by independent
experiments, and has established, in what seems to be a very
satisfactory manner, the existence of a peculiar action in the
interior of solid insulators when subjected to electrical influence. As far as can be gathered from the experiments which
have yet been made, it seems probable that a dielectric, subjected to electrical influence, becomes excited in such a manner
that every portion of it, however small, possesses *polarity*
exactly analogous to the magnetic polarity induced in the substance of a piece of soft iron under the influence of a magnet.
By means of a certain hypothesis regarding the nature of magnetic action,* Poisson has investigated the mathematical laws
of the distribution of magnetism, and of magnetic attractions
and repulsions. These laws seem to represent in the most
general manner the state of a body polarized by influence, and
therefore, without adopting any particular mechanical hypothesis, we may make use of them to form a mathematical
theory of electrical influence in dielectrics, the truth of which
can only be established by a rigorous comparison of its results
with experiment.

44. Let us therefore consider what would be the effect, according to this theory, which would be produced by the presence
of a solid dielectric, *C*, placed in the space between *A* and *B*,
the rest of which is occupied by air. The action of *C*, when
excited by the influence of the electricities on *A* and *B*, may
(as Poisson has shown for magnetism) be represented, whether

---

* Faraday adopts the corresponding hypothesis to explain the action of a
solid dielectric, which he states thus:—" If the space round a charged globe
were filled with a mixture of an insulating dielectric, as oil of turpentine or
air, and small globular conductors, as shot, the latter being at a little distance from each other, so as to be insulated, then these in their condition
and action exactly resemble what I consider to be the condition and action
of the particles of the insulating dielectric itself. If the globe were charged,
these little conductors would all be polar; if the globe were discharged, they
would all return to their normal state, to be polarized again upon the re-
charging of the globe."—(*Experimental Researches*, § 1679.) The results of
the mathematical analysis of such an action are given in the text. It may
be added that the value of the coefficient *k* will differ sensibly from unity if
the volume occupied by the small conducting balls bear a finite ratio to that
occupied by the insulating medium.

on points within or without C, by a certain distribution of positive electricity on one portion of the surface of C, and of an equal quantity of negative electricity on the remainder. The condition necessary and sufficient for determining this distribution may (as can be shown from Poisson's analysis) be expressed as follows. Let R be the resultant force on a point P without C, and R' on a point P' without C, due to the electrified surfaces A and B, and to the imagined distribution on C. If P and P' be taken infinitely near one another, and consequently each infinitely near the surface of C, the component of R' in the direction of the normal must bear to the component of R in the same direction a constant ratio $\left(\dfrac{1}{k}\right)$ depending on the capacity for dielectric induction of the matter of C.* The components of R and R' in the tangent plane will of course be equal and in the same direction, and, if ρ be the intensity of the imagined distribution on the surface of C, in the neighbourhood of P and P', the difference of the normal components will be $4\pi\rho$, as is evident from Coulomb's theorem, referred to above.

45.   Let us now suppose C to be a shell surrounding A, and let S and S', its interior and exterior surfaces, be *surfaces of equilibrium* in the system of forces due to the action of A and B, and of the polarity of C. It may be shown that the same surfaces S, S', would necessarily be surfaces of equilibrium, if C were removed and the whole space were filled with air; and consequently, that the whole series of surfaces of equi-

---

* From this it follows that, in the case of heat, C must be replaced by a body whose conducting power is k times as great as that of the matter occupying the remainder of the space between A and B.

[*Note added March* 1854.—The same demonstration, of course, is applicable to the influence of a piece of soft iron, or other " paramagnetic " (*i.e.*, substance of ferro-magnetic inductive capacity), or to the reverse influence of a diamagnetic on the magnetic force in any locality near a magnet in which it can be placed, and shows that the lines of magnetic force will be altered by it precisely as the lines of motion of heat in corresponding thermal circumstances would be altered by introducing a body of greater or of less conducting power for heat.   Hence we see how strict is the foundation for an analogy on which the *conducting power of a magnetic medium for lines of force* may be spoken of, and we have a perfect explanation of the condensing action of a paramagnetic, and the repulsive effect of a diamagnetic, upon the lines of force of a magnetic field, which have been described by Faraday.— (*Exp. Researches*, §§ 2807, 2808.)]

librium, commencing with A and ending with B, will be the same in the two cases. Hence the resultant force due to the excitation of the dielectric C (or to the imagined distributions of electricity on S and S' which produce it), on points within S or without S', must be such as not to alter the distributions on A and B when the quantity on A is given; and is therefore nothing. Accordingly, let Q be the total force on a point indefinitely near S, and within it; Q' the total force on a point without S', but indefinitely near it. Since the forces on points without S and within S' indefinitely near the former points are, according to the law stated above, $\dfrac{Q}{k}$ and $\dfrac{Q'}{k}$, it follows* that the intensities of the imagined distributions on S and S'. in the neighbourhood of the points considered, are

$$-\frac{1}{4\pi}\left(Q-\frac{Q}{k}\right) \text{ and } \frac{1}{4\pi}\left(Q'-\frac{Q'}{k}\right).$$

Hence, if U, U' be the potentials at S, S', due to A and B alone, and v the potential at any point P, it follows that the potential at P, due to the polarity of the dielectric, is

$$-\left(1-\frac{1}{k}\right)U+\left(1-\frac{1}{k}\right)U',$$

or
$$-\left(1-\frac{1}{k}\right)v+\left(1-\frac{1}{k}\right)U',$$

or
$$-\left(1-\frac{1}{k}\right)v+\left(1-\frac{1}{k}\right)v, \text{ that is, } 0,$$

according as P is within S, within S' and without S, or without S'. Hence the total potential will be, according to the position

of P,
$$v-\left(1-\frac{1}{k}\right)(U-U'),$$

or
$$\frac{v}{k}+\left(1-\frac{1}{k}\right)U',$$

or
$$v.$$

Hence the sole effect of the dielectric C, on the state of A and B, is to diminish the potential in the interior of the former by the quantity

$$\left(1-\frac{1}{k}\right)(U-U').$$

---

* See Green's *Essay*, Art. 12; or above, I. § 8.

If the whole space between A and B be occupied by the solid dielectric, the surfaces S and A will coincide, as also, S' and B, and therefore $U = V$, $U' = 0$. Hence the potential in the interior of A will be $\dfrac{V}{k}$,

or the fraction $\dfrac{1}{k}$ of the potential, with the same charge on A, and with a gaseous dielectric. From this it follows that, when the dielectric is solid, it would require, to produce a given potential in the interior of A, k times the charge which would be necessary to produce the same potential when the dielectric is gaseous, and therefore the body A in a given state, defined by the potential in its interior, produces on the interior surface of B, by induction, through the solid dielectric, a quantity of electricity k times as great as through a gaseous dielectric. On this account Faraday calls the property of a dielectric measured by k, its " specific inductive capacity."

46.   In Faraday's experiments an apparatus (which is in fact a Leyden phial, in which any solid or fluid may be substituted for the glass dielectric of an ordinary Leyden phial) is used, corresponding to the case we have been considering, in which A is a conducting sphere (2·33 inches in diameter), and B a concentric spherical shell surrounding it (the distance between the surfaces of A and B being ·62 of an inch). In the shell B there is an aperture into which a shell-lac stem is fixed; a wire, attached to A, passes through the centre of this stem to the outside of the shell, and supports a ball of metal, M, which is thus insulated and connected with A. It may be shown that in such an apparatus the state of the ball A and of the shell B will approximately be not affected by the aperture in the latter, or by the wire supporting M, and that the distribution of electricity on M will be approximately the same as if the wire supporting it and the conductors A and B were removed. Hence the sole relation between A and M will be that the *potentials* in their interiors are the same; and therefore the latter, which is accessible, may be taken as an index of the state of the former.

47.   To determine the specific inductive capacity of any dielectric, Faraday uses two apparatus of the kind just described,

precisely equal and similar, in one of which the space between A and B is filled with air, and in the other with the dielectric to be examined. One of these apparatus is charged, and the intensity measured: the balls M, M' in the two are then made to touch and separate again, and the remaining intensity on the first (which is equal to the intensity imparted to the second) is measured. If this be found to differ from half the original intensity, it will follow that the specific inductive capacity of the substance examined differs from that of air, which is unity, and its value may be determined by means of a simple expression from the experimental data. To investigate this, let us first suppose each apparatus to be charged, and let it be required to find the intensity on the balls after they are made to touch, and then removed from mutual influence; and let the dielectrics be any two substances, whose inductive capacities are k, k'. Let ρ, ρ' be the intensities before, and σ the common intensity after contact. Then, denoting by Q, Q' the quantities of electricity constituting the charges before, and q, q' after contact, we shall have, by the principles already

developed,
$$\frac{Q}{Q'} = \frac{k\rho}{k'\rho'}, \quad \frac{\sigma}{\rho} = \frac{q}{Q}, \quad \frac{\sigma}{\rho'} = \frac{q'}{Q'}.$$

Also
$$Q + Q' = q + q'.$$

Hence we deduce
$$\sigma = \frac{k\rho + k'\rho'}{k + k'}.$$

In the experiment described, one of the dielectrics is air. Hence, to obtain the required formula, we may put $k' = 1$, in this equation, and then resolve for k.

Thus we find
$$k = \frac{\sigma - \rho'}{\rho - \sigma}.$$

If only one of the apparatus be originally charged, according as it is the first or the second, we shall have

$$k = \frac{\sigma}{\rho - \sigma},$$

or
$$k = \frac{\rho' - \sigma}{\sigma}.$$

48. If the substance examined (the dielectric of the first apparatus) be any gas, or air in a different state as to pressure or temperature from the air of the second apparatus, Faraday

always finds the intensity after contact to be half the original intensity, and hence for every gaseous body $k = 1$.

49. If the dielectric of the first apparatus be solid, the intensity after contact is found to be greater than half the original intensity when the first, and less than half when the second is the apparatus originally charged. Hence for a solid dielectric, $k > 1$. For sulphur Faraday finds the value to be rather more than 2·2; for shell-lac, about 2; and for flint-glass, greater than 1·76.

50. The commonly received ideas of attraction and repulsion exercised at a distance, independently of any intervening medium, are quite consistent with all the phenomena of electrical action which have been here adduced. Thus we may consider the particles of air in the neighbourhood of electrified bodies to be entirely uninfluenced, and therefore to produce no effect in the resultant action on any point: but the particles of a solid non-conductor must be considered as assuming a polarized state when under the influence of free electricity, so as to exercise attractions or repulsions on points at a distance, which, with the action due to the charged surfaces, produce the resultant force at any point. It is, no doubt, possible that such forces at a distance may be discovered to be produced entirely by the action of contiguous particles of some intervening medium, and we have an analogy for this in the case of heat, where certain effects which follow the same laws are undoubtedly propagated from particle to particle. It might also be found that magnetic forces are propagated by means of a second medium, and the force of gravitation by means of a third. We know nothing, however, of the molecular action by which such effects could be produced, and in the present state of physical science it is necessary to admit the known facts in each theory as the foundation of the ultimate laws of action at a distance.

St Peter's College,
Nov. 22, 1845.

III. ON THE ELECTRO-STATICAL CAPACITY OF A LEYDEN PHIAL AND OF A TELEGRAPH WIRE INSULATED IN THE AXIS OF A CYLINDRICAL CONDUCTING SHEATH.*

[From the *Philosophical Magazine*, 1855, first half-year.]

51. The principles brought forward in the preceding articles On the Uniform Motion of Heat, etc., enable us with great ease to investigate the "capacity"† of a Leyden phial with either air, or any liquid or solid dielectric, and of other analogous arrangements, such as the copper wires in gutta-percha tubes under water, with which Faraday has recently performed such remarkable experiments.‡

52. Thus, for a Leyden phial, let us suppose a portion S of the surface of a conductor A to be everywhere so near the surface of a conductor A', that the distance between them at any point is a small fraction of the radii of curvature of each surface in the neighbourhood; and let z be the distance between them at a particular position, P. Then, by the analogy with heat, it is clear that if the two surfaces be kept at different electrical potentials, V and V', the potentials at equidistant points in any line across from one to the other will be in arithmetical progression. Hence $\dfrac{V - V'}{z}$ will be the rate of variation of the potential perpendicularly across in the position P. If, in the first place, the dielectric be air, the electric force in the air

---

* Communicated as an Additional Note to two papers (I. and II. above) "On the Uniform Motion of Heat in Homogeneous Solid Bodies, and its connexion with the Mathematical Theory of Electricity," and "On the Mathematical Theory of Electricity in Equilibrium;" only not in time to be appended to the reprints of those papers which appeared in the *Philosophical Magazine*, June and July 1854 (1854, I. and II.).

† Defined (*Philosophical Magazine*, June 1853) for any conductor (subject or not to the influence of other conductors), as the quantity of electricity which it takes to charge it to unit potential.

‡ Described in a lecture at the Royal Institution, Jan. 20, 1854, and subsequently published in the *Philosophical Magazine* (1854, I. p. 197).

between the two about the position P will consequently be $\dfrac{V-V'}{z}$, and therefore the electrical density (according to the theorem proved in the first article) on one surface must be $+\dfrac{1}{4\pi}\dfrac{V-V'}{z}$, and on the other $-\dfrac{1}{4\pi}\dfrac{V-V'}{z}$. The quantity of electricity in the position P, on an area ds of the surface S, is therefore $\dfrac{1}{4\pi}\dfrac{V-V'}{z}ds$, and therefore the whole quantity on S is

$$\frac{V-V'}{4\pi}\int\frac{ds}{z},$$

which is Green's general expression for the electrification of either coating of a Leyden phial. If the thickness of the dielectric be constant and equal to τ, it becomes

$$\frac{V-V'}{4\pi}\frac{S}{\tau}.$$

53. Now if A' be uninsulated, we have $V'=0$; and then, to charge S to the potential V, it takes the quantity $V\times\dfrac{S}{4\pi\tau}$. Hence the "capacity" of S is

$$\frac{S}{4\pi\tau}.$$

If instead of air there be a solid or liquid dielectric of inductive capacity, k, occupying the space between the two surfaces, the quantity of heat conducted across, in the analogous thermal circumstances, would be k times as great as in the case corresponding to the air dielectric, with the same difference of temperatures; and in the actual electrical arrangement, the quantity of electricity on each of the conducting surfaces would be k times as great as with air for dielectric and the same difference of potentials. The expression for the capacity of an actual Leyden phial is therefore

$$\frac{kS}{4\pi\tau},$$

k being the inductive capacity of the solid non-conductor of which it is formed, τ its thickness, and S the area of it which is coated on each side.

54. To investigate the capacity of a copper wire in the circumstances experimented on by Faraday, let us first consider the analogous circumstances regarding the conduction of heat; that is, let us consider the conduction of heat that would take place

across the gutta-percha, if the copper wire in its interior were
kept continually at a temperature a little above that of the
water which surrounds it.   Here the quantity of heat flowing
outwards from any length of the copper wire, the quantities
flowing across different surfaces surrounding it in the gutta-
percha, and the quantity flowing into the water from the same
length of gutta-percha tube, in the same time, must be equal.
But the areas of the same length of different cylindrical surfaces
are proportional to their radii, and therefore the flow of heat
across equal areas of different cylindrical surfaces in the gutta-
percha, coaxial with the wire, must be inversely as their radii.
Hence, in the corresponding electrical problem, with air as the
dielectric instead of gutta-percha, if R denote the resultant
electrical force at any point P in the air between an insulated,
electrified, infinitely long cylindrical conductor, and an uninsulated, coaxial, hollow cylindrical conductor surrounding it,
and if x be the distance of P from the axis, we have

$$R = \frac{A}{x},$$

where A denotes a constant.   But if v be the potential at P;
by the definition of "potential" we have

$$\frac{dv}{dx} = -R.$$

Hence

$$\frac{dv}{dx} = -\frac{A}{x};$$

and, by integration, $v = -A \log x + C.$

Assigning the constants A and C so that the potential may
have the value V at the surface of the wire, and may vanish
at the hollow conducting surface round it, if r and r' denote the
radii of these cylinders respectively, we have

$$v = V \frac{\log \frac{r'}{x}}{\log \frac{r'}{r}},$$

and

$$-\frac{dv}{dx} = R = \frac{V}{\log \frac{r'}{r}} \frac{1}{x}$$

55.   Taking $x = r$, we find by this the electric force in the air

infinitely near the inner electrified conductor; and dividing the
value found, by 4π (according to the general theorem), we have

$$\frac{1}{4\pi}\frac{V}{r\log\frac{r'}{r}}$$

for the electrical density on the surface of the conductor.
Multiplying this by $2\pi r l$, the area of a length l of the surface,
we find

$$\tfrac{1}{2}\frac{Vl}{\log\frac{r'}{r}},$$

for the whole quantity of electricity on that length. Hence, if
k be the specific inductive capacity of gutta-percha, the electricity resting on a length l of the wire in the actual circumstances
will amount to

$$\tfrac{1}{2}\frac{kl}{\log\frac{r'}{r}}V.$$

Or if S denote the surface of the wire, we have, for the quantity
of electricity which it holds,

$$V.\frac{kS}{4\pi r\log\frac{r'}{r}};$$

and therefore its capacity is the same as that of a Leyden
phial with an equal area of coated glass of thickness equal to
$\frac{I}{k}r\log\frac{r'}{r}$, if I denote the specific inductive capacity of the
glass.

56. In the case experimented on by Mr Faraday, the diameter
of the wire was $\frac{1}{16}$th of an inch, and the exterior diameter of the
gutta-percha covering was about four times as great. Hence
the thickness of the equivalent Leyden phial must have been

$$\frac{I}{k}\cdot\frac{1}{32}\log_e 4=\frac{I}{k}\cdot\frac{1}{23\cdot08}$$

As the surface of the wire amounted to 8300 square feet, we
may infer that if the gutta-percha had only the same inductive capacity as glass (and it probably has a little greater), the
insulated wire, when the outer surface of the gutta-percha was
uninsulated, would have had an electrical capacity equal to that
of an ordinary Leyden battery of 8300 square feet of coated
glass $\frac{1}{23}$d of an inch thick.

INVERCLOY, ARRAN, *June*, 1854.

IV. ON THE MATHEMATICAL THEORY OF ELECTRICITY
IN EQUILIBRIUM.

(Art. xxxviii. of complete list in *Mathematical and Physical Papers*, Vol. i.)

II.—A STATEMENT OF THE PRINCIPLES ON WHICH THE MATHE-MATICAL THEORY OF ELECTRICITY IS FOUNDED.

[*Cambridge and Dublin Mathematical Journal*, March, 1848.]

57. This paper may be regarded as introductory to some others which will follow, containing various investigations in the Theory of Electricity. The fundamental mathematical principles of the phenomena of Electricity in Equilibrium are stated and explained in as concise a manner as seems consistent with clearness. To avoid lengthening the paper and unnecessarily distracting the attention of the reader, no details are given with reference to the experiments which have been, or which might be, made for establishing the various propositions asserted; and, for the same reasons, scarcely any allusion is made to the history of the subject. With regard to the nature of the evidence on which the mathematical theory of electricity rests, the reader is referred to the preceding paper "On the Elementary Laws of Statical Electricity," where, besides some general explanations on the subject, the works containing accounts of the actual experimental researches of principal importance are indicated. That paper is marked as the first of a series which it was my intention to publish in this *Journal*, and of which the second now appears. In this series it will not be attempted to adhere to a systematic course of investigations such as might constitute a complete treatise on the subject; and my only reason for publishing this introductory article is for the sake of reference in other papers, there being no published work in which the principles are stated in a sufficiently concise and correct form, independently of any hypothesis, to be altogether satisfactory in the present state of science.

The Two Kinds of Electricity.

58. If a piece of glass and a piece of resin are rubbed together and then separated, it is found that they attract one another

mutually. The term *electricity** has been applied to the agency
developed in this operation; the excitation of the bodies, to
which the attractive force is due, is called *electrical,* and the
bodies so excited are said to be *electrified,* or to be *charged with
electricity.*

If second pieces of glass and resin be rubbed together and
then separated, and placed in the neighbourhood of the first pair
of electrified bodies, it may be observed—

(1)   *That the two pieces of glass repel one another.*

(2)   *That each piece of glass attracts each piece of resin.*

(3)   *That the two pieces of resin repel one another.*

Hence it is inferred that the two pieces of glass possess electrical properties which differ in their characteristics from those
of the resin; and the two kinds of electricity thus indicated are
called vitreous and resinous, after the substances on which they
are developed.. Bodies may in various ways be made electric;
but the characteristics presented are always those of either
vitreous electricity or *resinous electricity.*

59. An electrified body exerts no force, whether of attraction
or of repulsion, upon any non-electric matter. When in any case
bodies not previously electrified are observed to be attracted, or
urged in any direction, by an electrical mass, it is because the
bodies have become *electrically excited by influence.*

60. If a small piece of glass and a small piece of resin, which
have been electrified by mutual friction, be placed successively
in the same position in the neighbourhood of an electrified body,
they will be acted upon by equal forces, in the same line,
but in contrary directions. Hence the two bodies are said to be
equally charged with the two kinds of electricity respectively.

Electrical Quantity.

61. The force between two electrified bodies depends, *ceteris
paribus,* on the amounts of their charges, or on the quantities
of electricity which they possess.

If a small piece of glass and a small piece of resin be electrified
by mutual friction to such an extent that, when separated and
placed at a unit of distance, they attract one another with a
unit of force, the quantity of electricity possessed by the former

---

* From ἤλεκτρον, amber, on account of such phenomena having been first
observed with amber as one of the substances rubbed together.

is said to be unity; the latter possesses what may be called a unit of resinous electricity.

If m bodies, each possessing a unit of vitreous electricity, be incorporated together, the single body thus composed is charged with m units of the same kind of electricity: It is said to possess a quantity of electricity equal to m, or its *electrical mass* is m. A similar definition is applicable with reference to the measurement of resinous electricity.

62. If two bodies possessing equal quantities of vitreous and resinous electricity be incorporated, the single body thus composed will be found either to be non-electric, or to be in such a state that, without the removal of any electricity of either kind from it, it may, merely by an alteration in the distribution of what it already possesses, be deprived of all electrical symptoms. Thus it appears that a body either vitreously or resinously electrified, may be deprived of its charge merely by supplying it with an equal quantity of the other kind of electricity.

In consequence of this fact, we may establish a complete system of algebraic notation with reference to electrical quantity, whether of vitreous or resinous electricity, by adopting as universal the law that the total quantity of electricity possessed by two bodies, or the quantity possessed by one body made up of two, is equal to the sum of the quantities with which they are separately charged. Thus let m be the quantity of electricity with which a vitreously electrified body is charged, and let m' be the quantity contained by a body equally charged with resinous electricity. We must have

$$m + m' = 0,$$

and therefore m' is equal to $-m$. Now it is usual to regard vitreous electricity as positive; and we must therefore regard the other kind as negative; so that a body possessing m units of resinous electricity is to be considered as charged with a quantity $-m$ of electricity.

The Superposition of Electrical Forces.

63. If a body, electrified in a given invariable manner, be placed in the neighbourhood of any number of electrified bodies, it will experience a force which is the resultant of the forces that would be separately exerted upon it by the different bodies

if they were placed in succession in the positions which they actually occupy, without any alteration in their electrical conditions.

This law is true even if any number of the bodies considered be merely different parts of one continuous mass.

COR. 1. The total mechanical action between two electrified bodies, whether parts of one continuous mass or isolated bodies, is the resultant of the forces due to the mutual actions between all parts of either body and all parts of the other, if we conceive the two bodies to be arbitrarily divided each into parts in any manner whatever.

Cor. 2. We may, in any electrical problem, imagine the charge possessed by a body to be divided into two or more parts, each distributed arbitrarily with the sole condition that the sum of the quantities of electricity in any very small space of the body due to the different distributions shall be equal to the given quantity of electricity in that space, according to the actual distribution of electricity in the body; and we may consider the force actually exerted upon any other electrified body as equivalent to the resultant of the forces due to these partial distributions.

The Law of Force between Electrified Bodies.

64. The force between two small electrified bodies varies inversely as the square of the distance between them.

COR. If two small bodies be charged respectively with quantities m and m' of electricity, they will mutually repel with a force equal to $\dfrac{mm'}{\Delta^2}$;

(an action which will be really attractive when m and m' have unlike signs, as would be the case were the bodies dissimilarly electrified). For two units, placed at a distance unity, repel with a force equal to unity, and therefore if placed at a distance Δ, they will repel with a force $\dfrac{1}{\Delta^2}$; and the expression for the repulsion between m units and m' units is deduced from this, according to the principle of the superposition of forces, by multiplying by mm'.

Definition of the Resultant Electrical Force at a Point.

65. Let a unit of negative electricity be conceived to be concentrated at a point P in the neighbourhood of an electrified
body or group of bodies, without producing any alteration in
the previously existing electrical distribution. The force exerted
upon this electrical point is what we shall throughout understand *as the resultant force at P* due to the electricity of the
body or bodies considered.

Cor. If R *be the resultant force at P* in any case, then
the force actually exerted upon an electrical mass m, concentrated at P, will be equal to $-mR$.

Electrical Equilibrium.

66. When a body held at rest is electrified, and when, being
either subject to electrical action from other bodies, or entirely
isolated, the distribution of its charge remains permanently
unaltered, the electricity upon it is said to be in equilibrium.

Electrical equilibrium may be disturbed in various ways.
Thus if a body charged with electricity in equilibrium be
touched, or even approached by another electrified body, the
equilibrium may be broken, and can only be restored after a
different distribution has been effected, by a motion of electricity
through the body or along its surface : or if a body be initially
electrified in any arbitrary manner, whether by friction or otherwise, it may be that, as soon as the exciting cause is removed,
the electricity will either gradually become altered from its
initial distribution, by moving slowly through the body, or will
suddenly assume a certain definite distribution.

The laws which regulate the distribution of electricity in
equilibrium on bodies in various circumstances have been the
subject of most important experimental researches ; and having
been established with perfect precision by Coulomb, and placed
beyond all doubt by verifications afforded in subsequent experiments, they constitute the foundation of an extremely interesting branch of the Mathematical Theory of Electricity.   In
connexion with these laws, and before stating them, it will be
convenient to explain the nature of the distinction which is
drawn between the two great classes of bodies in nature, called
Conductors of Electricity, and Non-Conductors of Electricity.

Non-Conductors of Electricity.

67. A body which affords such a resistance to the transmission of electricity through it, or along its surface, that, if it be once electrified in any way, it retains permanently, without any change of distribution, the charge which it has received, is called a Non-Conductor of Electricity.

No body exists in nature which fulfils strictly the terms of this definition; but glass and resin, besides many other substances, are such that they may, within certain limits and subject to certain restrictions, be considered as non-conductors.

Conductors of Electricity.

68. A very extensive class of bodies in nature, including all the metals, many liquids, etc., are found to possess the property that, in all conceivable circumstances of electrical excitation, the resultant force at any point within their substance vanishes. Such bodies are called Conductors of Electricity, since they are destitute of the property, possessed by non-conductors, of retaining permanently, by a resistance to every change, any distribution of electricity arbitrarily imposed; the only kind of distribution which can exist unchanged for an instant on a conductor being such as satisfies the condition that the resultant force must vanish in the interior.

It is found by experiment that the electricity of a charged conductor rests entirely on its surface, and that the electrical circumstances are not at all affected by the nature of the interior, but depend solely upon the form of the external conducting surface. Thus the electrical properties of a solid conductor, of a hollow conducting shell, or of a non-conductor enclosed in an envelop, however thin (the finest gold leaf, for instance), are identical, provided the external forms be the same. A hollow conductor never shows symptoms of electricity on its interior surface, unless an electrified body be insulated within it; in which case the interior surface will become electrified by *influence* or by *induction,* in such a way as to make the total resultant force at any point in the conducting matter vanish, by balancing, for any such point, the force due to the electricity of the insulated body.

It has been frequently assumed that electricity penetrates to a finite depth below the surface of conductors; and, in accordance with certain hypothetical ideas regarding the nature of electricity, the "thickness of the stratum" at different points of the surface of a conductor has been considered as a suitable term with reference to the varying or uniform distribution of electricity over the body. All the conclusion with reference to this delicate subject which can as yet be drawn from experiment, is that the "thickness," if it exist at all, must be less than that of the finest gold leaf; and in the present state of science we must regard it as immeasurably small. It may be conceived that the actual thickness of the excited stratum at the surface of an electrified conductor is of the same order as the space through which the physical properties of the pervading matter change continuously from those of the solids to those which characterize the surrounding air.

Electrical Density at any Point of a Charged Surface.

69.   In this, and in all the papers which will follow, instead of the expression "the thickness of the stratum," Coulomb's far more philosophical term, *Electrical Density*, will be employed with reference to the distribution of electricity on the surface of a body; a term which is to be understood strictly in accordance to the following definitions, without involving even the idea of a hypothesis regarding the nature of electricity. The electrical density of a uniformly charged surface is the quantity of electricity distributed over a unit of surface.

The electrical density at any point of a surface, whether the distribution be uniform or not, is the quotient obtained by dividing the quantity of electricity distributed over an infinitely small element at this point, by the area of the element.

Exclusion of all Non-Conductors except Air.

70.   In the present paper, and in some others which will follow, no bodies will be considered except conductors; and the air surrounding them, which will be considered as offering a resistance to the transference of electricity between two detached conductors, but as otherwise destitute of electrical properties.   A full development of the mathematical theory, of the internal electrical polarization of solid or liquid non-con-

ductors, subject to the influence of electrified bodies, discovered by Faraday (in his Experimental Researches on the specific inductive capacities of non-conducting media), must be reserved for a later communication.*

Insulated Conductors.

71. A conductor separated from the ground, and touched only by air, is said to be insulated. Insulation may be practically effected by means of solid props of matter, such as glass, shell-lac, or gutta percha;† and if the props be sufficiently thin, it is found that their presence does not in any way alter or affect the electrical circumstances, and that their *resisting* power, as non-conductors of electricity, prevents any alteration in the quantity of electricity possessed by the insulated body; so that however the distribution may be affected by the influence of surrounding bodies, it is only by a temporary breaking of the insulation that the absolute charge can be increased or diminished.

If an insulated uncharged conductor be placed in the neighbourhood of bodies charged with electricity, it will become "electrified by influence," in such a manner that its resultant electrical force at every internal point shall counterbalance the force due to the exterior charged bodies: but, in accordance with what has been stated in the preceding paragraph, the total quantity of electricity will remain equal to nothing; that is to say, the two kinds of electricity produced upon it by influence will be equal to one another in amount.

Recapitulation of the Fundamental Laws.

72. The laws of electricity in equilibrium in relation with conductors may—if we tacitly take into account such principles

---

* The results of this Theory were explained briefly in a paper entitled "Note sur les Lois Élémentaires de l'Electricité Statique" (published, in 1845, in Liouville's *Journal*), and more fully in the first paper of the present series, on the "Mathematical Theory of Electricity" (II. above). A similar view of this subject has been taken by Mossotti, whose investigations are published in a paper entitled "Discussione Analitica sull' Influenza che l'Azione di un Mezzo Dielettrico ha sulla Distributione dell' Elettricità alla Superficie di piu Corpi Elettrici Disseminati in Esso" (Vol. xxiv. of the *Mémorie della Società Italiana delle Scienze Residente in Modena*, dated 1846).

† It has been recently discovered by Faraday that gutta percha is one of the best insulators among known substances (*Phil. Mag.*, March, 1848).

T. E.   4

as the superposition of electrical forces, and the invariableness
of the quantity of electricity on a body, except by addition or
subtraction (in the extended algebraic sense of these terms)—
be considered as fully expressed in the three following propositions :—

I.   The repulsion between two electrical points is inversely
proportional to the square of their distance.

II.   Electricity resides at the boundary of a charged conductor.

III.   The resultant force at any point in the substance of
a conductor, due to all existing electrified bodies, vanishes.

It has been proved by Green that the second of these laws
is a mathematical consequence of the first and third; and it has
been demonstrated by La Place* that the first law may be inferred from the truth, in a certain particular case, of the second
and third.   The three laws were, however, first announced by
Coulomb, as the result of his experimental researches on the
subject.

Objects of the Mathematical Theory of Electricity.

73.   The varied problems which occur in the mathematical
theory of electricity in equilibrium may be divided into the
two great classes of Synthetical and Analytical investigations.
In problems of the former class, the object is in each case the
determination either of a resultant force or of an aggregate
electrical mass, according to special data regarding distributions
of electricity: in the latter class, inverse problems, such as the
determination of the electrical density at each point of the
surface of a conductor in any circumstances, according to the
laws stated above, are the objects proposed.

It has been proved (by Green and Gauss) that there is a
determinate unique solution of every actual analytical problem
of the Theory of Electricity in relation with conductors.   The
demonstration of this with reference to the complete Theory of
Electricity (including the action of solid non-conducting media
discovered by Faraday), as well as with reference to the Theories
of Heat, Magnetism, and Hydrodynamics, may be deduced from
two theorems proved in the Cambridge and Dublin Mathematical Journal for 1847, "Regarding the Solution of certain Partial

---

* [Originally by Cavendish, as I learned after the first publication of this
paper.   See footnote of March 1854 on § 34 above.]

Differential Equations" (XIII. below, or Thomson and Tait's *Natural Philosophy*, App. A.).

The full investigation of any actual case of electrical equilibrium will generally involve both analytical and synthetical problems; as it may be desirable, besides determining the distribution, to find the resulting electrical force at points not in the interior of any conductor, or to find the total mechanical action due to the attractions or repulsions of the elements of two conductors, or of two portions of one conductor; and besides, it is frequently interesting to verify synthetically the solutions obtained for analytical problems.

Actual Progress in the Mathematical Theory of Electricity.

74. In Poisson's valuable memoirs on this subject, the distribution of electricity on two electrified spheres, uninfluenced by other electric matter, is considered; a complete solution of the analytical problem is arrived at; and various special cases of interest are examined in detail with great rigor. In a very elaborate memoir by Plana*, the solution given by Poisson is worked out much more fully, the excessive mathematical difficulties in the way of many actual numerical applications of interest being such as to render a work of this kind extremely important.

The distribution of electricity on an ellipsoid (including the extreme cases of elliptic and circular discs, and of a straight rod), and the results of consequent *synthetical* investigations are well known.

The analytical problem regarding an ellipsoid subject to the influence of given electrical masses, has been solved by M. Liouville, by the aid of a very refined mathematical method suggested by some investigations of M. Lamé with reference to corresponding problems in the Theory of Heat.

Green's Essay on Electricity and his other papers on allied subjects contain, besides the solution of several special problems of interest, most valuable discoveries with reference to the general Theory of Attraction, and open the way to much more extended investigations in the Theory of Electricity than any that have yet been published.

GLASGOW COLLEGE, *March* 4, 1848.

---

* *Turin Academy of Sciences*, tome VII. Serie II. published separately in a quarto volume of 333 pages: Turin, 1845.

V.—ON THE MATHEMATICAL THEORY OF ELECTRICITY IN EQUILIBRIUM.

(Art. xxxviii. of complete list in *Mathematical and Physical Papers*, Vol. i.)

III.—GEOMETRICAL INVESTIGATIONS WITH REFERENCE TO THE DISTRIBUTION OF ELECTRICITY ON SPHERICAL CONDUCTORS.*

[*Cambridge and Dublin Mathematical Journal*, March, May, and Nov. 1848, Nov. 1849, Feb. 1850.]

75. There is no branch of physical science which affords a surer foundation, or more definite objects for the application of mathematical reasoning, than the theory of electricity. The small amount of attention which this most attractive subject has obtained is no doubt owing to the extreme difficulty of the analysis by which even a very limited progress has as yet been made; and no other circumstance could have totally excluded from an elementary course of reading, a subject which, besides its great physical importance, abounds so much in beautiful illustrations of ordinary mechanical principles. This character of difficulty and impracticability is not however inseparable from the mathematical theory of electricity: by very elementary geometrical investigations we may arrive at the solution

---

* The investigations given in this paper (§§ 75—127) form the subject of the first part of a series of lectures on the Mathematical Theory of Electricity given in the University of Glasgow during the present session [1847—8]. They are adaptations of certain methods of proof which first occurred to me as applications of *the principle of electrical images*, made with a view to investigating the solutions of various problems regarding spherical conductors, without the explicit use of the differential or integral calculus. The spirit, if not the notation, of the *differential* calculus must enter into any investigations with reference to Green's theory of the potential, and therefore a more extended view of the subject is reserved for a second part of the course of lectures. A complete exposition of the *principle of electrical images* (of which a short account was read at the late meeting of the British Association at Oxford) has not yet been published; but an outline of it was communicated by me to M. Liouville in three letters, of which extracts are published in the *Journal de Mathématiques* (1845 and 1847, vols. x., xii.). [See xiv. below.] A full and elegant exposition of the method indicated, together with some highly interesting applications to problems in geometry not contemplated by me, are given by M. Liouville himself, in an article written with reference to those letters, and published along with the last of them. I cannot neglect the present opportunity of expressing my thanks for the honour which has thus been conferred upon me by so distinguished a mathematician, as well as for the kind manner in which he received those communications, imperfect as they were, and for the favourable mention made of them in his own valuable memoir.

of a great variety of interesting problems with reference to the distribution of electricity on spherical conductors, including Poisson's celebrated problem of the two spheres, and others which might at first sight be regarded as presenting difficulties of a far higher order. The object of the following paper is to present, in as simple a form as possible, some investigations of this kind. The methods followed, being for the most part *synthetical*, were suggested by a knowledge of results founded on a less restricted view of the theory of electricity; and it must not be considered either that they constitute the best or the easiest way of advancing towards a *complete* knowledge of the subject, or that they would be suitable as instruments of research in endeavouring to arrive at the solutions of new problems.

Insulated Conducting Sphere subject to no External Influence.

76. We may commence with the simplest possible case, that of a spherical conductor, charged with electricity and insulated in a position removed from all other bodies which could influence the distribution of its charge. In this, as in the other cases which will be considered, the various problems, of the analytical and synthetical classes, alluded to in a previous paper (IV. § 73), will be successively subjects of investigation. Thus let us first determine the density at any point of the surface, and then, after verifying the result by showing that the laws (§ 72) are satisfied, let us investigate the resultant force at an external point.

Determination of the Distribution.

77. Let a be the radius of the sphere, and E the amount of the charge.

According to Law II., the whole charge will reside on the surface, and, on account of the symmetry, it must be uniformly distributed. Hence, if ρ be the required density at any point, we have
$$\rho = \frac{E}{4\pi a^2}.$$

Verification of Law III.

78. The well-known theorem, that the resultant force due to a uniform spherical shell vanishes for any interior point, constitutes the verification required in this case. This theorem was

first given by Newton, and is to be found in the *Principia*; but as his demonstration is the foundation of every synthetical investigation which will be given in this paper, it may not be superfluous to insert it here; and accordingly the passage of the *Principia* in which it occurs, translated literally, is given here.

Newton, First Book, Twelfth Section, Prop. LXX. Theorem XXX.

If the different points of a spherical surface attract equally with forces varying inversely as the squares of the distances, a particle placed within the surface is not attracted in any direction.

Let $HIKL$ be the spherical surface, and P the particle within it. Let two lines HK, IL, intercepting very small arcs HI, KL, be drawn through P; then on account of the similar triangles HPI, KPL (Cor. 3, Lemma VII. Newton), those arcs will be proportional to the distances HP, LP; and any small elements of the spherical surface at HI and KL, each bounded all round by straight lines passing through P [and very nearly coinciding with HK], will be in the duplicate ratio of those lines. Hence the forces exercised by the matter of these elements on the particle P are equal; for they are as the quantities of matter directly, and the squares of the distances, inversely; and these two ratios compounded give that of equality. The attractions therefore, being equal and opposite, destroy one another: and a similar proof shows that all the attractions due to the whole spherical surface are destroyed by contrary attractions. Hence the particle P is not urged in any direction by these attractions. Q. E. D.

Digression on the Division of Surfaces into Elements.

79. The division of a spherical surface into infinitely small elements will frequently occur in the investigations which follow: and Newton's method, described in the preceding demonstration, in which the division is effected in such a manner that all the parts may be taken together in *pairs of opposite elements with reference to an internal point;* besides other

methods deduced from it, suitable to the special problems to be examined; will be repeatedly employed. The present digression, in which some definitions and elementary geometrical propositions regarding this subject are laid down, will simplify the subsequent demonstrations, both by enabling us, through the use of convenient terms, to avoid circumlocution, and by affording us convenient means of reference for elementary principles, regarding which repeated explanations might otherwise be necessary.

Explanations and Definitions regarding Cones.

80. If a straight line which constantly passes through a fixed point be moved in any manner, it is said to describe, or generate, a *conical surface* of which the fixed point is the vertex.

If the generating line be carried from a given position continuously through any series of positions, no two of which coincide, till it is brought back to the first, the entire line on the two sides of the fixed point will generate a complete conical surface, consisting of two sheets, which are called *vertical or opposite cones.* Thus the elements *HI* and *KL*, described in Newton's demonstration given above, may be considered as being cut from the spherical surface by two *opposite cones* having *P* for their common vertex.

The Solid Angle of a Cone, or of a complete Conical Surface.

81. If any number of spheres be described from the vertex of a cone as centre, the segments cut from the concentric spherical surfaces will be similar, and their areas will be as the squares of the radii. The quotient obtained by dividing the area of one of these segments by the square of the radius of the spherical surface from which it is cut, is taken as the measure of the *solid angle of the cone.* The segments of the same spherical surfaces made by the opposite cone, are respectively equal and similar to the former. Hence the solid angles of two vertical or opposite cones are equal: either may be taken as the solid angle of the complete conical surface, of which the opposite cones are the two sheets.

Sum of all the Solid Angles round a Point $= 4\pi$.

82. Since the area of a spherical surface is equal to the square of its radius multiplied by 4π, it follows that the sum of the solid angles of all the distinct cones which can be described with a given point as vertex, is equal to 4π.

Sum of the Solid Angles of all the complete Conical Surfaces $= 2\pi$.

83. The solid angles of vertical or opposite cones being equal, we may infer from what precedes that the sum of the solid angles of all the complete conical surfaces which can be described without mutual intersection, with a given point as vertex, is equal to 2π.

Solid Angle subtended at a Point by a Terminated Surface.

84. The solid angle subtended at a point by a superficial area of any kind, is the solid angle of the cone generated by a straight line passing through the point, and carried entirely round the boundary of the area.

Orthogonal and Oblique Sections of a Small Cone.

85. A very small cone, that is, a cone such that any two positions of the generating line contain but a very small angle, is said to be cut at right angles, or orthogonally, by a spherical surface described from its vertex as centre, or by any surface, whether plane or curved, which touches the spherical surface at the part where the cone is cut by it.

A very small cone is said to be cut obliquely, when the section is inclined at any finite angle to an orthogonal section; and this angle of inclination is called the *obliquity of the section.*

The area of an orthogonal section of a very small cone is equal to the area of an oblique section in the same position, multiplied by the cosine of the obliquity.

Hence the area of an oblique section of a small cone is equal to the quotient obtained by dividing the product of the square of its distance from the vertex, into the solid angle, by the cosine of the obliquity.

Area of the Segment cut from a Spherical Surface by a Small Cone.

86. Let E denote the area of a very small element of a

spherical surface at the point E (that is to say, an element every part of which is very near the point E), let ω denote the solid angle subtended by E at any point P, and let PE, produced if necessary, meet the surface again in E': then, a denoting the radius of the spherical surface, we have

$$E = \frac{2a \cdot \omega \cdot PE^2}{EE'}.$$

For, the obliquity of the element E, considered as a section of the cone of which P is the vertex and the element E, a section; being the angle between the given spherical surface and another de-scribed from P as centre, with PE as radius; is equal to the angle between the radii, EP and EC, of the two spheres. Hence, by con-sidering the isosceles triangle ECE', we find that the cosine of the obliquity is equal to $\dfrac{\frac{1}{2}EE'}{EC}$, or to $\dfrac{EE'}{2a}$, and we arrive at the preceding expression for E.

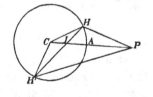

87. *Theorem.* [*] The attraction of a uniform spherical surface on an external point is the same as if the whole mass were collected at the centre.

Let P be the external point, C the centre of the sphere, and CAP a straight line cutting the spherical surface in A. Take I in CP, so that CP, CA, CI may be continual proportionals, and let the whole spherical surface be divided into *pairs of opposite elements with reference to the point I.*

Let H and H' denote the magnitudes of a pair of such

---

[*] This theorem, which is more comprehensive than that of Newton in his first proposition regarding attraction on an external point (Prop. LXXI.), is fully established as a corollary to a subsequent proposition (Prop. LXXIII Cor. 2). If we had considered the proportion of the forces exerted upon two external points at different distances, instead of, as in the text, investigating the absolute force on one point, and if besides we had taken together all the pairs of elements which would constitute two narrow annular portions of the surface, in planes perpendicular to PC, the theorem and its demonstration would have coincided precisely with Prop. LXXI. of the *Principia*.

elements, situated respectively at the extremities of. a chord HH'; and let ω denote the magnitude of the solid angle subtended by either of these elements at the point I.

We have (§ 85)

$$H = \frac{\omega \cdot IH^2}{\cos CHI}, \text{ and } H' = \frac{\omega \cdot IH'^2}{\cos CH'I}.$$

Hence, if ρ denote the density of the surface (§ 69), the attractions of the two elements H and H' on P are respectively

$$\rho \frac{\omega}{\cos CHI} \cdot \frac{IH^2}{PH^2}, \text{ and } \rho \frac{\omega}{\cos CH'I} \cdot \frac{IH'^2}{PH'^2}.$$

Now the two triangles PCH, HCI have a common angle at C, and, since $PC : CH :: CH : CI$, the sides about this angle are proportional. Hence the triangles are similar; so that the angles CPH and CHI are equal, and

$$\frac{IH}{HP} = \frac{CH}{CP} = \frac{a^*}{CP}.$$

In the same way it may be proved, by considering the triangles PCH', $H'CI$, that the angles CPH' and $CH'I$ are equal, and that

$$\frac{IH'}{H'P} = \frac{CH'}{CP} = \frac{a}{CP}.$$

Hence the expressions for the attractions of the elements H and H' on P become

$$\rho \frac{\omega}{\cos CHI} \cdot \frac{a^2}{CP^2}, \text{ and } \rho \frac{\omega}{\cos CH'I} \cdot \frac{a^2}{CP^2},$$

which are equal, since the triangle HCH' is isosceles; and, for the same reason, the angles CPH, CPH', which have been proved to be respectively equal to the angles CHI, $CH'I$, are equal. We infer that the resultant of the forces due to the two elements is in the direction PC, and is equal to

$$2\omega \cdot \rho \cdot \frac{a^2}{CP^2}.$$

To find the total force on P, we must take the sum of all

---

* From this we infer that the ratio of IH to HP is constant, whatever be the position of H on the spherical surface, a well-known proposition.—(Thomson's *Euclid*, vi. Prop. G.)

the forces along PC due to the pairs of opposite elements; and, since the multiplier of ω is the same for each pair, we must add all the values of ω, and we therefore obtain (§ 83), for the required resultant,

$$\frac{4\pi\rho a^2}{\overline{CP^2}}.$$

The numerator of this expression; being the product of the density into the area of the spherical surface; is equal to the mass of the entire charge; and therefore the force on P is the same as if the whole mass were collected at C. Q. E. D.

COR. The force on an external point, infinitely near the surface, is equal to $4\pi\rho$, and is in the direction of a normal at the point. The force on an internal point, however near the surface, is, by a preceding proposition, equal to nothing.

Repulsion on an element of the Electrified Surface.

88. Let σ be the area of an infinitely small element of the surface at any point P, and at any other point H of the surface let a small element subtending a solid angle ω, at P, be taken. The area of this element will be equal to

$$\frac{\omega . PH^2}{\cos \overline{CHP}},$$

and therefore the repulsion along HP, which it exerts on the element σ at P, will be equal to

$$\frac{\rho\omega . \rho\sigma}{\cos \overline{CHP}}, \text{ or } \frac{\omega}{\cos \overline{CHP}}\rho^2\sigma.$$

Now the total repulsion on the element at P is in the direction CP; the component in this direction of the repulsion due to the element H, is

$$\omega . \rho^2\sigma;$$

and, since all the cones corresponding to the different elements of the spherical surface lie on the same side of the tangent plane at P, we deduce, for the resultant repulsion on the element σ,

$$2\pi\rho^2\sigma.$$

From the corollary to the preceding proposition, it follows that this repulsion is half the force which would be exerted on an external point, possessing the same quantity of electricity as the element σ, and placed infinitely near the surface.

GLASGOW COLLEGE, *March* 14, 1848.

INSULATED SPHERE SUBJECTED TO THE INFLUENCE OF AN ELECTRICAL POINT—(§§ 89—95).

89. A conducting sphere placed in the neighbourhood of an electrified body must necessarily become itself electric, even if it were previously uncharged; since (Law III.) the entire resultant force at any point within it must vanish, and consequently there must be a distribution of electricity on its surface which will for internal points balance the force resulting from the external electrified body. If the sphere, being insulated, be previously charged with a given quantity of electricity, the whole amount will (§ 71) remain unaltered by the electrical influence, but its distribution cannot be uniform, since in that case, it would exert no force on an internal point, and there would remain the unbalanced resultant due to the external body. In what follows, it will be proved that the conditions are satisfied by a certain assumed distribution of electricity in each instance; but the proposition that no other distribution can satisfy the conditions, which is merely a case of a general theorem referred to above (§ 73), will not be specially demonstrated with reference to the particular problems; although we shall have to assume its truth when a certain distribution which is proved synthetically to satisfy the conditions is asserted to be the unique solution of the problem.

Attraction of a Spherical Surface of which the density varies inversely as the cube of the distance from a given point.

90. Let us first consider the case in which the given point S and the attracted point P are separated by the spherical surface. The two figures represent the varieties of this case in which the point S being without the sphere, P is within; and, S being within, the attracted point is external. The same demonstration is applicable literally with reference to the two

figures; but, for avoiding the consideration of negative quantities, some of the expressions may be conveniently modified to
suit the second figure. In such instances the two expressions
are given in a double line, the upper being that which is most
convenient for the first figure, and the lower for the second.

Let the radius of the sphere be denoted by a, and let f be the
distance of S from C, the centre of the sphere (not represented
in the figures).

Join SP and take T in this line (or its continuation) so that

$$\text{(fig. 1)} \quad SP \cdot ST = f^2 - a^2 \\ \text{(fig. 2)} \quad SP \cdot TS = a^2 - f^2 \Big\} \quad \dots\dots\dots\dots (1).*$$

Through T draw any line cutting the spherical surface at K, K'.
Join SK, SK', and let the lines so drawn cut the spherical surface again in EE'.

Let the whole spherical surface be divided into pairs of opposite elements with reference to the point T. Let K and K'
be a pair of such elements situated at the extremities of the
chord KK', and subtending the solid angle ω at the point T;
and let elements E and E' be taken subtending at S the same
solid angles respectively as the elements K and K'. By this
means we may divide the whole spherical surface into pairs of
conjugate elements, E, E', since it is easily seen that when we
have taken every pair of elements, K, K', the whole surface will

---

* If, in geometrical investigations in which diagrams are referred to, the
distinction of *positive* and *negative* quantities be observed, the order of the
letters expressing a straight line will determine the algebraic sign of the
quantity denoted: thus we should have, universally, if A, B be the extremities
of a straight line, $AB = -BA$, each member of this equation being positive
or negative according to the conventional direction in which positive quantities
are estimated. In the present instance, lengths measured along the line SP in
the direction from S towards P, or in corresponding directions in the continuation of this line on either side, are, in both figures, considered as positive.
Hence, in the first figure ST will be positive; but when f is less than a,
ST must be negative on account of the equation $SP \cdot ST = f^2 - a^2$. Hence the
second figure represents this case; and, if we wish to express the circumstances
without the use of negative quantities, we must change the signs of both
members of the equation, and substitute for the positive quantity $-ST$ its
equivalent TS, so that we have $SP \cdot TS = a^2 - f^2$, as the most convenient form
of the expression, when reference is made to the second figure. See above
(Symbolical Geometry, § 4), in volume of the *Cambridge and Dublin Mathematical Journal* for 1848, where the principles of interpretation of the sign − in
geometry are laid down by Sir William R. Hamilton [or Tait's *Quaternions*,
§ 20, 1868].

have been exhausted, without repetition, by the deduced ele-

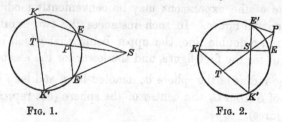

Fig. 1.   Fig. 2.

ments, E, E'. Hence the attraction on P will be the final resultant of the attractions of all the pairs of elements, $E\,E''$.

Now if ρ be the electrical density at E, and if F denote the
attraction of the element E on P, we have

$$F = \frac{\rho \cdot E}{EP^2}.$$

According to the given law of density we shall have

$$\rho = \frac{\lambda}{SE^3},$$

where λ is a constant. Again, since SEK is equally inclined
to the spherical surface at the two points of intersection, we
have (§§ 85, 86)

$$E = \frac{SE^2}{SK^2} \cdot K = \frac{SE^2}{SK^2} \cdot \frac{2a\omega \cdot TK^2}{KK'};$$

and hence

$$F = \frac{\dfrac{\lambda}{SE^3} \cdot \dfrac{SE^2}{SK^2} \cdot \dfrac{2a\omega \cdot TK^2}{KK'}}{EP^2} = \lambda \cdot \frac{2a}{KK'} \cdot \frac{TK^2}{SE \cdot SK^2 \cdot EP^2} \cdot \omega.$$

Now, by considering the great circle in which the sphere is cut
by a plane through the line SK, we find that

$$\left. \begin{array}{l} \text{(fig. 1)} \quad SK \cdot SE = f^2 - a^2 \\ \text{(fig. 2)} \quad KS \cdot SE = a^2 - f^2 \end{array} \right\} \quad \ldots\ldots\ldots\ldots(2),$$

and hence $SK \cdot SE = SP \cdot ST$, from which we infer that the
triangles KST, PSE are similar; so that $TK : SK :: PE : SP$.
Hence

$$\frac{TK^2}{SK^2 \cdot PE^2} = \frac{1}{SP^2},$$

and the expression for F becomes

$$F = \lambda \cdot \frac{2a}{KK'} \cdot \frac{1}{SE \cdot SP^2} \cdot \omega \quad \ldots\ldots\ldots\ldots(3).$$

Modifying this by (2) we have

$$\text{(fig. 1)} \quad F = \lambda \cdot \frac{2a}{KK'} \cdot \frac{\omega}{(f^2 - a^2)\, SP^2} \cdot SK$$

$$\text{(fig. 2)} \quad F = \lambda \cdot \frac{2a}{KK'} \cdot \frac{\omega}{(a^2 - f^2)\, SP^2} \cdot KS$$

$$\Bigg\} \quad \dots\dots (4).$$

Similarly, if F' denote the attraction of E' on P, we have

$$\text{(fig. 1)} \quad F' = \lambda \frac{2a}{KK'} \cdot \frac{\omega}{(f^2 - a^2)\, SP^2} \cdot SK',$$

$$\text{(fig. 2)} \quad F' = \lambda \frac{2a}{KK'} \cdot \frac{\omega}{(a^2 - f^2)\, SP^2} \cdot K'S.$$

Now in the triangles which have been shown to be similar, the angles TKS, EPS are equal; and the same may be proved of the angles $TK'S$, $E'PS$. Hence the two sides SK, SK' of the triangle KSK' are inclined to the third at the same angles as those between the line PS and directions PE, PE' of the two forces on the point P; and the sides SK, SK' are to one another as· the forces, F, F', in the directions PE, PE'. It follows, by "the triangle of forces," that the resultant of F and F' is along PS, and that it bears to the component forces the same ratios as the side KK' of the triangle bears to the other two sides. Hence the resultant force due to the two elements E and E', on the point P, is towards S, and is equal to

$$\lambda \cdot \frac{2a}{KK'} \cdot \frac{\omega}{(f^2 \sim a^2) \cdot SP^2} \cdot KK', \text{ or } \frac{\lambda \cdot 2a \cdot \omega}{(f^2 \sim a^2)\, SP^2}.$$

The total resultant force will consequently be towards S; and we find, by summation (§ 83) for its magnitude,

$$\frac{\lambda \cdot 4\pi a}{(f^2 \sim a^2)\, SP^2}.$$

Hence we infer that the resultant force at any point P, separated from S by the spherical surface, is the same as if a quantity of matter equal to $\dfrac{\lambda \cdot 4\pi a}{f^2 \sim a^2}$ were concentrated at the point S.

91. To find the attraction when S and P are either both without or both within the spherical surface.

Take in CS (fig. 3), or in CS produced through S (fig. 4), a point S_1, such that

$$CS \cdot CS_1 = a^2.$$

Then, by a well-known geometrical theorem (see note on § 87), if E be any point on the spherical surface, we have

$$\frac{SE}{S_1 E} = \frac{f}{a}.$$

Hence we have

$$\frac{\lambda}{SE^3} = \frac{\lambda a^3}{f^3 . S_1 E^3}.$$

Hence, ρ being the electrical density at E, we have

$$\rho = \frac{\dfrac{\lambda a^3}{f^3}}{S_1 E^3} = \frac{\lambda_1}{S_1 E^3},$$

if

$$\lambda_1 = \frac{\lambda a^3}{f^3}.$$

Hence, by the investigation in the preceding paragraph, the attraction on P is towards S_1, and is the same as if a quantity

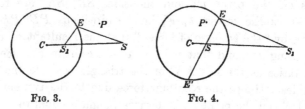

FIG. 3.   FIG. 4.

of matter equal to $\dfrac{\lambda_1 . 4\pi a}{f_1^2 \sim a^2}$ were concentrated at that point; f_1 being taken to denote CS_1. If for f_1 and λ_1 we substitute their values, $\dfrac{a^2}{f}$ and $\dfrac{\lambda a^3}{f^3}$, .we have the modified expression

$$\frac{\lambda \dfrac{a}{f} . 4\pi a}{a^2 \sim f^2}$$

for the quantity of matter which we must conceive to be collected at S.

92. PROP. If a spherical surface be electrified in such a way that the electrical density varies inversely as the cube of the distance from an internal point S (fig. 4), or from the corresponding external point S_1, it will attract any external point, as if its whole mass were concentrated at S, and any internal

point as if a quantity of matter greater than the whole mass in the ratio of a to f were concentrated at S_1.

Let the density at E be denoted, as before, by $\dfrac{\lambda}{SE^3}$. Then, if we consider two opposite elements at E and E' which subtend a solid angle ω at the point S, the areas of these elements being (§ 96) $\dfrac{\omega \cdot 2a \cdot SE^2}{E'E}$ and $\dfrac{\omega \cdot 2a \cdot SE'^2}{E'E}$, the quantity of electricity which they possess will be

$$\frac{\lambda \cdot 2a \cdot \omega}{E'E}\left(\frac{1}{SE}+\frac{1}{E'S}\right) \text{ or } \frac{\lambda \cdot 2a \cdot \omega}{SE \cdot E'S}.$$

Now $SE \cdot E'S$ is constant (Euc. III. 35), and its value is $a^2 - f^2$. Hence, by summation, we find for the total quantity of electricity on the spherical surface

$$\frac{\lambda \cdot 4\pi a}{a^2 - f^2}.$$

Hence, if this be denoted by m, the expressions in the preceding paragraphs, for the quantities of electricity which we must suppose to be concentrated at the point S or S_1, according as P is without or.within the spherical surface, become respectively

$$m, \text{ and } \frac{a}{f}m. \qquad\qquad \text{Q. E. D.}$$

Application of the preceding Theorems to the Problem of Electrical Influence.

93. PROB. To find the electrical density at any point of an insulated conducting sphere (radius a) charged with a quantity Q (either positive, or negative, or zero) of electricity, and placed with its centre at a given distance f from an electrical point M possessing m units of electricity.

If the expression for the electrical density at any point E of the surface be

$$\rho = \frac{\lambda}{ME^3} + k \dots\dots\dots(a),$$

λ and k being constants; the force exerted by the electrified surface on any internal point will be the same as if the constant distribution k, which (§ 78) exerts no force on an internal point, were removed; and therefore (§ 90) will be

T. E. 5

the same as if a quantity of matter equal to $\dfrac{\lambda \cdot 4\pi a}{f^2 - a^2}$ were collected at the point M.   Hence, if the condition

$$\frac{\lambda \cdot 4\pi a}{f^2 - a^2} = -m \dots\dots\dots\dots\dots\dots(b)$$

be satisfied, the total attraction on an internal point, due to the electrified surface and to the influencing point, will vanish.

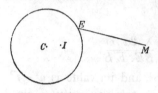

Hence this distribution satisfies the condition of equilibrium (§ 72); and to complete the solution of the proposed problem it only remains to determine the quantity k, so that the total quantity of electricity on the surface may have the given value Q. Now (§ 92) the total mass of the distribution, depending on the term $\dfrac{\lambda}{ME^3}$ in the expression for the density, since M is an *external* point, is equal to

$$\frac{a}{f} \cdot \frac{\lambda \cdot 4\pi a}{f^2 - a^2}.$$

Hence, adding $4\pi a^2 k$, the quantity depending on the constant term k, we obtain the entire quantity, which must be equal to Q; and we therefore have the equation

$$\frac{a}{f} \cdot \frac{\lambda \cdot 4\pi a}{f^2 - a^2} + 4\pi a^2 k = Q \dots\dots\dots\dots (c).$$

From equations (b) and (c) we deduce

$$\lambda = -\frac{(f^2 - a^2)\,m}{4\pi a} \quad \text{and} \quad k = \frac{Q + \dfrac{a}{f} m}{4\pi a^2}.$$

Hence, by substituting in (a), we have

$$\rho = -\frac{(f^2 - a^2)\,m}{4\pi a} \cdot \frac{1}{ME^3} + \frac{Q + \dfrac{a}{f} m}{4\pi a^2} \dots\dots\dots\dots(A),$$

as the expression of the required distribution of electricity. This agrees with the result obtained by Poisson, by means of an investigation in which the analysis known as that of "Laplace's coefficients," is employed.

94. To find the attraction exerted by the electrified conductor on any external point.

We may consider separately the distributions corresponding to the constant and the variable term in the expression for the electrical density at any point of the surface. The attraction of the first of these on an external point is (§ 87) the same as if its whole mass were collected at the centre of the sphere: the attraction of the second on an external point is (§ 92) the same as if its whole mass were collected at an interior point I, taken in MC so that $MI.MC = a^2$. Hence, according to the investigation in the preceding paragraph, we infer that the conductor attracts any external point with the same force as would be produced by quantities $Q + \dfrac{a}{f}m$, and $-\dfrac{a}{f}m$ of electricity, concentrated at the points C and I respectively.

COR. The resultant force at an external point infinitely near the surface is in the direction of the normal, and is equal to $4\pi\rho$, if ρ be the electrical density of the surface, in the neighbourhood.

95. To find the mutual attraction or repulsion between the influencing point, M, and the conducting sphere.

According to what precedes, the required attraction or repulsion will be the entire force exerted upon m units of electricity at the point M, by $Q + \dfrac{a}{f}m$ at C and $-\dfrac{a}{f}m$ at a point I, taken in CM, at a distance $\dfrac{a^2}{f}$ from C. Hence, if the required *attraction* be denoted by F (a quantity which will be negative if the actual force be of *repulsion*), we have

$$F = -\frac{\left(Q + \dfrac{a}{f}m\right)m}{f^2} + \frac{\dfrac{a}{f}m.m}{\left(f - \dfrac{a^2}{f}\right)^2} \dots\dots\dots\dots (B),$$

$$= \frac{a\{f^4 - (f^2 - a^2)^2\}m^2 - f(f^2 - a^2)^2 Qm}{f^3(f^2 - a^2)^2},$$

or $$F = \frac{a^3(2f^2 - a^2)m^2 - f(f^2 - a^2)^2 Qm}{f^3(f^2 - a^2)^2} \dots\dots (C).$$

COR. 1. If Q be zero or negative, the value of F is necessarily positive, since f must be greater than a; and therefore there is a force of attraction between the influencing point and the conducting sphere, whatever be the distance between them.

Cor. 2. If Q be positive, then for sufficiently large values of f, F is negative, while for values nearly equal to a, F is positive. Hence if an electrical point be brought into the neighbourhood of a similarly charged insulated sphere, and if it be held at a great distance, the mutual action will be repulsive; if it then be gradually moved towards the sphere, the repulsion, which will at first increase, will, after attaining a maximum value, begin to diminish till the electrical point is moved up to a certain distance where there will be no force either of attraction or repulsion; if it be brought still nearer to the conductor, the action will become attractive and will continually augment as the distance is diminished.

If the value of Q be positive, and sufficiently great, a spark will be produced between the nearest part of the conductor and the influencing point, before the force becomes changed from repulsion to attraction.

St Peter's College,
July 7, 1848.

EFFECTS OF ELECTRICAL INFLUENCE ON INTERNAL SPHERICAL, AND ON PLANE CONDUCTING SURFACES.

96.　In the preceding articles of this series certain problems with reference to conductors bounded externally by spherical surfaces have been considered. It is now proposed to exhibit the solutions of similar problems with reference to the distribution of electricity on concave spherical surfaces, and on planes.

The object of the following short digression is to define and explain the precise signification of certain technical terms and expressions which will be used in this and in subsequent papers on the Mathematical Theory of Electricity.

External and Internal Conducting Surfaces.

97.　Def. 1.　A closed surface separating conducting matter

within it from air* without it, is called an *external conducting surface.*

DEF. 2. A closed surface separating air within it from conducting matter without it is called an *internal conducting surface.*

Thus, according to these definitions, a solid conductor has only one " conducting surface," and that " an external conducting surface."

A conductor containing within it one or more hollow spaces filled with air, possesses two or more "conducting surfaces;" namely, one "external conducting surface," and one or more "internal conducting surfaces."

A complex arrangement, consisting of a hollow conductor and other conductors insulated within it, presents several external and internal conducting surfaces; namely, an "external conducting surface" for each individual conductor, and as many "internal conducting surfaces" as there are hollow spaces in the different conductors.

98. In any arrangement such as this, there are different masses of air which are completely separated from one another by conducting matter. Now among the General Theorems alluded to in § 73, it will be proved that the bounding surface or surfaces of any such mass of air cannot experience any electrical influence from the surfaces of the other masses of air, or from any electrified bodies within them. Hence any statical phenomena of electricity which may be produced in a hollow space surrounded continuously by conducting matter,— whether this conducting envelope be a sheet even as thin as gold leaf, or a massive conductor of any external form and dimensions,—will depend solely on the form of the internal conducting surface.

99. PROP. *An internal conducting surface cannot receive a charge of electricity independently of the influence of electrified bodies within it.*

100. The demonstration of this proposition depends on what precedes, and on one of the General Theorems, already alluded to (§ 73), by which it appears that it is impossible to distribute a charge of electricity on a closed surface in such a manner that

---

See § 70, excluding all non-conductors except air, or gases.

there may be no resultant force exerted on external points, and consequently impossible, with merely a distribution of electricity on an internal conducting surface, to satisfy the condition of electrical equilibrium with reference to the conducting matter which surrounds it.

The preceding proposition (§ 99) is fully confirmed by experiment (Faraday's *Experimental Researches*, §§ 1173, 1174). In fact, the certainty with which its truth has been practically demonstrated in a vast variety of cases, by all electrical experimenters, may be regarded as a very strong part of the evidence on which the Elementary Laws as stated above (§ 72) rest.

101. It might be further stated that the total quantity of electricity produced by influence on an internal conducting surface is necessarily equal in every case to the total quantity of electricity on the influencing electrified bodies insulated within it. This will also be demonstrated among the General Theorems ; but its truth in the special case which we are now to consider, will, as we shall see, be established by a special demonstration.

Electrical Influence on an Internal Spherical Conducting Surface.

102. In investigating the effects of electrical influence upon an external, or convex, spherical conducting surface (§§ 93, 94, 95), we have considered the conductor to be insulated and initially charged with a given amount of electricity. In the present investigation no such considerations are necessary, since, according to the statements in the preceding paragraphs, it is of no consequence, in the case now contemplated, whether the conductor containing the internal conducting surface be insulated or not; and it is impossible to charge this internal surface initially, or to charge it at all, independently of the influence of electrified bodies within it. With the modifications and omissions necessary on this account, the preceding investigations are applicable to the case now to be considered.

103. PROB. To find the electrical density at any point of an internal spherical conducting surface with an electrical point insulated within it.

Let m denote the quantity of electricity in the electrical point M; f its distance from C the centre of the sphere, and a the radius of the sphere.

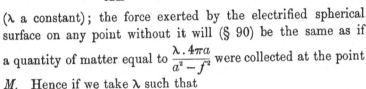

If the expression for the electrical density at any point E of the internal surface be

$$\rho = \frac{\lambda}{ME^3},^*$$

(λ a constant); the force exerted by the electrified spherical surface on any point without it will (§ 90) be the same as if a quantity of matter equal to $\dfrac{\lambda \cdot 4\pi a}{a^2 - f^2}$ were collected at the point M. Hence if we take λ such that

$$\frac{\lambda \cdot 4\pi a}{a^2 - f^2} = -m,$$

the total resultant force, due to the given electrical point and to the electrified surface, will vanish at every point external to the spherical surface, and consequently at every point within the substance of the conductor; so that the condition of electrical equilibrium (§ 72), in the prescribed circumstances, is satisfied. We conclude, therefore, that the required density at any point E, of the internal spherical surface is given by the equation

$$\rho = -\frac{(a^2 - f^2)\,m}{4\pi a} \cdot \frac{1}{ME^3} \quad \dots\dots\dots\dots\dots(A).$$

This solution of the problem is complete, since it satisfies all the conditions that can possibly be prescribed, and it is unique, as follows from the general Theorem referred to in § 73.†

---

* We cannot here, as in (a) of § 93, annex a constant term, since in this case there would result a force due to a corresponding quantity of electricity, concentrated at the centre of the sphere on all points of the conducting mass.

† For if there were two distinct solutions there would be two different distributions on the spherical surface, each balancing on external points the action of the internal influencing body, and therefore each producing the same force at external points. Hence a distribution, in which the electrical density at each point is equal to the difference of the electrical densities in those two, would produce no force at external points. But, by the theorem alluded to, no distribution on a closed surface of any form can have the property of producing no force on external points; and therefore the hypothesis that there are two distinct solutions is impossible.

The theorem made use of in this reasoning is susceptible of special *analytical*

COR. The total quantity of electricity produced by the influence of an electrical point within an internal spherical conducting surface is equal, but of the opposite kind to that of the influencing point.

This follows at once from the investigation of § 92; from which we also deduce the conclusion stated below in the next section.

104. The entire electrical force, which vanishes for all points external to the conducting surface, may, for points within it, be found by compounding the force due to the given influencing point M (charged, by hypothesis, with a quantity m of electricity) with that due to an imaginary point I, taken in CM produced, at such a distance from C that $CM . CI = a^2$, and charged with a quantity of electricity equal to $-\dfrac{a}{f} m$.

COR. The resultant force at an internal point infinitely near the surface, is in the direction of the normal, and is equal to $4\pi\rho$, if ρ be the electrical density of the surface in the neighbourhood.

105. The mutual attraction between the influencing point M, and the surface inductively electrified will be found as in § 95, provided the *uniform supplementary distribution* which was there introduced be omitted. Hence, omitting the term of (B) which depends on this supplementary distribution; or simply, without reference to (B), considering the mutual force between m at M and $-\dfrac{a}{f} m$ at I, a force which is necessarily attractive as the two electrical points M and I possess opposite kinds of electricity; we obtain

$$F = \frac{\dfrac{a}{f} m . m}{\left(\dfrac{a^2}{f} - f\right)^2} = \frac{a f m^2}{(a^2 - f^2)^2} \quad \dots\dots\dots\dots\dots(B)$$

as the expression for the required attraction.

---

demonstration (with the aid of the method in which "Laplace's coefficients" are employed) for the case of a spherical surface; but such an investigation would be inconsistent with the synthetical character of the present series of papers, and I therefore do no more at present than allude to the general theorem.

Electrical Influence on a Plane Conducting Surface of infinite extent.

106. If, in either the case of an external or the case of an internal spherical conducting surface, the radius of the sphere be taken infinitely great, the results will be applicable to the present case of an infinite plane; and it is clear that from either we may deduce the complete solution of the problem of determining the distribution of electricity, produced upon a conducting plane, by the influence of an electrical point. The "supplementary distribution," which, in the case of a convex spherical conducting surface, must in general be taken into account, will, in the case of a sphere of infinite radius, be a finite quantity of electricity uniformly distributed over a surface of infinite extent, and will therefore produce no effect; and the same results will be obtained whether we deduce them from the case of an external or of an internal spherical surface.

107. Let M be an electrical point possessing a quantity m of electricity placed in the neighbourhood of a conductor bounded on the side next M by a plane LL' which we must conceive to be indefinitely extended in every direction; it is required to determine the electrical density at any point E of the conducting surface.

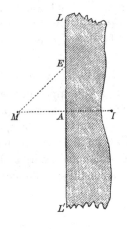

Draw MA perpendicular to the plane, and let its length be denoted by p. We may, in the first place, conceive that instead of the plane surface we have a spherical conducting surface entirely enclosing the air in which M is insulated; and, supposing the shortest line from M to the spherical surface to be equal to p, we should have, according to the notation of § 103, $f = a - p$.

Hence the expression (A) becomes

$$\rho = -\frac{2ap - p^2}{4\pi a} \cdot \frac{m}{ME^3} = -\left(\frac{p}{2\pi} - \frac{p^2}{4\pi a}\right)\frac{m}{ME^3}.$$

In this, let a be supposed to be infinitely great; the second term within the vinculum will vanish, and we shall have simply

$$\rho = -\frac{mp}{2\pi ME^3}\ldots\ldots\ldots\ldots\ldots(A)$$

for the required electrical density at the point E of the infinite plane electrified inductively through the influence of the point M.

Cor. The total amount of the electricity produced by induction is equal in quantity, but opposite in kind, to that of the influencing point M. We have seen already that the same proposition is true in general for internal spherical surfaces inductively electrified; but it does not hold for an external spherical surface, even if we neglect the "supplementary distribution," as it appears from the demonstration of § 92, that the amount of the distribution expressed by the first term (that which varies inversely as the cube of the distance from the influencing point) of the value of ρ in equation (A) of § 93, is equal to $-\frac{a}{f}m$. The infinite plane may, as we have seen, be regarded as an extreme case of either an external or an internal spherical surface; and the proposition which is in general true for internal, but not true for external spherical surfaces, holds in this limiting intermediate case.

108. To determine the resultant force at any point in the air, before the conducting plane, it will be only necessary, as in § 104, to compound the action of the given electrical point with that of an imaginary point I.

To find this point, we must produce MA beyond A to a distance AI, determined by the equation $CM . CI = a^2$; which, if we denote AI by p', becomes $(a-p)(a+p') = a^2$.

From this we deduce

$$p' = \frac{ap}{a-p} = \frac{p}{1-\frac{p}{a}},$$

and thence, in the case of $a = \infty$, we deduce $p' = p$.

Again, for the quantity of electricity to be concentrated at I, we have the expression

$$m' = -\frac{a}{a-p}m, \text{ or, when } a = \infty, m' = -m.$$

Hence the force at any point before the plane will be obtained by compounding that due to the given electrical point M, with a force due to an imaginary point I, possessing an equal quantity of the other hand of electricity, and placed at an equal distance behind the plane in the perpendicular MA produced.

109. If reference be made to the general demonstration (§ 90) on which all the special conclusions with reference to the effects of electrical influence on convex, concave, or plane conducting surfaces depend, we see that the geometrical construction employed fails in the case of a sphere of infinite radius, becoming nugatory in almost every step: we have however deduced *conclusions* which are not nugatory, but, on the contrary, assume a remarkably simple form for this case; and we may regard as rigorously established the solution of the problem of electrical influence on an infinite plane which has been thus obtained.

110. It is interesting to examine the nugatory forms which occur in attempting to apply the demonstrations of §§ 90 and 92, to the case of an infinite plane; and it is not difficult to derive a special demonstration, free from all nugatory steps, of the following proposition.

Let LL' be an infinite "material plane," of which the "density" in different positions varies inversely as the cube of the distance from a point S, or from an equidistant point S_1, on the other side of the plane. The resultant force at any point P is the same as if the whole matter of the plane were concentrated at S; and the resultant force at any point P_1, on the other side of the plane, is the same as if the whole matter were collected at S_1.

111. In the course of the demonstration (in that part which corresponds to the investigation in § 93) it would appear that, if the density at any point E of the plane is given by the expression

$$\rho = \frac{\lambda}{\overline{SE}^3},$$

the entire quantity of matter distributed over the infinite extent of the plane is given by the expression

$$-m = \frac{2\pi\lambda}{p}.$$

This proposition and that which precedes it * contain the simplest expression of the mathematical truths on which the solution of the problem of electrical influence on an infinite plane depends, and we might at once obtain from them the results given above. For an isolated investigation of this case of electrical equilibrium, this would be a better form of solution: but I have preferred the method given above, since the solution of the more general problem, of which it is a particular case, had been previously given.

112. The case of electrical influence which has been considered

---

* The two propositions may be analytically expressed as follows :—

Let O, the point in which SS_1 cuts the plane, be origin of co-ordinates, and let this line be axis of z. Then, taking OX, OY in the plane, let the co-ordinates of P be (x, y, z). Let also those of E be $(\xi, \eta, 0)$; so that we have

$$\rho = \frac{\lambda}{(\xi^2 + \eta^2 + p^2)^{\frac{1}{2}}}.$$

Hence the proposition stated in the text (§ 111), that the entire quantity of matter distributed over the infinite extent of the plane is equal to $\frac{2\pi\lambda}{p}$, is thus expressed:—

$$\int_{-\infty}^{\infty}\int_{-\infty}^{\infty} \frac{\lambda d\xi d\eta}{(\xi^2 + \eta^2 + p^2)^{\frac{3}{2}}} = \frac{2\pi\lambda}{p}.$$

This equation may be very easily verified, and so an extremely simple analytical demonstration of one of the theorems enunciated above is obtained.

Again, the proposition with reference to the attraction of the plane may, according to the well-known method, be expressed most simply by means of the potential. This must, in virtue of the enunciation in § 110, be equal to the potential due to the same quantity of matter, collected at the point S, or the point S_1, according as the attracted point is separated from the former or from the latter by the plane. Hence we must have

$$\int_{-\infty}^{\infty}\int_{-\infty}^{\infty} \frac{\lambda d\xi d\eta}{(\xi^2 + \eta^2 + p^2)^{\frac{3}{2}}\{(x-\xi)^2 + (y-\eta)^2 + z^2\}^{\frac{1}{2}}} = \frac{\frac{2\pi\lambda}{p}}{\{x^2 + y^2 + (\pm z + p)^2\}^{\frac{1}{2}}},$$

the positive or negative sign being attached to z in the denominator of the second member, according as z is given with a positive or negative value. This equation (of which a geometrical demonstration is included in §§ 107 and 108, in connexion with § 90) is included in a result (the evaluation of a certain multiple integral), of which three different analytical demonstrations were given in a paper *On certain Definite Integrals suggested by Problems in the Theory of Electricity*, published in March 1847 in this *Journal*, vol. ii. p. 109 (IX. below).

might at first sight appear to be of a singularly unpractical nature, since a conductor presenting on one side a plane surface of infinite extent in every direction would be required for fully realizing the prescribed circumstances. If, however, we have a plane table of conducting matter, or covered with a sheet of tinfoil, or if we have a wall presenting an uninterrupted plane surface of some extent, the imagined circumstances are, as we readily see, *approximately* realized with reference to the influence of any electrical point in the neighbourhood of such a conducting plane, provided the distance of the influencing point from the plane be small compared with its distance from the nearest part where the continuity of the plane surface is in any way broken.

FORTBREDA, BELFAST, *Oct.* 17, 1849.

INSULATED SPHERE SUBJECT TO THE INFLUENCE OF A BODY OF ANY FORM ELECTRIFIED IN ANY GIVEN MANNER.

113.   The problem of determining the distribution of electricity upon a sphere, or upon internal or plane spherical conducting surfaces, under the influence of an electrical point, was fully solved in §§ 89...112 of this series of papers. On the principle of the superposition of electrical forces (§ 63) we may apply the same method to the solution of corresponding problems with reference to the influence of any number of given electrical points.

114.   Thus let M, M', M'' be any number of electrical points possessing respectively m, m', m'' units of electricity, at distances f, f', f'' from C the centre of a sphere insulated and charged with a quantity Q of electricity. The actual distribution of electricity on the spherical surface must be such that the force due to it at any internal point shall be equal and opposite to the force due to the electricity at M, M', M''. Now if there were a distribution of electricity on the spherical surface such that the density at any point E would be $\dfrac{\lambda}{ME^3}$, the force due to this at any internal point

would (§ 90) be the same as that due to a quantity $\dfrac{\lambda \cdot 4\pi a}{f^2 - a^2}$ concentrated at the point M; and therefore if we take

$$\lambda = -\frac{(f^2 - a^2)m}{4\pi a},$$

the force at internal points due to this distribution would be equal and opposite to the force due to the actual electricity of M. We might similarly express distributions which would respectively balance the actions of M', M'', etc., upon points within the sphere; and thence, by supposing all those distributions to coexist on the surface, we infer that a single distribution such that the density at E is equal to

$$-\left\{ \frac{(f^2-a^2)m}{4\pi a} \frac{1}{ME^3} + \frac{(f'^2-a^2)m'}{4\pi a} \frac{1}{M'E^3} + \frac{(f''^2-a^2)m''}{4\pi a} \frac{1}{M''E^3} \right\}$$

would balance the joint action of all the electrical points M, M', M'', on points within the sphere. Again, from § 92, we infer that the total quantity of electricity in such a distribution is

$$-\left(\frac{a}{f}m + \frac{a}{f'}m' + \frac{a}{f''}m'' \right).$$

Hence, unless the data chance to be such that Q is equal to this quantity, a supplementary distribution will be necessary to constitute the actual distribution which it is required to find. The amount of this supplementary distribution will be

$$Q + \frac{a}{f}m + \frac{a}{f'}m' + \frac{a}{f''}m'';$$

which must be so distributed as to produce no force on internal points.

115. Taking then the distribution found above, which balances the action of the electricity at M, M', etc., on points within the sphere, and a uniform supplementary distribution; and superimposing one on the other, we obtain a resultant electrical distribution in which the density at any point E of the surface of the sphere is given by the equation

$$\rho = -\left\{ \frac{(f^2 - a^2)m}{4a\pi} \frac{1}{ME^3} + \frac{(f'^2 - a^2)}{4\pi a} \frac{1}{M'E^3} + \text{etc.} \right\}$$

$$+ \frac{Q + \frac{a}{f}m + \frac{a}{f'}m' + \text{etc.}}{4\pi a^2} \quad \dots\dots\dots\dots\dots(1);$$

and we draw the following conclusions:—

(1) The total force at any internal point, due to this distribution and to the electricity of M, M', etc., vanishes.

(2) The entire quantity of electricity on the spherical surface is equal to Q.

Hence this distribution of the given charge on the sphere satisfies the condition of electrical equilibrium under the influence of the given electrical points M, M', etc.; and (§ 73) it is therefore the distribution which actually exists upon the spherical conductor in the prescribed circumstances.

116. The resultant force at any external point may be found as in the particular case treated in § 94. Thus, if we join MC, $M'C$, $M''C$, and take in the lines so drawn, points I, I', I'' respectively, at distances from C such that

$$CI \cdot CM = CI' \cdot CM' = CI'' \cdot CM'' = a^2,$$

the resultant action due to the actual electricity of the spherical surface will, at any external point, be the same as if the sphere were removed, and electrical points I, I', etc., substituted in its stead, besides (except in the case when the supplementary distribution vanishes) an electrical point at C: and the quantities of electricity which must be conceived for this representation, to be concentrated at these points, are respectively

$$
\left.
\begin{aligned}
& -\frac{a}{f}m, && \text{at } I \\[2mm]
& -\frac{a}{f'}m', && \text{at } I' \\[2mm]
& \;\cdots\cdots\cdots\cdots\cdots \\[2mm]
& Q + \frac{a}{f} + \frac{a}{f'} + \text{etc.,} && \text{at } C
\end{aligned}
\right\} \quad \ldots\ldots\ldots\ldots(2).
$$

and

117. By means of these imaginary electrical points we may give another form to the expression for the distribution on the spherical surface, which in many important cases, especially that of two mutually influencing spherical surfaces, is extremely convenient. For (as in § 94, Cor.) it is readily seen that the first term, in the expression for ρ multiplied by 4π, or

$$-\frac{(f^2 - a^2)m}{a}\,\frac{1}{ME^3},$$

is the resultant force at E, due to M and I, and that this force is in the direction of a normal to the spherical surface through

E; and that similar conclusions hold with reference to the other similar terms of (2). Again, the last term,

$$\frac{Q + \frac{a}{f} m + \frac{a}{f'} m' + \text{etc.}}{4\pi a^2}$$

is the expression for the force at E, due to the imaginary electric point C, divided by 4π; and this force also is in the direction of the normal. Hence, with reference to the total resultant action at E, due to M, M', etc., and the spherical surface, or the imaginary electrical points within it, we infer

(1)   That this force is in the direction of the normal;

(2)   That if R be its magnitude considered as positive or negative according as it is from or towards the centre of the sphere, and ρ the electrical density at E, we have

$$\rho = \frac{1}{4\pi} R \ldots\ldots\ldots\ldots\ldots\ldots\ldots\ldots(3).$$

These two propositions constitute the expression, for the case of a spherical conductor subject to any electric influence, of *Coulomb's Theorem.**

118.   The total action exerted by the given electrical points, and by the sphere with its electricity disturbed by their influence upon a given electrified body placed anywhere in their neighbourhood, might, as we have seen, be found by substituting in place of the sphere the group of electrical points which represents its external action, provided there were no disturbance produced by the influence of this electrified body. This hypothesis, however, cannot be true unless the sphere, after experiencing as a conductor the influence of M, M', etc., were to become a non-conductor so as to preserve with *rigidity* the distribution of its electricity when the new electrified body is brought into its neighbourhood: and consequently, when it is asserted that the resultant force *at* any external point P is due to the group of electrical points determined in the preceding paragraphs, we must remember that the disturbing influence that would be actually exerted upon the distribution on the

---

* For a general demonstration of this theorem, virtually the same as the original demonstration given by Coulomb himself, see *Cambridge Mathematical Journal* (1842), vol. iii. p. 75 (or §§ 7, 8, above).

spherical surface by a unit of electricity at the point P, is excluded in the definition (§ 65) of the expression "*the resultant electrical force at a point.*"

119. The actual force exerted upon any one, M, of the influencing points may be determined by investigating the resultant force at M, due to all the others and to the conductor, and multiplying it by the quantity of electricity, m, situated at this point, since in this case the influence of the body on which the force is required has been actually taken into account.

120. It follows that the entire mutual action between all the given electrical points and the sphere under their influence is the same as the mutual action between the two systems of electrical points,

$$
\left.
\begin{array}{c}
m \text{ at } M \\[1ex]
m' \text{ at } M' \\[1ex]
\cdots\cdots \\[1ex]
\cdots\cdots
\end{array}
\right\}
\text{ and }
\left\{
\begin{array}{ll}
-\dfrac{a}{f}m & \text{at } I \\[2ex]
-\dfrac{a}{f'}m' & \text{at } I' \\[2ex]
\cdots\cdots\cdots\cdots\cdots\cdots \\[1ex]
\cdots\cdots\cdots\cdots\cdots\cdots \\[1ex]
Q+\dfrac{a}{f}m+\dfrac{a}{f'}m'+\text{etc., at } C.
\end{array}
\right.
$$

This action may be fully determined with any assigned *data*, by the elementary principles of statics.

121. There is a remarkable characteristic of this resultant action which ought not to be passed over, as it is related to a very important physical principle of symmetry, of which many other illustrations occur in the theories of electricity and magnetism. It is expressed in the following proposition :—

The mutual action between a spherical conductor and any given electrified body consists of a single force in a line through the centre of the sphere.

Let us conceive the given electrified body either to consist of a group of electrical points, or to be divided into infinitely small parts, each of which may be regarded as an electrical point. The mutual action between the given body and the conducting sphere under its influence is therefore to be found by compounding all the forces between the points M, M', etc., of the given body, and the points I, I', etc....and C, of the imaginary system within the sphere determined by the construction and formulæ of the preceding paragraphs. Of these, the forces between M and C, between M' and C', etc.; and again, between M and I, between M' and I', etc., are actually in lines passing through C; and, therefore, if there were no other forces to be taken into account the proposition would be proved. But we have also a set of forces between M and I', between M and I'', etc., none of which, except in particular cases, are in lines through C, and, therefore, it remains for us to determine the nature of the resultant action of all these

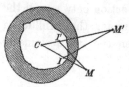

forces. For this purpose let us consider any two points M, M' of the given influencing body and the corresponding imaginary points I, I'; and let us take the force between M and I', and along with it the force between I and M'.

These two forces lie in the plane MCM', since, by the construction given above, I and I' are respectively in the lines CM and CM'; and hence they have a single resultant. Now the force in MI' is due to m units of electricity at M, and $-\dfrac{a}{f'}m'$ units at I'; and (§ 64) it is therefore a force of repulsion equal to

$$\frac{-\dfrac{a}{f'}m'.m}{I'M^2}, \text{ or a force of attraction equal to } \frac{\dfrac{a}{f'}m'.m}{I'M^2}.$$

Similarly, we find

$$\frac{\dfrac{a}{f}m.m'}{IM^2}$$

for the attraction between M and I. Now since, by construction, $CM.CI = CM'.CI'$, the triangles $I'MC$, $IM'C$, which have a common angle at C, are similar. Hence

$$\frac{I'M^2}{IM'^2} = \frac{CI'.CM}{CI.CM'} = \frac{\dfrac{a^2}{f'}.f}{\dfrac{a^2}{f}.f'},$$

from which we deduce

$$\frac{\dfrac{a}{f'}m'.m}{I'M^2}f = \frac{\dfrac{a}{f}m.m'}{IM'^2}f'.$$

Now if we multiply the first member of this equation by $\sin CMI'$, we obtain the moment round C of the force between

I' and M; and similarly, by multiplying the second member
by sin $CM'I$, we find the moment of the force between M' and
I; and, since the angle at M is equal to the angle at M', we
infer that the moments of the two forces round C are equal.
From this it follows that the resultant of the forces in MI' and
$M'I$ is a force in a line passing through C. Now the entire
group of forces between points of the given body and *non-
correspondent* imaginary points, consists of pairs such as that
which we have just been considering; and therefore the mutual
action is the resultant of a number of forces in lines passing
through C. This, compounded with the forces between M, M',
etc., and the *corresponding* imaginary points, and the forces
between M, M', etc., and the imaginary electrical point at C,
gives for the total mutual action a final resultant in a line
passing through C.

122. It follows from this theorem that if a spherical conductor be supported in such a manner as to be able to turn
freely round its centre, or round any axis passing through its
centre, it will remain in equilibrium when subjected to the
influence of any external electrified body or bodies. We may
arrive at the same conclusion by merely considering the perfect
symmetry of the sphere, round its centre or round any line
through its centre, without assuming any specific results with
reference to the distribution of electricity on spherical conductors. For if there were a tendency to turn round any
diameter through the influence of external electrified bodies,
the sphere would, on account of its symmetry, experience the
same tendency when turned into any other position, its centre
and the influencing bodies remaining fixed; and there would
therefore result a continually accelerated motion of rotation.
This being a physical impossibility, we conclude that the
sphere can have no tendency to move when its centre is fixed,
whatever be the electrical influence to which it is subjected.

123. It is very interesting to trace the different actions
which, according to the synthetical solution of the problem of
electrical influence investigated above, must balance to produce
this equilibrium round the centre of a spherical conductor
subjected to the influence of a group of electrical points. Let
us, for example, consider the case of two influencing points.

For fixing the ideas, let us conceive the sphere to be capable of turning round a vertical axis, and let the influencing points be situated in the horizontal plane of its centre, C. If at first there be only one electrical point, M, which we may suppose to be positive, the sphere under its influence will be electrified with a distribution symmetrical round the line MC, but with more negative, or, as the case may be, less positive, electricity, on the hemisphere of the surface next M than on the remote hemisphere. If another positive electrical point, M', be brought into the neighbourhood of the sphere, on a level with its centre, and on one side or the other of MC, and if for a moment we conceive the sphere to be a perfect non-conductor of electricity; this second point, acting on the electricity as distributed under the influence of the first, will make the sphere tend to turn round its vertical axis. Thus if AA_1 be a diameter of the sphere in the line $MACA_1$, the sphere would tend to turn from its primitive position so as to bring the point A of its surface nearer M'. If now the sphere be supposed to become a perfect conductor, the distribution of its electricity will be altered so as to be no longer symmetrical round AA_1. This alteration we may conceive to consist of the superposition of a distribution of equal quantities of positive and negative electricities symmetrically distributed round the line $M'C$, with the negative electricity preponderating on the hemisphere nearest to M'. To obtain the total action of the two points on the electrified sphere, it will now be necessary to compound the action of M', and the action of M, on this superimposed distribution with the action previously considered. Of these the former consists of a simple force of attraction in the line $M'C$; but the latter, if referred to C the centre of the sphere, will give, besides a simple force, a couple round a vertical axis, tending to turn the sphere in such a direction as to bring the point A' of its surface nearer M. Now, as we know *a priori* that there can be no resultant tendency to turn arising from the entire action upon the sphere, it follows that the moment of this couple must be equal to the moment of the contrary couple, which, as we have seen previously, results from the action of M' on the sphere as primitively electrified under the influence of M. This is precisely the proposition of which a synthetical demonstration was given in § 121, and we accordingly see that that demonstration is merely the verification of a proposition of which the truth is rendered certain by *a priori* reasoning founded on general physical principles.

124. When the influencing body, instead of being, as we have hitherto conceived it, a finite group of isolated electrical points, is a continuous mass continuously electrified, we must imagine it to be divided into an infinite number of electrical points; and then, by means of the integral calculus, the expressions investigated above may be modified so as to be applicable to any conceivable case.

125. It appears from the considerations adduced in §§ 99, 100, that it is impossible to have an internal spherical conducting surface, or an infinite plane conducting surface, insulated and charged with a given amount of electricity; and that consequently, there being no " uniform supplementary distributions " to be taken into account, the solutions of ordinary problems with reference to such surfaces are somewhat simpler than those in which it may be proposed to consider an insulated conducting sphere possessing initially a given electrical charge. All the investigations of the present article, except those which have reference to the "supplementary distribution " and which are not required, are at once applicable to cases of internal or of plane conducting surfaces.

126. The importance of considering the imaginary electrical points *I, I'*, etc. (and *C*, the centre of the sphere in the case of an external spherical surface), whether for solving problems with reference to the mutual forces called into action by the electrical excitation, or for determining the distribution of electricity on the spherical surface, has been shown in what precedes. Hence it will be useful, before going further in the subject, to examine the nature of such groups of imaginary points, when the influencing bodies are either finite groups of electrical points, or continuously electrified bodies. [See XIV. below, or Thomson and Tait's *Natural Philosophy*, §§ 512...518.]

127. The term *Electrical Images*, which will be applied to the imaginary electrical points or groups of electrical points, is suggested by the received language of Optics; and the close analogy of optical images will, it is hoped, be considered as a sufficient justification for the introduction of a new and extremely convenient mode of expression into the Theory of Electricity.

STOCKHOLM, *September* 20, 1849.

VI.—ON THE MUTUAL ATTRACTION OR REPULSION BETWEEN TWO ELECTRIFIED SPHERICAL CONDUCTORS.

(Art. LXIV. of *Mathematical and Physical Papers*, Vol. II.)

[Philosophical Magazine, April and August 1853.]

128. In a communication made to the British Association at Cambridge in 1845, I indicated a solution adapted for numerical calculation, of the problem of determining the mutual attraction between two electrified spherical conductors. A paper (II. above) published in November of the same year in the first Number of the *Cambridge and Dublin Mathematical Journal* contains a formula actually expressing the complete solution for the case of an insulated sphere and a non-insulated sphere of equal radius (§ 30, above), and numerical results calculated for four different distances for the sake of comparison with experimental results which had been published by Mr Snow Harris. The investigation by which I had arrived at this solution, which was equally applicable to the general problem of finding the attraction between any two electrified spherical conductors, has not hitherto been published; but it was communicated in July 1849 to M. Liouville, along with another very different method by which I had just succeeded in arriving at the same result, in a letter the substance of which constitutes the present communication. Formulæ marked (8) . . . . (18) in that letter expressed the details of the solution according to the two methods. They are reproduced here in terms of the same notation, and with the same numbers affixed. The first-mentioned method is expressed by the formulæ (16), (17), (18), and the other by (8) . . . . (15). The formulæ marked with letters (*a*), (*b*), etc., in the present paper, express details of which I had not preserved exact memoranda.

129. Let *A* and *B* designate the two spherical conductors; let *a* and *b* be their radii, respectively; and let *c* be the distance between their centres. Let them be charged with such quantities of electricity, that, when no other conductors and no

excited electrics are near them, the values of the potential*
within them may be u and v respectively.

130. The distribution of electricity on each surface may be
determined with great facility by applying the "principle
of successive influences" suggested by Murphy (Murphy's
Electricity, Cambridge, 1833, p. 93), and determining the effect
of each influence by the method of "electrical images," given
in a paper entitled "Geometrical Investigations regarding
Spherical Conductors."† The following statement shows as
much as is required of the results of this investigation for our
present purpose.

131. Let us imagine an electrical point containing a quantity
of electricity equal to ua to be placed at the centre of A, and
another vb at the centre of B. The image of the former in B
will be $-\dfrac{b}{c} \cdot ua$, at a point in the line joining the centres, and
distant by $\dfrac{b^2}{c}$ from the centre of B. The image of this in A will
be $\dfrac{a}{c-\dfrac{b^2}{c}} \cdot \dfrac{b}{c} \cdot ua$, in the same line, at a distance $\dfrac{a^2}{c-\dfrac{b^2}{c}}$ from the
centre of A; the image of this point in B will be $\dfrac{-b}{c-\dfrac{a^2}{c-\dfrac{b^2}{c}}}$
$\dfrac{a}{c-\dfrac{b^2}{c}} \cdot \dfrac{b}{c} \cdot ua$, at a distance $\dfrac{b^2}{c-\dfrac{a^2}{c-\dfrac{b^2}{c}}}$ from the centre of B; and

so on : and in a similar manner we may derive a series of
imaginary points from vb at the centre of B. To specify completely these two series of imaginary points, let p_1, p'_1, p_2, p'_2,
p_3, p'_3, etc., denote the masses of the series of which the first
is at the centre of A ; and let f_1, f'_1, f_2, f'_2, etc., denote the
distances of these points from the centres of A and B alter-

---

* The potential at any point in the neighbourhood of, or within, an electrified
body, is the quantity of work that would be required to bring a unit of positive
electricity from an infinite distance to that point, if the given distribution of
electricity were maintained unaltered. Since the electrical force vanishes at
every point within a conductor, the potential is constant throughout its interior.
† *Cambridge and Dublin Mathematical Journal*, Feb. 1850 (v. above, § 127).

nately; and, again, let q_1, q'_1, q_2, q'_2, ..., denote the masses, and g_1, g'_1, g_2, g'_2, ..., the distances of the successive points of the other series from the centres of B and A alternately. These quantities are determined by using the following equations, and giving n successively the values 1, 2, 3, ... :—

$$\left\{ \begin{aligned} &f_1 = 0, & &p_1 = ua \\ &f'_n = \frac{b^2}{c - f_n}, & &p'_n = - p_n \frac{f'_n}{b} \\ &f_{n+1} = \frac{a^2}{c - f'_n}, & &p_{n+1} = -p'_n \frac{f_{n+1}}{a} \end{aligned} \right\} \left\{ \begin{aligned} &g_1 = 0, & &q_1 = vb \\ &g'_n = \frac{a^2}{c - g_n}, & &q'_n = -q_n \frac{g'_n}{a} \\ &g_{n+1} = \frac{b^2}{c - g'_n}, & &q_{n+1} = -q'_n \frac{g_{n+1}}{b} \end{aligned} \right\} \quad (8).$$

The two series of imaginary electrical points thus specified, would, if they existed, produce the same action in all space external to the spherical surfaces as the actual distributions of electricity do, those (p_1, q'_1, p_2, q'_2, etc.) which lie within the surface A producing the effect of the distribution on A, and the others (q_1, p'_1, q_2, p'_2, etc.), all within the surface B, the effect of the actual distribution on B. Hence the resultant force between the two partial groups is the same as the resultant force due to the mutual action between the actual distributions of electricity on the two conductors; and if this force, considered as positive or negative according as repulsion or attraction preponderates, be denoted by F, we have

$$F = \sum_{s=1}^{s=\infty} \sum_{t=1}^{t=\infty} \left\{ \frac{p_s q_t}{(c - f_s - g_t)^2} + \frac{p'_s q'_t}{(c - f'_s - g'_t)^2} + \frac{p_s p'_t}{(c - f_s - f'_t)^2} + \frac{q_s q'_t}{(c - g_s - g'_t)^2} \right\} \quad (9),$$

where $\Sigma\Sigma$ denotes a double summation, with reference to all integral values of s and t. The following process reduces this double series to the form of a single infinite series, of which the successive terms may be successively calculated numerically in any particular case with great ease.

132. First, taking from (8) expressions for p_s and f_s in terms of inferior order, and for q_t and g_t in terms of higher order, and continuing the reduction successively, we have

$$\frac{p_s q_t}{c - f_s - g_t} = \frac{p'_{s-1} \dfrac{a}{c - f'_{s-1}} \cdot q'_t \dfrac{a}{g_t}}{c - \dfrac{a^2}{c - f'_{s-1}} - \left(c - \dfrac{a^2}{g'_t} \right)} = \frac{p'_{s-1} q'_t}{c - f'_{s-1} - g'_t}$$

$$= \frac{p_{s-1} \dfrac{b}{c - f_{s-1}} \cdot q_{t+1} \dfrac{b}{g_{t+1}}}{c - \dfrac{b^2}{c - f_{s-1}} - \left(c - \dfrac{b^2}{g_{t+1}} \right)} = \frac{p_{s-1} q_{t+1}}{c - f_{s-1} - g_{t+1}};$$

and therefore

$$\frac{p_s q_t}{c-f_s-g_t} = \frac{p_{s-1}q_{t+1}}{c-f_{s-1}-g_{t+1}} = \frac{p_{s-2}q_{t+2}}{c-f_{s-2}-g_{t+2}} \cdots = \frac{p_1 q_{t+s-1}}{c-f_1-g_{t+s-1}} = -uq'_{t+s-1}$$

and

$$\frac{p'_s q'_t}{c-f'_s-g'_t} = -uq'_{t+s}.$$

Similarly, we find

$$\frac{p_s p'_t}{c-f_s-f'_t} = \frac{p_{s-1}p'_{t+1}}{c-f_{s+1}-f'_{t+1}} = \cdots = \frac{p_1 p'_{t+s-1}}{c-f'_{t+s-1}} = -up_{t+s},$$

and

$$\frac{q_s q'_t}{c-g_s-g'_t} = -vq_{t+s}.$$

Now $\dfrac{p'_n}{u} = \dfrac{q'_n}{v}$; and $\dfrac{p_n}{u}$ and $\dfrac{q_n}{v}$ are each independent of u and v; hence the following notation may be adopted conveniently:

$$\left. \begin{aligned} p'_n &= -\frac{u}{S_n}, \quad q'_n = -\frac{v}{S_n} \\[2mm] p_n &= \frac{u}{P_n}, \quad q_n = \frac{v}{Q_n} \end{aligned} \right\} \ldots\ldots\ldots\ldots\ldots(13).$$

Then, taking n to denote $t+s$ in the preceding equations, we have

$$\left. \begin{aligned} \frac{p_{n-t}q_t}{c-f_{n-t}-g_t} &= \frac{p'_{n-t-1}q'_t}{c-f'_{n-t-1}-g'_t} = \frac{uv}{S_{n-1}}; \\[2mm] \frac{p_{n-t}p'_t}{c-f_{n-t}-f'_t} &= -\frac{u^2}{P_n}; \quad \frac{q_{n-t}q'_t}{c-g_{n-t}-g'_t} = -\frac{v^2}{Q_n} \end{aligned} \right\} \ldots (14).$$

Hence we have

$$\frac{p_{n-t+1}q_t}{(c-f_{n-t+1}-g_t)^2} = \left(\frac{uv}{S_n}\right)^2 \frac{1}{p_{n-t+1}q_t} = \frac{uv}{S^2_n} \cdot P_{n-t+1} Q_t;$$

from which we conclude that

$$\sum_{s=1}^{t=\infty} \sum_{t=1}^{s=\infty} \frac{p_s q_t}{(c-f_s-g_t)^2} = \sum_{n=1}^{n=\infty} \frac{uv}{S^2_n} \sum_{t=1}^{t=n} (P_{n-t+1}Q_t);$$

and, by using this and transformations similarly obtained for the other parts of the expression for F, we obtain

$$F = \sum_{n=1}^{n=\infty} \left\{ \frac{uv}{S^2_n} \left[\sum_{t=1}^{t=n} (P_{n-t+1}Q_t) + \sum_{t=1}^{t=n-1} (S_{n-t}S_t) \right] \right.$$
$$\left. - \frac{u^2}{P^2_n} \left[\sum_{t=1}^{t=n-1} (P_{n-t}S_t) \right] - \frac{v^2}{Q^2_n} \left[\sum_{t=1}^{t=n-1} (Q_{n-t}S_t) \right] \right\} \ldots(15).$$

133. The quantities P_n, Q_n, S_n which occur in this expression, may be determined successively for successive values of n in the following manner:—By substituting, in (8), for p_n, p'_n, q_n, q'_n their values by (13), and eliminating f_n, f'_n, g_n, g'_n, we find

$$cP_n = aS_{n-1} + bS_n, \quad cQ_n = bS_{n-1} + aS_n, \Big\}$$
$$cS_{n-1} = bP_{n-1} + aP_n = aQ_{n-1} + bQ_n \Big\} \dots\dots(a);$$

from which we derive

$$P_{n+1} = \frac{c^2 - a^2 - b^2}{ab} P_n - P_{n-1}$$

$$Q_{n+1} = \frac{c^2 - a^2 - b^2}{ab} Q_n - Q_{n-1} \Bigg\} \dots\dots\dots(b).$$

$$S_{n+1} = \frac{c^2 - a^2 - b^2}{ab} S_n - S_{n-1}$$

By giving n the values 1 and 2 in (13) and (8), we find

$$\left.
\begin{cases}
P_1 = \dfrac{1}{a}, \\[2mm]
P_2 = \dfrac{c^2 - b^2}{a^2 b} = \dfrac{c^2 - a^2 - b^2}{ab} P_1 + \dfrac{1}{b},
\end{cases}
\right.$$

$$\left.
\begin{cases}
Q_1 = \dfrac{1}{b}, \\[2mm]
Q_2 = \dfrac{c^2 - a^2}{ab^2} = \dfrac{c^2 - a^2 - b^2}{ab} Q_1 + \dfrac{1}{a},
\end{cases}
\right\} \dots\dots\dots(c).$$

$$\left.
\begin{cases}
S_1 = \dfrac{c}{ab}, \\[2mm]
S_2 = \dfrac{c^2 - a^2 - b^2}{ab} S_1
\end{cases}
\right.$$

By these equations we have directly the values of the first two terms of each of the sets of quantities P_1, P_2, P_3, etc., Q_1, Q_2, Q_3, etc., and S_1, S_2, S_3, etc.; and the others may be calculated successively by the preceding equations.

134. The polynomials which constitute the numerators of the successive terms of the second member of (15) may also be calculated successively, by means of equations obtained in the following manner. We have by (c), (b), and (a),

$$P_1 Q_n + P_2 Q_{n-1} + P_3 Q_{n-2} + \text{etc.} = \frac{1}{a} Q_n + \left(\frac{c^2 - a^2 - b^2}{ab} P_1 + \frac{1}{b} \right) Q_{n-1}$$
$$+ \left(\frac{c^2 - a^2 - b^2}{ab} P_2 - P_1 \right) Q_{n-2} + \text{etc.}$$
$$= \frac{c}{ab} S_{n-1} + \frac{c^2 - a^2 - b^2}{ab} (P_1 Q_{n-1} + P_2 Q_{n-2} + \text{etc.}) - (P_1 Q_{n-2} + P_2 Q_{n-3} + \text{etc.});$$

and similarly we find

$$S_1 S_{n-1} + S_2 S_{n-2} + S_3 S_{n-3} + \text{etc.}$$
$$= \frac{c}{ab} S_{n-1} + \frac{c^2 - a^2 - b^2}{ab} (S_1 S_{n-2} + S_2 S_{n-3} + \text{etc.}) - (S_1 S_{n-3} + S_2 S_{n-4} + \text{etc.});$$

$S_1 P_{n-1} + S_2 P_{n-2} + S_3 P_{n-3} + \text{etc.}$

$$= \frac{c}{ab} P_{n-1} + \frac{c^2 - a^2 - b^2}{ab} (S_1 P_{n-2} + S_2 P_{n-3} + \text{etc.}) - (S_1 P_{n-3} + S_2 P_{n-4} + \text{etc.});$$

and

$S_1 Q_{n-1} + S_2 Q_{n-2} + S_3 Q_{n-3} + \text{etc.}$

$$= \frac{c}{ab} Q_{n-1} + \frac{c^2 - a^2 - b^2}{ab} (S_1 Q_{n-2} + S_2 Q_{n-3} + \text{etc.}) - (S_1 Q_{n-3} + S_2 Q_{n-4} + \text{etc.}).$$

Hence, if we put

$$\left.\begin{array}{c} \sum_{t=1}^{t=n} (P_{n-t} Q_t) + \sum_{t=1}^{t=n-1} (S_{n-t} S_t) = 2S'_n \\[2mm] \sum_{t=1}^{t=n-1} (P_{n-t} S_t) = P'_n, \\[2mm] \text{and} \quad \sum_{t=1}^{t=n-1} (Q_{n-t} S_t) = Q'_n \end{array}\right\} \dots\dots(e),$$

in terms of which notation the expression (15) for F becomes

$$\left.\begin{array}{c} F = 2uv \left(\dfrac{S'_1}{S_1^2} + \dfrac{S'_2}{S_2^2} + \dfrac{S'_3}{S_3^2} + \text{etc.} \right) \\[3mm] - \left\{ u^2 \left(\dfrac{P'_1}{P_1^2} + \dfrac{P'_2}{P_2^2} + \dfrac{P'_3}{P_3^2} + \text{etc.} \right) + v^2 \left(\dfrac{Q'_1}{Q_1^2} + \dfrac{Q'_2}{Q_2^2} + \dfrac{Q'_3}{Q_3^2} + \text{etc.} \right) \right\} \end{array}\right\} (f);$$

we have

$$\left.\begin{array}{c} S'_{n+1} = \dfrac{c^2 - a^2 - b^2}{ab} S'_n - \left(S'_{n-1} - \dfrac{c}{ab} S_n \right) \\[3mm] P'_{n+1} = \dfrac{c^2 - a^2 - b^2}{ab} P'_n - \left(P'_{n-1} - \dfrac{c}{ab} P_n \right) \\[3mm] Q'_{n+1} = \dfrac{c^2 - a^2 - b^2}{ab} Q'_n - \left(Q'_{n-1} - \dfrac{c}{ab} Q_n \right) \end{array}\right\} \dots\dots(g).$$

Also we have directly from (e) and (c),

$$\left.\begin{array}{c} S'_1 = \tfrac{1}{2} \dfrac{1}{ab}, \quad S'_2 = \tfrac{1}{2} \dfrac{3c^2 - a^2 - b^2}{a^2 b^2} \\[3mm] P'_1 = 0, \quad P'_2 = \dfrac{c}{a^2 b} \\[3mm] Q'_1 = 0, \quad Q'_2 = \dfrac{c}{ab^2} \end{array}\right\} \dots\dots(h).$$

135. These equations enable us to calculate successively the values of S'_1, S'_2, S'_3, etc., P'_1, P'_2, P'_3, etc., and Q'_1, Q'_2, Q'_3, etc., after the values of S_1, S_2, etc., P_1, P_2, etc., and Q_1, Q_2, etc., have been found.

136.   The solution of (*b*) as equations of finite differences with reference to *n*, and the determination of the arbitrary constants of integration by (*c*), leads to general expressions for S_n, P_n, and Q_n; and by using these in (*g*), integrating the equations so obtained, and determining the arbitrary constants by means of (*h*), general expressions for S'_n, P'_n, and Q'_n are obtained. The expression for *F* may therefore be put in the form of an infinite series, with a finite expression for the general term. Further, the value of this series may be expressed, by means of analysis similar to that which Poisson has used for similar purposes, in terms of a definite integral.   I do not, however, in the present communication give any of this analysis, except for the case of two spheres in contact which is discussed below, because, except for cases in which the spheres are very near one another, the series for *F* is rapidly convergent, and the terms of it may be successively calculated with great ease, by regular arithmetical processes, for any set of values of *c*, *a*, and *b*, by using first the equations (*c*), to calculate S_1, S_2, P_1, P_2, Q_1, Q_2; then (*b*) with the values 2, 3, etc., successively substituted for *n*, to calculate S_3, S_4, etc., and P_3, P_4, etc., and Q_3, Q_4, etc.; then (*h*) and (*g*) to calculate by a similar succession of processes, the values of S'_1, S'_2, S'_3, etc., P'_1, P'_2, P'_3, etc., and Q'_1, Q'_2, Q'_3, etc.

137.   The following is the method, alluded to above, by which I first arrived at the solution of this problem in the year 1845.

138.   The "mechanical value" of a distribution of electricity on a group of insulated conductors, may be easily shown to be equal to half the sum of the products obtained by multiplying the quantity of electricity on each conductor into the potential within it.*   Hence, if *D* and *E* denote the quantities of electricity on the two spheres in the present case, and if *W* denote the mechanical value of the distribution of electricity on them, we have $$W = \tfrac{1}{2} (Du + Ev).$$

---

* This proposition occurred to me in thinking over the demonstration which Gauss gave of the theorem *that a given quantity of matter may be distributed in one and only one way over a given surface so as to produce a given potential at every point of the surface*, and considering the mechanical signification of the function on the rendering of which a minimum that demonstration is founded. It was published, I believe, by Helmholtz in 1847, in his treatise *Ueber die Erhaltung der Kraft*, by the translation of which, in the last number of the *New Scientific Memoirs*, a great benefit has been conferred on the British scientific public.

Now if the two spheres, kept insulated, be pushed towards one another, so as to diminish the distance between their centres from c to $c - dc$, the quantity of work that will have to be spent will be $F \cdot dc$, since F denotes the repulsive force against which this relative motion is affected. But the mechanical value of the distribution in the altered circumstances must be increased by an amount equal to the work spent in producing no other effect but this alteration. Hence $F \cdot dc = - dW$, and therefore

$$F = -\tfrac{1}{2} \frac{d\,(Du + Ev)}{dc} \quad\ldots\ldots\ldots\ldots\ldots (16),$$

where u and v are to be considered as varying with c, and D and E as constants. Now, according to the notation expressed in (13), we have

$$\left. \begin{aligned} \left(\frac{1}{P_1} + \frac{1}{P_2} + \text{etc.}\right) u - \left(\frac{1}{S_1} + \frac{1}{S_2} + \text{etc.}\right) v &= D \\ -\left(\frac{1}{S_1} + \frac{1}{S_2} + \text{etc.}\right) u + \left(\frac{1}{Q_1} + \frac{1}{Q_2} + \text{etc.}\right) v &= E \end{aligned} \right\} \ldots (17).$$

Determining $\dfrac{du}{dc}$ and $\dfrac{dv}{dc}$ by the differentiation of these equations, and using the results in (16), we find

$$F = \tfrac{1}{2}\left\{ u^2 \frac{d}{dc}\left(\frac{1}{P_1} + \frac{1}{P_2} + \text{etc.}\right) - 2uv \frac{d}{dc}\left(\frac{1}{S_1} + \frac{1}{S_2} + \text{etc.}\right) \right.$$
$$\left. + v^2 \frac{d}{dc}\left(\frac{1}{Q_1} + \frac{1}{Q_2} + \text{etc.}\right) \right\} \ldots\ldots (18).$$

This expression agrees perfectly with (f), given above; since, by differentiating the equations (b) and (c) with reference to c, we find that the quantities denoted above by S'_1, S'_2, S'_3, etc., P'_1, P'_2, P'_3, etc., Q'_1, Q'_2, Q'_3, etc., and expressed by the equations (g) and (h), are equal respectively to

$$\tfrac{1}{2}\frac{dS_1}{dc},\ \tfrac{1}{2}\frac{dS_2}{dc},\ \tfrac{1}{2}\frac{dS_3}{dc},\ \text{etc.,} \quad \tfrac{1}{2}\frac{dP_1}{dc},\ \tfrac{1}{2}\frac{dP_2}{dc},\ \tfrac{1}{2}\frac{dP_3}{dc},\ \text{etc.,}$$

$$\tfrac{1}{2}\frac{dQ_1}{dc},\ \tfrac{1}{2}\frac{dQ_2}{dc},\ \tfrac{1}{2}\frac{dQ_3}{dc},\ \text{etc.}$$

139. The series (f) or (18) for F becomes divergent for the case of two spheres in contact, but the doubly infinite series from which this was derived in the first of the two investigations given above, is convergent when the terms are properly grouped together; and its sum may be expressed by means of a definite integral in the following manner :—

140. Since the two spheres are in contact, the potentials within them must be equal, that is, we must have $u = v$. For the sake of simplicity, let us suppose the radii of the two spheres to be equal, and let each be taken as unity. Then we shall have $a = b = 1$, and $c = 2$; and the terms of doubly infinite series (9) in this case are easily expressed,* in very simple forms, by equations (8). Thus we find

$$F = v^2 \times \left\{ \begin{aligned} &\frac{1}{2^2} - \frac{1.2}{3^2} + \frac{1.3}{4^2} - \frac{1.4}{5^2} + \frac{1.5}{6^2} - \text{etc.} \\ &\quad - \frac{2.1}{3^2} + \frac{2.2}{4^2} - \frac{2.3}{5^2} + \frac{2.4}{6^2} - \text{etc.} \\ &\qquad + \frac{3.1}{4^2} - \frac{3.2}{5^2} + \frac{3.3}{6^2} - \text{etc.} \\ &\qquad\quad - \frac{4.1}{5^2} + \frac{4.2}{6^2} - \text{etc.} \\ &\qquad\qquad + \frac{5.1}{6^2} - \text{etc.} \end{aligned} \right\} \quad \dots\dots(k).$$

If we add the terms in the vertical columns, we find

$$F = v^2 \times \tfrac{1}{6} \left(\frac{1.2.3}{2^2} - \frac{2.3.4}{3^2} + \frac{3.4.5}{4^2} - \text{etc.} \right),$$

which is a diverging series, and is the same as we should have found by using the form (f) or (18). But if we add the terms in the horizontal lines, we find the following convergent series for F:—

$$F = v^2 \left\{ \int_0^1 \frac{\log\frac{1}{\theta} . \theta d\theta}{(1+\theta)^2} - 2 \int_0^1 \frac{\log\frac{1}{\theta} . \theta^2 d\theta}{(1+\theta)^2} + 3 \int_0^1 \frac{\log\frac{1}{\theta} . \theta^3 d\theta}{(1+\theta)^2} - \text{etc.} \right\}$$

---

* From equations (8) we find, in this case,

$$f'_n = g'_n = \frac{2n-1}{2n}; \quad f_n = g_n = \frac{2n-2}{2n-1}$$

$$p'_n = q'_n = -\frac{v}{2n}; \quad p_n = q_n = \frac{v}{2n-1}.$$

Hence
$$\frac{p_s q_t}{(c - f_s - g_t)^2} = \frac{(2s-1)(2t-1)}{\{2(s+t)-2\}^2}$$

$$\frac{p'_s q'_t}{(c - f'_s - g'_t)^2} = \frac{2s.2t}{\{2(s+t)\}^2}$$

$$\frac{p_s p'_t}{(c - f_s - f'_t)^2} = \frac{q_s q'_t}{(c - g_s - g'_t)^2} = -\frac{2t(2s-1)}{\{2(s+t)-1\}^2}$$

and then, by (9), we obtain the expression for F in this particular case, given in the text.

Hence, since $(1 + \theta)^{-2} = 1 - 2\theta + 3\theta^2 -$ etc., we have

$$F = v^2 \int_0^1 \frac{\log \frac{1}{\theta}\; \theta d\theta}{(1 + \theta)^4} \quad\ldots\ldots\ldots\ldots (l);$$

or, by actual integration,

$$F = v^2 \left[\frac{\log \frac{1}{\theta}}{(1 + \theta)^3} \frac{3\theta^2 + \theta^3}{6} + \tfrac{1}{6} \log (1 + \theta) - \tfrac{1}{6} \frac{\theta}{(1 + \theta)^2} \right]_0^1$$

$$= v^2 \cdot \tfrac{1}{6} \times (\log 2 - \tfrac{1}{4}) = v^2 \cdot \tfrac{1}{6} \times (\cdot 69315 - \cdot 25)$$

$$= v^2 \times \cdot 073858.$$

The quantity of electricity on each sphere being equal to the sum of the masses of the imaginary series of points within it, is, according to the formulæ for p_1, q'_1, p_2, q'_2, etc.,

$$v (1 - \tfrac{1}{2} + \tfrac{1}{3} - \tfrac{1}{4} + \text{etc.}), \text{ or } v \log 2.$$

Hence we have the following expression for the repulsion between the two spheres, in terms of Q the quantity of electricity on each,

$$F = Q^2 \cdot \frac{\tfrac{1}{6} \times (\log 2 - \tfrac{1}{4})}{(\log 2)^2}.$$

141. If x denote the distance at which two electrical points, containing quantities equal to the quantities on the two spheres, must be placed so as to repel one another with a force equal to the actual force of repulsion between the spheres, we have

$$\frac{(v \cdot \log 2)^2}{x^2} = F.$$

Using the value for F found above, we obtain

$$x = \frac{\log 2}{\sqrt{\{\tfrac{1}{6} \times (\log 2 - \tfrac{1}{4})\}}} = 2 \cdot 550.$$

If the electrical distribution on each surface were uniform, this distance would be equal to 2, the distance between the centres of the spheres; but it exceeds this amount, to the extent shown by the preceding result, because in reality the electrical density on each conductor increases gradually from the point of contact to the remotest points of the two surfaces.

P.S.—The calculation by the method shown in the preceding paper, of the various quantities required for determining the force between two *spheres of equal radii* (each unity), insulated with their centres at distances 2·1, 2·2, 2·3, etc., up to 4, has been undertaken, and is now nearly complete.

GLASGOW COLLEGE, *March* 21, 1853.

---

142.   The following numerical results have been calculated (by means of the formulæ established above) for application to the theory of a new electrometer which I have recently had constructed to determine electrical potentials in absolute measure, from the repulsions of uninsulated balls in the interior of a hollow insulated and electrified conductor, by means of a bifilar or torsion balance bearing a vertical shaft which passes through a small aperture to the outside of the conductor :—

TABLE I.—*Showing the Quantities of Electricity on two equal Spherical Conductors, of radius* r, *and the mutual force between them, when charged to potentials* u *and* v *respectively.*

Col. 1.	Cols. 2 and 3.		Cols. 4 and 5.		Col. 6.
Distance from centre to centre $=cr$.	For determining the quantities of electricity, $D=(Iu-Jv)r$ $E=(Iv-Ju)r.$		For determining the mutual force, $F=2Buv-A(u^2+v^2)$; being repulsion when positive, and attraction when negative.		Ratio of the potentials when there is neither attraction nor repulsion, $\rho=\frac{B}{A}-\left(\frac{B^2}{A^2}-1\right)^{\frac12}$ $\frac{1}{\rho}=\frac{B}{A}+\left(\frac{B^2}{A^2}-1\right)^{\frac12}.$
c.	I.	J.	A.	B.	ρ.
2·0	$J+$·693147	∞	∞	$A+\frac12\times$·073858	$1-\sqrt{\dfrac{·073858}{A}}$
2·1	1·58396	·88175	1·13844	1·17439	·77828
2·2	1·43131	·72378	·52852	·56350	·69637
2·3	1·34827	·63395	·32917	·36357	·63553
2·4	1·29316	·57202	·23159	·26464	·58975
2·5	1·25324	·52537	·17432	·20630	·55888
2·6	1·22218	·48819	·13696	·16787	·51699
2·7	1·19755	·45746	·11082	·14090	·47805
2·8	1·17738	·43140	·09174	·12073	·46049
2·9	1·16056	·40886	·07720	·10526	·43667
3·0	1·14629	·38908	·06592	·09299	·41567
3·1	1·13404	·37151	·05693	·08304	·39672
3·2	1·12340	·35571	·04963	·07481	·37947
3·3	1·11410	·34150	·04363	·06791	·36376
3·4	1·10588	·32852	·03863	·06203	·34939
3·5	1·09859	·31663	·03441	·05697	·33615
3·6	1·09208	·30569	·03084	·05257	·32418
3·7	1·08623	·29557	·02775	·04872	·31263
3·8	1·08095	·28617	·02509	·04531	·30211
3·9	1·07617	·27742	·02278	·04229	·29233
4·0	1·07182	·26924	·02075	·03958	·28318

TABLE II. —*Showing the Potentials in two equal Spherical Conductors, and the mutual force between them, when charged with quantities D and E of electricity respectively.*

Col. 1. Distance from centre to centre =or.	Cols. 2 and 3. For determining the potentials, $u=\left(\dfrac{I}{I^2-J^2}\cdot D+\dfrac{J}{I^2-J^2}\cdot E\right)\dfrac{1}{r}$ $v=\left(\dfrac{I}{I^2-J^2}\cdot E+\dfrac{J}{I^2-J^2}\cdot D\right)\dfrac{1}{r}$		Cols. 4 and 5. For determining the mutual force, $F=\left\{2\beta DE-\alpha(D^2+E^2)\right\}\dfrac{1}{r^2}$, where $\alpha=\dfrac{A(I^2+J^2)-2BIJ}{(I^2-J^2)^2}$ $\beta=\dfrac{B(I^2+J^2)-2AIJ}{(I^2-J^2)^2}$.		Col. 6. Ratio of the quantities when there is neither attraction nor repulsion, $\theta=\dfrac{I\rho-J}{I-J\rho}$.
$c.$	$\dfrac{I}{I^2-J^2}.$	$\dfrac{J}{I^2-J^2}.$	$\alpha.$	$\beta.$	$\theta.$
2·0	$\tfrac{1}{2}\times\dfrac{1}{·693147}$	$\tfrac{1}{2}\times\dfrac{1}{·693147}$	∞	$\alpha+\tfrac{1}{2}\times·153726$	$1-\sqrt{\dfrac{·153726}{a}}$
2·1	·91482	·50926	·15375	·22668	·39102
2·2	·93869	·47467	·08263	·15251	·29435
2·3	·95220	·44782	·05444	·12186	·23580
2·4	·96142	·42528	·03955	·10309	·19944
2·5	·96829	·40599	·02997	·09038	·16908
2·6	·97354	·38888	·02342	·08078	·14476
2·7	·97771	·37348	·01849	·07341	·12786
2·8	·98105	·35946	·01500	·06710	·11318
2·9	·98376	·34658	·01222	·06186	·09971
3·0	·98598	·33467	·01010	·05731	·08877
3·1	·98782	·32361	·00842	·05333	·07944
3·2	·98934	·31327	·00708	·04981	·07139
3·3	·99067	·30366	·00599	·04666	·06442
3·4	·99178	·29462	·00510	·04382	·05839
3·5	·99272	·28612	·00437	·04126	·05298
3·6	·99351	·27810	·00378	·03891	·04868
3·7	·99423	·27054	·00326	·03679	·04349
3·8	·99484	·26338	·00283	·03484	·04061
3·9	·99537	·25659	·00247	·03305	·03736
4·0	·99583	·25015	·00216	·03139	·03444

(Art. VII. of complete list in *Mathematical and Physical Papers*, Vol. I.)

[From the *Cambridge Mathematical Journal*, May 1843.]

144. In measuring the action exerted upon an electrified
body, by a quantity of free electricity distributed in any manner
over another body, the methods followed in the cases in which
the attracted body is conducting and non-conducting are
different. Now, the only difference between the state of a
conducting body and that of a non-conducting body is, that
the electricity is held upon a conducting body by the pressure
of the atmosphere (to a certain extent at least), while on a non-
conducting body it is held by the *friction* of the particles of the
body.

145. To find the attraction of an electrical mass E, on a non-
conducting electrified body A, the obvious way is to proceed as
in ordinary cases of attraction, considering the electricity on A
as the attracted mass.

In finding the action on a conducting body A, the method
followed is to consider its electricity as exerting no pressure
upon the particles of the body, but disturbing its equilibrium,
by making the pressure of the air unequal at different parts
of its surface. These two methods of measuring the action
of E on A *should* obviously lead to the same result, since the
action must be the same, whether A be conducting or non-
conducting, the distribution remaining the same. It is the
object of the following paper to show that they *do* lead to the
same result.

146. We must first find the pressure of an element of the
electricity of A, on the atmosphere.

Let ds be the area of the element, and ρds its electrical mass.
Let ds form part of another element σ, indefinitely larger than
ds in every direction, but so small that it may be considered
as plane. Now, if $\rho\sigma$ be a material plane, it can exercise no
attraction on ρds, in a direction perpendicular to the plane, and

it may be readily shown that this is also true if $\rho\sigma$ be a plate of matter of different densities, arranged in parallel planes, the thickness being either finite or indefinitely small, and the law of density being any whatever.

147. Hence, the force acting on ρds is due to the repulsion of all the electrical mass, except σ; and, since the electricity on A is in equilibrium under the influence of E, the repulsion acts along the normal through ds, and is in magnitude $2\pi\rho^2 ds$ (see I. above, § 7), which is therefore the pressure of ds on the air. Hence, if p be the barometric pressure of the atmosphere, the pressure on ds, perpendicular to the surface, is

$$(p - 2\pi\rho^2)\, ds.$$

Hence, if X be the whole pressure on A, resolved along a fixed line $X'X$, and if ν be the angle which the normal through ds makes with this line, we have

$$X = -\iint (p - 2\pi\rho^2) \cos\nu ds,$$

the integrals being extended over the surface of A. Now,

$$\iint p \cos\nu ds = 0,$$

since the pressure of the atmosphere does not disturb the equilibrium of A. Hence, we have

$$X = 2\pi\iint\rho^2 \cos\nu ds \quad\quad\quad\quad\quad\quad (a),$$

which is the expression for the attraction on a conducting body A, either separate from the body on which E is distributed, or connected with it.

148. To show that this is identical with the expression for the attraction of E on the electricity of A, let $R\rho ds$ and $R'\rho ds$ be the components of the repulsion on ρds, which are due to E, and to the electricity of A; and let α, α' be the angles which their directions make with XX'. Then we shall have

$$2\pi\rho \cos\nu = R \cos\alpha + R' \cos\alpha';$$

therefore $\quad\quad X = \iint (R \cos\alpha + R' \cos\alpha')\, ds.$

Now, $\iint R' \cos\alpha' ds$ is the attraction of the electricity of A on itself in the direction XX', and is therefore $= 0$. Hence,

$$X = \iint R \cos\alpha ds \quad\quad\quad\quad\quad\quad (b).$$

But this expression for X is the attraction of E on the electricity of A: [also, the *moment* round OX is the same for the diminution of air pressure as for the attraction of E on the electricity of A:] and hence the two methods of measuring the action lead to the same result.

VIII.—DEMONSTRATION OF A FUNDAMENTAL PROPOSITION IN THE MECHANICAL THEORY OF ELECTRICITY.

(Art. xiv. of complete list in *Mathematical and Physical Papers*, Vol. i.)

[From the *Cambridge Mathematical Journal*, Feb. 1845.]

149. If a material point be in a position of equilibrium when under the influence of any number of masses attracting it or repelling it with forces which are inversely proportional to the square of the distance, the equilibrium will be unstable.*

The first thing to be proved is, that if the material point receive a slight displacement, there will in general be a moving force called into action.

150. Let O be the position of equilibrium; P any adjacent point; V the potential of the influencing masses, μ, at P, which point we suppose not to be contained within any portion of μ; U the value of V at O. Now it is shown by Gauss, in his Mémoire on General Theorems in Attraction, (also in Thomson and Tait's *Natural Philosophy*, § 497,) that V cannot have the constant value U through any finite volume, however small, adjacent to O, without having it for every point external to μ. But this is impossible, as may be shown in the following manner.

Let σ be a closed surface containing within it a quantity of matter, μ_1, consisting of any number of detached portions of μ, or of the whole of μ, if μ be a continuous mass. Let $d\sigma$ be an element of σ, and P the force due to the total action of μ, resolved in a direction perpendicular to $d\sigma$, which may be considered positive when directed towards the space within σ. Then, by a theorem demonstrated in this *Journal* (see XII. below, § 20ϕ), we have
$$\iint P d\sigma = 4\pi\mu,$$
the integrations being extended over the whole of σ. Hence P cannot be $= 0$ for every point of the surface σ, and therefore V cannot be constant for all the space exterior to μ.

---

* This theorem was first given by Mr Earnshaw, in his Memoir on Molecular Forces, read at the Cambridge Philosophical Society, March 18, 1839. See Vol. vii. of the *Transactions*.

Hence V cannot have the constant value U for every point of any finite volume, however small, adjacent to O.

151. Now let a sphere S be described round O as centre, with any radius a, sufficiently small that no portion of μ shall be included, and let P be any point of the surface S, and ds an element of the surface at P.

In the equations (3) and (4) of the article already referred to (XII. below, § 199), let the sphere S be the surface there considered; let $v = V$, and $v_1 = \dfrac{1}{r}$, if $OP = r$.

Hence $P_1 = \dfrac{1}{a^2}$ and $v_1 = \dfrac{1}{a}$, at every point of S;

$$m = \mu, \quad m_1 = 1, \quad \iiint v\,dm_1 = U.$$

Also
$$\iint v_1 P ds = \frac{1}{a} \iint P ds = 0,$$

and $\iiint v_1 dm = 0$, since S does not contain any of the matter μ. We have therefore, by comparing (3) and (4) of § 199,

$$0 = 4\pi U - \frac{1}{a^2} \iint V ds.$$

Therefore $\iint V ds = 4\pi a^2 U,$

which shows that the mean value for the surface of a sphere, of the potential of any external masses, is equal to the value at the centre.   Let $V = U + u.$

Therefore $\iint u\,ds = 0.$

152. Now, as has already been shown, u cannot be $= 0$ for every point P adjacent to O, and therefore if the sphere pass through a point P' where u is negative, there must also be a point P'' in the surface, for which u is positive. But if we assume the potential of an attracting particle to be positive, the direction of the resultant force, resolved along any straight line, will be that in which V increases. Hence there will be a force towards O, for points displaced along OP', and from O, for points displaced along OP''. Hence if M, the material point in equilibrium at O, be displaced along OP'', the moving force generated will tend to remove it further from O, which is therefore an unstable position.

153. As an application of this theorem, let us consider the

case of any number of material points repelling one another according to the inverse square of the distance, and contained in the interior of a rigid closed envelope. Let the system be in equilibrium when acted upon by attracting or repelling masses distributed in any manner without the envelope.

It will generally be possible that there may be a position or positions of equilibrium, in which at least some of the particles are not in contact with the surface. If now we suppose all the particles fixed except one, not in contact with the surface, the equilibrium of this particle is, as has been shown, unstable. Hence, generally, the equilibrium of the system is unstable if any of the particles be not in contact with the surface, and therefore in nature the particles cannot remain in such a position. There must, however, be some stable position or positions in which the particles can rest, but in such, all the particles must be in contact with the surface of the envelope. The sole condition of equilibrium in this case will be that the resultant force on each particle shall be in the direction of a normal to the surface, and directed towards the exterior space. If the number of particles be infinite, and there be one position in which the whole surface is covered, there can be no other in which this is the case, as is shown in the paper in this *Journal* already quoted (XII. below, § 204) ; and it is also readily seen that this position will be stable, and that no other in which the surface is not entirely covered can be stable. In this case the particles will be distributed according to the law of the intensity of electricity on the surface, the space within being conducting matter, and the masses without being any electrified bodies. If a mechanical theory be adopted, *electricity* will actually be a number of material points without weight, which repel one another according to the inverse square of the distance. Thus the result we have arrived at is, that there can be permanently no free electricity in the interior of a conducting body under any circumstances whatever.

154. If, as may happen through the influence of the exterior masses, there cannot be a position of equilibrium of the particles covering the whole surface, there will be a permanent distribution, in which part of the surface is uncovered. This,

however, is never the case with electricity, as a certain quantity
of latent electricity is then decomposed, so that the whole
surface is covered with electricity, either positive or negative.
All the above reasoning would still apply, if we considered the
masses of some points to be negative, and of some positive, and
the force between any two to be a repulsion equal to the product of their masses divided by the square of their distance.

155. Since every particle is on the surface, the whole *medium*
(if it can be properly so called), will be an indefinitely thin
stratum, the thickness being in fact the ultimate breadth of an
atom or material point. If we suppose these atoms to be merely
centres of force, the thickness will therefore be absolutely
nothing, and thus the *fluid* will be absolutely compressible and
inelastic. Any thickness which the stratum can have must
depend on a force of elasticity, or on a force generated by the
contact of material points, and in either case will therefore
require * an ultimate law of repulsion more intense than that of
the inverse square,† when the distance is very small, and we
therefore conclude that this cannot be the ultimate law of
repulsion in any elastic fluid. As, however, all experiments
yet made serve to confirm the fact that there is no electricity
in the interior of conducting bodies, or that the stratum has
absolutely no thickness, we conclude that there is no elasticity
in the assumed electric fluid, and thus the law of force, deduced
independently by direct experiments, is confirmed.

St Peter's College, *Jan.* 16, 1845.

---

* [*Note added Jan.* 1869.—This was written without knowledge of Davy's
"repulsive motion," and without the slightest idea that elasticity of every kind
is most probably a result of motion. The conclusions of the text are, however,
not affected by these views.]

† This agrees with a result of Mr Earnshaw.

IX.—NOTE ON INDUCED MAGNETISM IN A PLATE.

(Art. xx. of complete list in *Mathematical and Physical Papers*, Vol. I.)

[From the *Cambridge and Dublin Mathematical Journal*, Nov. 1845.]

156. If a plate of soft iron be submitted to the action of a magnet of any kind, it immediately becomes magnetized "by induction;" and the effects of this are exhibited in the attraction or repulsion it exercises upon small magnetic bodies in its neighbourhood. The determination of these effects, from the elementary laws of magnetic induction, is a problem of considerable practical interest. In the case of a plate bounded by infinite parallel planes, I have succeeded in obtaining a complete solution of a very simple nature, by means of a principle which will be developed in a future paper (see above, §§ 127, 107, 108, 44). The object of the present note is to compare this solution with a formula given by Green in his *Essay on Electricity and Magnetism*, as an approximate result, but which appears to be inadmissible.

157. Let the influencing magnet, which may be of any form and size, and magnetized in any manner, be denoted by Q; and let us suppose it to be held *behind* the plate of soft iron. The solution which I have obtained enables us to find the total magnetic action on a point, P, situated in any position, either within or without the plate; but at present I shall only state the result when P is *before* the plate. In this case the actual magnetic effect on P may be produced by supposing Q and the plate to be removed, and a certain imaginary series of magnets Q', Q_1, Q_2, etc., to be substituted, the system being constructed thus. Each of the imaginary magnets is equal and similar to Q, and similarly magnetized; Q' occupies the place of Q, and the others are similarly placed behind it, along a line perpendicular to the plate, the distance between corresponding points of each consecutive pair being equal to twice the thickness of

the plate.   The intensities of the successive magnets decrease in a geometrical progression, of which the common ratio is m^2 (a quantity measuring (§ 45) the inductive capacity for magnetism of the plate), commencing with that of Q', which is equal to $1 - m^2$, if the intensity of Q be unity.   It is hardly necessary to point out the analogy between this and the corresponding result in optics, in which the illumination produced through a plate of glass, by a candle, is found to be due to the candle itself, with diminished brightness, and to a row of images behind it, with intensities decreasing in a geometrical progression, which arise from successive internal reflections.

158.   If the iron plate be infinitely thin, all the *images*, Q_1, Q_2, etc., will coincide with Q' ; and, since the sum of their intensities is unity, the total effect will be the same as that of Q, which will therefore be unaffected by the interposition of the screen. The same will be the case if the distance of Q be infinitely great, and the thickness of the screen finite; but in this case, at least as far as the present result can show us, the dimensions of the planes which bound the plate must be infinitely great compared with the distance of Q.

159.   The result which I have stated is applicable also to the imaginary case in which, instead of being a magnet, Q is a mass of positive or negative magnetism.*   Thus, let Q be a unit of positive magnetism collected in a point, which case is investigated by Green.   To express the action analytically, let Q be taken as origin of co-ordinates, a line perpendicular to the plate as axis of x, and the plane through this line, and P, as plane of (x, y).   Then denoting by a the thickness of the plate, and considering Q as a positive unit of matter, we shall have, for the total potential at P, due to Q and the plate,

$$F=(1-m^2)\left\{\frac{1}{(x^2+y^2)^{\frac{1}{2}}} + \frac{m^2}{\{(x+2a)^2+y^2\}^{\frac{1}{2}}} + \frac{m^4}{\{(x+4a)^2+y^2\}^{\frac{1}{2}}} + \text{etc.}\right\}(1).$$

---

* This expression does not imply any hypothesis of a magnetic matter or of a fluid or fluids, but it is merely used for brevity in consequence of the principle established by Coulomb, Poisson, and Ampère, that the action of a magnetized body of any kind, or of a collection of electric "closed currents," may always be represented by an imaginary positive and negative distribution of matter, of which the whole mass is algebraically nothing.   By an element of positive or negative magnitude, we merely mean a portion of this imagined matter.

160.   For all magnetic bodies m is between 0 and 1, the former limit being its value when the inductive capacity for magnetism is nothing, and the latter being never attained, though it is approached in such bodies as iron, of which the inductive capacity is great.   In the extreme case of $m = 1$, the laws of induction in a magnetic body degenerate into those of electrical equilibrium on the surface of a conductor of electricity.   If in the expression for F we put $m = 1$, one of the factors vanishes and the other becomes infinite, but the ultimate value of the product is nothing, which shows that the effect of the plate is to destroy all action behind it.   This we know to be the case when an infinite conducting screen of any form is placed before an electrified body.

161.   In the case when the plate is of iron, the value of m is nearly unity.   Hence, as the series is multiplied by $1 - m^2$, it might be imagined that, if we "neglect small quantities of the order $(1 - g)$ compared with those which are retained," $(1 - g$ being, in Green's notation, a quantity of the same order as $1 - m)$, an approximate result would be obtained by putting $m = 1$ in the successive terms of the series within the vinculum.   And it is thus that Green, having, in the investigation, neglected quantities multiplied by $(1 - g)^2$, arrives at the result,

$$F = \frac{4(1-g)}{3}\left\{ \frac{1}{(x^2+y^2)^{\frac{1}{2}}} + \frac{1}{\{(x+2a)^2+y^2\}^{\frac{1}{2}}} + \frac{1}{\{(x+4a)^2+y^2\}^{\frac{1}{2}}} + \text{etc.} \right\}.$$

As, however, this series has an infinite sum, it is clear that no value of m can be sufficiently near to unity to render the approximation admissible.   If instead of Q we were to substitute a magnet, or any collection of positive and negative particles, such that the sum of the masses is zero, the series for the potential, deduced from Green's expression, would converge : and the same remark is applicable to the series which would be found for the *attraction* of the system on a point beyond the screen, even when Q is a positive point, by differentiating the expression for F.   Notwithstanding this, the approximation is still inadmissible ; since, if we expand the rigorous expression in either case in ascending powers $(1 - m)$, we find that, though the first term is finite, the coefficients of all the terms which follow it are infinite.

162. Although the method by which I obtained the rigorous solution is quite distinct from that followed by Green, being independent of any mathematical process, it may be satisfactory to show that the result can be deduced from his own analysis, and even with greater ease than his solution is obtained after making unnecessary approximation.

By a very remarkable investigation, in which he extends Laplace's well-known analysis for spherical co-ordinates to the case when the radius of the sphere becomes infinite, Green arrives (*Essay on Electricity*, p. 64) at the following expression for the total potential at P, due to the positive unit of matter Q, and to the interposed plate, before making any approximation :—

$$F = \frac{8}{\pi}(1-g)(1+2g)\int_0^\infty \frac{d\gamma\epsilon^{-\gamma x}}{(2+g)^2 - 9g^2\epsilon^{-2\gamma a}}\int_0^1 \frac{d\beta}{(1-\beta^2)^{\frac{1}{2}}}\cos(\beta\gamma y).$$

Let $m = \dfrac{3g}{2+g}$. Then we have, by expansion, and by changing the order of the integration,

$$F = \frac{2}{\pi}(1-m^2)\int_0^1 \frac{d\beta}{(1-\beta^2)^{\frac{1}{2}}}$$

$$\int_0^\infty d\gamma . \epsilon^{-\gamma x}(1 + m^2\epsilon^{-2\gamma a} + m^4\epsilon^{-4\gamma a} + \text{etc.})\cos(\beta\gamma y)$$

$$= \frac{2}{\pi}(1-m^2)\int_0^1 \frac{d\beta}{(1-\beta^2)^{\frac{1}{2}}}$$

$$\left\{\frac{x}{x^2+\beta^2 y^2} + \frac{m^2(x+2a)}{(x+2a)^2+\beta^2 y^2} + \frac{m^4(x+4a)}{(x+4a)^2+\beta^2 y^2} + \text{etc.}\right\}$$

$$= \frac{2}{\pi}(1-m^2)\Sigma\int_0^{\frac{1}{2}\pi} \frac{m^{2i}x_i d\theta}{x_i^2 + y^2\sin^2\theta}, \quad \text{where } x_i = x + 2ia,$$

$$= (1-m^2)\Sigma \frac{m^{2i}}{(x_i^2 + y^2)^{\frac{1}{2}}},$$

which agrees with the expression given above.

St Peter's College, *Oct.* 1*4th*, 1845.

X.—SUR UNE PROPRIÉTÉ DE LA COUCHE ÉLECTRIQUE EN ÉQUILIBRE À LA SURFACE D'UN CORPS CONDUCTEUR.

Par M. J. LIOUVILLE.

(Art. xxiv. of complete list in *Mathematical and Physical Papers*, Vol. i.)

[From the *Cambridge and Dublin Mathematical Journal*, Nov. 1846.]

163. La méthode la plus générale que l'on connaisse pour former des couches électriques, en équilibre à la surface de corps conducteurs, consiste à considérer une masse M; et le potentiel,

$$V = \iiint \frac{f(x',\, y',\, z')\, dx'\, dy'\, dz'}{\Delta},$$

de cette masse, par rapport à un point quelconque (x, y, z), dont la distance au point (x', y', z'), ou à l'élément

$$f(x',\, y',\, z')\, dx'\, dy'\, dz',$$

est désignée par Δ. Prenons ensuite une surface de niveau ou d'équilibre relativement à l'attraction de la masse M, et qui entoure cette masse, c'est à dire prenons une surface fermée (A), contenant la masse M dans son intérieur, et pour tous les points de laquelle V conserve une valeur constante. En fin soit $\frac{dV}{ds} ds$ la variation infiniment petite que V éprouve lorsqu'on passe d'un point de cette surface à un point extérieur infiniment voisin situé sur la normale à une distance ds. C'est la dérivée $\frac{dV}{ds}$, multipliée si l'on veut par une constante, qui réglera la loi des densités de l'électricité en équilibre sur un corps conducteur terminé par la surface (A). Plusieurs géomètres sont parvenus, chacun de leur côté, à ce beau théorème; mais c'est George Green qui l'a, je crois, donné le premier dans un excellent mémoire publié en 1828, sous ce titre : *An Essay on the Application of Mathematical Analysis to the Theories of Electricity and Magnetism.* Je me propose de montrer que la couche électrique en équilibre ainsi obtenue a précisément le même centre de gravité que la masse M.

164. Plaçons l'origine des coordonnées x, y, z, au centre de

gravité de la masse M; et désignons par x_1 une quelconque des coordonnées du centre de gravité de la couche électrique, laquelle sera fournie par la formule

$$x_1 \iint \frac{dV}{ds}\, d\omega = \iint x\, \frac{dV}{ds}\, d\omega,$$

où les intégrations s'appliquent à la surface (A) dont l'élément est représenté par $d\omega$. Il s'agit de prouver que $x_1 = 0$.

D'après l'expression de V, on a

$$\frac{d^2V}{dx^2} + \frac{d^2V}{dy^2} + \frac{d^2V}{dz^2} = -4\pi f(x, y, z), \text{ ou } = 0,$$

suivant que le point (x, y, z) appartient ou non à la masse M.

Pour plus de simplicité, écrivons toujours

$$\frac{d^2V}{dx^2} + \frac{d^2V}{dy^2} + \frac{d^2V}{dz^2} = -4\pi f(x, y, z),$$

en regardant la fonction $f(x, y, z)$ comme nulle hors de la masse M; et combinons cette équation avec cette autre de forme analogue

$$\frac{d^2U}{dx^2} + \frac{d^2U}{dy^2} + \frac{d^2U}{dz^2} = 0,$$

où nous supposons que U est une fonction de x, y, z, qui reste finie et continue ainsi que ses dérivées dans tout l'espace intérieur à (A). Nous aurons

$$V\frac{d^2U}{dx^2} - U\frac{d^2V}{dx^2} + V\frac{d^2U}{dy^2} - U\frac{d^2V}{dy^2} + V\frac{d^2U}{dz^2} - U\frac{d^2V}{dz^2} = 4\pi\, Uf(x, y, z).$$

Multiplions par $dx\,dy\,dz$, et intégrons dans tout l'espace intérieur à (A). En conservant à ds et à $d\omega$ la même signification que ci-dessus, on trouve, après des transformations bien connues :

$$\iint V\frac{dU}{ds}\, d\omega - \iint U\frac{dV}{ds}\, d\omega = 4\pi \iiint Uf(x, y, z)\, dx\,dy\,dz.$$

Mais l'équation en U est satisfaite par $U = x$; nous avons donc :

$$\iint V\frac{dx}{ds}\, d\omega - \iint x\frac{dV}{ds}\, d\omega = 4\pi \iiint xf(x, y, z)\, dx\,dy\,dz.$$

L'intégrale triple du second membre, divisée par M, donne l'abscisse du centre de gravité de la masse M. Ce centre étant à l'origine des coordonnées, l'intégrale dont nous parlons est

nulle. Je vais prouver que l'intégrale $\iint V \frac{dx}{ds} d\omega$ l'est aussi.
D'abord on peut faire sortir V du signe $\int$, puisque, sur la surface (A), V est constant. Observons ensuite que $\frac{dx}{ds}$ a pour
valeur le cosinus de l'angle α que la normale ds fait avec l'axe
des x. Notre intégrale deviendra donc : $V\iint \cos \alpha d\omega$. Or
l'intégrale $\iint \cos \alpha d\omega$ est nulle, d'après un théorême connu,
comme composée d'éléments deux à deux égaux et de signes
contraires. Ainsi $\iint V \frac{dx}{ds} d\omega = 0$. Il·reste donc finalement

$$\iint x \frac{dV}{ds} d\omega = 0,$$

et l'on en conclut $x_1 = 0$, ce qu'il fallait démontrer.

Toul, 4 *Juillet* 1846.

NOTE ON THE PRECEDING PAPER.

By William Thomson.

[Extracted from a Letter to M. Liouville.]

165. "…The demonstration which you have given has led me
to this other theorem, that the mass M, and the shell surrounding it, have the same principal axes, through any point.

To demonstrate this, let $U = yz$ in the formula which you
have given. Then, since, if we denote by K the constant value
of V at the shell, we have

$$\iint V \frac{dU}{ds} d\omega = K \iint \frac{dU}{ds} = 0*,$$

we find

$$\iint yz \frac{-dV}{ds} d\omega = 4\pi \iiint yz \cdot f(x, y, z) \, dxdydz \ldots\ldots(1),$$

which proves the proposition enunciated.

If we take $U = x^2$, we find

$$V \frac{d^2U}{dx^2} - U \frac{d^2V}{dx^2} + V \frac{d^2U}{dy^2} - U \frac{d^2V}{dy^2} + V \frac{d^2U}{dz^2} - U \frac{d^2V}{dz^2} = 2V + 4\pi x^2 f(x, y, z):$$

---

* See xii. below, § 200, (8).

from which, observing that

$$\iint V\left(\frac{dU}{dx}dydz + \frac{dU}{dy}dzdx + \frac{dU}{dz}dxdy\right) = K\iiint\left(\frac{d^2U}{dx^2} + \frac{d^2U}{dy^2} + \frac{d^2U}{dz^2}\right)dxdydz$$

$$= 2K\iiint dxdydz;$$

we deduce

$$\frac{1}{4\pi}\iint x^2\frac{-dV}{ds}d\omega = \frac{1}{2\pi}\iiint(V-K)\,dxdydz + \iiint x^2 f(x,y,z)\,dxdydz.$$

Let A, B, C be the moments of inertia of the mass M round the axes of co-ordinates, and A_1, B_1, C_1, those of the shell, round the same axes, it being supposed that the quantity of matter of the shell is the same as that of M;* the preceding equation, and the two others which correspond relatively to the axes of y and z, are with this notation,

$$A_1 = Q + A, \quad B_1 = Q + B, \quad C_1 = Q + C \ldots\ldots\ldots(2),\dagger$$

where

$$Q, = \frac{1}{2\pi}\iiint(V-K)\,dxdydz,$$

is a quantity which is independent of the position of the origin.

From equations (2), we have

$$B - C = B_1 - C_1, \quad C - A = C_1 - A_1, \quad A - B = A_1 - B_1 \ldots(3).$$

A demonstration of your theorem and of the theorems expressed by the equations (1) and (3) may be arrived at by comparing the expressions for the equal potentials‡ produced by the mass M, and the shell at very distant points."‖

St Peter's College, *July* 15, 1846.

---

* In this case the "density" of the distribution at any point of the shell will be equal to $\frac{1}{4\pi}\cdot\frac{-dV}{ds}$. See I. above, § 7.

† If the origin be taken at the centre of gravity, and the axes of co-ordinates principal axes of M (and therefore of the shell, according to the proposition enunciated above), these equations show that the "central ellipsoid" (see note to p. 202 of *Cambridge and Dublin Mathematical Journal*, 1846) for the shell is confocal with that for the body M.

‡ A shell constructed round the mass M, in the manner described by M. Liouville, with a quantity of matter equal to M, exerts the same force upon points without the shell, as was proved first by Green (see also I. above, § 9); and since the potential of each vanishes at an infinite distance, it follows that the two bodies produce equal potentials at every point without the shell.

‖ [See Thomson and Tait's *Natural Philosophy*, § 539.]

XI.—ON CERTAIN DEFINITE INTEGRALS SUGGESTED BY PROBLEMS IN THE THEORY OF ELECTRICITY.

(Art. xxviii. of complete list in *Mathematical and Physical Papers*, Vol. i.)

[From the *Cambridge and Dublin Mathematical Journal*, March 1847.]

166.   It follows from the solution of the problem of the distribution of electricity on an infinite plane,* subject to the influence of an electrical ˙point, that the value of the double integral,

$$\int_{-\infty}^{\infty}\int_{-\infty}^{\infty} \frac{z\,d\xi\,d\eta}{\{(\xi-x)^2+(\eta-y)^2+z^2\}^{\frac{3}{2}}\{(\xi-x')^2+(\eta-y')^2+z'^2\}^{\frac{1}{2}}},$$

is

$$\frac{2\pi}{\{(x-x')^2+(y-y')^2+(z+z')^2\}^{\frac{1}{2}}}.$$

A direct analytical verification of this result is therefore interesting in connexion with the physical problem.   In the following paper the multiple integral

$$\int_{-\infty}^{\infty}\int_{-\infty}^{\infty}\cdots \frac{u\,d\xi_1 d\xi_2\ldots d\xi_s}{\{(\xi_1-x_1)^2+(\xi_2-x_2)^2+\ldots+u^2\}^{\frac{1}{2}(s+1)}\{(\xi_1-x_1')^2+(\xi_2-x_2')^2+\ldots+u'^2\}^{\frac{1}{2}(s-1)}}$$

is considered, and its value is shown to be

$$\frac{\pi^{\frac{1}{2}(s+1)}}{\Gamma\frac{1}{2}(s+1)}\frac{1}{\{(x_1-x_1')^2+(x_2-x_2')^2+\ldots+(u+u')^2\}^{\frac{1}{2}(s-1)}},$$

a result of which the one mentioned above is a particular case. Several distinct demonstrations of this theorem are given, and some other formulæ, which have occurred to me in connexion with it, are added.

167.   The first part of the following paper, which is a translation, with slight alterations, of a memoir in Liouville's *Journal*,† contains a demonstration suggested to me by a method followed by Green in proving the remarkable theorem in Art. (5) of his Essay on Electricity.   In the second part some formulæ are given which, in the case of two variables, are such as would

---

* See above, § 111, footnote.

† 1845, p. 137, "Démonstration d'un Théorème d'Analyse" (April 1845).

occur in the analysis of problems in heat and electricity, with reference to a body bounded in one direction by an infinite plane, if the methods indicated by Fourier were followed; and from them the value of the multiple integral mentioned above is deduced. In § III. the evaluation is effected by a direct process of reduction, suggested by geometrical considerations*.

PART I.

168. Let the value of the multiple integral, which, if we use a very convenient notation analogous to that of factorials, may be written thus,

$$\left[\int_{-\infty}^{\infty}\right]^n \frac{[d\xi]^s}{\{\Sigma(\xi-x)^2 + u^2\}^{\frac{1}{2}(s+1)}\{\Sigma(\xi-x')^2 + u'^2\}^{\frac{1}{2}(s-1)}},$$

be denoted by U.

Let $u + u' = a$, it being understood that u and u' are taken as positive. Then, if we assume

$$R = \frac{1}{\{\Sigma(\xi-x)^2 + v^2\}^{\frac{1}{2}(s-1)}} - \frac{1}{\{\Sigma(\xi-x)^2 + (2u-v)^2\}^{\frac{1}{2}(s-1)}} \quad (1),$$

$$R' = \frac{1}{\{\Sigma(\xi-x')^2 + (a-v)^2\}^{\frac{1}{2}(s-1)}} \quad \ldots\ldots\ldots\ldots(2),$$

we have

$$-2(s-1)uU = \left[\int_{-\infty}^{\infty}\right]^s R'\frac{dR}{dv}[d\xi]^s, \text{ when } v = u.$$

It is easily seen that the second member of this equatio.. vanishes when $v = \pm \infty$, and that it does not become infinite, even when one of the values 0, $2u$, or a is assigned to u. Hence the preceding equation may be written

$$-2(n-1)uU = \int_{-\infty}^{u}\left[\int_{-\infty}^{\infty}\right]\left(\frac{dR'}{dv}\frac{dR}{dv} + R'\frac{d^2R}{dv^2}\right)[d\xi]^s dv.$$

But we have

$$\int\left[\int_{-\infty}^{\infty}\right]^s \frac{dR'}{dv}\frac{dR}{dv}[d\xi]^s dv = \left[\int_{-\infty}^{\infty}\right]^s \int\frac{dR'}{dv}\frac{dR}{dv} dv[d\xi]^s$$

$$= \left[\int_{-\infty}^{\infty}\right]^s R\frac{dR'}{dv}[d\xi]^s - \int\left[\int_{-\infty}^{\infty}\right]^s R\frac{d^2R'}{dv^2}[d\xi]^s dv.$$

When we take the integral with respect to v between the

---

* See "Extrait d'une lettre à M. Liouville, etc." Liouville's *Journal*, 1845, p. 364 (XIV. § 210, below).

limits $-\infty$ and u, the first term vanishes, since at each limit $R = 0$. Thus the preceding equation is reduced to

$$-2(s-1)uU = \int_{-\infty}^{u}\left[\int_{-\infty}^{\infty}\right]^s\left(R'\frac{d^2R}{dv^2} - R\frac{d^2R'}{dv^2}\right)[d\xi]^s dv.$$

169. Now we have $\dfrac{d^2R'}{dv^2} + \Sigma\dfrac{d^2R'}{d\xi^2} = 0$,

for all values of ξ_1, $\xi_2 \ldots$, provided v be not equal to a. Hence this equation is satisfied for all the values of the variables between the limits of the integration in the preceding expression, and we may therefore employ it to eliminate $\dfrac{d^2R'}{dv^2}$: we thus obtain

$$-2(s-1)uU = \int_{-\infty}^{u}\left[\int_{-\infty}^{\infty}\right]^s\left(R'\frac{d^2R}{dv^2} + R\Sigma\frac{d^2R'}{d\xi^2}\right)[d\xi]^s dv.$$

Taking one of the terms of the second member, and integrating by parts, we have

$$\int_{-\infty}^{u}\left[\int_{-\infty}^{\infty}\right]^s R\frac{d^2R'}{d\xi_1^2}[d\xi]^s dv$$

$$= \int_{-\infty}^{u}\left[\int_{-\infty}^{\infty}\right]^{s-1}\left(\int_{-\infty}^{\infty}R\frac{d^2R'}{d\xi_1^2}d\xi_1\right)[d\xi]^{s-1}dv$$

$$= -\int_{-\infty}^{u}\left[\int_{-\infty}^{\infty}\right]^{s-1}\left(\int_{-\infty}^{\infty}\frac{dR'}{d\xi_1}\frac{dR}{d\xi_1}d\xi_1\right)[d\xi]^{s-1}dv$$

$$= \int_{-\infty}^{u}\left[\int_{-\infty}^{\infty}\right]^s R'\frac{d^2R}{d\xi_1^2}[d\xi]^s dv,$$

since the integrated parts vanish at each limit. By applying a similar process to each term under the sign Σ, we find

$$-2(s-1)uU = \int_{-\infty}^{u}\left[\int_{-\infty}^{\infty}\right]^s R'\left(\frac{d^2R}{dv^2} + \Sigma\frac{d^2R}{d\xi^2}\right)[d\xi]^s dv.$$

But, if we denote by Q and Q' the two parts of R, in equation (1), so that $R = Q - Q'$, we have

$$\frac{d^2Q'}{dv^2} + \Sigma\frac{d^2Q'}{d\xi^2} = 0$$

for all values of the variables v, ξ_1, etc., within the limits of integration ; hence there remains

$$-2(n-1)uU = \int_{-\infty}^{u}\left[\int_{-\infty}^{\infty}\right]^s R'\left(\frac{d^2Q}{dv^2} + \Sigma\frac{d^2Q}{d\xi^2}\right)[d\xi]^s dv.$$

To determine the value of this expression it may be remarked that the quantity under the integral signs vanishes for all values of the variables which differ sensibly from those expressed by

$$v = 0, \quad \xi_1 = x_1, \quad \xi_2 = x_2, \quad \text{etc.},$$

and moreover, that if we consider separately the terms of the second member, each is found to be a converging integral: it follows that, if we denote by P the value which R' receives when the variables have these values assigned, we have

$$- 2(s-1)uU = P \iiint \dots \left(\frac{d^2Q}{d^2v} + \Sigma \frac{d^2Q}{d\xi^2} \right) dv d\xi_1 d\xi_2 \dots d\xi_s \quad (3),$$

where the limits of integration must be such as to include the values 0, x_1, x_2, etc., but are otherwise entirely arbitrary. By considering separately the different terms of this expression, and integrating each with respect to the variable to which it is related, without yet assigning the limits of the integration, we find

$$- 2(s-1)uU = P \left(\iint \dots \frac{dQ}{dv} d\xi_1 d\xi_2 \dots + \iint \frac{dQ}{d\xi_1} dv d\xi_2 + \text{etc.} \right) \quad (4).$$

170.   Let us now assume

$$\xi_1 = v_1 + x_1, \quad \xi_2 = v_2 + x_2, \quad \text{etc.},$$

and

$$v^2 + v_1^2 + \dots + v_s^2 = r^2,$$

from which we have

$$Q = \frac{1}{r^{s-1}}, \quad \frac{dQ}{dv} = - \frac{s-1}{r^{s+1}} v, \quad \frac{dQ}{d\xi_1} = - \frac{s-1}{r^{s+1}} v_1, \quad \text{etc.}$$

The integrations in equation (3) may be extended to all the values of the variables which satisfy the condition

$$v^2 + v_1^2 + v_2^2 + \dots + v_s^2 \overset{=}{<} a^2;$$

and the limits in (4) will then be such as to include all the values which satisfy the equation

$$v^2 + v_1^2 + v_2^2 + \dots + v_s^2 = a^2, \quad \text{or} \quad r^2 = a^2, \quad \text{etc.}$$

If in the integrations we only take the positive values of the variables v, v_1, v_2, etc., which satisfy the limiting condition, we must multiply each integral by 2^{s+1}; and we may then simply take, in the successive terms the second member of (4),

$$\frac{dQ}{dv} = - \frac{s-1}{a^{s+1}} v, \quad \frac{dQ}{d\xi_1} = - \frac{s-1}{a^{s+1}} v_1, \quad \text{etc.}$$

Thus we have

$$uU = \frac{2^s P}{a^{s+1}} \left(\iint \ldots v\, dv_1 dv_2 \ldots dv_s + \iint \ldots v_1 dv\, dv_2 \ldots dv_s + \text{etc.} \right)^*$$

$$= \frac{2^s (s+1) P}{a^{s+1}} \iint \ldots (a^2 - v_1^2 - v_2^2 - \ldots - v_s^2)^{\frac12}\, dv_1 dv_2 \ldots dv_s$$

$$= (s+1) P \iint \ldots (1 - l_1 - l_2 \ldots - l_s)^{\frac12} l_1^{-\frac12} l_2^{-\frac12} \ldots l_s^{-\frac12} dl_1 dl_2 \ldots dl_s ;$$

in which last expression the limits include all positive values satisfying the condition

$$l_1 + l_2 + \ldots + l_s \lessgtr 1.$$

Hence, by Liouville's theorem†,

$$uU = (s+1) P \frac{\Gamma(\tfrac12)^s}{\Gamma(\tfrac12 s)} \int_0^1 (1-h)^{\frac12} h^{\frac12 s - 1} dh = \frac{\pi^{\frac12(s+1)}}{\Gamma\frac12(s+1)} P,$$

which gives the required value of the integral U.

171. If we denote by U' any integral corresponding to U, in which the system of variables $u, x_1, x_2 \ldots$ and $u', x'_1, x'_2 \ldots$ are inverted, we shall have $uU = u' U'$, since P is a function symmetrical with respect to the two systems; and we therefore deduce from the preceding result,

$$u \left[\int_{-\infty}^{\infty} \right]^s \frac{[d\xi]^s}{\{\Sigma(\xi-x)^2 + u^2\}^{\frac12(s+1)} \{\Sigma(\xi-x')^2 + u'^2\}^{\frac12(s-1)}}$$
$$= u' \left[\int_{-\infty}^{\infty} \right]^s \frac{[d\xi]^s}{\{\Sigma(\xi-x')^2 + u'^2\}^{\frac12(s+1)} \{\Sigma(\xi-x)^2 + u^2\}^{\frac12(s-1)}} \quad \ldots\ldots(5).$$
$$= \frac{\pi^{\frac12(s+1)}}{\Gamma\frac12(s+1)} \frac{1}{\{\Sigma(x-x')^2 + (u+u')^2\}^{\frac12(s-1)}}$$

172. I shall add another demonstration of this theorem, as an application of some remarkable analysis given by Mr Green in his memoir "On the determination of the exterior and interior attractions of ellipsoids of variable densities‡."

Let $$V = \left[\int_{-\infty}^{\infty} \right]^s \frac{u[d\xi]^s}{\{\Sigma(\xi-x)^2 + u^2\}^{\frac12(s+1)} \{\Sigma(\xi-x')^2 + u'^2\}^{\frac12(s-1)}} \ldots\ldots(6),$$

an integral which may also be expressed thus:

$$-\frac{1}{n-1} \frac{d}{du} \left\{ \left[\int_{-\infty}^{\infty} \right]^s \frac{[d\xi]^s}{\{\Sigma(\xi-x)^2 + u^2\}^{\frac12(s-1)} \{\Sigma(\xi-x')^2 + u'^2\}^{\frac12(s-1)}} \right\}.$$

---

* [By putting, in this, $v = \int dv$; $v_1 = \int dv_1$; etc., we have

$$uU = \frac{2^s P}{a^{s+1}} (s+1) \iiint \ldots dv\, dv_1 dv_2 \ldots :$$

whence immediately, by a simpler case of Liouville's theorem than in the text, or by Green's transformation (see § 186), the same result.]

† See Gregory's *Examples* (Ed. 1841), p. 469.

‡ Read at the Cambridge Phil. Soc., May 6, 1833. See *Trans.* of that date.

From this latter form, we see that the equation

$$\frac{d^2 V}{du^2} + \Sigma \frac{d^2 V}{dx^2} = 0 \quad \dots\dots\dots\dots(7)$$

is satisfied, provided u does not vanish. Hence V is a function which satisfies this equation for all values of x_1, x_2 ... and for all the values of u between 0 and ∞ At these limits the value of V may be easily determined, and the general value inferred in the following manner :—

173. When $u = 0$, the quantity under the signs of integration in the expression for V vanishes for all the values of ξ_1, ξ_2 ... which are not equal to x_1, x_2 ... respectively. Hence it follows that, when $u = 0$,

$$V = \frac{1}{\{\Sigma (x - x')^2 + u'^2\}^{\frac{1}{2}(s-1)}} \left[\int_{-\infty}^{\infty} \right]^s \frac{u \, [d\xi]^s}{\{\Sigma (\xi - x)^2 + u^2\}^{\frac{1}{2}(s+1)}}$$

$$\dagger = \frac{1}{\{\Sigma (x - x')^2 + u'^2\}^{\frac{1}{2}(s-1)}} \left[\int_{-\infty}^{\infty} \right]^s \frac{dz_1 dz_2 \dots dz_s}{(1 + z_1^2 + z_2^2 + \dots + z_s^2)^{\frac{1}{2}(s+1)}}$$

$$= \frac{1}{\{\Sigma (x - x')^2 + u'^2\}^{\frac{1}{2}(s-1)}} \cdot \int_0^{\infty} \int_0^{\infty} \cdots \frac{l_1^{-\frac{1}{2}} l_2^{-\frac{1}{2}} \dots dl_1 dl_2 \dots}{(1 + l_1 + l_2 + \dots + l_s)^{\frac{1}{2}(s+1)}}$$

$$= \frac{1}{\{\Sigma (x - x')^2 + u'^2\}^{\frac{1}{2}(s-1)}} \frac{\pi^{\frac{1}{2}s}}{\Gamma(\frac{1}{2}s)} \int_0^{\infty} \frac{h^{\frac{1}{2}s-1} dh}{(1 + h)^{\frac{1}{2}(s+1)}}$$

$$= \frac{\pi^{\frac{1}{2}(s+1)}}{\Gamma\frac{1}{2}(s+1)} \frac{1}{\{\Sigma (x - x')^2 + u'^2\}^{\frac{1}{2}(s-1)}}.$$

Also, when $u = \infty$, the value of V is nothing.

174. Thus we see that V has the same value as the expression

$$\frac{\pi^{\frac{1}{2}(s+1)}}{\Gamma\frac{1}{2}(s+1)} \cdot \frac{1}{\{\Sigma (x - x')^2 + (u + u')^2\}^{\frac{1}{2}(s-1)}},$$

when $u = 0$, and when $u = \infty$; which enables us to infer that

$$V = \frac{\pi^{\frac{1}{2}(s+1)}}{\Gamma\frac{1}{2}(s+1)} \cdot \frac{1}{\{\Sigma (x - x')^2 + (u + u')^2\}^{\frac{1}{2}(s-1)}},$$

for all positive values of u, provided u' be taken as positive; for the second member of this equation satisfies equation (7) for all positive values of u, and for any values of the other variables, and at the limits $u = 0$ and $u = \infty$ has the same value as V, and therefore, by a theorem of Green's*, in the memoir referred to, must be equal to V for all positive values of u.

---

* [Included in *Theorem* 2 of XIII. below.]

175.   From what has been proved above we may deduce the solution of the following problem :—

Having given for all values of ξ_1, ξ_2 ..., the value of the multiple integral

$$S\frac{\rho' dx_1' dx_2' \ldots dx_s'}{\{(x_1' - \xi_1)^2 + (x_2' - \xi_2)^2 + \ldots + (x_s' - \xi_s)^2 + u'^2\}^{\frac{1}{2}(s-1)}} \ldots (a),$$

where u' and ρ' are any unknown functions of x_1', $x_2' \ldots x_s'$, let it be required to find the value of

$$S\frac{\rho' dx_1' dx_2' \ldots dx_s'}{\{(x_1' - x_1)^2 + (x_2' - x_2)^2 + \ldots + (x_s' - x_s)^2 + (u' + u)^2\}^{\frac{1}{2}(s-1)}} \ldots (b),$$

where $x_1, x_2 \ldots x_s$ are any given quantities, and u a given positive quantity.

Denoting the expression (a) by Φ, and the expression (b) by ϕ, we have, from the theorem established above,

$$\phi = \frac{u\Gamma\frac{1}{2}(s+1)}{\pi^{\frac{1}{2}(s+1)}} S\rho' dx_1' dx_2' \ldots dx_s'.$$

$$\left[\int_{-\infty}^{\infty}\right]^s \frac{[d\xi]^s}{\{\Sigma(\xi - x)^2 + u^2\}^{\frac{1}{2}(s+1)} \{\Sigma(\xi - x')^2 + u'^2\}^{\frac{1}{2}(s-1)}}$$

$$= \frac{u\Gamma\frac{1}{2}(s+1)}{\pi^{\frac{1}{2}(s+1)}} \left[\int_{-\infty}^{\infty}\right]^s \frac{[d\xi]^s}{\{\Sigma(\xi - x)^2 + u^2\}^{\frac{1}{2}(s+1)}} S\frac{\rho' dx_1' dx_2' \ldots dx_s'}{\{\Sigma(\xi - x')^2 + u'^2\}^{\frac{1}{2}(s-1)}},$$

or $\qquad \phi = \frac{u\Gamma\frac{1}{2}(s+1)}{\pi^{\frac{1}{2}(s+1)}} \left[\int_{-\infty}^{\infty}\right]^s \frac{\Phi[d\xi]^s}{\{\Sigma(\xi - x)^2 + u^2\}^{\frac{1}{2}(s+1)}} \ldots \ldots (c).$

But, by hypothesis, Φ is given for all values of ξ_1, $\xi_2 \ldots \xi_s$; and therefore this equation expresses the solution of the problem. We may also deduce from the theorem (5) the expression

$$\phi = -\frac{\Gamma\frac{1}{2}(s+1)}{(s-1)\pi^{\frac{1}{2}(s+1)}} \left[\int_{-\infty}^{\infty}\right]^s \frac{\Psi[d\xi]^s}{\{\Sigma(\xi - x)^2 + u^2\}^{\frac{1}{2}(s-1)}} \ldots (d),$$

by means of which ϕ may be determined when the value, Ψ, of $\frac{d\phi}{du}$ corresponding to $u = 0$ is given.

176.   For the particular case of $u' = 0$, the theorem (d) is included in a theorem given by Green, in which the number n in the exponent of the denominator may differ from the number s of variables, the sole condition being that $n - s + 1$ must be positive ; but it is only in the case of $n = s$ that a general theorem such as (d), by means of which the general value of ϕ is obtained from the value $\frac{d\phi}{du}$ when $u = 0$, can be established.

177. Let us now apply these formulæ to the case of $s = 2$: we may in this case conveniently replace x_1, x_2, u by x, y, z, and ξ_1, ξ_2, by ξ, η. Equations (c) and (d) become

$$\phi = \frac{z}{2\pi} \int_{-\infty}^{\infty} \int_{-\infty}^{x} \frac{\Phi d\xi d\eta}{\{(\xi - x)^2 + (\eta - y)^2 + z^2\}^{\frac{3}{2}}} \quad \ldots\ldots(e),$$

and
$$\phi = -\frac{1}{2\pi} \int_{-\infty}^{\infty} \int_{-\infty}^{\infty} \frac{\Psi d\xi d\eta}{\{(\xi - x)^2 + (\eta - y)^2 + z^2\}^{\frac{1}{2}}} \quad \ldots(f),$$

where Ψ denotes the value of $\dfrac{d\phi}{dz}$ when $x = \xi$, $y = \eta$, $z = 0$.

178. The first of these theorems may be deduced from a very general theorem given by Green in his essay on Electricity and Magnetism [§ (5) eq. (6)]. The second may be demonstrated in the following manner:—

Let x', y', z' be considered as the co-ordinates of a point P', where there is situated a quantity of matter $\rho' dx' dy' dz'$, in the volume $dx' dy' dz'$. Then ϕ will be the potential on a point $P(x, y, z)$, above the plane of x, y which we may regard as horizontal, due to a quantity of matter,

$$M, (= \iiint \rho' dx' dy' dz')$$

situated below this plane. Now it follows from a theorem, first, so far as I am aware, given by Gauss, for a surface of any form, that there is a determinate distribution of matter upon the plane (xy) which will produce this same potential on points above the plane. Let k be the density of this distribution at a point Π (ξ, η) of the plane, so that

$$\phi = \int_{-\infty}^{\infty} \int_{-\infty}^{\infty} \frac{k d\xi d\eta}{\{(\xi - x)^2 + (\eta - y)^2 + z^2\}^{\frac{1}{2}}},$$

which gives
$$\frac{d\phi}{dz} = -z \int_{-\infty}^{\infty} \int_{-\infty}^{\infty} \frac{k d\xi d\eta}{\{(\xi - x)^2 + (\eta - y)^2 + z^2\}^{\frac{3}{2}}}.$$

Let $z = 0$; then denoting by k and $\left(\dfrac{d\phi}{dz}\right)_0$ the values of k and $\dfrac{d\phi}{dz}$ at the point $(x, y, 0)$, we find

$$\left(\frac{d\phi}{dz}\right)_0 = -k \int_{-\infty}^{\infty} \int_{-\infty}^{\infty} \frac{z d\xi d\eta}{\{(\xi - x)^2 + (\eta - y)^2 + z^2\}^{\frac{3}{2}}}$$
$$= -k \cdot 2\pi,$$

since the value of the integral in the second member is 2π, whatever be the value of z. Hence we conclude that

$$k = -\frac{1}{2\pi} \cdot \Psi,$$

and equation (f) is established.

179. It should be remarked that the total quantity of matter distributed over the plane xy must be equal to the mass M, which it represents : this is readily verified from the preceding formulæ.

180. The same formulæ admit of an interesting application in the theory of heat. Thus let ϕ be the permanent temperature of a point P in an infinite homogeneous solid, heated by constant sources distributed below the plane (xy), (the case in which some of the sources are in this plane being of course included). If the temperature Φ at any point Π in the plane (xy) be given, the formula (e) enables us to find the temperature at any point above the plane.

181. As an example, let us suppose that the sources of heat are such that the temperature of a portion A of the plane (xy), between two lines parallel to OY and at equal distances, a, on its two sides, has a constant value c, and the temperature of the remainder of the plane zero. In this case the formula (e) will give, for the temperature at a point (x, y, z) above the plane,

$$\phi = \frac{zc}{2\pi} \int_{-\infty}^{\infty} \int_{-a}^{a} \frac{d\xi d\eta}{\{(\xi - x)^2 + (\eta - y)^2 + z^2\}^{\frac{3}{2}}}$$

$$= \frac{c}{\pi} \left(\tan^{-1} \frac{x+a}{z} - \tan^{-1} \frac{x-a}{z} \right)$$

$$= \frac{c}{\pi} \tan^{-1} \frac{2ax}{x^2 + z^2 - a^2}.$$

From this we conclude that the isothermal surfaces which correspond to this case are circular cylinders, which intersect the plane (xy) in the two parallel lines bounding A.

The application to this example, and all others in which the isothermal surfaces are cylindrical, may be made directly by putting $s = 1$ in the general formulæ.

PART II.

182. I now proceed to find the values, which will be denoted by V and W, of the integrals

$$\left[\int_{-\infty}^{\infty}\right]^{s}\frac{[d\xi]^{s}[\cos m\xi]^{s}}{(\Sigma\xi^{2}+u^{2})^{\frac{1}{2}(s-1)}}$$

and

$$\left[\int_{-\infty}^{\infty}\right]^{s}[dm]^{s}[\cos mx]^{s}\frac{\epsilon^{-(\Sigma m^{2})^{\frac{1}{2}}u}}{(\Sigma m^{2})^{\frac{1}{2}}},$$

where the symbols $[\cos m\xi]^{s}$, $[\cos mx]^{s}$ denote the products

$$\cos m_{1}\xi_{1}.\cos m_{2}\xi_{2}..\cos m_{s}\xi_{s},$$
$$\cos m_{1}x_{1}.\cos m_{2}x_{2}..\cos m_{s}x_{s};$$

and the notation is in other respects the same as before.

By means of the formula

$$[\cos m\xi+\sin m\xi.\sqrt{(-1)}]^{s}=\cos(\Sigma m\xi)+\sin(\Sigma m\xi).\sqrt{(-1)},$$

it is easily shown that

$$V=\left[\int_{-\infty}^{\infty}\right]^{s}\frac{[d\xi]^{s}\cos\Sigma(m\xi)}{(\Sigma\xi^{2}+u^{2})^{\frac{1}{2}(s-1)}}\dots\dots\dots\dots(a).$$

Hence, by a suitable linear transformation, in which one of the assumptions is $\Sigma m\xi=\eta\,(\Sigma m^{2})^{\frac{1}{2}}$, we have [if μ denote $(\Sigma m^{2})^{\frac{1}{2}}$]

$$V=\int_{-\infty}^{\infty}\cos\mu\eta\,.d\eta\,.\left[\int_{-\infty}^{\infty}\right]^{s-1}\frac{[d\zeta]^{s-1}}{(u^{2}+\eta^{2}+\Sigma\zeta^{2})^{\frac{1}{2}(s-1)}}\dots\dots(b).$$

Now, by means of Liouville's theorem[*], we find

$$\left[\int_{-\infty}^{\infty}\right]^{s-1}\frac{[d\zeta]^{s-1}}{(u^{2}+\eta^{2}+\Sigma\zeta^{2})^{\frac{1}{2}(s-1)}}=\frac{\pi^{\frac{1}{2}(s-1)}}{\Gamma\frac{1}{2}(s-1)}\int_{0}^{\infty}\frac{2\xi d\xi\,.\xi^{s-3}}{(\xi^{2}+\eta^{2}+u^{2})^{\frac{1}{2}(s-1)}}.$$

Hence

$$V=\frac{4\pi^{\frac{1}{2}(s-1)}}{\Gamma\frac{1}{2}(s-1)}\int_{0}^{\infty}\int_{0}^{\infty}\frac{\xi^{s-2}\cos\mu\eta\,.d\xi d\eta}{(\xi^{2}+\eta^{2}+u^{2})^{\frac{1}{2}(s-1)}}\dots\dots\dots\dots(c).$$

Differentiating with respect to u, by which the further reduction of the integral will be facilitated, we have

$$-\frac{dV}{du}=(s-1)\frac{4\pi^{\frac{1}{2}(s-1)}}{\Gamma\frac{1}{2}(s-1)}\int_{0}^{\infty}\int_{0}^{\infty}\frac{u\,.\xi^{s-2}\cos\mu\eta\,.d\xi d\eta}{(\xi^{2}+\eta^{2}+u^{2})^{\frac{1}{2}(s+1)}}\dots(d).$$

Now

$$\int_{0}^{\infty}\frac{\xi^{s-2}d\xi}{(\xi^{2}+\eta^{2}+u^{2})^{\frac{1}{2}(s+1)}}=\int_{0}^{\infty}\frac{\xi^{s-3}\,d\xi}{\left(1+\dfrac{\eta^{2}+u^{2}}{\xi^{2}}\right)^{\frac{1}{2}(s+1)}}=\frac{1}{2}\int_{0}^{\infty}\frac{dt}{\{1+(\eta^{2}+u^{2})t\}^{\frac{1}{2}(s+1)}}$$

$$=\frac{1}{s-1}.\frac{1}{\eta^{2}+u^{2}}.$$

---

[*] See *Cambridge Mathematical Journal*, Feb. 1841, p. 221 [or Gregory's *Examples*, Ed. 1841, p. 469].

Hence
$$-\frac{dV}{du} = \frac{4\pi^{\frac{1}{2}(s-1)}}{\Gamma\frac{1}{2}(s-1)}\int_0^\infty \frac{u\cos\mu\eta d\eta}{\eta^2 + u^2} \quad\dots\dots\dots(e).$$
$$= \frac{2\pi^{\frac{1}{2}(s+1)}}{\Gamma\frac{1}{2}(s-1)}\cdot\epsilon^{-\mu u}.$$

From this, by integration with respect to u, we deduce the value of V: thus we have the result

$$\left[\int_{-\infty}^\infty\right]^s \frac{[d\xi]^s[\cos m\xi]^s}{(\Sigma\xi^2 + u^2)^{\frac{1}{2}(s-1)}} = \frac{2\pi^{\frac{1}{2}(s+1)}}{\Gamma\frac{1}{2}(s-1)}\frac{\epsilon^{-(\Sigma m^2)^{\frac{1}{2}}u}}{(\Sigma m^2)^{\frac{1}{2}}} \quad\dots\dots(V).$$

183.   To evaluate the integral W we may in the first place reduce it to a double integral by a process similar to that indicated above, for obtaining the expression (c); and we thus find

$$W = \frac{4\pi^{\frac{1}{2}(s-1)}}{\Gamma\frac{1}{2}(s-1)}\int_0^\infty\int_0^\infty dm\,dn\,.\,m^{s-2}\cos(nr)\,.\,\frac{\epsilon^{-(m^2+n^2)^{\frac{1}{2}}u}}{(m^2+n^2)^{\frac{1}{2}}} \quad\dots\dots(a),\checkmark$$

where r denotes $(\Sigma x^2)^{\frac{1}{2}}$. If we take $m = \rho\cos\vartheta$, $n = \rho\sin\vartheta$, this becomes

$$W = \frac{4\pi^{\frac{1}{2}(s-1)}}{\Gamma\frac{1}{2}(s-1)}\int_0^\infty\int_0^{\frac{1}{2}\pi} d\theta d\rho\rho^{s-2}\cos^{s-2}\vartheta\cos(r\rho\sin\vartheta)\epsilon^{-\rho u}\dots(b).$$

Now we have

$$\left(\frac{d^2}{du^2} + \frac{d^2}{dr^2}\right)^f\cos(r\rho\sin\vartheta)\,\epsilon^{-\rho u} = \rho^{2f}\cos^{2f}\vartheta\,.\,\cos(r\rho\cos\vartheta)\,\epsilon^{-\rho u}\dots(c).$$

Considering first the case where s is even, let $f = \frac{1}{2}s - 1$; we thus find

$$\rho^{s-2}\cos^{s-2}\vartheta\cos(r\rho\cos\vartheta)\epsilon^{-\rho u} = \left(\frac{d^2}{du^2} + \frac{d^2}{dr^2}\right)^{\frac{1}{2}s-1}\cos(r\rho\sin\vartheta)\epsilon^{-\rho u},$$

and, by substitution in (b), we have

$$W = \frac{4\pi^{\frac{1}{2}(s-1)}}{\Gamma\frac{1}{2}(s-1)}\int_0^\infty\int_0^{\frac{1}{2}\pi} d\vartheta d\rho\,.\,\left(\frac{d^2}{du^2} + \frac{d^2}{dr^2}\right)^{\frac{1}{2}s-1}\cos(r\rho\sin\vartheta)\,\epsilon^{-\rho u},$$
$$= \frac{4\pi^{\frac{1}{2}(s-1)}}{\Gamma\frac{1}{2}(s-1)}\left(\frac{d^2}{du^2} + \frac{d^2}{dr^2}\right)^{\frac{1}{2}s-1}\int_0^\infty\int_0^{\frac{1}{2}\pi} d\vartheta d\rho\cos(r\rho\sin\vartheta)\,\epsilon^{-\rho u}$$
$$= \frac{4\pi^{\frac{1}{2}(s-1)}}{\Gamma\frac{1}{2}(s-1)}\left(\frac{d^2}{du^2} + \frac{d^2}{dr^2}\right)^{\frac{1}{2}s-1}\int_0^{\frac{1}{2}\pi}\frac{ud\vartheta}{u^2 + r^2\sin^2\vartheta}$$
$$= \frac{4\pi^{\frac{1}{2}(s-1)}}{\Gamma\frac{1}{2}(s-1)}\left(\frac{d^2}{du^2} + \frac{d^2}{dr^2}\right)^{\frac{1}{2}s-1}\frac{\frac{1}{2}\pi}{(u^2+r^2)^{\frac{1}{2}}}$$
$$= \frac{4\pi^{\frac{1}{2}(s-1)}}{\Gamma\frac{1}{2}(s+1)}\frac{1^2.3^2\dots(s-1)^2}{(u^2+r^2)^{\frac{1}{2}(s-1)}} = 2^{s-1}\pi^{\frac{1}{2}(s-1)}\Gamma\frac{1}{2}(s-1)\frac{1}{(u^2+r^2)^{\frac{1}{2}(s-1)}}(d).$$

In the second case, when s is odd, let $f = \frac{1}{2}(s-1)$ in (c); then, making use of the result in (b), we have

$$W = \frac{4\pi^{\frac{1}{2}(s-1)}}{\Gamma\frac{1}{2}(s-1)} \left(\frac{d^2}{du^2} + \frac{d^2}{dr^2}\right)^{\frac{1}{2}(s-1)} \int_0^\infty \int_0^{\frac{1}{2}\pi} d\vartheta d\rho\rho \cos\vartheta . \cos(r\rho\sin\vartheta)\epsilon^{-\rho u}$$

$$= \frac{4\pi^{\frac{1}{2}(s-1)}}{\Gamma\frac{1}{2}(s-1)} \left(\frac{d^2}{du^2} + \frac{d^2}{dr^2}\right)^{\frac{1}{2}(s-1)} \int_0^\infty d\rho . \frac{\sin(r\rho)}{r} \epsilon^{-\rho u}$$

$$= \frac{4\pi^{\frac{1}{2}(s-1)}}{\Gamma\frac{1}{2}(s-1)} \left(\frac{d^2}{du^2} + \frac{d^2}{dr^2}\right)^{\frac{1}{2}(s-1)} \frac{1}{r^2 + 0^2} = 2^{s-1}\pi^{\frac{1}{2}(s-1)}\Gamma\frac{1}{2}(s-1)\frac{1}{(u^2+r^2)^{\frac{1}{2}(s-1)}}.$$

Hence, whether s be odd or even, we conclude that

$$\left[\int_0^\infty\right]^s [dm]^s [\cos mx]^s \frac{\epsilon^{-(\Sigma m^2)^{\frac{1}{2}}u}}{(\Sigma m^2)^{\frac{1}{2}}} = 2^{s-1}\pi^{\frac{1}{2}(s-1)}\Gamma\frac{1}{2}(s-1)\frac{1}{(u^2+\Sigma x^2)^{\frac{1}{2}(s-1)}} \quad (W).$$

184. The investigation which we have just gone through, of the integrals (V), (W) constitutes the verification of "Fourier's theorem" in a particular case. For, by this theorem, we have, if $F(x_1, x_2 ...)$ be a function which remains the same when the signs of any of the variables are changed,

$$2^s\pi^s F(x_1, x_2 ...) =$$
$$\left[\int_{-\infty}^\infty\right]^s [dm]^s [\cos mx]^s \int_{-\infty}^\infty [d\xi]^s [\cos m\xi]^s F(\xi_1, \xi_2 ...) ...(e):$$

and if we take

$$F(\xi_1, \xi_2 ...) = \frac{1}{(\Sigma \xi^2 + u^2)^{\frac{1}{2}(s-1)}},$$

the result of the integrations with respect to $\xi_1, \xi_2 ...$, is given by (V), and the second member thus becomes a multiple integral with respect to $m_1, m_2 ...$, which is shown by (W) to be equal to the first member. Conversely, if we assume Fourier's theorem, we may deduce the value W, by means of it, from that of V. The integrals V and W are also connected by means of another case of Fourier's theorem, found by taking, in (e),

$$F(\xi_1, \xi_2 ...) = \frac{\epsilon^{-(\Sigma\xi^2)^{\frac{1}{2}}u}}{(\Sigma\xi)^{\frac{1}{2}}} .$$

In this way, after the value of W has been found, that of V may be deduced.

185. The formulæ (V) and (W) may be applied to evaluate the multiple integral u, and we shall thus obtain the result of the investigation in § I. in a different manner.

By means of the equation obtained by differentiating (W) with respect to u, we find

$$\frac{u}{\{\Sigma(\xi-x)^2+u^2\}^{\frac{1}{2}(s+1)}} = \frac{1}{2^{s-1}(s-1)\,\pi^{\frac{1}{2}(s-1)}\,\Gamma^{\frac{1}{2}}(s-1)}$$

$$\left[\int_{-\infty}^{\infty}\right]^s [dm]^s[\cos m\,(\xi-x)]^s\epsilon^{-(\Sigma m^2)^{\frac{1}{2}}u};$$

Making this substitution, for one of the factors of the expression under the integral signs U, we have

$$Uu = \frac{1}{2^{s-1}(s-1)\,\pi^{\frac{1}{2}(s-1)}\,\Gamma^{\frac{1}{2}}(s-1)}\left[\int_{-\infty}^{\infty}\right]^s \frac{[d\xi]^s}{\{\Sigma(\xi-x')^2+u'^2\}^{\frac{1}{2}(s-1)}}$$

$$\left[\int_{-\infty}^{\infty}\right]^s [dm]^s[\cos m\,(\xi-x)]^s\,\epsilon^{-(\Sigma m^2)^{\frac{1}{2}}u}$$

$$= \frac{1}{2^{s-1}(s-1)\pi^{\frac{1}{2}(s-1)}\Gamma^{\frac{1}{2}}(s-1)}$$

$$\left[\int_{-\infty}^{\infty}\right]^s [dm]^s[\cos m\,(x-x')]^s\epsilon^{-(\Sigma m^2)\frac{1}{2}u}\left[\int_{-\infty}^{\infty}\right]^s \frac{[d\xi]^s[\cos m\,(\xi-x')]^s}{\{\Sigma(\xi-x')^2+u'^2\}^{\frac{1}{2}(s-1)}}$$

$$= \frac{\pi}{2^{s-2}(s-1)\{\Gamma^{\frac{1}{2}}(s-1)\}^2}$$

$$\left[\int_{-\infty}^{\infty}\right]^s [dm]^s[\cos m\,(x-x')]^s\epsilon^{-(\Sigma m^2)\frac{1}{2}u}\frac{\epsilon^{-(\Sigma m^2)^{\frac{1}{2}}u'}}{(\Sigma m^2)^{\frac{1}{2}}},\ \text{by } (V),$$

$$= \frac{2\pi^{\frac{1}{2}(s+1)}}{(s-1)\Gamma^{\frac{1}{2}}(s-1)}\frac{1}{\{\Sigma(x-x')^2+(u+u')^2\}^{\frac{1}{2}(s-1)}},\ \text{by } (W),$$

which agrees with the value obtained above.

PART III.

186. The value of the integral U may also be obtained by a direct process of reduction, as follows:—

By a suitable linear transformation, in which assumptions such as

$$\xi_1 - x_1 = \Sigma a\zeta$$

are made, we find

$$U = \left[\int_{-\infty}^{\infty}\right]^s \frac{[d\zeta]^s}{(\Sigma\zeta^2+u^2)^{\frac{1}{2}(s+1)}(\Sigma\zeta^2-2f\zeta_1+f^2+u'^2)^{\frac{1}{2}(s-1)}} \quad\ldots\ldots(a),$$

where

$$f^2 = \Sigma(x-x')^2.$$

Let us now assume

$$\zeta_1 = \rho\cos\phi,\ \zeta_2 = \rho\sin\phi\cos\theta_1,\ \zeta_3 = \rho\sin\phi\sin\theta_1\cos\theta_2\ldots,$$
$$\zeta_{s-1} = \rho\sin\phi\sin\theta_1\sin\theta_2\ldots\ldots\cos\theta_{s-2},$$
$$\zeta_s = \rho\sin\phi\sin\theta_1\sin\theta_2\ldots\ldots\sin\theta_{s-2},$$

from which we deduce*

$$[d\xi]' = \rho'^{-1} \sin^{r-2}\phi \sin^{r-3}\theta_1 \sin^{-4}\theta_2 \ldots \sin\theta_{s-3}[d\theta]^{r-2}d\phi d\rho;$$

a transformation given first by Green. Equation (a) is thus reduced to

$$U = H_{s-2}\int_0^\infty \int_0^\pi \frac{\rho'^{-1}\sin^{s-2}\phi\, d\phi d\rho}{(\rho^2+u^2)^{\frac12(s+1)}(\rho^2 - 2\rho f\cos\phi + f^2 + u'^2)^{\frac12(s-1)}} \ldots(b),$$

where H_{s-2} denotes the product

$$\int_0^\pi \sin^{s-3}\theta d\theta . \int_0^\pi \sin^{s-4}\theta d\theta \ldots\ldots \int_0^{2\pi} d\theta.$$

Let $\rho = u\tan\frac12\vartheta$; we thus get

$$Uu = \frac12 H_{s-2}\int_0^\pi \int_0^\pi$$

$$\frac{\sin^{s-1}\vartheta \sin^{s-2}\phi d\phi d\vartheta}{\{2(f^2+u'^2+u^2)+2(f^2+u'^2-u^2)\cos\vartheta - 4uf\sin\vartheta\cos\phi\}^{\frac12(s-1)}}$$

and we may now conveniently assume

$$2(f^2 + u'^2 + u^2) = h^2 + k^2,$$

$$2(f^2 + u'^2 - u^2)\cos\vartheta - 4uf\sin\vartheta\cos\phi$$
$$= 2\{(f^2 + u'^2 - u^2)^2 + 4u^2 f^2\}^{\frac12}\cos\theta = 2hk\cos\theta,$$

and $\sin\phi\sin\vartheta = \sin\varphi\sin\theta,$

from which we deduce

$$h^2 = (u'+u)^2 + f^2, \quad k^2 = (u'-u)^2 + f^2,$$

$$\sin\vartheta d\phi d\vartheta = \sin\theta d\varphi d\theta;$$

the expression for U becomes

$$Uu = \frac12 H_{s-2}\int_0^\pi \int_0^\pi \frac{\sin^{s-1}\theta \sin^{s-2}\varphi d\varphi d\theta}{(h^2 - 2hk\cos\theta - k^2)^{\frac12(s-1)}}$$

$$= \frac12 H_{s-1}\int_0^\pi \frac{\sin^{s-1}\theta d\theta}{(h^2 - 2hk\cos\theta + k^2)^{\frac12(s-1)}}$$

Let $h\sin(\psi - \theta) = k\sin\psi;$

by means of this transformation, observing that $h > k$, we readily find

$$Uu = \frac{\frac12 H_s}{h^{s-1}}$$

or $$Uu = \frac{\pi^{\frac12(s+1)}}{\Gamma\frac12(s+1)}\frac{1}{\{\Sigma(x-x')^2 + (u+u')^2\}^{\frac12(s-1)}},$$

which is the same as the result previously obtained.

St Peter's College, *Oct.* 3, 1846.

---

* See *Cambridge Mathematical Journal*, Nov. 1843, p. 24, First Series; [or Green, "Attraction of Ellipsoids," § 6, *Camb. Phil. Trans.*, May, 1833.]

(Art. VI. of complete list in *Mathematical and Physical Papers*, Vol. I.)

[From the *Camb. Math. Jour.*, Nov. 1842 and Feb. 1843.]

187. Let x, y, z be the co-ordinates of any point P in an attracting or repelling body M; let dm be an element of the mass, at the point P, which will be positive or negative according as it is attractive or repulsive; let x', y', z' be the co-ordinates of an attracted point P'; let

$$\Delta = \{(x' - x)^2 + (y' - y)^2 + (z' - z)^2\}^{\frac{1}{2}};$$

and let

$$v' = \int \frac{dm}{\Delta},$$

the integral including the whole of M. This expression has been called by Green the potential* of the body M, on the point P', and the same name has been employed by Gauss (in a Mémoire on "General Theorems relating to Attractive and Repulsive Forces, in the *Resultate aus den Beobachtungen des magnetischen Vereins im Jahre* 1839, Leipsic 1840, edited by M. Gauss and Weber)†. By a known theorem, the components of the attraction of M on P', in the directions of x, y, z, are

$$-\frac{dv'}{dx'}, \quad -\frac{dv'}{dy'}, \quad -\frac{dv'}{dz'},$$

and if $d\gamma'$ be the element of any line, straight or curved, which passes through P', the attraction in the direction of this element is $\dfrac{dv'}{d\gamma'}$. Hence it follows that if a surface be drawn through any point P' for every point of which the potential has the same value, the attraction on every point in the surface is wholly in the direction of the normal. Surfaces for which the potential is constant are therefore called, by Gauss, *surfaces of equilibrium*. It has been shown in a former paper (I. above),

---

* ["This I found in a reference to his memoirs, in Murphy's first memoir on "definite integrals. Ever since I have been trying to see Green's memoir, but "could not hear of it from anybody till to-day, when I have got a copy from "Mr Hopkins. Jan. 25, 1845." (Private note which I find written on p. 190 of vol. iii. of my copy of the *Camb. Math. Jour.*)]

† Translations of this paper have been published in Taylor's *Scientific Memoirs* for April, 1842, and in the Numbers of Liouville's *Journal* for July and August, 1842.

that if M, instead of an attractive mass, were a group of sources of heat or cold in the interior of an infinite homogeneous solid, v' would be the permanent temperature produced by them at P'. In that case, the surfaces of equilibrium would be *isothermal surfaces*.

188. When the attraction of (positive or negative) matter, as for instance electricity, spread over a surface is considered, the density of the matter at any point is measured by the quantity of matter on an element of the surface, divided by that element.

The principal object of this paper is to prove the following theorems :—

If upon E, one of the surfaces of equilibrium enclosing an attracting mass, its matter be distributed in such a manner that its density at any point P is equal to the attraction of M on P; then—

(1) The attraction of the matter spread over E, on an external point, is equal to the attraction of M on the same point multiplied by 4π.

(2) The attraction of the matter on E, on an internal point, is nothing.

189. These theorems were proved in a previous paper (I. §§ 5, 9), from considerations relative to the uniform motion of heat; but in the following they are proved by direct integration :—

Let u be the potential of M, on the point P, (xyz) in E. The components of the attraction of M on P, in the directions of x, y, z, are

$$-\frac{du}{dx}, \quad -\frac{du}{dy}, \quad -\frac{du}{dz};$$

and hence, if α, β, γ be the angles which a normal to E at P makes with these directions, the total attraction on P is

$$-\left(\frac{du}{dx}\cos\alpha + \frac{du}{dy}\cos\beta + \frac{du}{dz}\cos\gamma\right) \text{ or } -\frac{du}{dn},$$

if dn be an element of the normal through P.

This is therefore the expression for the density at P of the matter we have supposed to be spread over E. Let ds be an element of E at P; let v' be the potential of E, on a point P', $(x'y'z')$, either within or without E; and let Δ be the distance from P to P'. Then

$$v' = -\left\{ \int \frac{\left(\frac{du}{dx}\cos\alpha + \frac{du}{dy}\cos\beta + \frac{du}{dz}\cos\gamma\right)ds}{\Delta} \right\} = -\left\{ \int \frac{\frac{du}{dn}ds}{\Delta} \right\} \ (a),$$

the brackets enclosing the integrals denoting that the integrations are to be extended over the whole surface E. Now for ds, we may choose any one of the expressions,

$$ds = \frac{dydz}{\cos\alpha}, \quad ds = \frac{dxdz}{\cos\beta}, \quad ds = \frac{dxdy}{\cos\gamma}.$$

Hence any integral of the form
$$\{\int (A\cos\alpha + B\cos\beta + C\cos\gamma)ds\}$$
may be transformed into the sum of the three integrals,
$$(\iint Adydz), \quad (\iint Bdxdz), \quad (\iint Cdxdy),$$
by using the first, second, and third of the expressions for ds in the first, second, and third terms of the integral respectively.

Hence, if $\quad A = \dfrac{d\phi}{dx}\psi, \quad B = \dfrac{d\phi}{dy}\psi, \quad C = \dfrac{d\phi}{dz}\psi,$

$$\left(\int \frac{d\phi}{dn}\psi ds\right) \text{ or } \left\{\int\left(\frac{d\phi}{dx}\cos\alpha + \frac{d\phi}{dy}\cos\beta + \frac{d\phi}{dz}\cos\gamma\right)\psi ds\right\},$$

$$= \left\{\iint \psi\left(\frac{d\phi}{dx}dydz + \frac{d\phi}{dy}dxdz + \frac{d\phi}{dz}dxdy\right)\right\}\ldots\ldots(b),$$

the limits of the integrations relative to y and z, x and z, x and y, being so chosen as to include the whole of the surface considered.

190.　Making use of this transformation in (a) we have

$$v' = -\left\{\iint\left(\frac{du}{dx}\frac{dydz}{\Delta} + \frac{du}{dy}\frac{dxdz}{\Delta} + \frac{du}{dz}\frac{dxdy}{\Delta}\right)\right\}\ldots\ldots(a').$$

Now $\quad \displaystyle\iint \frac{du}{dx}\frac{dydz}{\Delta} = \iint dydz \int dx\left(\frac{d^2u}{dx^2}\frac{1}{\Delta} + \frac{du}{dx}\frac{d}{dx}\frac{1}{\Delta}\right)$

$$= \iiint dxdydz\left(\frac{d^2u}{dx^2}\frac{1}{\Delta} + \frac{du}{dx}\frac{d}{dx}\frac{1}{\Delta}\right).$$

191.　Hence, if the integrals in the second member include every point in the space contained between E, and another surface of equilibrium, $E_{,}$ without E, and which we shall suppose to be also without P', we have

$$\left\{\iint\frac{du}{dx}\frac{dydz}{\Delta}\right\}_{,} - \left\{\iint\frac{du}{dx}\frac{dydz}{\Delta}\right\} = \iiint\left(\frac{d^2u}{dx^2}\frac{1}{\Delta} + \frac{du}{dx}\frac{d}{dx}\frac{1}{\Delta}\right)dxdydz,$$

the accent denoting that, in the term accented, the integrals are to be extended over the surface E,. Modifying in a similar manner the second and third terms of v', we have

$$\left\{\int \frac{\frac{du}{dn}ds}{\Delta}\right\}, - \left\{\int \frac{\frac{du}{dn}ds}{\Delta}\right\} \text{ or } \left\{\int \frac{\frac{du}{dn}ds}{\Delta}\right\}, + v'$$

$$= \iiint \left(\frac{d^2u}{dx^2} + \frac{d^2u}{dy^2} + \frac{d^2u}{dz^2} + \frac{du}{dx}\frac{d}{dx}\cdot\frac{1}{\Delta} + \frac{du}{dy}\frac{d}{dy}\cdot\frac{1}{\Delta} + \frac{du}{dz}\frac{d}{dz}\cdot\frac{1}{\Delta}\right) dxdydz \ (c).$$

Now, for all points without M,

$$\frac{d^2u}{dx^2} + \frac{d^2u}{dy^2} + \frac{d^2u}{dz^2} = 0,$$

by a known theorem; and such points only are included in the integrals in the second member of (c).

Also, by integration by parts,

$$\iiint \frac{du}{dx}\frac{d}{dx}\frac{1}{\Delta}dxdydz = \iint u\frac{d}{dx}\frac{1}{\Delta}dydz - \iiint u\frac{d^2}{dx^2}\frac{1}{\Delta}dxdydz$$

$$= \left\{\iint u\frac{d}{dx}\frac{1}{\Delta}dydz\right\}, - \left\{\iint u\frac{d}{dx}\frac{1}{\Delta}dydz\right\} - \iiint u\frac{d^2}{dx^2}\frac{1}{\Delta}dxdydz.$$

Modifying similarly the two remaining terms of the second member of (c), we have

$$\left\{\int \frac{\frac{du}{dn}ds}{\Delta}\right\}, + v' = \left\{\iint u\left(\frac{d}{dx}\frac{1}{\Delta}dydz + \frac{d}{dy}\frac{1}{\Delta}dxdz + \frac{d}{dz}\frac{1}{\Delta}dxdy\right)\right\}$$

$$- \left\{\iint u\left(\frac{d}{dx}\frac{1}{\Delta}dydz + \frac{d}{dy}\frac{1}{\Delta}dxdz + \frac{d}{dz}\frac{1}{\Delta}dxdy\right)\right\}$$

$$- \iiint u\left(\frac{d^2}{dx^2}\frac{1}{\Delta} + \frac{d^2}{dy^2}\frac{1}{\Delta} + \frac{d^2}{dz^2}\frac{1}{\Delta}\right)dxdydz\ldots\ldots(c').$$

Now, since E and E, are surfaces of equilibrium, u is constant for each. Again,

$$\frac{d^2}{dx^2}\frac{1}{\Delta} + \frac{d^2}{dy^2}\frac{1}{\Delta} + \frac{d^2}{dz^2}\frac{1}{\Delta} = 0,$$

except when P coincides with P', at which point u has the value u'. Hence, the value of the integrals,

$$\iiint u\left(\frac{d^2}{dx^2}\frac{1}{\Delta} + \frac{d^2}{dy^2}\frac{1}{\Delta} + \frac{d^2}{dz^2}\frac{1}{\Delta}\right) dxdydz$$

is only affected by these elements, for which $u = u'$, and hence u may be taken without the integral sign, as being constant and equal to u'. If, therefore, for brevity, we put

T. E.   9.

$$\iint\left(\frac{d}{dx}\frac{1}{\Delta}dydz + \frac{d}{dy}\frac{1}{\Delta}dxdz + \frac{d}{dz}\frac{1}{\Delta}dxdy\right) = (h)\text{ or }(h),\quad(d),$$

according as the integrals refer to E, or to $E_{\prime}$, and

$$\iiint\left(\frac{d^2}{dx^2}\frac{1}{\Delta} + \frac{d^2}{dy^2}\frac{1}{\Delta} + \frac{d^2}{dz^2}\frac{1}{\Delta}\right)dxdydz = k\ldots\ldots(e),$$

the integrations including every point between E and E''; equation (c) becomes

$$\left\{\int\frac{\frac{du}{dn}ds}{\Delta}\right\} + v' = (u),(h), -(u)(h) - u'k\ldots\ldots(c'').$$

Now it is obvious that, at a great distance from M, the surfaces of equilibrium are very nearly spherical. Let E' be taken so far off that it may be considered as spherical, without sensible error, and let γ be the distance of any point in E' from the centre, a fixed point in M, or, which is the same, the radius of the sphere. Then $-\frac{du}{dn}$, or $-\frac{du}{d\gamma}$, is the attraction of M, on a point in E', and is therefore equal to $\frac{M}{\gamma^2}$, and therefore, by the known expression for the potential of a uniform spherical shell, on an interior point,

$$-\left\{\int\frac{\frac{du}{dn}ds}{\Delta}\right\},\text{ or }\frac{M}{\gamma^2}\left\{\int\frac{ds}{\Delta}\right\}, = \frac{M}{\gamma^2}4\pi\gamma = 4\pi(u),\ldots\ldots(f).$$

It now only remains to determine the integrals (h), $(h)_{\prime}$ and k.

By putting, in (b), $\psi = 1$, $\phi = \frac{1}{\Delta}$, we find the following transformation, for (h),

$$h = \int\frac{d\frac{1}{\Delta}}{dn}ds = -\int\frac{d\Delta}{dn}\frac{ds}{\Delta^2}.$$

Now let the point (xyz) be referred to the polar co-ordinates, γ, θ, ϕ. Then, if P' be pole, $\gamma = \Delta$. Also, if ψ be the angle between Δ and dn, the expression for ds is

$$ds = \frac{\Delta^2\sin\theta d\theta d\phi}{\cos\psi},\text{ or, since }\cos\psi = \frac{d\Delta}{dn},$$

$$ds = \frac{\Delta^2\sin\theta d\theta d\phi}{\frac{d\Delta}{dn}}.$$

Hence $h = -\iint\sin\theta d\theta d\phi.$

If P' be within the surface to which the integrals refer, the limits for θ are 0 and π, and for ϕ, 0 and 2π, and in that case, $h = -4\pi$; therefore, since P' is always within $E_{,}$,

$$(h)_{,} = -4\pi \quad\quad\quad\quad\quad\quad\quad (g).$$

If P' be without the surface considered, then, for each value of θ, we must take the sum of the expressions

$$-\sin\theta\, d\theta d\phi, \text{ and } -\sin\theta\,(-d\theta)\, d\phi,$$

and, therefore, each element of the integral is destroyed by another equal to it, but with a contrary sign, and the value of the complete integral is therefore zero.

Hence, according as P' is without or within E,

$$(h) = 0, \text{ or } (h) = -4\pi \quad\quad\quad\quad\quad (h).$$

Again, to find the value of k, we have, by dividing it into three terms, and integrating each once,

$$k = \left\{ \iint \left(\frac{d}{dx}\frac{1}{\Delta}\, dydz + \frac{d}{dy}\frac{1}{\Delta}\, dxdz + \frac{d}{dz}\frac{1}{\Delta}\, dxdy \right) \right\}_{,}$$

$$-\left\{ \iint \left(\frac{d}{dx}\frac{1}{\Delta}\, dydz + \frac{d}{dy}\frac{1}{\Delta}\, dxdz + \frac{d}{dz}\frac{1}{\Delta}\, dxdy \right) \right\}$$

$$= (h)_{,} - (h) = -4\pi - 0, \text{ or } = -4\pi + 4\pi ;$$

and, therefore, according as P' is without or within E,

$$k = -4\pi, \text{ or } k = 0 \quad\quad\quad\quad\quad\quad (k).$$

Hence, making use of (f), (g), (h), (k), in (c''), we have

$$v' = 4\pi u', \text{ when } P' \text{ is without } E \quad\quad\quad (1),$$

$$v' = 4\pi (u), \text{ when } P' \text{ is within } E \quad\quad\quad (2).$$

From the first of these equations it follows that the attraction of E, on a point without it, is the same as that of M, multiplied by 4π; and since the second shows that the potential of E on internal points is constant, we infer that the attraction of E on internal points is nothing.

These theorems, along with some others which were also proved in the previous paper in this Journal, already referred to, had, I have since found, been given previously by Gauss. One of the most important of these is the following:—If a mass M be wholly within or wholly without a surface, an equal mass may be distributed over this surface [in the former case, or a certain less mass may be distributed over it in the latter case] in such a manner that its attraction, in the former case on external points, and in the latter on internal, will be equal to the attraction of M on the same points. This theorem, which was proved from physical considerations in the paper *On the Uniform Motion of Heat, etc.*, is proved analytically in Gauss's *Mémoire*, but the same method is used in both to infer from it the truth of propositions (1) and (2).

From Prop. (2) it follows that, if E be the surface of an electrified conducting body, the intensity of the electricity at any point will be proportional to the attraction of M on the point. Hence we have the means of finding an infinite number of forms for conducting bodies, on which the distribution of electricity can be determined.

Thus, if M consists of a group of material points, m_1, m_2, etc., whose co-ordinates are x_1, y_1, z_1, ; x_2, y_2, z_2, etc.: the general equation to the surfaces of equilibrium is

$$\frac{m_1}{\{(x-x_1)^2+(y-y_1)^2+(z-z_1)^2\}^{\frac{1}{2}}} + \frac{m_2}{\{(x-x_2)^2+(y-y_2)^2+(z-z_2)^2\}^{\frac{1}{2}}} + \text{etc.} = \lambda,$$

and the intensity of electricity at any point of a solid body, bounded by one of them, will be the value of

$$\left\{\left(\frac{d\lambda}{dx}\right)^2 + \left(\frac{d\lambda}{dy}\right)^2 + \left(\frac{d\lambda}{dz}\right)^2\right\}^{\frac{1}{2}},$$

at the point.

To take a simple case :—Let there be only two material points, of equal intensity. The surface will then be a surface of revolution, and will be symmetrical with regard to a plane perpendicular, through its point of bisection, to the line joining the two points, and would probably very easily be constructed in practice. We should thus have a simple method of verifying numerically the mathematical theory of electricity.

PART II.

[From the *Cambridge Mathematical Journal*, February 1843.]

199. I shall now prove a general theorem, which comprehends the propositions demonstrated in Part I., along with several others of importance in the theories of electricity and heat.

Let M and M_1 be two bodies, or groups or attracting or repelling points; and let v and v_1 be their potentials on xyz; let R and R_1 be their total attractions on the same point; and let θ be the angle between the directions of R and R_1, and $\alpha\beta\gamma$, $\alpha_1\beta_1\gamma_1$, the angles which they make with xyz. Let S be a closed surface, ds an element, corresponding to the co-ordinates xyz; and P and P_1 the components of R, R_1, in a direction perpendicular to the surface at ds. Then we have

$$R\cos\alpha = -\frac{dv}{dx}, \quad R\cos\beta = -\frac{dv}{dy}, \quad R\cos\gamma = -\frac{dv}{dz},$$

$$R_1\cos\alpha_1 = -\frac{dv_1}{dx}, \quad R_1\cos\beta_1 = -\frac{dv_1}{dy}, \quad R_1\cos\gamma_1 = -\frac{dv_1}{dz},$$

$$\cos\theta = \cos\alpha\cos\alpha_1 + \cos\beta\cos\beta_1 + \cos\gamma\cos\gamma_1;$$

hence, $\quad \dfrac{dv}{dx}\dfrac{dv_1}{dx} + \dfrac{dv}{dy}\dfrac{dv_1}{dy} + \dfrac{dv}{dz}\dfrac{dv_1}{dz} = RR_1\cos\theta.$

Hence

$$\iiint RR_1\cos\theta\,dxdydz = \iiint\left(\frac{dv}{dx}\frac{dv_1}{dx} + \frac{dv}{dy}\frac{dv_1}{dy} + \frac{dv}{dz}\frac{dv_1}{dz}\right)dxdydz \ldots(a),$$

where we shall suppose the integrals to include every point in the interior of S. Now, by integration by parts, the second member may be put under the form,

$$\iint v_1\left(\frac{dv}{dx}\,dydz + \frac{dv}{dy}\,dxdz + \frac{dv}{dz}\,dxdy\right)$$

$$- \iiint v_1\left(\frac{d^2v}{dx^2} + \frac{d^2v}{dy^2} + \frac{d^2v}{dz^2}\right)dxdydz \ldots(b);$$

where the double integrals are extended over the surface S, and the triple integrals, as before, over every point in its interior. If we transform the first term of this by (b), Part I., and observe that $-\dfrac{dv}{dn} = P$, it becomes

$$- \iint v_1 P\,ds.$$

Again, $\quad \dfrac{d^2v}{dx^2} + \dfrac{d^2v}{dy^2} + \dfrac{d^2v}{dz^2} = 0 \ldots\ldots\ldots\ldots(c),$

except when xyz is a point of the attracting mass.

If this be the case, and if k be the density of the matter at the point, we have

$$\frac{d^2v}{dx^2} + \frac{d^2v}{dy^2} + \frac{d^2v}{dz^2} + 4\pi k = 0$$

therefore $\quad \left(\dfrac{d^2v}{dx^2} + \dfrac{d^2v}{dy^2} + \dfrac{d^2v}{dz^2}\right)dxdydz + 4\pi dm = 0 \left.\right\} \ldots\ldots(d).$

Hence (*a*) is transformed into

$$\iiint RR_1 \cos\theta dxdydz = 4\pi\iiint v_1 dm - \iint v_1 Pds \,\ldots\ldots(3);$$

similarly, by performing the integration in (*a*), on the terms

$$\frac{dv}{dx}, \ \frac{dv}{dy}, \ \frac{dv}{dz}, \ \text{instead of} \ \frac{dv_1}{dx}, \ \frac{dv_1}{dy}, \ \frac{dv_1}{dz},$$

we should have found

$$\iiint RR_1 \cos\theta dxdydz = 4\pi\iiint vdm_1 - \iint vP_1 ds \,\ldots\ldots(4).$$

200. If the triple integrals in (*a*) were extended over all the space without *S*, or over every point between *S*, and another surface, *S,*, enclosing it, at an infinite distance, it may be shown, as in Part I., that the superior values of the double integrals in (*b*), corresponding to *S,*, vanish. Hence, the inferior values being those which correspond to *S*, we have, instead of (3) and (4),

$$\iiint RR_1 \cos\theta dxdydz = 4\pi\iiint v_1 dm + \iint v_1 Pds \,\ldots\ldots(5),$$
$$\iiint RR_1 \cos\theta dxdydz = 4\pi\iiint vdm_1 + \iint vP_1 ds \,\ldots\ldots(6).$$

It is obvious that *v* and *v₁* in these equations may be any functions, each of which satisfy equations (*c*) and (*d*), whether we consider them as potentials or temperatures, or as mere analytical functions with the restriction that, in (5) and (6), *v* and *v₁* must be such as to make $\iint v_1 Pds$ and $\iint vP_1 ds$ vanish at *S,* [and (a condition the necessity for which has been discovered by Helmholtz),* that, in (3) and (4), if *S* be multiply continuous, *v* and *v₁* must be single-valued functions throughout it]. If each of them satisfy (*c*) for all the points within the limits of the triple integrals considered, *dm* and *dm₁* will each vanish; but if there be any points within the limits, for which either *v* or *v₁* does not satisfy (*c*), the value of *dm* or *dm₁* at those points will be found from (*d*).

201. Thus let $v_1 = 1$, for every point. Then we must have $dm_1 = 0$. Also $R_1 = 0$, $P_1 = 0$.

Hence (3) becomes

$$\iint Pds = 4\pi\iiint dm = 4\pi m \,\ldots\ldots\ldots\ldots(7),$$

---

* [See Helmholtz; Crelle's *Journal*, 1858 (Wirbelbewegung), translated by Tait, *Phil. Mag.* 1867, I. (Vortex-Motion); or Thomson (Vortex-Motion, §§ 54...58), *Trans. Royal Society of Edinburgh*, 1868.]

if m be the part of M within S. This expression is independent of the quantity of matter without S, and if $m = 0$ it becomes

$$\iint P ds = 0 \dots\dots\dots\dots\dots\dots(8).$$

If M be a group of sources of heat in a solid body, P will be the flux across a unit of surface, at the point xyz. Hence the total flux of heat across S is equal to the sum of the expenditures from all the sources in the interior; and if there be no sources in the interior, the whole flux is nothing. Both these results, though our physical ideas of heat would readily lead us to anticipate them, are by no means axiomatic when considered analytically. In exactly a similar manner, Poisson* proves that the total flux of heat out of a body during an instant of time is equal to the sum of the diminutions of heat of each particle of the body, during the same time. This follows at once from (7). For if we suppose there to be no sources of heat within S, but the temperature of interior points to vary with the time, on account of a non-uniform initial distribution of heat, we have

$$\frac{d^2v}{dx^2} + \frac{d^2v}{dy^2} + \frac{d^2v}{dz^2} = \frac{dv}{dt}.$$

Hence, by (d), we must use $-\dfrac{dv}{dt}\, dx dy dz$, instead of $4\pi dm$, and therefore (7) becomes

$$\iint P ds = \iiint \frac{-dv}{dt}\, dx dy dz.$$

It was the analysis used by Poisson, in the demonstration of this theorem, that suggested the demonstrations given in Part I. of propositions (1) and (2).

202.   As another example of the application of the theorem expressed by (3) and (4), let v_1 be the potential of a unit of mass, concentrated at a fixed point, $x'y'z'$. Hence, $M_1 = 1$ and $dm_1 = 0$, except when xyz, at which dm_1 is supposed to be situated, coincides with $x'y'z'$; and, if Δ be the distance of xyz from $x'y'z'$, $v_1 = \dfrac{1}{\Delta}$.

Hence, according as $x'y'z'$ is without or within S,

$$\iiint v\, dm_1 = 0, \text{ or } \iint v\, dm_1 = v' \iiint dm_1 = v' \dots\dots\dots(e),$$

---

* See *Théorie de la Chaleur*, p. 177.

the triple integrals being extended over the space within S. Now let us suppose M to be such, that v has a constant value (v) at S. Then $\iint vP_1 ds = (v)\iint P_1 ds$, which, by (7), is $= 0$, or to $4\pi\,(v)$, according as $x'y'z'$ is without or within S. Hence, by comparing (3) and (4), we have, in the two cases,

$$4\pi\iiint\frac{dm}{\Delta} - \iint\frac{Pds}{\Delta} = 0,\ \text{or}\ \iint\frac{Pds}{\Delta} = 4\pi\iiint\frac{dm}{\Delta} = 4\pi v'\ (9),$$

and $4\pi\iiint\dfrac{dm}{\Delta} - \iint\dfrac{Pds}{\Delta} = -4\pi\,(v) + 4\pi v'$;

therefore $\qquad\qquad \iint\dfrac{Pds}{\Delta} = 4\pi\,(v)$ ....(10).

These are the two propositions (1) and (2) proved in Part I., which are therefore, as we see, particular cases of the general theorem expressed by (3) and (4).*

203. If $v = v_1$, and if both arise from sources situated without S, (3) becomes

$$\iiint R^2 dx dy dz = \iint vP ds \ \ldots\ldots\ldots\ldots(11),$$

a proposition given by Gauss. If v have a constant value (v) over S, we have

$$\iint vP ds = (v)\iint P ds = 0,\ \text{by (8)},$$

hence $\qquad\qquad \iiint R^2 dx dy dz = 0.$

Therefore $R = 0$ and $v = (v)$ for interior points. Hence, if the potential produced by any number of sources have the same value over every point of a surface which contains none of them, it will have the same value for every interior point also. If we consider the sources to be spread over S, it follows that $v = (v)$ at the surface is a condition which implies that the attraction on an interior point will be nothing. Hence the sole condition for the distribution of electricity over a conducting surface, is that its attraction shall be everywhere perpendicular to the surface, a proposition which was proved from indirect considerations, relative to heat, in a former paper.†

---

* It may be here proper to state that these theorems, which were first demonstrated by Gauss, are the subject of a Mémoire by M. Chasles, in the Additions to the *Connaissance des Temps* for 1845, published in June, 1842. In this Mémoire he refers to an announcement of them, without a demonstration, in the *Comptes Rendus des Séances de l'Académie des Sciences*, Feb. 11, 1839, a date earlier than that of M. Gauss's Mémoire, which was read at the Royal Society of Göttingen in March, 1840.

† See I. above, § 5.

204. In exactly a similar manner, if none of the sources be without S, by means of (5) and (7), it may be shown that

$$\iiint R^2 dx dy dz = 4\pi M(v)\dots\dots\dots\dots(12);$$

the triple integrals being extended over all the space without S. Hence a quantity of matter μ can only be distributed in one way on S, so as to make (v) be constant. For if there were two distributions of μ, each making (v) constant, there would be a third, corresponding to their difference, which would also make (v) constant. The whole mass in the third case would be nothing. Hence, by (12), we must have $\iiint R^2 dx dy dz = 0$, and therefore $R = 0$ for external points; and, since (v) is constant at the surface, R must be $= 0$ for interior points also. Now this cannot be the case unless the density at each point of the surface be nothing, on account of the theorem of Laplace, that, if ρ be the density at any point of a stratum which exerts no attraction on interior points, its attraction on an interior point close to the surface will be $4\pi\rho$. This important theorem, which shows that there is only one distribution of electricity on a body that satisfies the condition of equilibrium, was first given by Gauss. It may be readily extended, as has been done by Liouville,[*] to the case of any number of electrified bodies, influencing one another, by supposing S to consist of a number of isolated portions, which will obviously not affect the truth of (5) and (6).

Then, if we suppose v to have the constant values, (v), $(v)'$, etc., at the different surfaces, and the quantities of matter on these surfaces to be M, M', etc., we should have, instead of (11),

$$\iiint R^2 dx dy dz = 4\pi \{M(v) + M'(v)' + \text{etc.}\}\dots\dots(13),$$

and from this it may be shown, as above, that there is only one distribution of the same quantities of matter, M, M', etc., which satisfies the conditions of equilibrium.

205. If both M and M_1 be wholly within S, by comparing (5) and (6), or if both be without S, by comparing (3) and (4), we have

$$\iint P v_1 ds = \iint P_1 v ds\dots\dots\dots\dots\dots(14).$$

---

[*] See Note to M. Chasles' Mémoire in the *Connaissance des Temps* for 1845.

Now let S be a sphere, and let $r\theta\phi$ be the polar co-ordinates, from the centre as pole, of any point in the surface to which the potentials v and v_1 correspond. Then we shall have $P = -\dfrac{dv}{dr}$, $P_1 = -\dfrac{dv_1}{dr}$, and we may assume $ds = r^2 \sin\theta d\theta d\phi$. Hence (14) becomes

$$\int_0^\pi \int_0^{2\pi} v_1 \frac{dv}{dr} \sin\theta d\theta d\phi = \int_0^\pi \int_0^{2\pi} v \frac{dv_1}{dr} \sin\theta d\theta d\phi \dots (15).$$

This equation leads at once to the fundamental property of Laplace's coefficients. For if v and v_1 be of the forms $Y_m r^m$, $Y_n r^n$, m and n being any positive or negative integers, zero included, and Y_m and Y_n being independent of r, we have, by substitution in (15),

$$m \int_0^\pi \int_0^{2\pi} Y_m Y_n \sin\theta d\theta d\phi = n \int_0^\pi \int_0^{2\pi} Y_m Y_n \sin\theta d\theta d\phi.$$

If m be not $= n$, this cannot be satisfied unless

$$\int_0^\pi \int_0^{2\pi} Y_m Y_n \sin\theta d\theta d\phi = 0 \dots (16).$$

This is the* fundamental property of Laplace's coefficients.

There are some other applications of the general theorem which has been established, especially to the Theory of Electricity, which must, however, be left for a future opportunity.

---

* [For a justification of this use of the definite article, see Murphy's *Electricity*, Chap. I. Props. I. and II., Cambridge 1833.]

XIII. THEOREMS WITH REFERENCE TO THE SOLUTION OF CERTAIN PARTIAL DIFFERENTIAL EQUATIONS.

(Art. xxxvi. of complete list in *Mathematical and Physical Papers*, Vol. i.)

[From the *Cambridge and Dublin Mathematical Journal*, Jan. 1848.]

206. *Theorem* 1. It is possible to find a function V, of x, y, z,* which shall satisfy, for all real values of these variables, the differential equation

$$\frac{d\left(\alpha^2\frac{dV}{dx}\right)}{dx} + \frac{d\left(\alpha^2\frac{dV}{dy}\right)}{dy} + \frac{d\left(\alpha^2\frac{dV}{dz}\right)}{dz} = -4\pi\rho\ldots\ldots(A),$$

α being any real continuous or discontinuous function of x, y, z, and ρ a function which vanishes for all values of x, y, z, exceeding certain finite limits (such as may be represented geometrically by a finite closed surface), within which its value is finite, but entirely arbitrary.

Theorem 2. There cannot be two different solutions of equation (A) for all real values of the variables.

1. (*Demonstration*).—Let U be a function of x, y, z, given by the equation

$$U = \iiint \frac{\rho'dx'dy'dz'}{\{(x-x')^2 + (y-y')^2 + (z-z')^2\}^{\frac{1}{2}}} \ldots\ldots(a),$$

the integrations in the second member including all the space for which ρ' is finite; so that, if we please, we may conceive the limits of each integration to be $-\infty$ and $+\infty$, as thus all the values of the variables for which ρ' is finite will be included, and the amount of the integral will not be affected by those values of the variables for which ρ' vanishes, being included. Again, V being any real function of x, y, z, let

---

* The case of three variables, which includes the applications to physical problems, is alone considered here; although the analysis is equally applicable whatever be the number of variables.

$$Q = \int_{-\infty}^{\infty}\int_{-\infty}^{\infty}\int_{-\infty}^{\infty} \left\{ \left(\alpha\frac{dV}{dx} - \frac{1}{\alpha}\frac{dU}{dx}\right)^2 + \left(\alpha\frac{dV}{dy} - \frac{1}{\alpha}\frac{dU}{dy}\right)^2 \right.$$
$$\left. + \left(\alpha\frac{dV}{dz} - \frac{1}{\alpha}\frac{dU}{dz}\right)^2 \right\} dxdydz \ldots(b).$$

It is obvious that, although V may be assigned so as to make Q as great as we please, it is impossible to make the value of Q less than a certain limit, since we see at once that it cannot be negative. Hence Q, considered as depending on the arbitrary function V, is susceptible of a minimum value; and the calculus of variations will lead us to the assigning of V according to this condition.

Thus we have

$$\tfrac{1}{2}\delta Q = \iiint \left\{ \left(\alpha\frac{dV}{dx} - \frac{1}{\alpha}\frac{dU}{dy}\right).\alpha\frac{d\delta V}{dx} + \left(\alpha\frac{dV}{dy} - \frac{1}{\alpha}\frac{dU}{dy}\right).\alpha\frac{d\delta V}{dy} \right.$$
$$\left. + \left(\alpha\frac{dV}{dz} - \frac{1}{\alpha}\frac{dU}{dz}\right).\alpha\frac{d\delta V}{dz} \right\} dxdydz.$$

Hence, by the ordinary process of integration by parts, the integrated terms vanishing at each limit,[*] we deduce

$$-\tfrac{1}{2}\delta Q = \iiint \delta V \left\{ \frac{d}{dx}\left(\alpha^2\frac{dV}{dx} - \frac{dU}{dx}\right) + \frac{d}{dy}\left(\alpha^2\frac{dV}{dy} - \frac{dU}{dy}\right) \right.$$
$$\left. + \frac{d}{dz}\left(\alpha^2\frac{dV}{dz} - \frac{dU}{dz}\right) \right\} dxdydz.$$

But by a well-known theorem (proved in Pratt's *Mechanics*, and in the treatise on Attraction in Earnshaw's *Dynamics*), we have
$$\frac{d^2U}{dx^2} + \frac{d^2U}{dy^2} + \frac{d^2U}{dz^2} = -4\pi\rho.$$

Hence the preceding expression becomes

$$-\tfrac{1}{2}\delta Q = \iiint \delta V . \left\{ \frac{d}{dx}\left(\alpha^2\frac{dV}{dx}\right) + \frac{d}{dy}\left(\alpha^2\frac{dV}{dy}\right) \right.$$
$$\left. + \frac{d}{dz}\left(\alpha^2\frac{dV}{dz}\right) + 4\pi\rho \right\} dxdydz.$$

We have, therefore, for the condition that Q may be a maximum or minimum, the equation,

$$\frac{d}{dx}\left(\alpha^2\frac{dV}{dx}\right) + \frac{d}{dy}\left(\alpha^2\frac{dV}{dy}\right) + \frac{d}{dz}\left(\alpha^2\frac{dV}{dz}\right) = -4\pi\rho,$$

to be satisfied for all values of the variables.

---

[*] All the functions of x, y, z contemplated in this paper are supposed to vanish for infinite values of the variables.

Now it is possible to assign V so that Q may be a minimum, and therefore there exists a function, V, which satisfies equation (A).

2. (*Demonstration*).—Let V be a solution of (A), and let V_1 be any different function of x, y, z, that is to say, any function such that $V_1 - V$, which we may denote by ϕ, does not vanish for all values of x, y, z. Let us consider the integral Q_1, obtained by substituting V_1 for V in the expression for Q. Since

$$\left(\alpha\frac{dV_1}{dx} - \frac{1}{\alpha}\frac{dU}{dx}\right)^2 = \left(\alpha\frac{dV}{dx} - \frac{1}{\alpha}\frac{dU}{dx}\right)^2 + 2\left(\alpha\frac{dV}{dx} - \frac{1}{\alpha}\frac{dU}{dx}\right)\alpha\frac{d\phi}{dx} + \alpha^2\frac{d\phi^2}{dx^2},$$

we have

$$Q_1 = Q + 2\iiint\left\{\left(\alpha\frac{dV}{dx} - \frac{1}{\alpha}\frac{dU}{dx}\right)\alpha\frac{d\phi}{dx} + \left(\alpha\frac{dV}{dy} - \frac{1}{\alpha}\frac{dU}{dy}\right)\alpha\frac{d\phi}{dy}\right.$$
$$\left. + \left(\alpha\frac{dV}{dz} - \frac{1}{\alpha}\frac{d\phi}{dz}\right)\alpha\frac{d\phi}{dz}\right\}dxdydz$$
$$+ \iiint\alpha^2\left(\frac{d\phi^2}{dx^2} + \frac{d\phi^2}{dy^2} + \frac{d\phi^2}{dz^2}\right)dxdydz.$$

Now, by integration by parts, we find

$$\int_{-\infty}^{\infty}\int_{-\infty}^{\infty}\int_{-\infty}^{\infty}\left(\alpha\frac{dV}{dx} - \frac{1}{\alpha}\frac{dU}{dx}\right)\alpha\frac{d\phi}{dx}.dxdydz$$
$$= -\iiint\phi.\frac{d}{dx}\left(\alpha^2\frac{dV}{dx} - \frac{dU}{dx}\right)dxdydz,$$

the integrated term vanishing at each limit. Applying this and similar processes with reference to y and z, we find an expression for the second term of Q_1, which, on account of equation (A), vanishes. Hence

$$Q_1 = Q + \iiint\alpha^2\left(\frac{d\phi^2}{dx^2} + \frac{d\phi^2}{dy^2} + \frac{d\phi^2}{dz^2}\right)dxdydz\ldots\ldots(c),$$

which shows that Q_1 is greater than Q. Now the only peculiarity of Q is, that V, from which it is obtained, satisfies the equation (A), and therefore V_1 cannot be a solution of (A). Hence no function different from V can be a solution of (A).

The analysis given above, especially when interpreted in various cases of abrupt variations in the value of α, and of infinite or evanescent values, through finite spaces, possesses very important applications in the theories of heat, electricity, magnetism, and hydrodynamics, which may form the subject of future communications.

EDINBARNET, DUMBARTONSHIRE, *Oct.* 9, 1847.

[From Liouville's *Journal de Mathématiques*, 1847.]

207. Dans les applications qui présentent le plus d'intérêt, il faut considérer des transitions subites dans la valeur de α. Par exemple, si α a une valeur constante dans tout l'espace extérieur à une surface fermée S, dans l'intérieur de laquelle α est infinie, notre analyse convient au cas d'un corps conducteur S soumis à l'influence d'une masse électrique donnée ($\iiint \rho dx dy dz$), et cette application ne présente aucune difficulté. On en tire, en effet, les démonstrations données par Green, que la solution analytique du problème de la distribution d'électricité dans ces circonstances est possible et qu'elle est unique.

Dans une application à l'hydrodynamique, ou à un certain problème de magnétisme, il faut considérer un éspace dans lequel la valeur de α soit zéro. L'interprétation du résultat ne présente aucune difficulté, mais il est plus difficile de bien comprendre comment la démonstration telle que je l'ai donnée plus haut se prête à ce cas. En essayant de l'expliquer nettement, j'ai trouvé une démonstration directe du théorème suivant, qui renferme le résultat dont il s'agit :

"Il est possible de trouver une fonction V qui s'évanouisse pour les valeurs infiniment grandes des variables x, y, z, et satisfasse à l'équation

$$\frac{d^2V}{dx^2} + \frac{d^2V}{dy^2} + \frac{d^2V}{dz^2} = 0,$$

pour tous les points extérieurs à une surface fermée S, avec cette condition

$$\frac{dV}{dn} = F,$$

dans laquelle F est une fonction arbitraire des coordonnées d'un point sur la surface S, et dn est l'élément d'une normale extérieure à la surface en ce point."

Pour le démontrer, considérons l'intégrale

$$\iiint \left[\left(\frac{dV}{dx} \right)^2 + \left(\frac{dV}{dy} \right)^2 + \left(\frac{dV}{dz} \right)^2 \right] dx dy dz = Q,$$

relative à l'espace extérieur à *S*.   Parmi toutes les fonctions *V* qui vérifient la condition

$$\iint V F dS = A,$$

où *A* est une quantité quelconque, il y en a une pour laquelle l'intégrale *Q* est un minimum.   Une fonction *V*, ainsi déterminée, satisfait aux équations

$$\frac{d^2 V}{dx^2} + \frac{d^2 V}{dy^2} + \frac{d^2 V}{dz^2} = 0,$$

$$\frac{d V}{dn} = cF$$

(où *c* est une constante), comme on s'en assure par le calcul des variations.   Suivant les valeurs de *A*, *c* aura des valeurs proportionnelles; on peut prendre *A* telle que *c* = 1.   De là on conclut le théorème énoncé.   Il serait facile d'ajouter une démonstration, que la solution du problème de la détermination de *V* sous ces conditions est unique.*

---

* [Provided *S* is a simply continuous surface.   If *S* be a multiply continuous surface, as, for instance, the inner boundary of an endless tube (a finite tube with its ends united, so as to constitute a circuit), we may add to *V* the velocity-potential of a liquid moving through it irrotationally (Thomson and Tait's *Natural Philosophy*, §§ 184—190 ; Thomson, *Vortex Motion*, §§ 54...58) without violating the conditions prescribed in the text.   Compare above, § 200, footnote.]

EXTRAIT D'UNE LETTRE DE M. WILLIAM THOMSON
À M. LIOUVILLE.

(Art. xix. of complete list in *Mathematical and Physical Papers*, Vol. i.)

[From Liouville's *Journal de Mathématiques*, 1845.]

" CAMBRIDGE, 8 *Octobre* 1845.

208.   "...Pendant mon séjour à Paris, je vous ai parlé du principe des *images* pour la solution de quelques problèmes relatifs à la distribution de l'électricité.   Il y a une foule de problèmes auxquels je ne pensais pas alors, et où j'ai trouvé plus tard qu'on peut l'appliquer.   Par exemple, on parvient ainsi à exprimer algébriquement la distribution d'électricité sur deux plans conducteurs qui se coupent sous un angle $\frac{\pi}{i}$, quand un point électrique est posé dans l'espace entre les deux plans.   (L'idée est analogue à celle du *kaleidoscope* de Brewster.)   Quand il y a trois plans qui se coupent perpendiculairement, ou quand il y a un plan qui coupe perpendiculairement deux plans qui se coupent sous un angle $\frac{\pi}{i}$, on peut également trouver la distribution sous l'influence d'un point électrique donné.   On peut aussi exprimer très-facilement la distribution sur les parois intérieures d'un parallélipipède rectangulaire creux, soumis à l'influence d'un point électrique posé en dedans, en se servant des intégrales définies.

"Soient C le centre d'une sphère S; Q, Q' deux points pris sur un même rayon CA et sur son prolongement, de telle manière que $\qquad CQ \cdot CQ' = CA^2;$
et P un point quelconque sur la surface S.   On a, comme on sait,

$$\frac{PQ}{PQ'} = \frac{AQ}{AQ'}.$$

On peut, à cause de ce théorème, appeler Q et Q' *points réciproques relatifs à la sphère S*, dont chacun est *l'image* de l'autre dans la sphère. Suivant cette définition, l'image d'une ligne ou surface sera le lieu des images de points pris sur cette ligne ou surface. Ainsi, on trouve que l'image d'un plan ou d'une sphère est toujours une sphère (le plan étant compris sous cette désignation). Les images de deux sphères se coupent sous le même angle, réel ou imaginaire, que les surfaces données.

"Soient Q, Q' deux points réciproques, relativement à une sphère S, et q, q', s leurs images et l'image de la sphère S dans une autre sphère donnée. Les points q, q' seront réciproques relativement à la sphère s.

209. "A l'aide de ces théorèmes, je parviens facilement à déterminer les images successives d'un point quelconque (qui n'est pas nécessairement dans la ligne qui passe par leurs centres), dans deux sphères qui se coupent sous un angle donné. Quand cet angle est imaginaire, je parviens ainsi à exprimer la distribution de l'électricité sur les deux sphères, sous l'influence d'un point quelconque, chargé d'électricité, au moyen des séries de M. Poisson (qui convergent comme des séries géométriques). Quand l'angle d'intersection est réel et compris dans l'expression $\frac{\pi}{i}$, on parvient ainsi à exprimer algébriquement la distribution d'une quantité donnée d'électricité sur la surface extérieure des sphères, qui n'est soumise à aucune influence ou qui l'est à celle d'un point donné. S'il y a trois surfaces sphériques qui se coupent perpendiculairement, on exprime algébriquement, par les mêmes principes, la distribution sur la surface extérieure. Je parviens aussi à déterminer les températures stationnaires dans l'intérieur d'une lentille dont les deux surfaces se coupent sous un angle $\frac{\pi}{i}$, la température de chaque point de ces surfaces étant donnée.

210. "Si l'on veut déterminer la distribution d'électricité sur une surface donnée S, sous l'influence d'un point quelconque Q, on réduit, par les mêmes principes, le problème à la détermination de la distribution, sans aucune influence, sur l'image de S dans une sphère décrite du centre Q, avec un rayon quelconque. Une application générale de ce théorème conduit à une démonstration rigoureuse du théorème de M. Gauss, qu'on peut produire, au moyen d'une distribution déterminée de matière sur

une surface fermée quelconque, une valeur donnée du potentiel à chaque point de la surface. Il y a aussi beaucoup d'applications spéciales [see below, §§ 218...220] qu'on peut faire de ce théorème aux cas dans lesquels *S* est une sphère, un disque circulaire, ou un segment d'une surface sphérique fait par un plan. J'en ai aussi déduit une démonstration géométrique du théorème que vous avez publié dans le numéro d'avril 1845 de votre Journal (*voir* page 137), dont voici l'expression analytique * * *" [see above, XI. §§ 167, 186].

EXTRAITS DE DEUX LETTRES ADRESSÉES À M. LIOUVILLE, PAR M. WILLIAM THOMSON.

[From Liouville's *Journal de Mathématiques*, 1847.]

"Cambridge, 26 *juin* 1846.

211. "...Les recherches sur lesquelles je vous ai écrit, le 8 octobre 1845, m'ont conduit à l'emploi d'un système nouveau de coordonnées orthogonales très-commode dans quelques problèmes des théories de la chaleur et de l'électricité. Les surfaces coordonnées dans ce système sont les surfaces engendrées par la rotation, autour d'un axe convenable, d'un système de coordonnées curvilignes dans un plan, et les plans méridiens. En effet, soit *M* un plan méridien quelconque ; les coordonnées d'un point *P* dans ce plan sont deux cercles qui se coupent à angle droit en ce point, et dont le premier passe par deux points fixes *A*, *A'*, dans l'axe de révolution *X'X*, tandis que le second est la courbe orthogonale de la série entière des cercles qui passent par les points *A*, *A'*. On démontre facilement que cette courbe est un cercle qui passe par deux points imaginaires *B*, *B'*, dans la droite *Y'OY* perpendiculaire à *X'OX*, à des distances aux deux côtés de *O* dont chacune est égale à $a\sqrt{-1}$, *a* étant la valeur des distances égales *A'O*, *OA*. En effet, la première série est exprimée par l'équation

$$(1) \qquad x^2 + y^2 - 2uy = a^2,$$

u étant un paramètre variable, et l'on en déduit

$$(2) \qquad x^2 + y^2 - 2vx = -a^2,$$

pour l'équation de la courbe orthogonale.

212. " Posons

$$u = a \cot \theta, \quad v = a \sqrt{-1} \cdot \cot \psi ;$$

θ sera l'angle que la tangente du cercle (1), au point A ou A', fait avec l'axe $X'X$, et ψ sera l'angle imaginaire que la tangente du cercle (2), au point B ou B', fait avec $Y'Y$. Pour avoir la série entière des cercles (1), il faudrait donner à u toutes les valeurs réelles de $-\infty$ à ∞, ou à θ toutes les valeurs de 0 à π; et, pour la série (2), il faudrait donner à v toutes les valeurs de a à ∞, et de $-\infty$ à $-a$. On peut considérer un point P comme déterminé sans ambiguïté par les coordonnées θ, ψ (en prenant $\theta + \pi$ au lieu de θ pour l'autre point d'intersection des mêmes cercles). Les équations de transformation, entre les coordonneés (x, y) et (θ, ψ) d'un même point P, sont

(3) $\qquad x^2 + y^2 - 2ay \cot \theta = a^2,$

(4) $\qquad x^2 + y^2 - 2ax \cot \psi \sqrt{-1} = -a^2.$

On en déduit

$$x = -a \frac{\sin \psi \sqrt{-1}}{\cos \psi - \cos \theta},$$

$$y = a \frac{\sin \theta}{\cos \psi - \cos \theta},$$

$$x^2 + y^2 = a^2 \frac{\cos \psi + \cos \theta}{\cos \psi - \cos \theta}.$$

Dans les applications physiques, il s'agit d'exprimer la distance Δ, entre deux points P, P' en fonction des nouvelles coordonnées. On trouve facilement, à l'aide des formules données ci-dessus, dans le cas de P et P' dans un même plan méridien M,

$$\Delta^2 = 2a^2 \frac{\cos(\psi - \psi') - \cos(\theta - \theta')}{(\cos \psi - \cos \theta)(\cos \psi' - \cos \theta')}.$$

Pour le trois coordonnées d'un point dans l'espace, je prends θ, ψ qui fixent sa position dans un plan méridien, et l'angle ϕ que ce plan fait avec un plan méridien fixe. Je trouve maintenant, pour la distance entre deux points quelconques P, P',

$$\Delta^2 = 2a^2 \frac{\cos(\psi - \psi') - [\cos \theta \cos \theta' + \sin \theta \sin \theta' \cos(\phi - \phi')]}{(\cos \psi - \cos \theta)(\cos \psi' - \cos \theta')}.$$

Pour éviter l'emploi de quantités imaginaires, je pose

$$2 \cos \psi = r + \frac{1}{r}, \quad 2 \cos \psi' = r' + \frac{1}{r'},$$

10—2

d'où l'on déduit

$$2\cos(\psi - \psi') = \frac{r}{r'} + \frac{r'}{r},$$

et l'expression précédente se réduit à

$$\Delta^2 = a^2 \frac{r^2 - 2rr'\left[\cos\theta\cos\theta' + \sin\theta\sin\theta'\cos(\phi - \phi')\right] + r'^2}{(r^2 - 2r\cos\theta + 1)(r'^2 - 2r'\cos\theta' + 1)}.$$

A l'aide de cette expression, on trouve

$$r\frac{d^2(s^{-1}vr)}{dr^2} + \frac{1}{\sin\theta}\frac{d\left[\sin\theta\dfrac{d(s^{-1}v)}{d\theta}\right]}{d\theta} + \frac{1}{\sin^2\theta}\frac{d^2(s^{-1}v)}{d\phi^2} = 0,$$

où
$$s = (r^2 - 2r\cos\theta + 1)^{\frac{1}{2}},$$

pour l'équation du mouvement uniforme de la chaleur exprimée par les coordonnées r, θ, ϕ.

" Les surfaces représentées par l'équation

$$r = \text{constante}$$

sont des sphères engendrées par la révolution d'une série de cercles autour de la droite qui contient leurs centres. Supposons que l'espace entre deux de ces sphères (quand chaque sphère est en dehors de l'autre, cet espace sera l'espace infini en dehors des deux sphères), dont les équations sont

$$r = \alpha, \quad r = \alpha_1,$$

soit rempli d'un milieu solide homogène, que les températures de tous les points de chaque surface soient données, et qu'il s'agisse de déterminer la température stationnaire d'un point quelconque dans le solide; on résoudra ce problème avec beaucoup de facilité au moyen de l'analyse de Laplace, en employant les coordonnées que j'ai indiquées. Dans le cas particulier d'une température constante pour chaque sphère, on parvient, après quelques réductions, à trouver la solution que Poisson a donnée pour le problème correspondant de deux sphères électrisées.

213. " Il y a un système nouveau et très-remarquable de coordonnées, qu'on trouve en posant

$$r\cos\theta = \xi, \quad r\sin\theta\cos\phi = \eta, \quad r\sin\theta\sin\phi = \zeta,$$

r, θ, ϕ appartenant au système expliqué ci-dessus. Dans ce système (ξ, η, ζ), les surfaces coordonnées sont des sphères orthogonales qui passent par un point fixe, et qui touchent, par conséquent, trois plans orthogonaux menés par ce point. Je suis parvenu à considérer ces systèmes de coordonnées en cherchant les *images* des séries de surfaces des systèmes (polaire et rectangulaire) ordinaires, dans des sphères convenablement disposées.

" L'application du système (ξ, η, ζ) aux problèmes de physique, pour le cas de deux systèmes qui se touchent l'un l'autre, en donne les solutions avec beaucoup de facilité ; mais il est plus simple de faire directement la recherche de ces coordonnées, que de les déduire du système (r, θ, ϕ). En effet, soient

$$x^2 + y^2 + z^2 - \frac{x}{\xi} = 0,$$

$$x^2 + y^2 + z^2 - \frac{y}{\eta} = 0,$$

$$x^2 + y^2 + z^2 - \frac{z}{\zeta} = 0,$$

les équations de trois sphères qui se coupent à un point P (elles se coupent aussi à l'origine O). Je prends ξ, η, ζ pour les coordonnées de ce point (il faudrait substituer $\dfrac{\xi - 1}{a}, \dfrac{\eta}{a}, \dfrac{\zeta}{a}$ dans ces équations, au lieu de ξ, η, ζ, pour retrouver les coordonnées ξ, η, ζ indiquées ci-dessus). De ces équations on tire

$$x^2 + y^2 + z^2 = \frac{1}{\xi^2 + \eta^2 + \zeta^2},$$

$$x = \frac{\xi}{\xi^2 + \eta^2 + \zeta^2}, \quad y = \frac{\eta}{\xi^2 + \eta^2 + \zeta^2}, \quad z = \frac{\zeta}{\xi^2 + \eta^2 + \zeta^2},$$

$$(x - x')^2 + (y - y')^2 + (z - z')^2 = \frac{(\xi - \xi')^2 + (\eta - \eta')^2 + (\zeta - \zeta')^2}{(\xi^2 + \eta^2 + \zeta^2)(\xi'^2 + \eta'^2 + \zeta'^2)},$$

et l'équation

$$\frac{d^2 v}{dx^2} + \frac{d^2 v}{dy^2} + \frac{d^2 v}{dz^2} = 0$$

devient, pour les nouvelles coordonnées,

$$(a) \qquad \frac{d^2 (\rho^{-1} v)}{d\xi^2} + \frac{d^2 (\rho^{-1} v)}{d\eta^2} + \frac{d^2 (\rho^{-1} v)}{d\zeta^2} = 0,$$

où

$$\rho = (\xi^2 + \eta^2 + \zeta^2)^{\frac{1}{2}}.$$

Pour exemple de l'emploi qu'on peut faire de ce système de coordonnées, supposons que la température d'un point (a, η, ζ) est une fonction donnée $F(\eta, \zeta)$ des coordonnées η, ζ de sa position sur la sphère α, et que la température d'un point (α_1, η, ζ) est $F_1(\eta, \zeta)$, et qu'il s'agit de déterminer la température permanente d'un point quelconque P (ξ, η, ζ) dans l'espace entre les sphères α, α_1 (c'est-à-dire l'espace entier pour lequel ξ a une valeur intermédiaire à α et α_1), que nous supposerons rempli d'un solide homogène. Suivant la méthode de Fourier, en observant que les valeurs

$$\cos m\eta . \cos \eta\zeta . \epsilon^{h\xi},$$
$$\cos m\eta . \sin \eta\zeta . \epsilon^{h\xi},$$
$$\dotsc\dotsc\dotsc\dotsc\dotsc$$

substituées pour $\rho^{-1}v$, sont des solutions particulières de l'équation (a), pourvu que $h^2 = m^2 + n^2$, je trouve, pour la solution du problème proposé,

$$(b) \begin{cases} v = \dfrac{\rho}{4\pi^2} \displaystyle\int_{-\infty}^{\infty}\int_{-\infty}^{\infty} dm\,dn \int_{-\infty}^{\infty}\int_{-\infty}^{\infty} d\eta'd\zeta' \dfrac{\cos m(\eta - \eta')\cos n(\zeta - \zeta')}{\epsilon^{h(a-a_1)} - \epsilon^{-h(a-a_1)}} \\[2ex] \times \begin{cases} \dfrac{F(\eta', \zeta')}{(a^2 + \eta'^2 + \zeta'^2)^{\frac{1}{2}}} \left[\epsilon^{h(\xi - a_1)} - \epsilon^{-h(\xi - a_1)} \right] \\[2ex] - \dfrac{F_1(\eta', \zeta')}{(a_1^2 + \eta'^2 + \zeta'^2)^{\frac{1}{2}}} \left[\epsilon^{h(\xi - a)} - \epsilon^{-h(\xi - a)} \right] \end{cases} \end{cases},$$

où ϵ est la base des logarithmes népériens, et

$$h = (m^2 + n^2)^{\frac{1}{2}}.$$

214. "Comme exemple de l'usage de cette formule, je ferai

$$F(\eta, \zeta) = F_1(\eta, \zeta) = V,$$

V étant une constante. Pour la réduction de l'expression, dans ce cas, j'observe que

$$(c) \qquad \int_{-\infty}^{\infty}\int_{-\infty}^{\infty} dp\,dq \frac{\cos mp \cos nq}{(\lambda + p^2 + q^2)^{\frac{1}{2}}} = 2\pi \frac{\epsilon^{-(m^2 + n^2)^{\frac{1}{2}}\lambda^{\frac{1}{2}}}}{(m^2 + n^2)^{\frac{1}{2}}}$$

d'où l'on déduit

$$\int_{-\infty}^{\infty}\int_{-\infty}^{\infty} d\eta'd\zeta' \frac{\cos m(\eta - \eta')\cos(\zeta - \zeta')}{(a^2 + \eta'^2 + \zeta'^2)^{\frac{1}{2}}} = 2\pi \cos m\eta . \cos n\zeta . \frac{\epsilon^{-ha}}{h},$$

et

$$\int_{-\infty}^{\infty}\int_{-\infty}^{\infty} d\eta'd\zeta' \frac{\cos m(\eta - \eta')\cos n(\zeta - \zeta')}{(a_1^2 + \eta'^2 + \zeta'^2)^{\frac{1}{2}}} = 2\pi \cos m\eta . \cos n\zeta . \frac{\epsilon^{\mp ha_1}}{h}$$

le signe supérieur ou inférieur étant pris, dans la seconde expression, selon que α_1 est positif ou négatif (je prends α toujours positif et $> \alpha_1$). Ces réductions faites, l'expression (b) se trouve réduite à

(I) $\qquad v = \dfrac{V\rho}{2\pi} \displaystyle\int_{-\infty}^{\infty} \int_{-\infty}^{\infty} dm\,dn \cos m\eta \,.\, \cos n\zeta \,.\, \dfrac{\epsilon^{-(m^2+n^2)^{\frac12}\xi}}{(m^2+n^2)^{\frac12}}$

et

(II) $\qquad \begin{cases} v = \dfrac{V\rho}{2\pi} \displaystyle\int_{-\infty}^{\infty} \int_{-\infty}^{\infty} dm\,dn \cos m\eta \cos n\zeta \\[2mm] \times \dfrac{\left(\epsilon^{-ha_1} - \epsilon^{ha_1}\right)\epsilon^{h(\xi-a)} + \left(\epsilon^{-ha} - \epsilon_{ha}\right)\epsilon^{h(\xi-a_1)}}{h\left[\epsilon^{h(a-a_1)} - \epsilon^{-h(a-a_1)}\right]} \end{cases}$

suivant les deux cas. L'équation (I) se réduit à

$$v = V,$$

à cause de la valeur qu'on trouve pour l'intégrale définie qui y est contenue*.

215. "L'expression pour v, dans le second cas, se trouve réduite en série convergente, si l'on substitue pour

$$\frac{1}{\epsilon^{h(a-a_1)} - \epsilon^{-h(a-a_1)}} \quad \text{la série}$$

$$\epsilon^{-h(a-a_1)} \left[1 + \epsilon^{-2h(a-a_1)} + \epsilon^{-4h(a-a_1)} + \dots\right],$$

et puis, pour chaque terme, sa valeur, suivant la formule citée dans le cas (I). On trouve ainsi

$$v = V\rho \left\{ \begin{aligned} &\frac{1}{[(2a-\xi)^2+\eta^2+\zeta^2]^{\frac12}} + \frac{1}{[(\gamma+2a-\xi)^2+\eta^2+\zeta^2]^{\frac12}} + \frac{1}{[(2\gamma+2a-\xi)^2+\eta^2+\zeta^2]^{\frac12}} + \dots\dots \\ &- \frac{1}{[(\gamma-\xi)^2+\eta^2+\zeta^2]^{\frac12}} - \frac{1}{[(2\gamma-\xi)^2+\eta^2+\zeta^2]^{\frac12}} - \frac{1}{[(3\gamma-\xi)^2+\eta^2+\zeta^2]^{\frac12}} - \dots\dots \\ &+ \frac{1}{[(\xi-2a_1)^2+\eta^2+\zeta^2]^{\frac12}} + \frac{1}{[(\gamma+\xi-2a_1)^2+\eta^2+\zeta^2]^{\frac12}} + \frac{1}{[(2\gamma+\xi-2a_1)^2+\eta^2+\zeta^2]^{\frac12}} + \dots\dots \\ &- \frac{1}{[(\gamma+\xi)^2+\eta^2+\xi^2]^{\frac12}} - \frac{1}{[(2\gamma+\xi)^2+\eta^2+\zeta^2]^{\frac12}} - \frac{1}{[(3\gamma+\xi)^2+\eta^2+\zeta^2]^{\frac12}} - \dots\dots \end{aligned} \right\},$$

où $\qquad\qquad\qquad\qquad \gamma = 2(a - a_1).$

---

* Les intégrales définies (c) et (I) sont des cas particuliers de deux intégrales multiples dont j'ai trouvé les valeurs en cherchant une démonstration de la formule (5), tome X de votre Journal, page 141. J'ai trouvé, [above, § 182, formula (V)], en effet,

$$\int_{-\infty}^{\infty} \int_{-\infty}^{\infty} \dots \frac{dp_1\,dp_2\dots dh_n \cos m_1 p_1 \cos m_2 p_2 \dots}{(p_1^2 + p_2^2 + \dots + p_n^2 + u^2)^{\frac{n-1}{2}}} = \frac{(n-1)\pi^{\frac{n-1}{2}}}{\Gamma\left(\frac{n+1}{2}\right)} \frac{\epsilon^{-(m_1^2+m_2^2+\dots)^{\frac12}u^{\frac12}}}{(m_1^2+m_2^2+\dots)^{\frac12}}$$

et

$$\int_{-\infty}^{\infty} \int_{-\infty}^{\infty} \dots \frac{dm_1\,dm_2 \dots \cos m_1 x_1 \cos m_2 x_2 \dots \epsilon^{-(m_1^2+m_2^2+\dots)^{\frac12}u}}{(m_1^2+m_2^2\dots)^{\frac12}}$$

$$= \frac{2^n \pi^{\frac{n-1}{2}} \Gamma\left(\frac{n+1}{2}\right)}{(n-1)(x_1^2 + x_2^2 + \dots + x_n + u^2)^{\frac{n-1}{2}}}$$

d'où l'on déduit immédiatement les intégrales citées.

De cette expression on déduit facilement la distribution d'électricité sur deux sphères qui se touchent.

216. "Le cas (I) correspond à deux sphères dont l'une, (a), est en dedans de l'autre, (a_1). Dans le cas (II), le solide considéré remplit l'espace entier en dehors des deux sphères, et la température est zéro à une distance infinie.

217. "Il y a une interprétation pour le nouveau système de coordonnées (r, θ) dans un plan, qui est très-simple. En effet, soient A, A' deux points fixes, et P un point quelconque dont il s'agit d'exprimer la position. Cela peut se faire au moyen de l'angle APA', que j'appelle θ, et de la raison r de AP à AP'. Quand θ a une valeur constante, le lieu de P est un cercle qui passe par les points A, A'; et quand r a une valeur constante, le lieu de P est un cercle, dont le centre est dans le prolongement de AA', d'un côté ou de l'autre, suivant que cette valeur est plus grande ou plus petite que l'unité, et qui a la propriété de couper à angle droit tout cercle décrit par les points A, A'.

"Posons maintenant, pour expliquer le second système,

$$r \cos \theta = \xi, \ r \sin \theta = \eta.$$

Le lieu de P, quand ξ a une valeur constante, sera tel que, si l'on mène, de A, AD perpendiculaire à $A'P$, la raison $DP \div AP$ sera constante, et l'on trouve ainsi que ce lieu est un cercle qui touche en A' une droite perpendiculaire à $A'A$; et l'on trouve semblablement que le lieu de P, quand η a une valeur constante, est un cercle qui touche $A'A$ au point A'."

"Knock, le 16 septembre 1846.

218. "...Depuis que je vous ai écrit la dernière fois, j'ai considéré le problème de la distribution d'électricité sur le segment d'une couche sphérique infiniment mince, fait par un plan, ce corps étant composé de matière conductrice, et j'ai trouvé, en expression finie, la solution complète, en supposant que le corps possède une quantité donnée d'électricité et que la distribution se fait sous l'influence de masses électriques données. J'avais l'intention de rédiger de suite pour vous un petit Mémoire sur ces recherches, mais j'ai rencontré quelque difficulté dans l'exposition de la méthode suivie, et comme je suis à présent très-occupé (les cours à Glasgow commencent le 1er novembre, et il me faudra beaucoup de

préparation), il me faut différer cette tâche*. Je me bornerai pour le moment aux énoncés de quelques-uns des résultats.

219. "Soit S le corps conducteur sur lequel il s'agit de déterminer la distribution. Pour premier cas, soit Q un point en dehors de S, sur la même surface sphérique dont S fait partie, et supposons que S soit mis en communication avec le sol par un fil conducteur infiniment mince (ainsi le potentiel dans S sera toujours zéro, quels que soient les corps électrisés qui en soient voisins). Il s'agit de déterminer la distribution d'électricité sur S sous l'influence d'une quantité donnée d'électricité négative Q, concentrée au point Q. Je démontre que l'intensité d'électricité a la même valeur aux points voisins des deux côtés de la couche S, et, en dénotant par σ cette valeur, pour un point quelconque P de S, je trouve

$$\sigma = \frac{Q}{2\pi^2} \cdot \frac{(s^2 - a^2)^{\frac{1}{2}}}{(a^2 - r^2)^{\frac{1}{2}} \cdot \Delta^2},$$

où a, s et r sont les distances du bord de S, du point Q et du point P, à un point C de S qu'on peut appeler son centre, et Δ est la distance entre Q et P. Il est remarquable que cette expression ne contient pas le rayon de la sphère dont S fait partie. En supposant que ce rayon soit infini, on a l'expression pour la distribution d'électricité sur un disque circulaire, sous l'influence d'un point dans son plan, qui est, en effet, la même que celle que Green a donnée pour ce cas.

220. "Pour trouver la distribution dans le cas de S isolé et électrisé, je remarque que, si la quantité d'électricité sur S est telle que le potentiel qui en résulte a une valeur donnée V, la distribution sur S sera la même que celle qui aurait lieu si S était situé dans l'intérieur d'une couche électrique qui produit le potentiel $- V$, S étant dans l'état d'un corps qui n'est pas isolé. On peut prendre pour cette couche une sphère concentrique avec celle dont S fait partie; en supposant l'excès du rayon de la première sphère sur le rayon de la seconde infiniment petit, on réduit le problème à la détermination de la distribution sur S, sous l'influence d'une distribution donnée d'électricité sur la sphère dont S fait partie, ce corps S n'étant

---

* It has, in fact, been delayed till December 1868 and January 1869. See xv. below.

pas isolé. Ainsi, par intégration, je déduis du résultat donné
ci-dessus les expressions

$$\sigma = \frac{V}{2\pi^2 f}\left[\left(\frac{f^2 - a^2}{a^2 - r^2}\right)^{\frac{1}{2}} - \operatorname{arc\,tang}\left(\frac{f^2 - a^2}{a^2 - r^2}\right)^{\frac{1}{2}}\right] + \frac{V}{2\pi f},$$

$$\sigma' = \frac{V}{2\pi^2 f}\left[\left(\frac{f^2 - a^2}{a^2 - r^2}\right)^{\frac{1}{2}} - \operatorname{arc\,tang}\left(\frac{f^2 - a^2}{a^2 - r^2}\right)^{\frac{1}{2}}\right]$$

(où f est le diamètre de la sphère dont S fait partie), pour les
intensités sur les deux côtés, convexe et concave, de S en un
point P."

NOTE AU SUJET DE L'ARTICLE PRÉCÉDENT ;
PAR J. LIOUVILLE.

221*.　La Lettre de M. Thomson m'a suggéré quelques remarques que je crois devoir présenter ici, parce qu'elles montre-
ront, ce me semble, plus clairement encore toute l'importance
du travail dont le jeune géomètre de Glasgow nous a donné un
extrait rapide.

Nous résoudrons d'abord le problème suivant :

Problème.—Soient x, y,..., z et ξ, η,..., ζ deux groupes contenant un nombre égal ou inégal de variables, les premières
x, y,..., z, indépendantes, les autres ξ, η,..., ζ fonctions des premières, en sorte que

$$\xi = f(x, y,..., z), \quad \eta = F(x, y,..., z),..., \quad \zeta = \phi(x, y,..., z) ;$$

soit encore　　　　　$p = \psi(x, y,..., z).$

Désignons d'ailleurs par ξ', η',..., ζ', p' ce que deviennent les
fonctions ξ, η,..., ζ, p, quand on y remplace x, y,..., z par
x', y',..., z'.　Cela posé, on demande de déterminer les fonctions
f, F,..., ϕ, ψ, de manière à avoir généralement

$$(\xi' - \xi)^2 + (\eta' - \eta)^2 + ... + (\zeta' - \zeta)^2 = \frac{(x' - x)^2 + (y' - y)^2 + ... + (z' - z)^2}{p^2 p'^2}.$$

Pour fixer les idées, nous nous bornerons au cas de trois
variables x, y, z, et de trois variables ξ, η, ζ ; et la question sera
de vérifier l'équation

---

* [The original numbering of M. Liouville's sections has been altered by the
addition of 220, for more convenient reference in the present volume.]

$$(1) \quad (\xi' - \xi)^2 + (\eta' - \eta)^2 + (\zeta' - \zeta)^2 = \frac{(x' - x)^2 + (y' - y)^2 + (z' - z)^2}{p^2 p'^2}.$$

La même méthode réussirait pour deux groupes $x, y, \ldots, z$ et $\xi, \eta, \ldots, \zeta$ quelconques. Il n'y aurait de changement que dans quelques détails, et seulement si le nombre des variables était différent dans les deux groupes. Au surplus, nous n'aurons besoin plus tard que du cas où ce nombre est le même de part et d'autre, et ne surpasse pas trois, ce qui nous permettra d'interpréter géométriquement les résultats de notre analyse.

Donnons à x', y', z' des valeurs particulières x_0, y_0, z_0 à volonté, et représentons par p_0, ξ_0, η_0, ζ_0 les valeurs correspondantes de p', ξ', η', ζ'. L'équation (1) nous donnera

$$p^2 = \frac{(x - x_0)^2 + (y - y_0)^2 + (z - z_0)^2}{p_0^2 [(\xi - \xi_0)^2 + (\eta - \eta_0)^2 + (\zeta - \zeta_0)^2]}.$$

Mais, pour plus de simplicité, nous mettrons partout $\xi + \xi_0$, $\eta + \eta_0$, $\zeta + \zeta_0$, $x + x_0$, $y + y_0$, $z + z_0$, au lieu de ξ, η, ζ, x, y, z, et de même $\xi' + \xi_0$, $x' + x_0$, etc., au lieu de ξ', x', etc., ce qui ne change rien aux différences $\xi' - \xi$, $x' - x$, etc. La valeur de p^2 deviendra

$$p^2 = \frac{x^2 + y^2 + z^2}{p_0^2 (\xi^2 + \eta^2 + \zeta^2)},$$

et l'équation (1) subsistera telle qu'elle est.

En faisant

$$x^2 + y^2 + z^2 = r^2, \quad \xi^2 + \eta^2 + \zeta^2 = \rho^2,$$
$$x'^2 + y'^2 + z'^2 = r'^2, \quad \xi'^2 + \eta'^2 + \zeta'^2 = \rho'^2,$$

on aura

$$p^2 = \frac{r^2}{p_0^2 \rho^2}, \quad p'^2 = \frac{r'^2}{p_0^2 \rho'^2},$$

et en portant ces valeurs dans l'équation (1), on trouvera aisément

$$\frac{1}{\rho^2} + \frac{1}{\rho'^2} - 2\left(\frac{\xi}{\rho^2} \frac{\xi'}{\rho'^2} + \frac{\eta}{\rho^2} \frac{\eta'}{\rho'^2} + \frac{\zeta}{\rho^2} \frac{\zeta'}{\rho'^2}\right)$$
$$= p_0^4 \left[\frac{1}{r^2} + \frac{1}{r'^2} - 2\left(\frac{x}{r^2} \frac{x'}{r'^2} + \frac{y}{r^2} \frac{y'}{r'^2} + \frac{z}{r^2} \frac{z'}{r'^2}\right)\right].$$

Maintenant donnons à x', y', z' quatre systèmes de valeurs connues à volonté, à chacun desquels répondront des valeurs déterminées de r', ξ', η', ζ', ρ', et nous aurons ainsi quatre équations du premier degré qui fourniront les valeurs de

$$\frac{\xi}{\rho^2}, \quad \frac{\eta}{\rho^2}, \quad \frac{\zeta}{\rho^2}, \quad \frac{1}{\rho^2},$$

considérées comme quatre inconnues, en fonction linéaire de

$$\frac{x}{r^2},\ \frac{y}{r^2},\ \frac{z}{r^2},\ \frac{1}{r^2}.$$

En désignant donc par A, B, C, D des constantes, et par P, Q, R, S des polynômes du premier degré en x, y, z, ces valeurs seront de la forme

$$\frac{\xi}{\rho^2}=A+\frac{P}{r^2},\ \ \frac{\eta}{\rho^2}=B+\frac{Q}{r^2},\ \ \frac{\zeta}{\rho^2}=C+\frac{R}{r^2},\ \ \frac{1}{\rho^2}=D+\frac{S}{r^2}.$$

En faisant la somme des carrés des trois premières, on trouve une valeur de $\frac{1}{\rho^2}$ qui doit être égale à celle que donne la quatrième équation. Ainsi les deux fonctions

$$D+\frac{S}{r^2}$$

et $$A^2+B^2+C^2+\frac{2\,(AP+BQ+CR)}{r^2}+\frac{P^2+Q^2+R^2}{r^4}$$

doivent être égales. Mais la première devient une fonction entière quand on la multiplie par r^2. Il faut donc que la seconde le devienne aussi, et que, par conséquent, $P^2+Q^2+R^2$ soit également divisible par r^2. Le quotient ne peut évidemment être qu'une constante, puisque le numérateur et le dénominateur sont du même degré. Soit m^2 cette constante, et

$$P^2+Q^2+R^2=m^2r^2=m^2\,(x^2+y^2+z^2).$$

P, Q, R étant des polynômes du premier degré, je fais
$$P=m\,(ax+by+cz+g),$$
$$Q=m(a'x+b'y+c'z+g'),$$
$$R=m(a''x+b''y+c''z+g''),$$

et j'en conclus par la comparaison des deux membres, d'une part,
$$a^2+a'^2+a''^2=1,\quad ab+a'b'+a''b''=0,$$
$$b^2+b'^2+b''^2=1,\quad ac+a'c'+a''c''=0,$$
$$c^2+c'^2+c''^2=1,\quad bc+b'c'+b''c''=0,$$

équations d'où résultent, comme on sait, les équations inverses
$$a^2+b^2+c^2=1,\quad aa'+bb'+cc'=0,$$
$$a'^2+b'^2+c'^2=1,\quad aa''+bb''+cc''=0,$$
$$a''^2+b''^2+c''^2=1,\quad a'a''+b'b''+c'c''=0;$$

et, d'autre part,
$$ag+a'g'+a''g''=0,\quad cg+c'g'+c''g''=0,$$
$$bg+b'g'+b''g''=0,\quad g^2+g'^2+g''^2=0.$$

Si nous admettions que g, g', g'' sont des constantes réelles, l'équation $g^2 + g'^2 + g''^2 = 0$ nous donnerait $g = 0$, $g' = 0$, $g'' = 0$. Mais, dans tous les cas, on arrivera au même résultat à l'aide des trois précédentes, en ayant égard aux équations de condition entre a, b, c, etc. Pour prouver, par exemple, que $g = 0$, il suffira d'ajouter entre elles les trois équations dont nous parlons après les avoir multipliées par les facteurs respectifs a, b, c. Il nous reste donc

$$P = m\,(ax + by + cz),$$
$$Q = m\,(a'x + b'y + c'z),$$
$$R = m\,(a''x + b''y + c''z),$$

a, b, c, etc., satisfaisant aux équations de condition ci-dessus, les mêmes qu'on rencontre dans la transformation de coordonnées rectangulaires en d'autres rectangulaires aussi. Et comme les équations

$$\frac{\xi}{\rho^2} = A + \frac{P}{r^2}, \quad \frac{\eta}{\rho^2} = B + \frac{Q}{r^2}, \quad \frac{\zeta}{\rho^2} = C + \frac{R}{r^2}$$

donnent

$$\frac{1}{\rho^2} = \left(A + \frac{P}{r^2}\right)^2 + \left(B + \frac{Q}{r^2}\right)^2 + \left(C + \frac{R}{r^2}\right)^2,$$

on en conclut les formules suivantes :

$$\xi = \frac{A + \dfrac{P}{r^2}}{\left(A + \dfrac{P}{r^2}\right)^2 + \left(B + \dfrac{Q}{r^2}\right)^2 + \left(C + \dfrac{R}{r^2}\right)^2},$$

$$\eta = \frac{B + \dfrac{Q}{r^2}}{\left(A + \dfrac{P}{r^2}\right)^2 + \left(B + \dfrac{Q}{r^2}\right)^2 + \left(C + \dfrac{R}{r^2}\right)^2},$$

$$\zeta = \frac{C + \dfrac{R}{r^2}}{\left(A + \dfrac{P}{r^2}\right)^2 + \left(B + \dfrac{Q}{r^2}\right)^2 + \left(C + \dfrac{R}{r^2}\right)^2}.$$

Mais il faut à présent rétablir $\xi - \xi_0$, $\eta - \eta_0$, $\zeta - \zeta_0$ au lieu de ξ, η, ζ, et $x - x_0$, $y - y_0$, $z - z_0$ au lieu de x, y, z. Ce changement fait, on aura les formules les plus générales qui puissent satisfaire à l'équation (1). Nous avons donc le théorème suivant :

Les formules générales qui peuvent satisfaire à l'équation (1) s'obtiendront en posant d'abord

$$X = a\ (x - x_0) + b\ (y - y_0) + c\ (z - z_0),$$
$$Y = a'\ (x - x_0) + b'\ (y - y_0) + c'\ (z - z_0),$$
$$Z = a''(x - x_0) + b''(y - y_0) + c''(z - z_0),$$

les coefficients a, b, etc., vérifiant les équations de condition

$$a^2 + a'^2 + a''^2 = 1, \quad ab + a'b' + a''b'' = 0,$$
$$b^2 + b'^2 + b''^2 = 1, \quad ac + a'c' + a''c'' = 0,$$
$$c^2 + c'^2 + c''^2 = 1, \quad bc + b'c' + b''c'' = 0,$$

puis prenant

$$u = A + \frac{mX}{X^2 + Y^2 + Z^2}, \quad v = B + \frac{mY}{X^2 + Y^2 + Z^2}, \quad w = C + \frac{mZ}{X^2 + Y^2 + Z^2},$$

et enfin

$$\xi - \xi_0 = \frac{u}{u^2 + v^2 + w^2}, \quad \eta - \eta_0 = \frac{v}{u^2 + v^2 + w^2}, \quad \zeta - \zeta_0 = \frac{w}{u^2 + v^2 + w^2}.$$

Réciproquement, on peut démontrer que l'équation (1) est satisfaite de cette manière, et trouver la valeur de p qui convient.

D'abord, des trois dernières formules on conclut facilement

$$(\xi' - \xi)^2 + (\eta' - \eta)^2 + (\zeta' - \zeta)^2 = \frac{(u' - u)^2 + (v' - v)^2 + (w' - w)^2}{(u^2 + v^2 + w^2)\ (u'^2 + v'^2 + w'^2)} :$$

les trois précédentes donnent de même

$$(u' - u)^2 + (v' - v)^2 + (w' - w)^2 = m^2 \frac{(X' - X)^2 + (Y' - Y)^2 + (Z' - Z)^2}{(X^2 + Y^2 + Z^2)\ (X'^2 + Y'^2 + Z'^2)} ;$$

enfin, à cause des équations de condition entre a, b, etc., on trouve

$$(X' - X)^2 + (Y' - Y)^2 + (Z' - Z)^2 = (x' - x)^2 + (y' - y)^2 + (z' - z)^2.$$

Il vient donc, en effet,

$$(\xi' - \xi)^2 + (\eta' - \eta)^2 + (\zeta' - \zeta)^2 = \frac{(x' - x)^2 + (y' - y)^2 + (z' - z)^2}{p^2 p'^2},$$

la valeur de p^2 étant

$$p^2 = \frac{(X^2 + Y^2 + Z^2)\ (u^2 + v^2 + w^2)}{m},$$

valeur qu'on pourra aisément exprimer en x, y, z, en observant que le produit $(X^2 + Y^2 + Z^2)(u^2 + v^2 + w^2)$ est égal à

$$(A^2 + B^2 + C^2)(X^2 + Y^2 + Z^2) + 2AmX + 2BmY + 2CmZ + m^2,$$

et que X, Y, Z sont connus en fonction de x, y, z. La valeur qu'on trouvera ainsi peut se mettre sous la forme

$$mp^2 = (A^2 + B^2 + C^2)\,[(x - x_1)^2 + (y - y_1)^2 + (z - z_1)^2],$$

x_1, y_1, z_1 étant des constantes dont voici les valeurs:

$$x_1 = x_0 - \frac{m\,(Aa + Ba' + Ca'')}{A^2 + B^2 + C^2},$$

$$y_1 = y_0 - \frac{m\,(Ab + Bb' + Cb'')}{A^2 + B^2 + C^2},$$

$$z_1 = z_0 - \frac{m\,(Ac + Bc' + Cc'')}{A^2 + B^2 + C^2}.$$

Si donc nous regardons plus tard x, y, z comme étant les coordonnées rectangulaires d'un point quelconque, on voit que la quantité p sera proportionnelle à la distance de ce point (x, y, z) à un point fixe (x_1, y_1, z_1). Il est aisé aussi de s'assurer que

$$\frac{d^2\frac{1}{p}}{dx^2} + \frac{d^2\frac{1}{p}}{dy^2} + \frac{d^2\frac{1}{p}}{dz^2} = 0.$$

222. Pour avoir explicitement ξ, η, ζ en x, y, z, il suffira de remplacer u, v, w, X, Y, Z par leurs valeurs. La première substitution fournit

$$\xi - \xi_0 = \frac{A\,(\text{X}^2 + \text{Y}^2 + \text{Z}^2) + m\text{X}}{(A^2 + B^2 + C^2)(\text{X}^2 + \text{Y}^2 + \text{Z}^2) + 2Am\text{X} + 2Bm\text{Y} + 2Cm\text{Z} + m^2}.$$

Le dénominateur est précisément la valeur de mp^2 dont on vient de donner l'expression en x, y, z, savoir,

$$mp^2 = (A^2 + B^2 + C^2)\,[(x - x_1)^2 + (y - y_1)^2 + (z - z_1)^2].$$

Il ne reste donc plus qu'à chercher le numérateur. Le calcul deviendra d'ailleurs fort simple si l'on retranche des deux membres la quantité

$$\frac{A}{A^2 + B^2 + C^2},$$

car alors le second membre pourra se réduire à une fraction ayant pour numérateur un polynôme du premier degré en X, Y, Z, et, par conséquent aussi, en x, y, z. En désignant donc par X un tel polynôme, et posant, pour abréger,

$$\xi_0 + \frac{A}{A^2 + B^2 + C^2} = \xi^0,$$

on pourra écrire $$\xi - \xi^0 = \frac{X}{p^2},$$

et de même $\qquad \eta - \eta^0 = \dfrac{Y}{p^2}, \quad \zeta - \zeta^0 = \dfrac{Z}{p^2},$

η^0, ζ^0 étant des constantes, et Y, Z des fonctions linéaires de x, y, z. Les polynômes X, Y, Z s'obtiendraient sans peine par ce qu'on vient de dire ; mais on les trouve sous une forme plus commode en opérant comme il suit. Il est aisé de voir qu'en attribuant une valeur infinie à une ou plusieurs des quantités x, y, z, ou, si l'on veut, en faisant

$$x^2 + y^2 + z^2 = \infty,$$

on a $\quad \xi = \xi^0, \quad \eta = \eta^0, \quad \zeta = \zeta^0, \quad \dfrac{p^2}{x^2 + y^2 + z^2} = \dfrac{A^2 + B^2 + C^2}{m}.$

Si donc on introduit cette hypothèse de $x^2 + y^2 + z^2 = \infty$ dans l'équation générale

$$(\xi' - \xi)^2 + (\eta' - \eta)^2 + (\zeta' - \zeta)^2 = \dfrac{(x' - x)^2 + (y' - y)^2 + (z' - z)^2}{p^2 p'^2},$$

il viendra

$$(\xi' - \xi^0)^2 + (\eta' - \eta^0)^2 + (\zeta' - \zeta^0)^2 = \dfrac{m}{(A^2 + B^2 + C^2) p'^2},$$

d'où, en effaçant les accents,

$$(\xi - \xi^0)^2 + (\eta - \eta^0)^2 + (\zeta - \zeta^0)^2 = \dfrac{m}{(A^2 + B^2 + C^2) p^2}.$$

Mais, d'un autre côté,

$$(\xi - \xi^0)^2 + (\eta - \eta^0)^2 + (\zeta - \zeta^0)^2 = \dfrac{X^2 + Y^2 + Z^2}{p^4};$$

donc

$$X^2 + Y^2 + Z^2 = \dfrac{m p^2}{A^2 + B^2 + C^2},$$

c'est-à-dire

$$X^2 + Y^2 + Z^2 = (x - x_1)^2 + (y - y_1)^2 + (z - z_1)^2.$$

De là, par un calcul tout semblable à celui qu'on a effectué dans le numéro précédent pour l'équation

$$P^2 + Q^2 + R^2 = m^2 (x^2 + y^2 + z^2),$$

on conclut qu'en représentant par α, β, γ, α', etc., des constantes assujetties aux équations de condition

$$\alpha^2 + \alpha'^2 + \alpha''^2 = 1, \qquad \alpha\beta + \alpha'\beta' + \alpha''\beta'' = 0,$$
$$\beta^2 + \beta'^2 + \beta''^2 = 1, \qquad \alpha\gamma + \alpha'\gamma' + \alpha''\gamma'' = 0,$$
$$\gamma^2 + \gamma'^2 + \gamma''^2 = 1, \qquad \beta\gamma + \beta'\gamma' + \beta''\gamma'' = 0,$$

du même genre que celles entre a, b, etc., on devra prendre

$$X = \alpha \ (x - x_1) + \beta \ (y - y_1) + \gamma \ (z - z_1),$$
$$Y = \alpha' \ (x - x_1) + \beta' \ (y - y_1) + \gamma' \ (z - z_1),$$
$$Z = \alpha''(x - x_1) + \beta''(y - y_1) + \gamma''(z - z_1).$$

Et, réciproquement, il est facile de vérifier qu'en adoptant ces valeurs de X, Y, Z, les formules

$$\xi - \xi^0 = \frac{nX}{X^2 + Y^2 + Z^2}, \quad \eta - \eta^0 = \frac{nY}{X^2 + Y^2 + Z^2}, \quad \zeta - \zeta^0 = \frac{nZ}{X^2 + Y^2 + Z^2},$$

qui résultent de notre analyse en faisant, pour abréger,

$$\frac{m}{A^2 + B^2 + C^2} = n, \ \text{d'où} \ p^2 = \frac{X^2 + Y^2 + Z^2}{n},$$

entraîneront l'équation demandée (1) dont la solution générale est exprimée ainsi d'une manière nouvelle et plus simple. En effet, on trouve d'abord

$$(\xi' - \xi)^2 + (\eta' - \eta)^2 + (\zeta' - \zeta)^2 = \frac{n^2[(X' - X)^2 + (Y' - Y)^2 + (Z' - Z)^2]}{(X^2 + Y^2 + Z^2)(X'^2 + Y'^2 + Z'^2)},$$

puis

$$(X' - X)^2 + (Y' - Y)^2 + (Z' - Z)^2 = (x' - x)^2 + (y' - y)^2 + (z' - z)^2,$$

à cause des équations de condition entre α, β, etc. Et de là on tire

$$(\xi' - \xi)^2 + (\eta' - \eta)^2 + (\zeta' - \zeta)^2 = \frac{n^2[(x' - x)^2 + (y' - y)^2 + (z' - z)^2]}{(X^2 + Y^2 + Z^2)(X'^2 + Y'^2 + Z'^2)},$$

c'est-à-dire l'équation (1), en prenant

$$p^2 = \frac{X^2 + Y^2 + Z^2}{n} = \frac{(x - x_1)^2 + (y - y_1)^2 + (z - z_1)^2}{n}.$$

223. On pourrait former inversement les valeurs de x, y, z en ξ, η, ζ; mais il est clair sans calcul, et à priori, que ces valeurs doivent s'exprimer par des formules du même genre que celles qui donnent ξ, η, ζ en x, y, z. En effet, p étant une fonction de x, y, z, on peut concevoir cette quantité comme fonction de ξ, η, ζ. Soit donc

$$p = \frac{1}{\varpi}, \quad p' = \frac{1}{\varpi'},$$

ϖ étant une certaine fonction de ξ, η, ζ, et ϖ' la même fonction de ξ', η', ζ'. L'équation (1) se changera dans l'équation nouvelle

$$(x' - x)^2 + (y' - y)^2 + (z' - z)^2 = \frac{(\xi' - \xi)^2 + (\eta' - \eta)^2 + (\zeta' - \zeta)^2}{\varpi \varpi'},$$

d'une forme toute semblable à l'équation (1) elle-même, et qui, par conséquent, donnera x, y, z en ξ, η, ζ de la même manière · que l'équation (1) a donné ξ, η, ζ en x, y, z.

224. On voit que, par l'échange des lettres x, y, z et ξ, η, ζ les unes dans les autres, une solution particulière de l'équation (1), je veux dire une solution dans laquelle les constantes auraient des valeurs particulières, en donnera une autre, la plupart du temps différente, quoique rentrant toujours, bien entendu, dans le type général indiqué tout à l'heure. Il est aisé aussi de voir que deux solutions données en fournissent une troisième. Supposons, en effet, qu'en prenant pour ξ, η, ζ, q des fonctions de U, V, W, on ait

$$(\xi'-\xi)^2+(\eta'-\eta)^2+(\zeta'-\zeta)^2=\frac{(U'-U)^2+(V'-V)^2+(W'-W)^2}{q^2 q'^2},$$

et que, de même, en prenant pour U, V, W, p, des fonctions de x, y, z, on ait

$$(U'-U)^2+(V'-V)^2+(W'-W)^2=\frac{(x'-x)^2+(y'-y)^2+(z'-z)^2}{p^2 p'^2},$$

il est clair qu'on pourra exprimer aussi q, ξ, η, ζ en x, y, z, et qu'il viendra

$$(\xi'-\xi)^2+(\eta'-\eta)^2+(\zeta'-\zeta)^2=\frac{(x'-x)^2+(y'-y)^2+(z'-z)^2}{p^2 q^2 \cdot p'^2 q'^2},$$

d'où une solution nouvelle de notre problème.

On peut dire, en d'autres termes, que diverses transformations qui résolvent ce problème étant opérées successivement, la transformation unique composée de cet ensemble le résout aussi. Et par la manière dont nous avons vérifié ci-dessus notre solution générale, il est manifeste que cette solution n'est que le résultat d'une suite de solutions particulières ainsi ajoutées entre elles pour ainsi dire.

225. Il y a une solution particulière de l'équation (1) que nous devons étudier spécialement parce qu'elle constitue, à proprement parler, l'élément essentiel de nos formules générales, et qu'elle nous servira d'ailleurs à en bien montrer le sens géométrique. Elle a été employée par M. Thomson, et consiste à poser

$$\xi=\frac{nx}{x^2+y^2+z^2}, \quad \eta=\frac{ny}{x^2+y^2+z^2}, \quad \zeta=\frac{nz}{x^2+y^2+z^2},$$

d'où résulte, en effet, l'équation

$$(\xi'-\xi)^2+(\eta'-\eta)^2+(\zeta'-\zeta)^2=n^2\frac{(x'-x)^2+(y'-y)^2+(z'-z)^2}{(x^2+y^2+z^2)(x'^2+y'^2+z'^2)},$$

c'est-à-dire l'équation (1), en prenant

$$p^2 = \frac{x^2 + y^2 + z^2}{n}.$$

On a alors
$$\xi^2 + \eta^2 + \zeta^2 = \frac{n^2}{x^2 + y^2 + z^2},$$

et, par conséquent,

$$x = \frac{n\xi}{\xi^2 + \eta^2 + \zeta^2}, \quad y = \frac{n\eta}{\xi^2 + \eta^2 + \zeta^2}, \quad z = \frac{n\zeta}{\xi^2 + \eta^2 + \zeta^2},$$

valeurs de même composition en ξ, η, ζ que les précédentes en x, y, z.

On peut interpréter géométriquement ces formules en regardant x, y, z, par exemple, comme des coordonnées rectangulaires, et ξ, η, ζ comme des paramètres. Les surfaces (ξ), (η), (ζ), pour lesquelles un de ces paramètres conserve même valeur, sont des sphères qui se coupent deux à deux orthogonalement, et par l'intersection de trois desquelles M. Thomson détermine la position de chaque point (x, y, z) ou (ξ, η, ζ). Sous ce point de vue, ξ, η, ζ sont des coordonnées curvilignes qui se rapportent à la même figure que les coordonnées rectilignes x, y, z. Mais il est plus commode, je crois, d'introduire dans nos recherches une de ces transmutations de figures si familières aux géomètres, et qui ont tant contribué aux progrès de la science dans ces derniers temps. La transformation dont il s'agit est bien connue, du reste, et des plus simples; c'est celle que M. Thomson lui-même a jadis employée sous le nom de principe des *images**. Considérez x, y, z comme les coordonnées d'un point quelconque m d'une figure rapportée à trois axes rectangulaires Ox, Oy, Oz, ξ, η, ζ comme celles d'un point μ d'une autre figure rapportée à trois axes $O\xi$, $O\eta$, $O\zeta$, rectangulaires aussi, et auxquels nous donnons la même origine O, et respectivement les mêmes directions, une de ces figures dérivant de l'autre, et le point μ, en particulier, correspondant au point m, en vertu des relations par lesquelles ξ, η, ζ s'expriment en x, y, z, ou x, y, z en ξ, η, ζ. Il est évident que les deux points correspondants m, μ sont en ligne droite avec l'origine O, et que le produit $Om . O\mu$ des rayons vecteurs Om, $O\mu$ est constant

---

* Tome x. de ce Journal, page 364 [above, § 207].

et $= n$. Une des figures se déduit donc de l'autre en prenant
sur chacun des rayons vecteurs menés du point O à un point
quelconque de la première figure d'autres rayons vecteurs en
raison inverse des premiers; les extrémités de ces nouveaux
rayons vecteurs déterminent la seconde figure. Nous donne-
rons à cette transformation le nom de transformation *par rayons
vecteurs réciproques*, relativement à l'origine O. Si, pour un
point m, on a $Om = \sqrt{n}$, on aura aussi $O\mu = \sqrt{n}$, et les points
m et μ qui se correspondent ainsi dans les deux figures coïn-
cideront. En disposant de n, on peut faire en sorte qu'un point
donné m reste fixe dans la transformation; il suffit de prendre
$n = \overline{Om}^2$, et alors tous les points situés sur la sphère dont O est
le centre et Om le rayon, resteront fixes aussi, mais tous les
autres seront déplacés.

226. A l'aide de cette transformation *par rayons vecteurs réciproques*, on déduira d'une figure donnée une infinité d'autres
figures, soit en changeant l'origine O d'où partent les rayons
vecteurs, soit en prenant diverses valeurs de n avec une même
origine O, ce qui ne donne, au surplus, lieu qu'à des figures
transformées toutes semblables entre elles, du moins tant que
n garde le même signe; car les figures qui répondent à deux
valeurs de n égales et de signes contraires sont symétriques.
On peut d'ailleurs effectuer, l'une après l'autre, des transformations relatives à des origines différentes. Mais je dis que nos
formules générales de n° 222 peuvent toujours s'interpréter à
l'aide d'une seule transformation de cette espèce, en sorte qu'on
n'obtiendrait rien de vraiment nouveau en ajoutant d'autres
transformations à celle-là.

En effet, dans le cas le plus général, nous pouvons encore
considérer x, y, z et ξ, η, ζ comme les coordonnées de deux points
m, μ appartenant à deux figures différentes et rapportés à deux
systèmes d'axes rectangulaires des x, y, z et ξ, η, ζ. Et voici
comment s'opère la transformation de l'une des figures dans
l'autre.

D'abord on passe de x, y, z, à X, Y, Z par les formules
$$X = \alpha(x-x_0) + \beta(y-y_0) + \gamma(z-z_0),$$
$$Y = \alpha'(x-x_0) + \beta'(y-y_0) + \gamma'(z-z_0),$$
$$Z = \alpha''(x-x_0) + \beta''(y-y_0) + \gamma''(z-z_0).$$

Or, à cause des équations de condition entre α, β, etc., ce passage n'est qu'un changement de coordonnées rectangulaires en d'autres coordonnées rectangulaires, qui n'altère en rien la première figure à laquelle il est appliqué ; on peut le supposer opéré d'avance, et confondre dès lors X, Y, Z avec x, y, z.

De là nous irons aux formules

$$\xi - \xi^0 = \frac{nX}{X^2 + Y^2 + Z^2}, \quad \eta - \eta^0 = \frac{nY}{X^2 + Y^2 + Z^2}, \quad \zeta - \zeta^0 = \frac{nZ}{X^2 + Y^2 + Z^2},$$

et nous aurons ainsi une transformation de X, Y, Z en $\xi - \xi_0$, $\eta - \eta_0$, $\zeta - \zeta^0$, que nous regarderons comme des coordonnées rectangulaires prises par rapport aux mêmes axes. Cette transformation est à rayons vecteurs réciproques, comme nous l'avons vu nᵘ 225. Elle s'opère en portant sur les rayons vecteurs menés de l'origine actuelle des longueurs inversement proportionnelles à ces rayons vecteurs ; l'ancienne figure se trouve ainsi changée en celle qui résulte des extrémités de toutes ces longueurs. Passer ensuite de $\xi - \xi^0$, $\eta - \eta^0$, $\zeta - \zeta^0$ à ξ, η, ζ, n'est qu'un simple déplacement de l'origine, les axes restant parallèles à eux-mêmes ; cela ne produit dans la figure transformée aucune altération.

Nos formules du n° 222 résultent donc d'une transformation par rayons vecteurs réciproques, combinée avec des changements ordinaires de coordonnées. De telles transformations en nombre quelconque donnent toujours naissance à une équation de la forme (1), et l'interprétation géométrique des formules par lesquelles nous avions d'abord lié (n° 221) x, y, z et ξ, η, ζ semblait en demander deux, relatives à deux origines différentes, l'une pour le passage de x, y, z à u, v, w, l'autre pour le passage de u, v, w à ξ, η, ζ ; mais on voit, par ce qui précède, et grâce aux formules plus simples du n° 222, qu'une seule transformation suffit pour conduire au résultat le plus général ; il était important de le démontrer.

227. Les considérations géométriques dont nous venons de faire usage, pour interpréter les formules qui conduisent à l'équation (1), donnent lieu à des conséquences remarquables dont nous allons dire quelques mots. Dans les deux figures que déterminent respectivement les coordonnées x, y, z et les coordonnées ξ, η, ζ, considérons, d'une part, deux points quel-

conques m, m', et, d'autre part, les points correspondants μ, μ'.
Soient D la distance des deux premiers, Δ celle des deux autres,
en sorte que

$$D^2 = (x' - x)^2 + (y' - y)^2 + (z' - z)^2,$$

$$\Delta^2 = (\xi' - \xi)^2 + (\eta' - \eta)^2 + (\zeta' - \zeta)^2.$$

L'équation (1), qui pourra s'écrire

$$\Delta^2 = \frac{D^2}{p^2 p'^2}, \quad \Delta = \frac{D}{pp'}, \quad \frac{1}{\Delta} = \frac{pp'}{D},$$

fournit une relation entre la distance Δ de deux points μ, μ'
dans l'une des figures et les quantités D, p, p'. Nous venons
de dire que D est la distance des deux points m, m' correspondants dans l'autre figure; quant à p et p', ce sont, à un facteur
constant près, les distances des points m, m' à un certain point
fixe. Toute relation métrique entre deux ou plusieurs distances Δ dans l'une des figures fournira donc immédiatement
une relation analogue dans l'autre figure. Mais il ne faut pas
croire que les divers points correspondants à ceux de la droite
Δ soient sur la droite D; cela arrive pour les points extrêmes
par la définition même de ces droites, mais n'a pas lieu, en
général, pour les points intermédiaires. En général, la suite
des points correspondants à ceux d'une droite de la première
figure forme dans la seconde figure une circonférence de cercle,
laquelle ne se réduit à une ligne droite que dans un cas particulier, celui où son rayon est infini.

Ayant en ξ, η, ζ l'équation d'une surface ou les équations
d'une ligne appartenant à la première figure, il suffit de substituer à ξ, η, ζ leurs valeurs pour former en x, y, z l'équation de
la surface ou les équations de la ligne correspondante. On
trouve bien facilement, de cette manière, que les plans et des
sphères se transforment en des sphères qui peuvent se réduire
à des plans quand le rayon devient infini; que, de même, des
droites et des circonférences de cercle se transforment en des
circonférences de cercle, etc. Mais, pour suivre le mécanisme
de ces transformations, il suffit de considérer la transformation
par rayons vecteurs réciproques, qui combinée avec des changements de coordonnées donne, comme on l'a vu, la transformation la plus générale. Soit donc

$$\xi = \frac{nx}{x^2+y^2+z^2} = \frac{nx}{r^2}, \quad x = \frac{n\xi}{\xi^2+\eta^2+\zeta^2} = \frac{n\xi}{\rho^2},$$

$$\eta = \frac{ny}{x^2+y^2+z^2} = \frac{ny}{r^2}, \quad y = \frac{n\eta}{\xi^2+\eta^2+\zeta^2} = \frac{n\eta}{\rho^2},$$

$$\zeta = \frac{nz}{x^2+y^2+z^2} = \frac{nz}{r^2}, \quad z = \frac{n\zeta}{\xi^2+\eta^2+\zeta^2} = \frac{n\zeta}{\rho^2},$$

$$\Delta = \frac{D}{pp'} = \frac{nD}{\sqrt{(x^2+y^2+z^2)(x'^2+y'^2+z'^2)}} = \frac{nD}{rr'},$$

l'ensemble des formules relatives à la transformation par rayons vecteurs réciproques. On en conclut immédiatement ce que nous venons d'avancer, concernant les plans et les sphères, les droites et les circonférences de cercle. Mais on voit, de plus, et même sans calcul, que les plans qui passent par le point O, origine des rayons vecteurs, sont les seuls qui restent des plans dans la transformation; avant et après, leur position est la même, quoique leurs divers points, bien entendu, se soient déplacés pour se substituer les uns aux autres, ceux qui étaient loin de l'origine en étant à présent devenus voisins, et *vice versâ*. Tout autre plan se transforme en une sphère passant par le point O (où la transformation amène tous les points situés à l'infini) et ayant son centre sur la perpendiculaire au plan menée du point O; la perpendiculaire et le diamètre de la sphère ont un produit égal à la constante n, et se déduisent ainsi facilement l'une de l'autre. Il est inutile d'ajouter que deux sphères qui correspondent à deux plans parallèles se touchent au point O. De même, deux sphères ainsi posées se transformeraient en deux plans parallèles. Mais une sphère qui ne passe pas par le point O doit rester une sphère, puisqu'elle ne peut acquérir aucun point à l'infini. Les droites passant par le point O restent des droites, et conservent leur position invariable. Toute autre droite donne lieu à une circonférence de cercle dont le plan est déterminé par la droite et par le point O, et dont le centre est situé sur la perpendiculaire abaissée du point O sur la droite; le diamètre est le quotient de la constante n par cette perpendiculaire. Les circonférences provenant de droites parallèles sont toutes tangentes à une parallèle menée par le point O à ces droites. On peut voir, enfin, que la transformée d'une circonférence est une droite quand la circonférence passe par le point O, et, dans tout autre cas, reste une circonférence.

Une propriété remarquable de ce genre de transformation consiste en ce que les deux triangles formés par trois points infiniment voisins quelconques de la figure primitive et les trois points correspondants de sa transformée sont semblables l'un à l'autre, en sorte que si deux lignes se coupent dans l'une des deux figures sous un certain angle, les lignes correspondantes de l'autre figure se couperont sous le même angle*. La démonstration de cette propriété repose sur l'équation (1), à laquelle nous avons donné la forme

$$\Delta = \frac{D}{pp'}.$$

Supposons, en effet, que les deux points m, m', ou (x, y, z), (x', y', z'), soient infiniment voisins, et que leur distance D soit représentée par ds. Représentons par $d\sigma$ celle des deux points correspondants μ, μ'. Comme p et p' n'auront pas de différence sensible, il nous viendra

$$d\sigma = \frac{ds}{p^2}.$$

Les éléments $d\sigma$, ds ont donc en chaque lieu un rapport constant qui dépend de p et change, en général, d'un lieu à l'autre. Considérons un troisième point m'' infiniment voisin des deux premiers, et désignons par ds' et ds'' ses distances à m et à m'; $d\sigma'$, $d\sigma''$ étant les distances correspondantes dans la seconde figure, on aura encore

$$d\sigma' = \frac{ds'}{p^2},$$

$$d\sigma'' = \frac{ds''}{p^2}.$$

Donc $\qquad d\sigma : d\sigma' : d\sigma'' :: ds : ds' : ds''.$

Ainsi, le triangle infinitésimal $mm'm''$ est semblable au triangle

---

* De la similitude des triangles infiniment petits correspondants, il résulte encore que la figure transformée est semblable à la figure primitive, ou à sa symétrique, dans ses éléments infiniment petits. En s'en tenant au premier cas, qui est proprement celui de nos formules, où nous prenons naturellement la constante n positive, on aura, à trois dimensions, une sorte de représentations des corps, analogue au tracé des cartes géographiques [those according to the "stereographic projection"], pour lesquelles le rapport de similitude des éléments correspondants est variable aussi d'un lieu à l'autre.

correspondant $\mu\mu'\mu''$. L'angle de ds avec ds' est, par conséquent, le même que celui de $d\sigma$ avec $d\sigma'$. Cette démonstration,
on le voit, n'exige pas même que l'équation (1) ait lieu pour
deux points situés à une distance finie ; elle demande seulement que cette équation ait toujours lieu pour deux points
infiniment voisins. On doit en dire autant d'un théorème que
je vais établir, et qui n'est qu'un corollaire de la proposition
précédente.

Une surface appartenant à l'une des deux figures étant
donnée, représentez-vous les lignes de courbure de cette surface, et les deux séries de surfaces développables, orthogonales
entre elles et à la surface donnée, qui sont formées par les
normales successives. Dans la seconde figure, les séries de
surfaces correspondantes resteront orthogonales entre elles et à
la transformée de la surface donnée ; par suite, en vertu du
beau théorème de M. Ch. Dupin, elles traceront encore sur
cette transformée des lignes de courbure. Ces lignes de courbure résulteront ainsi des lignes de courbure de la première
surface donnée, et seront immédiatement connues si les autres
le sont. Il sera aisé d'appliquer ce théorème aux surfaces du
second degré, comme aussi aux systèmes triples de surfaces
orthogonales que M. Serret a indiqués dans une Note récente[*],
et qui, par notre transformation, en donneront d'autres non
moins curieux, etc.

Proposons-nous, par exemple, de trouver les lignes de courbure de la surface enveloppe des sphères qui touchent trois
sphères données, problème que M. Ch. Dupin a résolu jadis
dans la *Correspondance sur l'École Polytechnique*, tome I, page
22. Soient O et P les points d'intersection de ces trois sphères ;
prenons le point O pour origine, et opérons une transformation
par rayons vecteurs réciproques, ce qui nous fournira une
seconde figure d'où nous reviendrons aisément à la première.
Dans la seconde figure, les trois sphères données seront remplacées par trois plans qui se couperont en un point Π
correspondant au second point P d'intersection de nos trois
sphères. La surface enveloppe des sphères tangentes à ces
trois plans sera (en se bornant à un des angles solides et à son

---

[*] Page 241 du présent volume [Liouville's *Journal*, 1847].

opposé) celle d'un cône droit à base circulaire ayant son sommet
au point Π, et circonscrit à une quelconque des sphères tangentes aux trois plans.   Les lignes de courbure de cette surface
conique sont : 1° les génératrices rectilignes qui passent toutes
par le point Π : dans le retour à la première figure, ces droites
deviendront des cercles passant tous par le point P, dont les
tangentes en P feront toutes le même angle avec la tangente
au cercle dans lequel se transforme l'axe du cône, d'où résultera
un nouveau cône droit, et passant toutes aussi avec des circonstances semblables par le point O ; 2° des cercles, dont les
plans sont tous parallèles entre eux et perpendiculaires à l'axe
du cône, et qui, lors du retour à la première figure, deviendront
des cercles coupant à angle droit ceux qui résultent de
génératrices rectilignes.   Les lignes de courbure de la surface
enveloppe des sphères tangentes à trois sphères données sont
donc des circonférences de cercle.

On démontre avec la même facilité le théorème de M. Dupin
concernant la courbe que trace sur chacune des trois sphères
données la sphère variable qui les touche.   En effet, quand les
trois sphères données sont remplacées par trois plans, il est
clair que la suite des points suivant lesquels la sphère variable
touche un quelconque des plans est une ligne droite passant
par le point d'intersection Π.   Donc, en revenant aux trois
sphères données, la courbe demandée est une circonférence de
cercle qui passe par les points O et P.   Il peut arriver, bien
entendu, que les points O et P soient imaginaires ; mais il n'y
a alors aucun changement essentiel à faire dans ce que nous
venons de dire, et nos conclusions subsistent.

La circonstance d'une origine O imaginaire aurait plus
d'inconvénient s'il s'agissait de résoudre le problème d'une
sphère tangente à quatre autres, en le ramenant au problème
très-simple de trouver une sphère tangente à une sphère donnée
et à trois plans donnés ; mais on y remédierait en augmentant
d'une même quantité les rayons des quatre sphères données, ce
qui ne change pas la position du centre de la sphère tangente.
De même, en se bornant à considérer des points tous situés
dans un plan passant par l'origine O, on ramènera la détermination du cercle tangent à trois autres à celle d'un cercle qui
touche un cercle donné et deux droites données.

En général, les systèmes de sphères ou de cercles, et spécialement de sphères ou de cercles passant par un point donné,
jouissent de propriétés curieuses dont beaucoup deviennent
intuitives par la transformation dont nous venons de nous
occuper. On peut appliquer en particulier cette remarque aux
théorèmes que M. Miquel a donnés dans son Mémoire sur les
angles curvilignes*. Pour nous borner au cas le plus simple,
il est évident que, dans un triangle ABC formé par trois arcs
de cercles passant tous par un même point O, la somme des
angles vaut 2 droits, puisque notre transformation rend ce
triangle rectiligne sans altérer ses angles.

228. Le passage des relations métriques d'une figure à l'autre,
dans la transformation par rayons vecteurs réciproques, en
allant des coordonnées ξ, η, ζ aux coordonnées x, y, z, s'opère à
l'aide de la formule

$$\Delta = \frac{nD}{rr'},$$

ou simplement

$$\Delta = \frac{D}{rr'},$$

en posant $n = 1$, ce qui n'a aucun inconvénient. Mais en
désignant par O l'origine, dans la seconde figure seulement, et
en employant les autres lettres A, B, etc., pour représenter à la
fois les points de la première figure et les points correspondants
de la seconde figure, cette formule revient à dire que, dans toute
relation entre des distances AB, BD, etc., il faut remplacer
chaque distance telle que AB par $\dfrac{AB}{OA.OB}$. Voilà donc une
règle pratique très-commode ; cette règle convient aussi bien au
cas du plan qu'à celui de l'espace. Deux exemples suffiront.

Que des droites partant d'un point fixe A coupent chacune
un cercle en deux points B et C, B' et C', etc., on aura

$$AB \times AC = AB' \times AC' = \text{constante.}$$

Donc, dans la figure transformée,

$$\frac{AB}{OA.OB} \times \frac{AC}{OA.OC} \times \frac{AB'}{OA.OB'} \times \frac{AC'}{OA.OC'},$$

et par conséquent,

$$\frac{AB}{OB} \times \frac{AC}{OC} = \text{constante.}$$

---

* Tome IX. de ce Journal page 20 [Liouville's *Journal*, 1844].

D'ailleurs les points A, B, C, qui étaient en ligne droite, se trouvent à présent sur une circonférence de cercle passant par le point O. Nous voyons par là que les cercles passant par deux points fixes A, O coupent un cercle donné en deux points B, C tels, que le rapport des produits des distances $AB \times AC$ et $OB \times OC$ a une valeur constante pour tous ces cercles.

Que les côtés BC, AC, AB d'un triangle rectiligne ABC soient coupés en trois points A', B', C' par une transversale, on aura

$$AC' \times BA' \times CB' = BC'' \times CA' \times AB'.$$

Donc, dans la figure transformée,

$$\frac{AC''}{OA.OC'} \times \frac{BA'}{OB.OA'} \times \frac{CB'}{OC.OB'} = \frac{BC''}{OB.OC'} \times \frac{CA'}{OC.OA'} \times \frac{AB'}{OA.OB'},$$

ce qui redonne

$$AC' \times BA' \times CB' = BC'' \times CA' \times AB'.$$

Mais cette relation s'applique à présent à un triangle curviligne ABC formé par trois cercles qui passent tous au point O et dont les côtés sont coupés en A', B', C' par un quatrième cercle passant aussi au point O. Il est, du reste, inutile d'ajouter que AC', BA', etc., sont les plus courtes distances des points A et C', B et A', etc., et non des segments mesurés sur les côtés du triangle curviligne.

On généraliserait aisément de la même manière le théorème relatif à un polygone gauche coupé par un plan. Mais en voilà assez sur ce sujet.

229. Etant données deux sphères qui ne se coupent pas, on peut toujours placer l'origine O sur la droite qui joint leurs centres, en un point réel tel, qu'après la transformation par rayons vecteurs réciproques, ces deux sphères seront concentriques. Prenons la droite des centres pour axe des x; désignons par h la distance inconnue du point O au centre de la première sphère, et par $h + l$ sa distance au centre de la seconde sphère; soient k, k' les rayons. Les équations des deux sphères seront, avant la transformation,

$$(x-h)^2 + y^2 + z^2 = k^2,$$
$$(x-h-l)^2 + y^2 + z^2 = k'^2,$$

et après la transformation, qui consistera à remplacer x, y, z par

$$\frac{x}{x^2+y^2+z^2}, \quad \frac{y}{x^2+y^2+z^2}, \quad \frac{z}{x^2+y^2+z^2},$$

elles deviendront

$$\left(x - \frac{h}{h^2-k^2}\right)^2 + y^2 + z^2 = \frac{k^2}{(h^2-k^2)^2},$$

$$\left[x - \frac{h+l}{(h+l)^2-k'^2}\right]^2 + y^2 + z^2 = \frac{k'^2}{[(h+l)^2-k'^2)]^2}.$$

Pour que le centre soit le même à présent, il faut et il suffit
que

$$\frac{h}{h^2-k^2} = \frac{h+l}{(h+l)^2-k'^2},$$

d'où
$$lh^2 + (l^2 + k^2 - k'^2) h + lk^2 = 0,$$

équation du second degré qui donnera pour h deux valeurs,

$$h = -\frac{l^2 + k^2 - k'^2}{2l} \pm \frac{1}{2l}\sqrt{G},$$

en posant

$$G = (l-k-k')(l-k+k')(l+k-k')(l+k+k');$$

et il est aisé de voir que G sera positive si les deux sphères
qu'on a données d'abord ne se coupent pas.

230. Ce théorème pourra être utile en géométrie; mais il
aura surtout une application importante dans les questions de
physique mathématique. Essayons ici d'indiquer rapidement
l'usage, en ce genre de questions, de la transformation générale
qui donne l'équation (1). La Lettre de M. Thomson nous
servira de guide; nous y ajouterons quelques développements.
La généralité plus ou moins grande de la solution par laquelle
on satisfait à l'équation (1) ne change en rien la marche à
suivre, qui reste la même dans tous les cas.

Et d'abord de l'équation

$$\frac{1}{\Delta} = \frac{pp'}{D}$$

on peut conclure, avec M. Thomson, que, si une fonction U de
ξ, η, ζ satisfait à l'équation

$$\frac{d^2U}{d\xi^2} + \frac{d^2U}{d\eta^2} + \frac{d^2U}{d\zeta^2} = 0,$$

cette même fonction, divisée par p et exprimée en x, y, z,
vérifiera l'équation de même forme

$$\frac{d^2 \, p^{-1}U}{dx^2} + \frac{d^2 \, p^{-1}U}{dy^2} + \frac{d^2 . \, p^{-1}U}{dz^2} = 0.$$

De là une liaison entre deux problèmes distincts concernant tous deux l'équilibre de température dans les corps homogènes, mais relatifs à deux systèmes dont l'un résulte de l'autre par la transformation qui lie ξ, η, ζ à x, y, z.

Que le premier système soit formé de deux sphères qui ne se coupent pas, que la température soit donnée en chaque point de leurs surfaces, et demandons quelle est la loi des températures permanentes dans l'espace compris entre elles, si l'une est intérieure à l'autre, ou dans l'espace infini extérieur à toutes deux, si l'une est en dehors de l'autre, en ajoutant dans ce dernier cas la condition que la température soit nulle à l'infini. On ramènera cette question au cas très-facile de deux sphères concentriques. Cela résulte du théorème établi ci-dessus et en montre toute l'importance. En indiquant cette application à la théorie de la chaleur, M. Thomson ajoute, du reste, avec raison qu'elle s'étend d'elle-même à la théorie de l'électricité.

Dans la théorie de l'électricité ou du magnétisme, et, en général, dans la théorie de l'attraction, la quantité que G. Green et M. Gauss nomment *potentiel*, c'est-à-dire la quantité qu'on obtient en faisant la somme des éléments attractifs ou répulsifs d'une masse divisés par leurs distances à un point, joue un rôle capital. On connaît le problème de M. Gauss : " Distribuer sur une surface donnée une masse attractive ou répulsive, de telle sorte que le potentiel ait en chaque point de la surface une valeur donnée." On a résolu ce problème pour différentes surfaces, en particulier pour l'ellipsoïde. Or la solution relative à une surface quelconque donne la solution pour toutes les surfaces qui se déduisent de celle-là par une transformation pour laquelle l'équation (1) ait lieu. Ayant, en effet, l'équation

$$\iint \frac{\lambda' d\omega'}{\Delta} = Q$$

pour la première surface, on aura pour la seconde surface une équation du même genre, remplaçant par leurs nouvelles valeurs Δ et $d\omega'$. On a

$$\Delta = \frac{D}{pp'}.$$

Quant à $d\omega'$, j'observe que les éléments linéaires correspondants $d\sigma$ et ds sont liés par la formule

$$d\sigma = \frac{ds}{p^2}.$$

Donc entre deux éléments superficiels correspondants $d\omega$, da,

on aura $$d\omega = \frac{da}{p^4}, \quad d\omega' = \frac{da'}{p'^4};$$

par suite, $$\iint \frac{\lambda'}{p'^3} \frac{da'}{D} = \frac{Q}{p},$$

ce qui résout le problème de M. Gauss pour la surface transformée.

On peut voir aussi que les équations désignées par (A), (B), (C) dans mes Lettres à M. Blanchet*, et qui sont d'un si grand usage dans la plupart des questions physico-mathématiques concernant l'ellipsoïde, ont leurs analogues, qu'on en déduit immédiatement pour les surfaces transformées de l'ellipsoïde†.

On peut considérer encore l'équation

$$\frac{d^2U}{dt^2} = \frac{d^2U}{d\xi^2} + \frac{d^2U}{d\eta^2} + \frac{d^2U}{d\zeta^2},$$

et lui faire subir la transformation de ξ, η, ζ en x, y, z.

A cause de l'équation

$$d\sigma = \frac{ds}{p^2},$$

qui peut s'écrire

$$d\xi^2 + d\eta^2 + p\zeta^2 = \frac{dx^2 + dy^2 + dz^2}{p^4},$$

on trouve, par des formules connues, que la quantité

$$\frac{d^2U}{d\xi^2} + \frac{d^2U}{d\eta^2} + \frac{d^2U}{d\zeta^2}$$

est égale à

$$p^6 \left(\frac{d \cdot \frac{1}{p^2} \frac{dU}{dx}}{dx} + \frac{d \cdot \frac{1}{p^2} \frac{dU}{dy}}{dy} + \frac{d \cdot \frac{1}{p^2} \frac{dU}{dz}}{dz} \right),$$

---

* *Voyez* le tome XI de ce Journal.

† Parmi ces surfaces, il faut distinguer celle que donne la transformation *par rayons vecteurs réciproques*, en mettant l'origine au centre même de l'ellipsoïde. On sait qu'elle est aussi le lieu des pieds des perpendiculaires abaissées du centre sur les plans tangents à un autre ellipsoïde dont les axes ont pour valeurs les inverses des valeurs des axes de l'ellipsoïde donné. Une propriété analogue a lieu dans le plan, pour la lemniscate par exemple, qui peut ainsi être engendrée de deux manières différentes au moyen d'une hyperbole équilatère, circonstance dont M. Chasles a tiré un heureux parti dans ses recherches *sur les arcs égaux de la lemniscate* (*Comptes Rendus de l'Académie des Sciences*, tome XXI, séance du 21 juillet 1845).

c'est-à-dire à

$$p^4\left(\frac{d^2U}{dx^2} + \frac{d^2U}{dy^2} + \frac{d^2U}{dz^2}\right) - 2p^3\left(\frac{dU}{dx}\frac{dp}{dx} + \frac{dU}{dy}\frac{dp}{dy} + \frac{dU}{dz}\frac{dp}{dz}\right)$$

ou enfin à

$$p^5\left(\frac{d^2 \cdot p^{-1}U}{dx^2} + \frac{d^2 \cdot p^{-1}U}{dy^2} + \frac{d^2 \cdot p^{-1}U}{dz^2}\right),$$

en se rappelant que

$$\frac{d^2\frac{1}{p}}{dx^2} + \frac{d^2\frac{1}{p}}{dy^2} + \frac{d^2\frac{1}{p}}{dz^2} = 0.$$

Par là on voit d'abord que l'équation

$$\frac{d^2U}{d\xi^2} + \frac{d^2U}{d\eta^2} + \frac{d^2U}{d\zeta^2} = 0$$

revient à celle-ci :

$$\frac{d^2 \cdot p^{-1}U}{dx^2} + \frac{d^2 \cdot p^{-1}U}{dy^2} + \frac{d^2 \cdot p^{-1}U}{dz^2} = 0,$$

ce que nous savions déjà.   On voit ensuite que l'équation

$$\frac{d^2U}{dt^2} = \frac{d^2U}{d\xi^2} + \frac{d^2U}{d\eta^2} + \frac{d^2U}{d\zeta^2}$$

se transforme en

$$\frac{d^2U}{dt^2} = p^5\left(\frac{d^2 \cdot p^{-1}U}{dx^2} + \frac{d^2 \cdot p^{-1}U}{dy^2} + \frac{d^2 \cdot p^{-1}U}{dz^2}\right),$$

ou, mieux encore, en

$$\frac{d^2 \cdot p^{-1}U}{dt^2} = p^4\left(\frac{d^2 \cdot p^{-1}U}{dx^2} + \frac{d^2 \cdot p^{-1}U}{dy^2} + \frac{d^2 \cdot p^{-1}U}{dz^2}\right).$$

Réciproquement, cette dernière équation, où le coefficient p varie proportionnellement à la distance du point (x, y, z) à un point fixe, se ramène à l'équation

$$\frac{d^2U}{dt^2} = \frac{d^2U}{d\xi^2} + \frac{d^2U}{d\eta^2} + \frac{d^2U}{d\zeta^2},$$

qui est à coefficients constants, résultat qui trouve une application utile dans la théorie du son.

On peut enfin ajouter que les équations aux différences partielles

$$\left(\frac{dU}{d\xi}\right)^2 + \left(\frac{dU}{d\eta}\right)^2 + \left(\frac{dU}{d\zeta}\right)^2 = Q$$

et

$$\left(\frac{dU}{dx}\right)^2 + \left(\frac{dU}{dy}\right)^2 + \left(\frac{dU}{dz}\right)^2 = \frac{Q}{p^4}$$

sont des transformées l'une de l'autre, ce qui pourra servir dans les questions de dynamique, où MM. Hamilton et Jacobi ont introduit de telles équations aux différences partielles.

On me pardonnera, je l'espère, ces développements que j'ai cru pouvoir donner, à la suite des deux Lettres si intéressantes de M. Thomson, sans le gêner dans ses recherches. Mon but sera rempli, je le répète, s'ils peuvent aider à bien faire comprendre la haute importance du travail de ce jeune géomètre, et si M. Thomson lui-même veut bien y voir une preuve nouvelle de l'amitié que je lui porte et de l'estime que j'ai pour son talent.

XV. DETERMINATION OF THE DISTRIBUTION OF ELECTRI-CITY ON A CIRCULAR SEGMENT OF PLANE OR SPHERICAL CONDUCTING SURFACE, UNDER ANY GIVEN INFLUENCE.

[Jan. 1869. *Not hitherto published.*]

231. The electric density at any point of the surface of an insulated conducting ellipsoid, electrified and left undisturbed by external influence, is (§ 11) simply proportional to the distance of the tangent plane from the centre. If we take $\rho = kp$ as the expression of this law, and call q the whole quantity of electricity communicated, we have (§ 14) $4\pi kabc = q$; so that the formula for the electric density, ρ, at any point P of the surface in terms of p, the distance of the tangent plane from the centre, and a, b, c the three semi-axes, is

$$\rho = \frac{q}{4\pi abc}\, p \quad\dots\dots\dots\dots\dots\dots\dots(1);$$

or, in terms of rectangular co-ordinates of the point P,

$$\rho = \frac{q}{4\pi abc \left(\dfrac{x^2}{a^4} + \dfrac{y^2}{b^4} + \dfrac{z^2}{c^4}\right)^{\frac{1}{2}}} \quad\dots\dots\dots\dots\dots\dots(2).$$

232. To find the "electrostatic capacity" (§ 51, footnote) of the charged ellipsoid, let V denote the potential at its surface.

We have, by § 15 (*e*),

$$V = q \int_a^\infty \frac{du}{\sqrt{(u^2 - a^2 + b^2)}\,\sqrt{(u^2 - a^2 + c^2)}} \quad\dots\dots\dots(3);$$

and therefore the capacity is the reciprocal of the definite integral which appears in this formula.

233. By taking $c = 0$ we fall on the case of an infinitely thin plane elliptic disc: for which we have

$$c^2 \left(\frac{x^2}{a^4} + \frac{y^2}{b^4} + \frac{z^2}{c^4}\right) = c^2 \left(\frac{x^2}{a^4} + \frac{y^2}{b^4}\right) + \left(1 - \frac{x^2}{a^2} - \frac{y^2}{b^2}\right) = 1 - \frac{x^2}{a^2} - \frac{y^2}{b^2};$$

and therefore
$$\rho = \frac{q}{4\pi ab\left(1 - \dfrac{x^2}{a^2} - \dfrac{y^2}{b^2}\right)^{\frac{1}{2}}} \quad\ldots\ldots\ldots\ldots\ldots\ldots\ldots(4).$$

Putting $b = a$ in this, we have, for an infinitely thin circular disc,

$$\rho = \frac{q}{4\pi a\,(a^2 - r^2)^{\frac{1}{2}}} \quad\ldots\ldots\ldots\ldots\ldots\ldots\ldots(5),$$

where a denotes the radius of the disc, and ρ the electric density on either side of it, at a distance r from the centre. This result was first given by Green, near the conclusion of his paper " On the Laws of the Equilibrium of Fluids analogous to the Electric Fluid" (*Transactions of the Cambridge Philosophical Society* for Nov. 12, 1832); from which I make the following extract :—

234. " Biot (*Traité de Physique*, tome ii. p. 277) has related " the results of some experiments made by Coulomb on the " distribution of the electric fluid when in equilibrium upon a " plate of copper 10 inches in diameter, but of which the thick- " ness is not specified. If we conceive this thickness to be " very small compared with the diameter of the plate, which " was undoubtedly the case, the formula just found ought to be " applicable to it, provided we except those parts of the plate " which are in the immediate vicinity of its exterior edge. As " the comparison of any results mathematically deduced from " the received theory of electricity with those of the experi- " ments of so accurate an observer as Coulomb must always be " interesting, we will here give a table of the values of the " density at different points on the surface of the plate, calcu- " lated by means of the formula (29), together with the cor- " responding values found from experiment :—

Distances from the Plate's edge.	Observed Densities.	Calculated Densities.
5 in.	1,	1
4	1,001	1,020
3	1,005	1,090
2	1,17	1,250
1	1,52	1,667
, 5	2,07	2,294
0	2,90	infinite

" We thus see that the differences between the calculated " and observed densities are trifling ; and, moreover, that the

12—2

" observed are all something smaller than the calculated ones,
" which, it is evident, ought to be the case, since the latter
" have been determined by considering the thickness of the
" plate as infinitely small, and consequently they will be some-
" what greater than when this thickness is a finite quantity, as
" it necessarily was in Coulomb's experiments."

235.   In this case (3) of § 232 becomes

$$V = q \int_a^\infty \frac{du}{u \sqrt{(u^2 - a^2)}} = q \, \frac{\pi}{2a} \, \dots\dots\dots\dots\dots(6).$$

Hence the capacity is $\dfrac{2a}{\pi}$.   But [§ 232 (3)] the capacity of a
globe is numerically equal to its radius; and therefore the
capacity of an infinitely thin disc is less than that of a globe
of equal radius, in the ratio of 1 to $\dfrac{\pi}{2}$, or 1 to 1·571. Cavendish found the ratio 1 to 1·57, by experiment *!

236.   The expression (5), § 233, for the electric density at
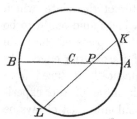
any point P on either side of an
infinitely thin circular disc of conducting material electrified and left
free from disturbing influence, may
be put into a form more convenient
for geometrical investigation, thus :—
Let C be the centre of the disc, so
that $CA = a$, $CP = r$, according to
previous notation.   Hence $BP = a + r$; $PA = a - r$;

$$a^2 - r^2 = BP.PA = KP.PL ;$$

if KL be any chord through P; and (5), with $\dfrac{2aV}{\pi}$ substituted

for q according to (6), becomes

$$\rho = \frac{V}{2\pi^2 \sqrt{KP.PL}} \quad\dots\dots\dots\dots\dots(7).$$

237. Consider a plane disc, S', thus electrified, to a potential
which we shall, for a moment, denote by V'; and, following the
suggestion of § 210, take its image relatively to a spherical
surface of radius R described from any point Q as centre.
This image will (§ 207) be a spherical segment, S, electrified
(§§ 210 and 238) as an infinitely thin conducting surface under
the influence of a quantity $V'R$ of electricity concentrated at
Q; and (compare § 213) the spherical surface of which S is a
part will pass through Q. The reader will have no difficulty
in verifying these statements for himself; but if he desires it,
he will find some further information and examples in Thomson and Tait's *Natural Philosophy*, §§ 512...518. Thus (§ 515
of that work) if ρ' be the electric density on either side of the
disc at P'_1, and ρ that on either side of its image at P, we have

$$\rho = \frac{-R^3}{QP^3} \rho' \quad\dots\dots\dots\dots\dots(8) ;$$

and if v' and v be the potentials at any point Π', and Π the
image of Π', due respectively to the disc S' and its image, we
have (Thomson and Tait, § 516)

$$v = \frac{R}{Q\Pi} v' \quad\dots\dots\dots\dots\dots(9).$$

This shows that, as the potential due to S' has a constant value,
V', at all points of S', the potential due to S will be, at different
points of S, inversely as their distances from Q; and if we take
$q = -RV'$, and denote by V the potential due to electricity
distributed over the two sides of S, we have

$$V = \frac{-q}{QP} \quad\dots\dots\dots\dots\dots(10),$$

and so see that S is electrified as an infinitely thin conducting
sheet of the same figure would be if connected with the earth
by an infinitely fine wire, and inductively electrified only by

the influence of a quantity $-q$ of electricity insulated at Q. Now, with our present notation (7) gives

$$\rho' = \frac{V'}{2\pi^2 \sqrt{L'P' . P'K'}} = -\frac{1}{2\pi^2 \sqrt{L'P' . P'K'}} \cdot \frac{q}{R} \dots(11),$$

if $K'L'$ be any chord through P' of the circle bounding the plane disc S'.

238.   Let K, P, and L be the images of K', P', L', so that KL, the image of $K'L'$, is the arc in which S is cut by the plane through Q and $K'L'$. We have (§ 207)

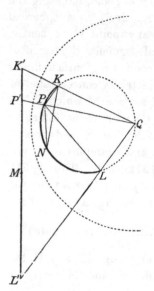

$$K'Q = \frac{R^2}{KQ}, \quad P'Q = \frac{R^2}{PQ}.$$

Hence $K'Q : P'Q :: PQ : KQ$; and therefore the triangles $K'P'Q$, PKQ are similar; and therefore

$$P'K' = KP \cdot \sqrt{\frac{P'Q . K'Q}{KQ . PQ}}$$

$$= KP \frac{R^2}{KQ \cdot PQ} \dots\dots\dots(12).$$

(Compare §§ 213, 227.) From this, and the corresponding expression for LP, we have

$$L'P' . P'K' = \frac{R^4}{PQ^2} \cdot \frac{LP . KP}{LQ . KQ} \dots(13);$$

an expression in which, as the first member has the same value for all lines such as $L'K'$ through P', the second must have the same value for all planes through PQ, cutting one circle on one spherical surface through P and Q, in K and L. As $\frac{R^4}{PQ^2}$ is constant, it follows that $\frac{LP . KP}{LQ . KQ}$ is constant*; a theorem of geometry given above (§ 228) by M. Liouville. Each mem-

---

* As a particular case let Q be either pole of the fixed circle. In this case $LQ = KQ$, a constant. Hence $LP . KP$ is constant; that is, the product of the two chords from any fixed point P on a spherical surface to the two points in which any fixed circle on the surface is cut by a plane through P and one of the poles of that circle, is constant, however the plane be varied. This is the simplest extension to spherical surfaces of the elementary geometrical theorem (Euc. III. 35) for the constancy of the rectangle under the two parts of a varying chord of a fixed circle through a fixed point, already used in the text (§ 236).

ber of (12) may be altered in form thus: bisect $L'K'$ in M, and arc LK in N. We have

$$\left.\begin{array}{l} L'P'.\,P'K' = MK'^2 - MP'^2 \\ LP.\,KP = NK^2 - NP^2 \\ LQ.\,KQ = NQ^2 - NK^2 \end{array}\right\}\quad\quad\quad(14),$$

equations of which the last two are very easily proved from the formula $\sin(\alpha - \beta)\sin(\alpha + \beta) = \sin^2\alpha - \sin^2\beta$,

by taking for α and β the angles subtended by $\tfrac{1}{2}NK$ and $\tfrac{1}{2}NP$ at the centre of the circle $QKPL$; and again, by taking for α and β the angles subtended by $\tfrac{1}{2}NQ$ and $\tfrac{1}{2}NK$ at the same point.

239. Using (13) in (11), and the result in (8), we find

$$\rho = \frac{1}{2\pi^2}\frac{q}{QP^2}\sqrt{\frac{LQ.\,KQ}{LP.\,KP}}\quad\quad\quad(15);$$

and modifying by (14),

$$\rho = \frac{1}{2\pi^2}\frac{q}{QP^2}\sqrt{\frac{NQ^2 - NK^2}{NK^2 - NP^2}}\quad\quad\quad(16).$$

If we take for $PKQL$ the plane through PQ and C the central point (or pole) of the spherical segment S, so that N becomes C, NK becomes equal to the chord of any arc from C to the lip; and (15) becomes

$$\rho = \frac{1}{2\pi^2}\frac{q}{QP^2}\sqrt{\frac{CQ^2 - a^2}{a^2 - CP^2}}\quad\quad\quad(17),$$

which is the result stated in § 219, above. It is remarkable that this expression is independent of the radius of the spherical surface of which the bowl is a part. Hence, if we suppose the radius infinite, we have the same expression (17) for the electric density at any point P on either side of an infinitely thin circular disc of radius a, connected with the earth by an infinitely fine wire, and influenced by a quantity Q of electricity collected at any point Q in the plane of the disc, but outside its bounding circle. It agrees with the solution previously given by Green for this case in his paper referred to in § 234, above.

240. (Compare § 220.) To find the distribution for the case in which S is insulated, electrified, and removed from all disturbing influence, let V be the constant potential produced throughout S by this distribution. Remark that the same

distribution of electricity on S would be produced inductively

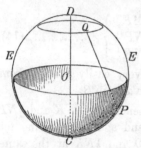

if it were connected by an infinitely fine wire with the earth, and enclosed by any surface, EE, rigidly electrified with such a quantity and distribution of electricity as (§§ 5, 73, 206, 207; also Thomson and Tait, § 499) to produce a uniform potential $-V$ through its interior.

Now take this enclosing surface, EE, to be spherical, concentric with that of which S is a part, and of radius greater than that of the last mentioned by an infinitely small excess. The electric density of the inducing distribution will be uniform all over EE, and equal to $-\dfrac{V}{2\pi f}$ if f be the diameter of the surface. The portion of EE which lies infinitely near to the convex surface of S will clearly induce on this convex surface an equal electric density of contrary sign, that is, $+\dfrac{V}{2\pi f}$. The remainder of EE will induce equal electric densities on the concave and convex sides of S, the amount of either of which at any point P is to be obtained by integration from (15), (16), or (17), thus:—

241. Let $d\sigma$ be an infinitesimal element of E, situated at a point Q anywhere on it. The quantity of electricity on this

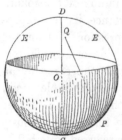

element is $-\dfrac{Vd\sigma}{2\pi f}$; and using this for $-q$ in (17), we find, for the density on either side at P, of the electrification induced by it, the following expression :—

$$\frac{V}{4\pi^3 f}\frac{d\sigma}{PQ^2}\sqrt{\frac{CQ^2-a^2}{a^2-CP^2}}.$$

Now calling O the centre of the spherical surface, let COP be denoted by η; COQ by θ; the value of either of these when P or Q is at the lip of the bowl, by α; and the angle between the planes of COP and COQ, by ϕ: so that we have

$$a^2=\tfrac{1}{2}f^2(1-\cos\alpha),\quad CP^2=\tfrac{1}{2}f^2(1-\cos\eta),\quad CQ^2=\tfrac{1}{2}f^2(1-\cos\theta)$$

and $\qquad PQ^2=\tfrac{1}{2}f^2(1-\cos\eta\cos\theta-\sin\eta\sin\theta\cos\phi);$

and we may take
$$d\sigma = \tfrac{1}{4}f^2 \sin\theta d\theta d\phi.$$

Hence if, lastly, ρ denote the electric density at P, on the concave side of the segment, we have

$$\rho = \frac{V}{8\pi^3 f \sqrt{(\cos\eta - \cos\alpha)}} \int_\alpha^\pi d\theta \sin\theta \sqrt{(\cos\alpha - \cos\theta)} \int_0^{2\pi} \frac{d\phi}{1 - \cos\eta\cos\theta - \sin\eta\sin\theta\cos\phi}.$$

But putting $\tan\tfrac{1}{2}\phi = t$, $A = 1 - \cos\eta\cos\theta$, and $B = \sin\eta\sin\theta$, we find

$$\int_0^{2\pi} \frac{d\phi}{1 - \cos\eta\cos\theta - \sin\eta\sin\theta\cos\phi} = 4\int_0^\infty \frac{dt}{A - B + (A+B)t^2} = \frac{2\pi}{\sqrt{(A^2 - B^2)}} = \frac{2\pi}{\cos\eta - \cos\phi},$$

and therefore

$$\rho = \frac{V}{4\pi^2 f \sqrt{(\cos\eta - \cos\alpha)}} \int_\alpha^\pi \frac{d\theta \sin\theta \sqrt{(\cos\alpha - \cos\theta)}}{\cos\eta - \cos\theta}$$

Lastly, putting $\sqrt{(\cos\alpha - \cos\theta)} = z$, we find

$$\int_\alpha^\pi \frac{d\theta \sin\theta \sqrt{(\cos\alpha - \cos\theta)}}{\cos\eta - \cos\theta} = 2\int_0^{\sqrt{(\cos\alpha + 1)}} \frac{z^2 dz}{\cos\eta - \cos\alpha + z^2}$$

$$= 2\left\{ \sqrt{(\cos\alpha + 1)} \div \sqrt{\cos\eta - \cos\alpha} \tan^{-1} \sqrt{\frac{\cos\alpha + 1}{\cos\eta - \cos\alpha}} \right\}$$

Hence we have, in conclusion,

$$\rho = \frac{V}{2\pi^2 f}\left\{ \sqrt{\frac{\cos\alpha + 1}{\cos\eta - \cos\alpha}} - \tan^{-1} \sqrt{\frac{\cos\alpha + 1}{\cos\eta - \cos\alpha}} \right\} \dots(18),$$

or, with f and α as above, and r to denote the chord CP,

$$\rho = \frac{V}{2\pi^2 f}\left\{ \sqrt{\frac{f^2 - a^2}{a^2 - r^2}} - \tan^{-1} \sqrt{\frac{f^2 - a^2}{a^2 - r^2}} \right\} \dots\dots(19);$$

and the same, with the addition of

$$\frac{V}{2\pi f} \dots\dots\dots\dots\dots\dots\dots\dots(20),$$

gives (§ 240) the electric density on the convex side; which are exactly the results stated above in § 220. Twenty-two years ago these and the very simple formula (17) were communicated by me to M. Liouville without proof, and were published in his Journal. From that time till now they have not been proved, or even noticed, so far as I am aware, by any other writer.

242. Numerical results, calculated from the preceding formulæ (19) and (20), are shown in the following tables :—

Plane Disc.	Curved Disc. Arc 10°.		Curved Disc. Arc 20°.		Bason. Arc 90°.	
	Concave.	Convex.	Concave.	Convex.	Concave.	Convex.
	·9136	1·0685	·8636	1·1364	·4459	1·5541
	·9457	1·0826	·8776	1·1504	·4469	1·5551
	·9920	1·1289	·9236	1·1964	·4828	1·5910
	1·0858	1·2227	1·0165	1·2893	·5566	1·6648
	1·2722	1·4091	1·2884	1·5611	·7065	1·8147
	1·7386	1·8755	1·6652	1·9379	1·0933	2 2015
	Mean.		Mean.		Mean.	
1·0000	1·0000		1·0000		1·0000	
1·0142	1·0141		1·0140		1·0010	
1·0607	1·0605		1·0600		1·0369	
1·1547	1·1542		1·1529		1·1106	
1·3416	1·3407		1·4247		1·2606	
1·8091	1·8071		1·8016		1·6474	

Bowl. Arc 180°.		Bowl. Arc 270°.		Bowl. Arc 340°.	
Concave.	Convex.	Concave.	Convex.	Concave.	Convex.
·1202	1·8798	·0135	1·9865	·0001	1·9999
·1266	1·8862	·0144	1·9874	·0002	1·9999
·1418	1·9014	·0176	1·9906	·0002	2·0000
·1779	1·9375	·0253	1·9983	·0004	2·0001
·2570	2·0166	·0451	2·0181	·0009	2·0006
·4959	2·2555	·1195	2·0925	·0042	2·0040
Mean.		Mean.		Mean.	
1·0000		1·0000		1·0000	
1·0064		1·0009		1·0000	
1·0216		1·0041		1·0001	
1·0577		1·0118		1·0002	
1·1366		1·0316		1·0007	
1·3757		1·1060		1·0041	

It is remarkable how slight an amount of curvature produces a very sensible excess of electric density on the convex side in the first two cases (10° and 20°) of curved discs; yet how nearly the mean of the densities on the convex and concave sides at any point agrees with that at the corresponding point on a plane disc shown in the first column. The results for bowls of 270° and 340° illustrate the tendency of the whole charge to the convex surface, as the case of a thin spherical conducting surface with an infinitely small aperture is approached.

The constant coefficient for each case has been taken so as
to make the mean of the electric densities on the convex and
concave sides unity at the middle point (as in Green's numbers,
§ 234 above, for the plane disc). The six points for which
the electric densities are shown in the tables below are (not
the six points to which Coulomb's observations and Green's
numbers quoted in § 234 refer, but) the middle point, and the
five points dividing the arc from the middle to the edge or lip
into six equal parts.

243. A second application of the principle stated in § 210,
and used in §§ 237...239, allows us to proceed from the solution now found for the electrification of an uninfluenced bowl
to determine the electrification of a bowl or disc under the
influence of electricity insulated at a point Q (not, as in the
solution of § 239, necessarily in the spherical surface or plane
of the bowl or disc, but) *anywhere in the neighbourhood.* Consider the image, S, of an uninfluenced electrified bowl, S',
relatively to a spherical surface described from any point Q in
its neighbourhood, as centre, with radius R. Let D' be the
point on the spherical surface of S' continued, which is equidistant from the lip (so that D' and the middle point of the
conducting surface S' are the two poles of the circle constituting the lip) ; $D'K'P'L'$ the circle in which S', and the continuation of its spherical surface, are cut by the plane through
D', Q, and any point P of S at which it is ·desired to find the
electric density ; and $DKPL$ the image of $D'K'P'L'$.

In the annexed diagrams two cases are illustrated ; in one
of which S is spherical and concave towards the influencing

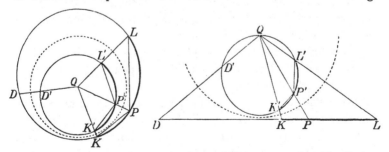

point, Q ; in the other, S is plane. Using now for S' all the
notation of §§ 240, 241, but with accents added, and taking
advantage of § 238, footnote, we see that

$$a'^2 - r'^2 = K'P' \cdot P'L'$$

and $$f'^2 - a'^2 = D'K^2 = D'L'^2 = D'K' \cdot D'L'.$$

Hence (19) becomes

$$\rho' = \frac{V'}{2\pi^2 f'} \left\{ \sqrt{\frac{D'K' \cdot D'L'}{P'K' \cdot P'L'}} - \tan^{-1} \sqrt{\frac{D'K' \cdot D'L'}{P'K' \cdot P'L'}} \right\} \quad \dots (21),$$

for the electric density at P' on the concave side of S'. But, as in § 238, we find

$$P'K' = PK \cdot \frac{R^2}{QP \cdot QK}, \quad P'L' = PL \cdot \frac{R^2}{QP \cdot QL} \quad \dots (22),$$

and $$D'K' = D'L' = DK \frac{R^2}{QD \cdot QK} = DL \frac{R^2}{QD \cdot QL} \dots (23).$$

Also, if h denote the shortest distance from Q to the spherical or plane surface of S, and f the diameter of this surface (infinite of course when the surface is plane, or negative if the convexity be towards Q), we have

$$f' = \frac{R^2}{h} + \frac{R^2}{f - h} = \frac{R^2 f}{h(f - h)} \quad \dots\dots\dots\dots\dots(24).$$

Using these in (21), putting $V' = \frac{q}{R}$, and substituting the expression so obtained for ρ' in (8) of § 237, we find

$$\rho = \frac{qh(f - h)}{2\pi^2 f \cdot PQ^3} \left\{ \frac{PQ}{DQ} \cdot \sqrt{\frac{DK \cdot DL}{PK \cdot PL}} - \tan^{-1} \left[\frac{PQ}{DQ} \sqrt{\frac{DK \cdot DL}{PK \cdot PL}} \right] \right\} (25),$$

for the electric density on the side of S *remote* from Q (that is, the convex or concave side, when S is spherical, according as Q is within or without the completed spherical surface). The electric density on the side next Q is [§ 241 (20)] the same, with the addition of $$\frac{qh(f - h)}{2\pi f \cdot PQ^3} \quad \dots\dots\dots\dots\dots\dots(26).$$

These formulæ, (25) and (26), express the electric density on the two sides of a circular segment or disc of infinitely thin spherical or plane conducting surface connected with the earth by an infinitely fine wire, and electrified by the influence of a quantity $-q$ of electricity insulated at a point Q anywhere in its neighbourhood.

244.   The position of the auxiliary point D (which appears in the diagrams as the image of D', the unoccupied pole of the lip of the original bowl S') may be found, without reference to S', by construction from S and Q supposed given; thus :—From (22) of § 243 we have

$$KD : DL :: KQ : QL \quad \dots\dots\dots\dots(27),$$

where K and L may be the points in which the lip of the bowl S is cut by *any* plane through $QD'D$. Let, for instance, this plane pass through the centre of one of the spherical surfaces. It must also pass through the centre of the other, and bisect each bowl; and if E, F be the points in which it cuts the lip of S, (26) applied to the present case gives

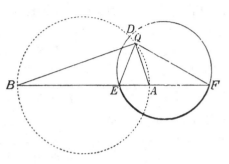

$ED : DF :: EQ : QF.$

Hence (Euclid, vi. 3) the lines bisecting the angles EDF, EQF cut the base EF in the same point; and D must be in the circle which is the locus of all points in the plane EFQ fulfilling this condition, being found by the well-known construction, thus :—Bisect the angle EQC by QA, meeting EF in A. Draw QB perpendicular to QA, and let it meet EF produced, in B. On BA as diameter describe a circle, which is the required locus; and D is the point in which this circle cuts the unoccupied part of the spherical or plane surface of S.

245. D being found by this simple construction, the solution of the problem is complete, without reference to S', thus :—To find the electric density at any point P, draw a plane through QDP, and let it meet the lip in K and L. Measure DK, DL, PK, PL, PQ, and DQ, and calculate by (25) and (26). But we have an important simplification from the geometrical theorem of § 238, which shows that

$$\frac{DK . DL}{PK . PL} = \frac{Dk . Dl}{Pk . Pl} \quad \dots\dots\dots\dots\dots(28)$$

if k, l be points in which the lip is cut by any plane whatever through PD. Choose, for instance, the plane through PD, and C the middle point of S. Then, as D, k, P, C, l lie all on one circle, and C is the middle point of the arc kPl, we have (as above, in § 238)

$$Dk . Dl = CD^2 - Ck^2 = CD^2 - a^2 ,$$
$$Pk . Pl = Ck^2 - CP^2 = a^2 - CP^2 ;$$

where, as before, a denotes the chord from the middle point to the lip. Using this in (28) and (25) we have, finally,

$$\rho = \frac{qh(f-h)}{2\pi^2 fPQ^3}\left\{\frac{PQ}{DQ}\sqrt{\frac{CD^2-a^2}{a^2-CP^2}} - \tan^{-1}\left[\frac{PQ}{DQ}\sqrt{\frac{CD^2-a^2}{a^2-CP^2}}\right]\right\} \quad (29),$$

for the density on the side remote from Q; h and $f-h$ being the shorter and longer distance.

246. For the case in which S is a plane disc, or $f = \infty$, this becomes

$$\rho = \frac{qh}{2\pi^2 PQ^3}\left\{\frac{PQ}{DQ}\sqrt{\frac{CD^2-a^2}{a^2-CP^2}} - \tan^{-1}\left[\frac{PQ}{DQ}\sqrt{\frac{CD^2-a^2}{a^2-CP^2}}\right]\right\} \quad (30),$$

and the addition (26) to it to give the electric density on the side next to Q,

$$\frac{qh}{2\pi PQ^3} \quad\dots\dots\dots\dots\dots\dots\dots\dots(31).$$

Also, as EFD is a straight line in this case, (27) gives

$$CD = \tfrac{1}{2}EF\,\frac{QF+QE}{QF-QE} \quad\dots\dots\dots\dots\dots\dots(32).$$

QE, QF are to be calculated immediately from data of whatever form, specifying the position of Q; and from them and CD found by this formula, DQ is to be calculated. Thus explicitly we have every element required for calculating electric densities by (30).

247. For the case of Q in the axis of the disc, D is infinitely distant, so that $CD = \infty$, $DQ = \infty$, and $\dfrac{CD}{DQ} = 1$.   And if for CP we put r, (30) and (31) give, for the density on remote side,

$$\rho = \frac{qh}{2\pi^2 PQ^3}\left\{\frac{PQ}{(a^2-r^2)^{\frac{1}{2}}} - \tan^{-1}\frac{PQ}{(a^2-r^2)^{\frac{1}{2}}}\right\} \quad\dots\dots(33),$$

and for the density on near side,

$$\rho + \frac{qh}{2\pi PQ^3} \quad\dots\dots\dots\dots\dots\dots\dots(34).$$

If P be at the centre of the disc, and if we take $q = 2\pi^2$, these become

$$\left.\begin{array}{l}\text{for remote side, } \rho = \dfrac{1}{h^2}\left(\dfrac{h}{a} - \tan^{-1}\dfrac{h}{a}\right) \\[2mm] \text{for near side, }\quad \rho + \dfrac{\pi}{h^2}\end{array}\right\} \quad\dots\dots\dots\dots(35);$$

from which the following numerical results have been calculated, with a, the radius of the disc taken as unity :—

Distance of Influencing Point h.	Induced Electric Density at middle of Disc	
	On remote side $\rho = h^{-1} - \cot^{-1}(h^{-1})$.	On near side $\rho + \pi h^{-2}$.
·2	·1651	78·7049
·4	·1218	19·7567
·6	·1655	8·8921
·8	·1957	5·1044
1·0	·2146	3·3562
1·2	·2250	2·4067
1·4	·2293	1·8322
1·6	·2296	1·4568
1·8	·2273	1·1969
2·0	·2232	1·0086
3·0	·1946	·5437

248. These numbers show that the distance at which the influencing point, if restricted to the axis of the disc, must be held to render the induced electric density at the middle on the far side a maximum is about 1·5 times the radius. But the characteristics (1) of the zero electric density on the far side, and infinite on the near side, when the influencing point is infinitely near the disc ; (2) the proportionality of the latter to h^{-2} for very small distances; and (3) the ultimate vanishing of the difference between the two sides as the influencing point is removed to an infinite distance, and the approximation of each to Green's result for a plane uninfluenced disc electrified to a potential equal to qh (§ 234, above), is better illustrated by the formulæ themselves, (35) ; (33), (34) ; and (30), (31) ; than by any numerical results calculated from them, however elaborately. It would be interesting to continue the analytical investigation far enough to determine the electric potential at any point in the neighbourhood of a disc electrified under influence, and so to illustrate further than is done by the numbers and formulæ already obtained, the theory of *electric screens*, and of Faraday's celebrated "induction in curved lines" (*Experimental Researches in Electricity*, §§ 1161, 1232; Dec. 21, 1837) ; but I am obliged to leave the subject for the present, in the hope that others may be induced to take it up.

XVI. ATMOSPHERIC ELECTRICITY.

[From Nichol's *Cyclopædia*, 2d Ed. (1860).]

249. It may be premised, to avoid circumlocution in this article, that every body in communication with the earth by means of matter possessing electric conductivity enough to prevent its electric potential* from differing sensibly from that of the earth, will be called part of the earth. Moist stone, and rock of all kinds, and all vegetable and animal bodies, in their natural conditions, except in circumstances of extraordinary dryness, possess, either superficially or throughout their substance, the requisite conductivity to fulfil that condition. On the other hand, various natural minerals and artificial compounds, such as glass,—various vegetable gums, such as India-rubber, gutta percha, rosin,—and various animal products, such as silk and gossamer fibre,—when either in a very dry natural or in an artificially dried atmosphere, resist electrical conduction so strongly that they may support a body, or otherwise form a material communication between it and the earth, and yet allow it to remain charged with electricity to a potential sensibly differing from the earth's, for fractions of a second, for minutes, for hours, for days, or even for years, without any fresh excitation or continued source of electricity. Again, air, whether dry or saturated with vapour of water, and probably all gases and vapours, unless ruptured by too strong an electromotive force, are very thoroughly destitute of conductivity—that is to say, are very perfectly endowed with the property of resisting the tendency of electricity to pass and establish

---

* Two conducting bodies are said to be of the same electric potential when, if put in conducting communication with the two electrodes of an electrometer, no electric effect is produced. When, on the other hand, the electrometer shows an effect, the amount of this effect measures the difference of potentials between the two bodies thus tested. Difference of potentials is also called electromotive force.

equality of potential between two bodies not otherwise materially connected.

250.   Hence, when "the surface of the earth" is spoken of, the surface separating the solids and liquids of the earth from the air will be meant; and when the more qualified expression "outer surface of the earth" is used, inner surfaces of vesicles, or the surfaces bounding completely enclosed spaces of air, must be understood to be excluded.   Thus, the surface of a mountain peak; the surface of a cave, up to the inmost recesses of the most intricate passages; the surface of a tunnel; the surface of the sea, or of a lake or river; all the surface of a sheet of unbroken water in such a fall as that of Niagara; the surface of blades of grass and flowers, and of soil below; in a wood, the surface of soil, and of trunks and leaves of trees; the surface of any animal resting on the earth; the outside of the roof of a house; the whole inside surface of a room with an open window; all belong to the outer surface of the earth.

251.   On the other hand, the moon, meteoric stones, birds or insects flying, leaves or fruit falling, seed wafted through the air, spray breaking away from a cascade or from waves of the sea, the liquid particles of a cloud or a fog, present surfaces not belonging to the earth, and between which and the earth's surface differences of potential, and lines of electric force, may and generally do exist.

252.   The whole surface of the earth, as defined above (§ 250), is at every moment electrified in every part, with the exception of neutral lines dividing portions which are negatively (resinously) from portions which are positively (vitreously) electrified. The negatively electrified portions are of very much greater extent, at all times, than those positively electrified; and there may be times when the whole surface is negatively electrified, because in all localities in which electrical observations have been hitherto made, with possibly one remarkable exception*, the earth's surface is always found negative, day and night,

---

* At Guajara station, on the Peak of Teneriffe, "During the whole period of "observation, by day and night, the electricity was moderate in quantity, and "always resinous.   This was during the period of N.E. trade wind, and within "its influence, though above its clouds."—[Professor Piazzi Smyth's Account of the Teneriffe Astronomical Experiment, *Philosophical Transactions*, 1858, and separate publication ordered by the Lords of the Admiralty.]   The "electricity" here referred to was that acquired by an insulated conductor carrying a burning

during fair weather, and only occasionally positive in broken
weather, or during an actual fall of rain in the immediate
neighbourhood, if not exactly on the place of observation. If,
then, at any one time there chances to be fair weather over the
whole earth, it may be presumed that the whole outer surface
of the earth is then negatively electrified, unless, judging from
the possible exception above alluded to, we are still to expect
positive electrification in same extreme positions.

253. As yet nothing is known regarding the electrification
of air itself*, or of clouds or other matter suspended in the air,
except what can be inferred (see below, § 254) from the electrification of the earth's surface, and its variations, with which
alone, as Peltier has remarked, the observations of "atmospheric electricity" hitherto published have dealt (see below,
§§ 265, 266). It is impossible, in the nature of things, to
investigate the bodily electrification of a non-conductor by any
observation whatever of electric action without it†, or in any
way whatever, except by something equivalent to a determination of the magnitude and direction of the resultant force at
every point of its mass‡. Towards this thorough investigation

match in the air at some distance from the earth. If it were really negative,
the earth's electrification at the place must have been positive; but the test as
to quality may have been deceptive, owing to the highly insulating condition of
both outer and inner surfaces of the glass shade enclosing the gold leaves,
and to the circumstance of the testing piece of rubbed sealing wax having been
applied possibly too near the gold leaves, instead of beside a remote part of the
insulated rod. Professor Smyth assures the writer, that he considers the
electrical experiment as not sufficiently complete or confirmed to allow any
conclusion to be built on it, and regards it rather as an indication of the
importance of making electrical observations with better apparatus, and
more available time for using it, than the first Teneriffe scientific expedition
afforded.

* For knowledge gained since this article was written see §§ 296—301 below.

† According to Green's remarkable theorems, triply rediscovered by Gauss,
Chasles, and the writer of this article, all different distributions of electricity
within a solid, which produce the same potential at its surface, produce the
same force at every point without it, and the problem of finding a distribution
of electricity within the interior, to produce a given distribution of potential at
the surface, is indeterminate.

‡ Let X, Y, Z be the components of the resultant force on a unit of electricity, if placed at any point x, y, z in a mass of air or other non-conductor;
and let ρ denote the electrical density of the substance, that is to say, the
quantity of electricity per unit of bulk actually possessed by the air in the
neighbourhood of this point. Then, by a well-known proposition of the mathematical theory of attraction, we have

$$\rho = \frac{1}{4\pi}\left(\frac{dX}{dx} + \frac{dY}{dy} + \frac{dZ}{dz}\right).$$

of the distribution of electricity within a non-conducting mass,
it may be remarked, that a determination of the normal component of the force all round a closed surface is just sufficient to
show the aggregate quantity of electricity possessed by all the
matter situated within it*. Hence observation in positions all
round a mass of air is necessary for determining the quantity
of electricity which it contains; and, therefore, the balloon
must be put in requisition if knowledge of the distribution of
electricity through the atmosphere is to be sought for.

254. Without leaving the earth, however, although we cannot
thoroughly investigate the electrification of the air, we can
make important inferences about it from observations of the
electric density over the earth's surface, by a principle of judging which may be thus explained :—If the earth were simply
an electrified body, placed in a perfectly insulating medium of
indefinite extent, and not sensibly influenced by any other
electrified matter, or by reflex influence from any conductor or
dielectric in its vicinity, its electricity would be distributed
over its surface according to a perfectly definite law, depending solely on the form of the surface, and deducible by a
sufficiently powerful mathematical analysis from sufficiently
perfect data of "geometry" (in the primitive sense of the term),
or of what, in more modern language, is called geodesy. If
the surface of the earth were truly spherical, this law would
simply be uniform distribution. A truly elliptic oblateness of
the earth would give, instead of uniformity, a distribution of
electric density in simple proportion to the perpendicular
distance between a tangent (that is horizontal) plane through
any point and the earth's centre; according to which the electric
density at the equator would be greatest, and would exceed
that at either pole, where it would be least, by $\frac{1}{305}$: a difference which, for the present, we may disregard.

255. The whole amount of electricity over the surface of
any great region of mountainous country, or of forest land,

---

* Let N be the normal component of the force at any point of a closed
surface, ds an element of the surface, $\int$ the sign of integration for the whole
surface, and Q the whole quantity of electricity within it. Then, by a well-
known theorem of Green's, rediscovered as alluded to in a preceding note,
we have
$$Q = \frac{1}{4\pi} \int N ds.$$

or of soil and vegetation of any kind, or of streets and houses in
a town, or of rough sea, would be very approximately the same
as that on an area of unruffled ocean, equal to the "reduced"
area of the irregular surface; but the distribution of the electricity over hill and valley, over the leaves and trunks of trees,
and the surfaces of plants generally, and on the soil beneath
them, over the roofs, perpendicular walls, and overhanging or
overshaded surfaces of buildings, and the surfaces of streets and
enclosed courts between them, and over the hollows and crests
of waves in a stormy sea, would be extremely irregular, with,
in general, greater electric density on the more prominent and
convex portions of surfaces, and less on the more covered and
concave—quite insensible, indeed, in any such position as the
interior of a cave, or the soil below trees in a forest even where
considerable angular openings of sky are presented,—or the
roof or floor of a tunnel, or covered chamber, even although
open to a considerable angle of sky.

256.   If thus a perfect electro-geodesy gave a "reduced"
electric density equal over the whole earth, we might infer that
the electrification of the earth is not influenced by any electricity in the air.   According to what has been stated above,
there might in that case be either no electricity in the air, from
the earth's atmosphere to the remotest star, and the lines of
electric force rising from the earth might either be infinite or
terminate in the surfaces of the moon, meteoric stones, sun,
planets, and stars; or there might be, at any distance considerably exceeding the height of the highest mountain, a uniformly electrified stratum of equal quantity and opposite kind
to the earth's, balancing through all the exterior space the force
due to the terrestrial electricity, and limiting the manifestations
of electric force to the atmosphere within it; or there might be
any of the infinite variety of distributions of electricity in space
round the earth, by which the electric density at the earth's
surface would be uninfluenced.

257.   But, in reality, the electric density varies greatly, even
in serene weather, over the earth's surface at any one time,
as we may infer from (1.) the facts (established for Europe,
and probably true in all the temperate zones of both hemispheres), that in any one place the electric density of the

surface observed during serene weather is much greater in winter than in summer, and that it varies according to something of a regular periodicity with the hours of the day and night; and (2.) the consideration that there is often serene weather of day and night, and of summer and winter, at one and the same time, in different temperate portions of the earth. We may, therefore, consider it as quite established that, even in serene weather, the electrification of the earth's surface is largely influenced by external electrified matter. Although we cannot (§ 253) discover the exact locality and distribution of this influencing electricity from its effects at the earth's surface alone, yet it is possible, from the character of the distribution of the terrestrial electric density as influenced by it, to assign a superior limit to its height*. If at any one instant the electric density reduced to the sea level were distributed according to a simple "harmonic" law, or, more generally, according to a certain definite character of non-abruptness of variation easily specified in mathematical language†, the external influencing electricity might be at any distance, however great, for all we could discover by observations near the earth's surface. But, little as we know yet regarding the diurnal law of electric variation in serene weather, it is, we may say with almost perfect certainty, not such as could give at any instant a distribution over the whole earth possessing any such gradual character as that referred to; and, therefore, we may, in all probability, from the character of the diurnal variation itself, say that its electric origin is not at a distance of many radii from the surface. On the other hand, when we consider that in temperate regions the velocity with which the earth's surface

---

* If at any instant the co-efficients of the series of "Laplace's functions," expressing the terrestrial electric density reduced to the sea level, converged ultimately with less rapidity than the geometrical series $1, \frac{1}{m}, \frac{1}{m^2} \ldots$ we might be sure that there is electricity in the air at some distance from the centre of the earth, not exceeding m times the radius of the earth's surface. For the principles on which this assertion is founded, see a short article, entitled "Note on Certain Points in the Theory of Heat," *Cambridge Mathematical Journal*, November 1843.

† For instance, if in simple proportion to the cosine of the angular distance from any point of the earth's surface, or more generally, if expressible by any finite number of "Laplace's functions," or still more generally, if expressible by a series of "Laplace's functions," with co-efficients converging ultimately more rapidly than any geometrical series.

is carried round in its diurnal course is from 500 to 900 miles per hour, we see clearly that any law of diurnal electric variation, established on observations even so frequent as once every hour, could not possibly fix the locality of the origin to within 100 miles of the surface; and as we have as yet nothing to go upon in the way of published observations more frequent than three or four times a day, towards establishing either the existence or the character of the diurnal law, we cannot consider it as proved by observation that the influencing electricity which produces it is even as near as the 50 or 100 miles limit which is commonly (but in the opinion of the writer of this article, most unreasonably) assigned as an end to the earth's atmosphere.

258.   The great suddenness of the electric variations during broken weather, and their close correspondence with beginnings, changes, and cessations of rain, hail, or snow, compel us (by a *common sense* estimate founded on an unconscious application of the mathematical law stated in the footnotes to the preceding § 257) to believe that their origin agrees in position with that of the showers, and to give it a "local habitation" and a name—Thundercloud.

259.   The writer of this article has observed extremely rapid variations of terrestrial electrification during perfectly serene weather.   Thus, in a calm summer night, with an unvarying cloudless sky overhead, and not the faintest appearance of auroral light to be seen, he has, in a temporary electric observatory in the Island of Arran, found large variations (as much as from a certain degree to double and back) in the course of a minute of time.   The influencing electricity by which these variations were produced, cannot possibly (unless on the extremely improbable hypothesis of their being due to highly electrified extra-terrestrial matter moving very rapidly with reference to the earth) have been very far removed from the earth's surface.   It is not impossible, and we have as yet nothing to make it decidedly improbable, that they were due to fluctuations up and down of aërial strata, perhaps those of the great atmospheric currents, in high regions of the atmosphere.   Judging, however, from still more recent observations referred to below (§ 262), we may think it more probable that these remarkable variations in the observed electric force were

due chiefly to positively or negatively electrified masses moving along within a few miles of the locality of observation.

260. Returning to the subject of the distribution of electricity over the earth's surface at any instant, we may remark, that if over an area of several miles in diameter, of perfectly level bare country, or of sea, the electrical density is sensibly uniform, we could not, without going up in a balloon, and observing the electric force at points in the air above, form any judgment whatever as to the distance from the earth at which the influencing electricity is situated. If, on the other hand, we find a very sensible variation in the electric density between two points of a piece of level open country, or at sea, not many miles apart, we may infer as quite certain that there is influencing electricity not many miles up in the air, and not uniformly distributed in level strata. Nothing can be easier than to make this trial—only to observe simultaneously with similar instruments, similarly placed, at two neighbouring stations, in a suitable locality—and most interesting and important results are to be derived from it, as soon as arrangements can be made for continuing the requisite observations day and night, during various vicissitudes of weather, especially during a time of perfect serenity.

261. Corresponding statements apply to a mountainous country, with this modification, that a very varied, instead of a uniform distribution of electric density, is, in such a locality, as explained above in § 255, the natural consequence of freedom from the disturbing influence of near electrified masses of air or cloud. The problem of accurately determining, from purely geometric data (§ 256), this undisturbed distribution over even the smoothest hillside, would infinitely transcend human mathematical power, although an approximate solution may be readily given for any piece of country over the whole of which both the inclination and the ratio of the height above the general level to the radius of curvature of the surface are small. For a rugged mountainous country, the most perfect geometric data, and the most strenuous mathematical efforts, could scarcely lead us towards an approximate estimate of the inequalities of electric density which different localities must present without any disturbance from near electrified atmosphere. Hence, in a

mountainous country—unless we find electricity strong in some
locality where from the configuration of the surface, we correctly
judge it ought to be weak if undisturbed, or weak where it ought
to be strong, or unless, at least, we find some very decided deviation from any such amount of difference between two stations
as, without being able to make a precise calculation, we can
estimate for the difference due to figure—we cannot judge as
to the influence of aërial electrification from simultaneous
absolute determinations at any one instant alone.   But of one
thing we may be sure, that although the absolute amounts of
the electrification at any two stations not far apart may differ
largely, they must remain in an absolutely constant proportion to one another, if there is no electrified air or cloud near.

262.   Hence, if we find observations made simultaneously by
two electrometers in neighbouring positions, in a mountainous
country, to bear always the same mutual proportion, we may
not be able to draw any inference as to electrified air ; but if,
on the contrary, we find their proportion varying, we may be
perfectly certain that there are varying electrified masses of air
or cloud not far off.   A first application of this test is described
in the following extract from the Proceedings of the Literary
and Philosophical Society of Manchester for October 18, 1859:—

"The following extract of a letter received from Professor W.
" Thomson, F.R.S., Glasgow, Honorary Member of the Society,
" etc., was read by Dr Joule :—

' I have a very simple "domestic" apparatus by which I can
' observe atmospheric electricity in an easy way.   It consists
' merely of an insulated can of water set on a table or window
' sill *inside,* and discharging by a small pipe through a fine nozzle
' two or three feet from the wall.   With only about ten inches
' head of water and a discharge so slow as to give no trouble in
' replenishing the can with water, the atmospheric effect is
' collected so quickly that any difference of potentials between
' the insulated conductor and the air at the place where the
' stream from the nozzle breaks into drops is done away with at
' the rate of five per cent. per half second, or even faster.   Hence
' a very moderate degree of insulation is sensibly as good as
' perfect, so far as observing the atmospheric effect is concerned.
' It is easy, by my plan of drawing the atmosphere round the
' insulating stems by means of pumice-stone moistened with

' sulphuric acid, to insure a degree of insulation in all weathers,
' by which there need not be more than five per cent. per hour
' lost by it from the atmospheric apparatus at any time. A little
' attention to keep the outer part of the conductor clear of
' spider lines is necessary. The
' apparatus I employed at In-
' vercloy stood on a table beside
' a window on the second floor,
' which was kept open about
' an inch to let the discharg-
' ing tube project out without
' coming in contact with the

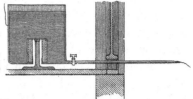

Fig. 1.

' frame. The nozzle was only about two feet and a half from
' the wall, and nearly on a level with the window sill. The
' divided ring electrometer stood on the table beside it, and
' acted in a very satisfactory way (as I had supplied it with a
' Leyden phial, consisting of a common thin white glass shade
' which insulated remark-
' ably well, instead of the
' German glass jar—the
' second of the kind which
' I had tried, and which
' would not hold its charge
' for half a day). I found
' from $13\frac{1}{4}°$ to 14° of torsion
' required to bring the index
' to zero, when urged aside
' by the electromotive force
' of ten zinc-copper water
' cells. The Leyden phial
' held so well, that the sensi-
' bility of the electrometer,
' measured in that way, did
' not fall more than from
' $13\frac{1}{2}°$ to $13\frac{1}{4}°$ in three days.
' The atmospheric effect
' ranged from 30° to above
' 420° during the four days
' which I had to test it; that

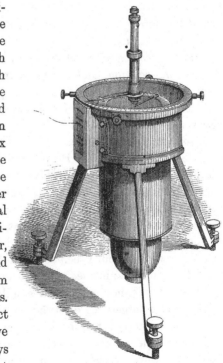

Fig. 2.

'is to say, the electromotive force per foot of air, measured hori-
'zontally from the side of the house, was from 9 to above 126 zinc-
'copper water cells. The weather was almost perfectly settled,
'either calm, or with slight east wind, and in general an easterly
'haze in the air. The electrometer twice within half an hour went
'above 420°, there being at the time a fresh temporary breeze
'from the east. What I had previously observed regarding the
'effect of east wind was amply confirmed. Invariably the
'electrometer showed very high positive in fine weather, before
'and during east wind. It generally rose very much shortly
'before a slight puff of wind from that quarter, and continued
'high till the breeze would begin to abate. I never once
'observed the electrometer going up unusually high during fair
'weather without east wind following immediately. One even-
'ing in August I did not perceive the east wind at all, when
'warned by the electrometer to expect it; but I took the
'precaution of bringing my boat up to a safe part of the beach,
'and immediately found by waves coming in that the wind
'must be blowing a short distance out at sea, although it did
'not get so far as the shore. I made a slight commencement
'of the *electrogeodesy* which I pointed out as desirable at the
'British Association, and in the course of two days, namely;
'October 10th and 11th, got some very decided results. Mac-
'farlane, and one of my former laboratory and Agamemnon
'assistants, Russell, came down to Arran for that purpose. Mr
'Russell and I went up Goatfell on the 10th instant, with the
'portable electrometer (see Fig. 3), and made observations, while
'Mr Macfarlane remained at Invercloy, constantly observing
'and recording the indications of the house electrometer. On
'the 11th instant the same process was continued, to observe
'simultaneously at the house and at one or other of several
'stations on the way up Goatfell. I have not yet reduced all
'the observations, but I see enough to leave no doubt whatever
'but that cloudless masses of air at no great distance from the
'earth, certainly not more than a mile or two, influence the
'electrometer largely by electricity which they carry. This I
'conclude because I find no constancy in the relation between
'the simultaneous electrometric indications at the different
'stations. Between the house and the nearest station the rela-

'tive variation was least. Between the house and a station about
'half way up Goatfell, at a distance estimated at two miles and
'a half in a right line, the number expressing the ratio varied
'from about 113 to 360 in the course of about three hours. On
'two different mornings the ratio of a house to a station about
'sixty yards distant on the road beside the sea was 97 and 96
'respectively. On the afternoon of the 11th instant, during a
'fresh temporary breeze of east wind, blowing up a little spray as
'far as the road station, most of which would fall short of the
'house, the ratio was 108 in favour of the house electrometer
'—both standing at the time very high—the house about 350°.
'I have little doubt but that this was owing to the negative
'electricity carried by the spray from the sea, which would
'diminish relatively the indications of the road electrometer'."

263. The electrometers referred to in the preceding extract
were on two different plans. The first, or "divided ring
electrometer," consists of—(1.) A ring of metal divided into
sectors, of which some—one or more—are insulated and connected with the conductor to be electrically tested, and the
remainder connected with the earth. (2.) An index of metal
supported by a glass fibre, or a wire, stretched in the line of
the axis of the ring, and capable of having its fixed end turned
through angles measured by a circle and pointer. (3.) A
Leyden phial, with its insulated coating electrically connected
with the index. (4.) A case to protect the index from currents
of air, and to keep an artificially dried atmosphere round the
insulating supports—glazed to allow the index to be seen from
without, but with the inner surface of the glass screened
(electrically) by wire cloth, perforated metal, or tinfoil, to do
away with irregular reflections on the index. In the instrument represented in the drawing (No. 2) above, the ring is
divided only into two parts, which are equal, and separated by
a space of air about $\frac{1}{20}$ of an inch. Each of these half rings is
supported on two glass pillars; and by means of screws acting
on a foot which bears these pillars, it is adjusted and fixed in
its proper position. The index is of thin sheet aluminium, and
projects in only one direction from the glass fibre bearing it.
A stiff vertical wire, rigidly connected with it, nearly in the
prolongation of the fibre, bears a counterpoise considerably

below the level of the index, and heavy enough to keep the index horizontal. A thin platinum wire hooked to the lower end of this vertical wire, dips in sulphuric acid in the bottom of the Leyden phial. The Leyden phial is charged either positively or negatively; and is found to retain its charge for months, losing, however, gradually, at some low rate, less generally than one per cent. per day of its amount. The index is thus, when the instrument is in use, kept in a state of charge corresponding to the potential of the inside coating of the phial. When one of the half rings is connected with the earth, and a charge of electricity communicated to the other, the index moves from or towards the latter, according as the charge communicated to it is of the same or the opposite kind to that of the index. This instrument, as an electroscope, possesses extreme sensibility—much greater than that of any other hitherto constructed; and by the aid of the torsion arrangement, it may be made to give accurate metrical results. There are some difficulties in the use of it, especially as regards the comparison of the indications obtained with different degrees of electrification of the index, and the reduction of the results to absolute measure, hitherto obviated only by a daily application of Delmann's method of reference to a zinc-copper water battery, which Delmann himself applies once for all, to one of his electrometers (unless his glass fibre breaks, when he must make a fresh determination of the sensibility of the instrument with its new fibre). The high sensibility of the divided ring

Fig. 3.—Portable Atmospheric Electrometer.

electrometer renders this test really very easy, as not more than
from ten to twenty cells are required ; and a comparison with a
few good cells of Daniell's may be made by its aid, to ascertain
the absolute value and the constancy of the water cells.   The
difficulty thus met is altogether done away with in another
kind of electrometer, also "heterostatic," of which only one has
yet been constructed—the electrometer of the portable apparatus
shown in the third drawing.   In it the index is attached at
right angles to the middle of a fine platinum wire, firmly
stretched between the inside coatings of two Leyden phials,
and consists simply of a very light bar of aluminium, extending equally on the two sides of the supporting wire.   It is
repelled by two short bars of metal, fixed on the two sides of
the top of a metal tube, which is supported by the inside coating of the lower phial, and has the fine wire in its axis.   A
conductor of suitable shape, bearing an electrode, to connect
with the body to be tested, insulated inside the case of the
instrument, in the neighbourhood of the index, and when electrified in the same way, or the contrary way, to the inside
coatings of the Leyden phials, causes, by its influence, the
repulsion between the index and the fixed bars to be diminished
or increased.   The upper Leyden phial is moveable about a
fixed axis, through angles measured by a pointer and circle,
and thus the amount of torsion, in one-half of the bearing
wire, required to bring the index to a constant position, in any
case, is measured.   The square root of the number of degrees
of torsion measures the difference of potentials between the
conductor tested and the inner coating of the Leyden phial.
In using the instrument, the conductor tested is first put in
connexion with the earth, and the torsion required to bring the
index to its fixed position is read off.   This is called the zero,
or earth reading.   The tested conductor is then electrified, and
the torsion reading taken.   In the atmospheric application, this
is called the air reading.   The excess—positive or negative—
of its square root, above that of the zero reading, measures the
electromotive force between the earth and the point of air
tested.   This result, when positive shows vitreous, when negative resinous potential in the air ; if the index is resinous.   By
the aid of Barlow's table of square roots, the indications of the

instrument may thus be reduced to definite measure of potential, almost as quickly as they can be written down. Once for all, the sensibility of the instrument can be determined by comparison with an absolute electrometer, or a galvanic battery. In the portable apparatus a burning match is used—instead of the water-dropping system, which the writer finds more convenient than any other for a fixed apparatus—to reduce the insulated conductor to the same potential as the air at its end.

264. As has been remarked above (§ 252), it is the electrification of the earth's surface which has either directly or virtually been the subject of measurement in all observations on atmospheric electricity hitherto made. The methods which have been followed may be divided into two classes—(1.) Those in which means are taken to reduce the potential of an insulated conductor to the same as that of the air, at some point, a few feet or yards distant from the earth. (2.) Those in which a portion of the earth (see above, § 253) is insulated, removed from its position, and tested by an electrometer, in a different position, or under cover. The first method was very imperfectly carried out by Beccaria with his long " exploring wire," stretched between insulating supports, or elevated portions of buildings, tree tops, or other prominent positions of the earth (see above, § 249) ; also, very imperfectly by means of " Volta's lantern "— an enclosed flame, supported on the top of an insulated conductor. On the other hand, it is put in practice very perfectly, by means of a match, or flame burning in the open air, on the top of a well insulated conductor—a plan adopted, after Volta's suggestion, by many observers ; also, even more decidedly, by means of the water-dropping system—described in the preceding extract—which has recently occurred to the writer, and has been found by him both to be very satisfactory in its action, and extremely easy and convenient in practice. The principle of each of these methods of the first class may be explained best by first considering the methods of the second class, as follows :—

265. If a large sheet of metal were laid on the earth in a perfectly level district, and if a circular area of the same metal were laid upon it, and, after the manner of Coulomb's proof plane, were lifted by an insulated handle, and removed

to an electrometer within doors, a measure of the earth's electrification, at the time, would be obtained; or, if a ball, placed
on the top of a conducting rod in the open air, were lifted from
that position by an insulating support, and carried to an
electrometer within doors, we should also have, on precisely the
same principle, a measure of the earth's electrification at the time.
If the height of the ball in this second plan were equal to one-
sixteenth of the circumference of the disc (compare § 235) used
in the first plan, the electrometric indications would be the same,
provided the diameter of the ball is small, in comparison with
the height to which it is raised in the air, and the electrostatic
capacity of the electrometer is small enough not to take any
considerable proportion of the electricity from the ball in its
application.   The idea of experimenting by means of a disc laid
flat on the earth, is merely suggested for the sake of illustration, and would obviously be most inconvenient in practice.
On the other hand, the method, by a carrier ball, instead of a
proof plane, is precisely the method by which, on a small scale,
Faraday investigated the distribution of electricity induced on
the earth's surface (see above, § 249), by a piece of rubbed shell-
lac; and the same method, applied on a suitable scale, for testing the natural electrification of the earth in the open air, has
given, in the hands of Delmann of Creuznach, the most accurate
results hitherto published in the way of electro-meteorological
observation*.

266.   If, now, we conceive an elevated conductor, first belonging to the earth (§ 249), to become insulated, and to be made
to throw off, and to continue throwing off, portions from an
exposed position of its own surface, this part of its surface will
quickly be reduced to a state of no electrification, and the whole
conductor will be brought to such a potential as will allow it to
remain in electrical equilibrium in the air, with that portion of
its surface neutral.   In other words, the potential throughout
the insulated conductor is brought to be the same as that of the

---

* Through some misapprehension, Mr Delmann himself has not perceived
that his own method of observation really consists in removing a portion of the
earth, and bringing it insulated with the electricity which it possessed *in situ*,
to be tested within doors, otherwise, he could not have objected, as he has,
to Peltier's view.

particular equi-potential surface in the air, which passes through
the point of it from which matter breaks away. A flame, or
the heated gas passing from a burning match, does precisely
this : the flame itself, or the highly-heated gas close to the
match being a conductor which is constantly extending out,
and gradually becoming a non-conductor. The drops into
which the jet issuing from the insulated conductor, on the plan
introduced by the writer, produce the same effects, with more
pointed decision, and with more of dynamical energy to remove
the rejected matter with the electricity which it carries from
the neighbourhood of the fixed conductor.

ROYAL INSTITUTION FRIDAY EVENING LECTURE,

May 18, 1860.

267. Stephen Gray, a pensioner of the Charter-house, after
many years of enthusiastic and persevering devotion to electric
science, closed his philosophical labours, about one hundred
and thirty years ago, with the following remarkable conjecture :—"That there may be found a way to collect a greater
"quantity of the electrical fire, and consequently to increase
"the force of that power, which, by several of these experi-
"ments, *si licet magna componere parvis,* seems to be of the
"same nature with that of thunder and lightning."

The inventions of the electrical machine and the Leyden
phial immediately fulfilled these expectations as to collecting
greater quantities of electric fire; and the surprise and delight
which they elicited by their mimic lightnings and thunders,
and above all by the terrible electric shock, had scarcely sub-
sided when Franklin sent his kite messenger to the clouds, and
demonstrated that the imagination had been a true guide to
this great scientific discovery—the identity of the natural agent
in the thunderstorm with the mysterious influence produced
by the simple operation of rubbing a piece of amber, which,
two thousand years before, had attracted the attention of those

philosophers among the ancients who did not despise the small things of nature.

268. The investigation of atmospheric electricity immediately became a very popular branch of natural science; and the discovery of remarkable and most interesting phenomena quickly rewarded its cultivators. The foundation of all we now know was completed by Beccaria, in his observations on "the mild electricity of serene weather," nearly a hundred years ago. It was not until comparatively recent years that definite quantitative comparisons from time to time of the electric quality manifested by the atmosphere in one locality were first obtained by the application of Peltier's mode of observation with his metrical electroscope. The much more accurate electrometer, and the greatly improved mode of observation, invented by Delmann, have given for the electric intensity, at any instant, still more precise results; but have left something to desire in point of simplicity and convenience for general use, and have not afforded any means for continuous observation, or for the introduction of self-recording apparatus. The speaker had attempted to supply some of these wants, and he explained the construction and use of instruments, now exhibited to the meeting, which he had planned for this purpose.

269. Apparatus for the observation of atmospheric electricity has essentially two functions to perform; to electrify a body with some of the natural electricity, or with electricity produced by its influence; and to measure the electrification thus obtained.

270. The measuring apparatus exhibited, consisted of three electrometers, which were referred to under the designations of (I.) The divided ring reflecting electrometer; (II.) The common house electrometer; and (III.) The portable electrometer.

(I.) The divided ring reflecting electrometer [compare § 263, above, and §§ 444...456, below] consists of :—

(1) A ring of metal divided into two equal parts, of which one is insulated, and the other connected with the metal case (5) of the instrument.

(2) A very light needle of sheet aluminium hung by a fine glass fibre, and counterpoised so as to make it project only to one side of this axis of suspension.

(3) A Leyden phial, consisting of an open glass jar, coated outside and inside in the usual manner, with the exception that the tinfoil of the inner coating does not extend to the bottom of the jar, which is occupied instead by a small quantity of sulphuric acid [connected with the tinfoil by means of a platinum wire].

(4) A stiff straight wire rigidly attached to the aluminium needle, as nearly as may be in the line of the suspending fibre, bearing a light platinum wire linked to its lower end, and hanging down so as to dip into the sulphuric acid.

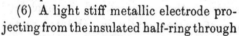

(5) A case protecting the needle from currents of air, and from irregular electric actions, and maintaining an artificially dried atmosphere round the glass pillar or pillars supporting the insulated half-ring and the uncoated portion of the glass of the phial.

(6) A light stiff metallic electrode projecting from the insulated half-ring through the middle of a small aperture in the metal case, to the outside.

(7) A wide metal tube of somewhat less diameter than the Leyden jar, attached to a metal ring borne by its inside coating, and standing up vertically to a few inches above the level of the mouth of the jar.

(8) A stiff wire projecting horizontally from this metal tube above the edge of the Leyden jar, and out through a wide hole in the case of the instrument to a convenient position for applying electricity to charge the jar with.

(9) A very light glass mirror, about three-quarters of an inch diameter, attached by its back to the wire (4), and therefore rigidly connected with the aluminium needle.

(10) A circular aperture in the case shut by a convex lens, and a long horizontal slit shut by plate glass, with its centre immediately above or below that of the lens, one of them above, and the other equally below the level of the centre of the mirror.

(11) A large aperture in the wide metal tube (7), on a level with the mirror (9), to allow light from a lamp outside the case, entering through the lens, to fall upon the mirror, and be

reflected out through the plate-glass window; and three or four fine metal wires stretched across this aperture to screen the mirror from irregular electric influences, without sensibly diminishing the amount of light falling on and reflected off it.

271. The divided ring (1) is cut out of thick strong sheet metal (generally brass). Its outer diameter is about 4 inches, its inner diameter $2\frac{1}{4}$; and it is divided into two equal parts by cutting it along a diameter with a saw. The two halves are fixed horizontally; one of them on a firm metal support, and the other on glass, so as to retain as nearly as may be their original relative position, with just the saw cut, from $\frac{1}{10}$ to $\frac{1}{20}$ of an inch broad, vacant between them. They are placed with their common centre as nearly as may be in the axis of the case (5), which is cylindrical, and placed vertically. The Leyden jar (3), and the tube (7), carried by its inside coating, have their common axis fixed to coincide as nearly as may be with that of the case and divided ring. The glass fibre hangs down from above in the direction of this axis, and supports the needle about an inch above the level of the divided ring. The stiff wire (4), attached to the needle, hangs down as nearly as may be along the axis of the tube (7).

[The following diagrams, placed here to facilitate comparison, represent the arrangement of "needle" and quadrants described below in § 345, as substituted in the modern instrument for the bisected ring and narrow needle of the old electrometer here described] :—

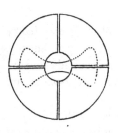 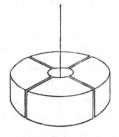

272. Before using the instrument, the Leyden phial (3) is charged by means of its projecting electrode (8). When an electrical machine is not available, this is very easily done by the aid of a stick of vulcanite, rubbed by a piece of chamois leather. The potential of the charge thus communicated to the phial, is

14—2

to be kept as nearly constant as is required for the accuracy of
the investigation for which the instrument is used. Two or
three rubs of the stick of vulcanite once a day, or twice a day,
are sufficient when the phial is of good glass, well kept dry.
The most convenient test for the charge of the phial is a
proper electrometer or electroscope, of any convenient kind,
kept constantly in communication with the charging electrode (8). [Compare § 353, below.]

The electrometer (II.) is to be ordinarily used for that purpose in the Kew apparatus. Failing any such gauge electrometer or electroscope, a zinc-copper-water battery of ten, twenty,
or more small cells may be very conveniently used (after the
manner of Delmann) to test directly the sensibility of the reflecting electrometer, which is to be brought to its proper degree
by charging its Leyden phial as much as is required.

273. In the use of this electrometer, the two bodies of which
the difference of potentials is to be tested are connected, one of
them, which is generally the earth, with the metal case of the
instrument, and the other with the insulated half ring. The
needle being, let us suppose, negatively electrified, will move
towards or from the insulated half ring, according as the potential of the conductor connected with this half ring differs positively or negatively from that of the other conductor (earth)
connected with the case. The mirror turns accordingly in one
direction or the other through a small angle from its zero position, and produces a corresponding motion in the image of the
lamp on the screen on which it is thrown.

274. (II.) The common house electrometer [compare § 263,
above, and §§ 374...377, below].—This instrument consists of:—

(1) A thin flint-glass bell, coated outside and inside like a
Leyden phial, with the exception of the bottom inside, which
contains a little sulphuric acid.

(2) A cylindrical metal case, enclosing the glass jar, cemented
to it round its mouth outside, extending upwards about an inch
and a half above the mouth, and downwards to a metal base
supporting the whole instrument, and protecting the glass
against the danger of breakage.

(3) A cover of plate glass, with a metal rim, closing the top
of the cylindrical case of the instrument.

(4) A torsion head, after the manner of Coulomb's balance, supported in the centre of the glass cover, and bearing a glass fibre which hangs down through an aperture in its centre.

(5) A light aluminium needle attached across the lower end of the fibre (which is somewhat above the centre of the glass bell), and a stiff platinum wire attached to it at right angles, and hanging down to near the bottom of the jar.

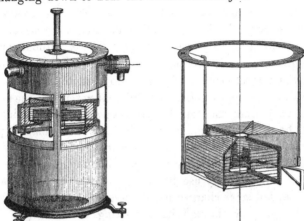

(6) A very light platinum wire, long enough to hang within one-eighth of an inch or so of the bottom of the jar, and to dip in the sulphuric acid.

(7) A metal ring, attached to the inner coating of the jar, bearing two plates in proper positions for repelling the two ends of the aluminium needle when similarly electrified, and proper stops to limit the angular motion of the needle to within about 45° from these plates.

(8) A cage of fine brass wire, stretched on brass framework, supported from the main case above by two glass pillars, and partially enclosing the two ends of the needle, and the repelling plates, from all of which it is separated by clear spaces, of nowhere less than one-fourth of an inch of air.

(9) A charging electrode, attached to the ring (7), and projecting over the mouth of the jar to the outside of the metal case (2), through a wide aperture, which is commonly kept closed by a metal cap, leaving at least one quarter of an inch of air round the projecting end of the electrode.

(10) An electrode attached to the cage (8), and projecting over

the mouth of the jar to the outside of the metal case (2), through the centre of an aperture, about a quarter of an inch diameter.

275. This instrument is adapted to measure differences of potential between two conducting systems, namely; as one, the aluminium needle (5), the repelling plates (7), and the inner coating of the jar; and, as the other, the insulated cage (8). This latter is commonly connected by means of its projecting electrode (10), with the conductor to be tested. The two conducting systems, if through their projecting electrodes connected by a metallic wire, may be electrified to any degree, without causing the slightest sensible motion in the needle. If, on the other hand, the two electrodes of these two systems are connected with two conductors, electrified to different potentials, the needle moves away from the repelling plates; and if, by turning the torsion head, it is brought back to one accurately marked position, the number of degrees of torsion required is proportional to the square of the difference of potentials thus tested.

276. In the ordinary use of the instrument, the inner coating of the Leyden jar is charged negatively, by an external application of electricity through its projecting electrode (9). The degree of the charge thus communicated, is determined by putting the cage in connexion with the earth through its electrode (10), and bringing the needle by torsion to its marked position. The square root of the number of degrees of torsion required to effect this, measures the potential of the Leyden charge. This result is called the reduced earth reading. When the atmosphere inside the jar is kept sufficiently dry,—this charge is retained from day to day with little loss; not more, often, than one per cent. in the twenty-four hours.

In using the instrument the charging electrode (9) of the jar is left untouched, with the aperture through which it projects closed over it by the metal cap referred to above. The electrode (10) of the cage, when an observation is to be made, is connected with the conductor to be tested, and the needle is brought by torsion to its marked position. The square root of the number of degrees of torsion now required measures the difference of potentials between the conductor tested and the interior coating of the Leyden jar. The excess, positive or negative, of this result above the reduced earth reading, measures the excess of the potential, positive or negative, of the conductor tested above that of the earth; or simply the potential of the conductor tested, if we regard that of the earth as zero.

277.   (III.)   The portable electrometer [compare § 263, above, and §§ 363...373, below] is constructed on the same electrical principles as the house electrometer just described. The mode of suspension of the needle is, however, essentially different; and a varied plan of connexion between the different electrical parts has been consequently adopted as more convenient.   In the portable electrometer, the needle is firmly attached at right angles to the middle of a fine platinum wire, tightly stretched in the axis of a brass tube with apertures in its middle to allow the needle to project on the two sides. One end of the platinum wire is rigidly connected with this tube; the other is attached to a graduated torsion head.   The brass tube carries two metal plates in suitable positions to repel the two ends of the needle in contrary directions, and metal stops to limit its angular motion within a convenient range.   The conducting system composed of these different parts is supported from the metal cover, or roof of the jar, by three glass stems.   The torsion head is carried round by means of a stout glass bar, projecting down from a pinion centered on the lower side of this cover, and turned by the action of a tangent screw presenting a milled head, to the hand of the operator outside.   The conducting system thus borne by insulating supports is connected with the outside conductor to be tested by means of an electrode passing out through the centre of the top of the case by a wide aperture in the centre of the pinion. A wire cage, surrounding the central part of the tube and the needle and repelling plates, is rigidly attached to the interior coating of the Leyden jar.   It carries two metal sectors, or "bulkheads," in suitable positions to attract the two ends of the needle, which, however, is prevented from touching them by the limiting stops referred to above.   The effect of these attracting plates, as they will be called, is to increase very much the sensibility of the instrument.   The square root of the number of degrees of torsion required to bring the needle to a sighted position near the repelling plates, measures the

difference of potentials between the cage and the conducting
system, consisting of tube, torsion-head, repelling plates, and
needle. The metal roof of the jar is attached to a strong metal
case, cemented round the outside of the top of the jar, and
enclosing it all round and below, to protect it from breakage
when being carried about. There are sufficient apertures in
this case, opened by means of a sliding piece, to allow the
observer to see the needle and graduated circle (torsion-head),
when using the instrument. On the outside of the roof of the
jar a stout glass stem is attached, which supports a light stiff
metallic conductor, by means of which a burning match is
supported, at the height of two or three feet above the observer.
This conductor is connected by means of a fine wire with the
electrometer, in the manner described above, through the centre
of the aperture in the roof. An artificially dried atmosphere
is maintained around this glass stem, by means of a metal case
surrounding it, and containing receptacles of gutta percha, or
lead, holding suitably shaped pieces of pumice-stone moistened
with sulphuric acid. The conductor which bears the match
projects upwards through the centre of a sufficiently wide aperture, and bears a small umbrella, which both stops rain from
falling into this aperture, and diminishes the circulation of air,
owing to wind blowing round the instrument, from taking place,
to so great a degree as to do away with the dryness of the interior atmosphere required to allow the glass stem to insulate
sufficiently. The instrument may be held by the observer in
his hand in the open air without the assistance of any fixed
stand. A sling attached to the instrument and passing over
his left shoulder, much facilitates operations, and renders it
easy to carry the apparatus to the place of observation, even if
up a rugged hill side, with little risk of accident.

278. The burning match in the apparatus which has just been
described, performs the collecting function referred to above.
The collector employed for the station apparatus, whether the
reflecting electrometer or the common house electrometer is
used, is an insulated vessel of water, allowed to flow out in a
fine stream through a small aperture at the end of a pipe projecting to a distance of several feet from the wall of the building in which the observations are made.

279. The principle of collecting, whether by fire or by water, in the observation of atmospheric electricity, was explained by the speaker thus :—The earth's surface is, except at instants, always found electrified, in general negatively, but sometimes positively. [Quotation from Nichol's *Cyclopædia*, viz., § 265, above, comes here in the original.]

After having given so much of these explanations as seemed necessary to convey a general idea of the principles on which the construction of the instruments of investigation depended, the speaker proceeded to call attention to the special subject proposed for consideration this evening.

280. What is terrestrial atmospheric electricity? Is it electricity of earth, or electricity of air, or electricity of watery or other particles in the air? An endeavour to answer these questions was all that was offered; abstinence from speculation as to the origin of this electric condition of our atmosphere, and its physical relations with earth, air, and water, having been painfully learned by repeated and varied failure in every attempt to see beyond facts of observation. In serene weather, the earth's surface is generally, in most localities hitherto examined, found negatively or resinously electrified; and when this fact alone is known, it might be supposed that the globe is merely electrified as a whole with a resinous charge, and left insulated in space.

281. But it is to be remarked that the earth, although insulated in its atmospheric envelope, being in fact a conductor, touched only by air one of the best although not the strongest of insulators, cannot with its atmosphere be supposed to be insulated so as to hold an electric charge in interplanetary space. It has been supposed, indeed, that outside the earth's recognised atmosphere there exists something or nothing in space which constitutes a perfect insulator; but this supposition seems to have no other foundation than a strange idea that electric conductivity is a strength or a power of matter rather than a mere *non-resistance*. In reality we know that air highly rarefied by the air-pump, or by other processes, as in the construction of the "vacuum tubes," by which such admirable phenomena of electric light have recently been seen in this place, becomes extremely weak in its resistance to the transference of elec-

tricity through it, and begins to appear rather as a conductor than an insulator. One hundred miles or upwards from the earth's surface, the air in space cannot in all probability have resisting power enough to bear any such electric forces as those which we generally find even in serene weather in the lower strata. Hence we cannot, with Peltier, regard the earth as a resinously charged conductor, insulated in space, and subject only to accidental influences from temporary electric deposits in clouds, or air round it; but we must suppose that there is always *essentially* in the higher aërial regions a distribution arising from the self-relief of the outer highly rarefied air by disruptive discharge. This electric stratum must constitute very nearly the electro-polar complement to all the electricity that exists on the earth's surface, and in the lower strata of the atmosphere; in other words, the total quantity of electricity, reckoned as excess of positive above negative, or of negative above positive, in any large portion of the atmosphere, and on the portion of the earth's surface below it, must be very nearly zero. The quality of non-resistance to electric force of the thin interplanetary air being duly considered, we might regard the earth, its atmosphere, and the surrounding medium as constituting respectively the inner coating, the di-electric (as it were glass), and the outer coating of a great Leyden phial, charged negatively; and even if we were to neglect the consideration of possible deposits of electricity through the body of the di-electric itself, we should arrive at a correct view of the electric indications discoverable at any one time and place of the earth's surface. In fact, any kind of "collector," or plan for collecting electricity from or in virtue of the natural "terrestrial atmospheric electricity," gives an effect simply proportional to the electrification of the earth's surface then and there. The methods of collecting by fire and water which the speaker exhibited, gave definitively, in the language of the mathematical theory, the "electric potential" of the air at the point occupied by the burning end of the match, or by the portion of the stream of water where it breaks into drops. If the apparatus is used in an open plane, and care be taken to eliminate all disturbance due to the presence of the electrometer itself and of the observer above the ground, the indicated effect, if expressed in absolute electrostatic measure, and divided by the height of the point tested above the ground, has only to be [according to an old theorem of Coulomb's (see footnote on § 25, above), corrected by Laplace] divided by four times the ratio of the circumference of a circle to its diameter, to reduce it to an expression of the number of units, in absolute electrostatic measure, of the electricity per unit of area of the earth's surface at the time and place. The mathematical theory does away with every difficulty in explaining the various and seemingly irreconcilable views which different writers have expressed, and explanations which different observers have given of the functions of their testing apparatus. In the present state of electric science, the most convenient and generally intelligible way to state the result of an observation of terrestrial atmospheric electricity, in absolute measure, is in terms of the number of elements of a constant galvanic battery, required to produce the same difference of potentials as exists between the earth and a point in the air at a stated height above an open level plane of ground. Observations with the portable electrometer had given, in ordinary fair weather, in the island of Arran, on a flat open sea beach, readings varying from 200 to 400, Daniel's elements, as the difference of potentials between the earth and the match, at a height of 9 feet above it. Hence, the intensity of electric force perpendicular to the earth's surface must have amounted to from 22 to 44 Daniel's elements per foot of air. In fair weather, with breezes from the east or north-east, he had often found from 6 to 10 times the higher of these intensities.

282. Even in fair weather, the intensity of the electric force in the air near the earth's surface is perpetually fluctuating. The speaker had often observed it, especially during calms or very light breezes from the east, varying from 40 Daniel's elements per foot to three or four times that amount during a few minutes; and returning again as rapidly to the lower amount. More frequently he had observed variations from about 30 to about 40, and back again, recurring in uncertain periods of perhaps about two minutes. These gradual variations cannot but be produced by electrified masses of air or cloud, floating by the locality of observation. Again, it is well known that

during storms of rain, hail, or snow, there are great and sometimes sudden variations of electric force in the air close to the
earth. These are undoubtedly produced, partly as those of fair
weather, by motions of electrified masses of air and cloud;
partly by the fall of vitreously or resinously electrified rain,
leaving a corresponding deficiency in the air or cloud from
which it falls ; and partly by disruptive discharges (flashes of
lightning) between masses of air or cloud, or between either
and the earth. The consideration of these various phenomena
suggested the following questions, and modes of observation for
answering them :—

283. *Question* 1. How is electricity distributed through the
different strata of the atmosphere to a height of five or six
miles above the earth's surface in ordinary fair weather ? To be
answered by electrical observations in balloons at all heights
up to the highest limit, and simultaneous observations at the
earth's surface.

Q. 2. Does electrification of air close to the earth's surface,
or within a few hundred feet of it, sensibly influence the
observed electric force ? and if so, how does it vary with the
weather, and with the time of day or year ? The first part of
this question has been answered very decidedly in the affirmative, first, for large masses of air within a few hundred yards
of the earth's surface, by means of observations made simultaneously at a station near the seashore in the island of Arran,
and at one or other of several stations at different distances,
within six miles of it, on the sides and summit of Goatfell.
After that it was found, by simultaneous observations made at
a window in the Natural Philosophy Lecture-Room, and on the
College Tower of the University of Glasgow, that the influence
of the air within 100 feet of the earth's surface was always
sensible at both stations, and often paramount at the lower.
Thus, for example, when, in broken weather, the superficial
electrification of the outside of the lecture-room, about 20 feet
above the ground, in a quadrangle of buildings, was found
positive, the superficial electrification of the sides of the tower,
about 70 feet higher, was often found negative, or nearly zero ;
and this sometimes even when the positive electrification of the
sides of the building at the lower station equalled in amount

an ordinary fair weather negative. This state of things could only exist in virtue of a negative electrification of the circumambient air, inducing a positive electrification on the ground and sides of the quadrangle, but not sufficient to counterbalance the influence, on the higher parts of the tower, of more distant positively electrified aërial masses.

A long continuation of such systems of simultaneous observation—not in a town only, but in various situations of flat and of mountainous country, on the sea coast as well as far inland, in various regions of the world—will be required to obtain the information asked for in the second part of this question.

Q. 3. Do the particles of rain, hail, and snow in falling through the air possess absolute charges of electricity? and if so, whether positive or negative, and of what amounts in different conditions as to place and weather? Attempts to answer this question have been made by various observers, but as yet without success; as, for instance, by an "electro-pluviometer," tried at Kew many years ago. By using a sufficiently well-insulated vessel to collect the falling particles, it is quite certain that a decided answer may be obtained with ease for the cases of hail and snow. Inductive effects produced by drops splashing away from the collecting vessel, if exposed to the electric force of the air in an open position, or inductive effects of the opposite kind produced by drops splashing away from surrounding walls or screens and falling into the collecting vessel, if not in an exposed position, make it less easy to ascertain the electrical quality of rain; but, by taking means to obviate the disturbing effects of these influences, the speaker hoped to arrive at definite results.

284. It would have been more satisfactory to have been able to conclude a discourse on atmospheric electricity otherwise than in questions, but no other form of conclusion would have been at all consistent with the present state of knowledge.

285. The discourse was illustrated by the use of the mirror electrometer reflecting a beam of light from the electric lamp, and throwing it on a white screen, where its motions were measured by a divided scale. The principle of the water-dropping collector was illustrated by allowing a jet of water to flow by a fine nozzle into the middle of the lecture-room, from

an uninsulated metal vessel of water and compressed air, and collecting the drops in an insulated vessel on the floor. This vessel was connected with the testing electrode of the reflecting electrometer; and it was then found to experience a continually increasing negative electrification, when fixed positively electrified bodies were in the neighbourhood of the nozzle. If the same experiment were made in ordinary fair weather in the open air, instead of under the roof and within the walls of the lecture-room, the same result would be observed, without the presence of any artificially electrified body. The vessel from which the water was discharged was next insulated; and other circumstances remaining unvaried, it was shown that this vessel became rapidly electrified to a certain degree of positive potential, and the falling drops ceased to communicate any more electricity to the vessel in which they were gathered.

286. The influence of electrified masses of air was illustrated by carrying about the portable electrometer, with its match burning, to different parts of the lecture-room, while insulated spirit-lamps connected with the positive and negative conductor of an electrical machine, burned on the two sides. The speaker observed the indications on the portable electrometer; but the potentials thus measured were seen by the audience marked on the scale by the spot of light; the reflecting electrometer being kept connected with the portable electrometer in all its positions, by means of a long fine wire. It was found that, when the burning match was on one side of a certain surface dividing the air of the lecture-room, the potential indicated was positive, and on the other side negative.

287. The water-dropping collector constructed for the self-registering apparatus to be used at Kew, had been previously set upon the roof of the Royal Institution, and an insulated wire (Beccaria's "Deferent Wire") led down to the reflecting electrometer on the lecture-room table. The electric force in the air above the roof was thus tested several times during the meeting; and it was at first found to be, as it had been during several days preceding, somewhat feeble positive (corresponding to a feeble negative electrification of the earth's surface, or rather housetops, in the neighbourhood). This was a not unfrequent electrical condition of days, such as these had been

of dull rain, with occasional intervals of heavier rain and of cessation. The natural electricity was again observed by means of the reflecting electrometer during several minutes near the end of the discourse; and was found, instead of the weak positive which had been previously observed, to be strong positive of three or four times the amount. Upon this the speaker quoted* an answer which Prior Ceca had given to a question Beccaria had put to him "concerning the state of "electricity when the weather clears up." " 'If, when the rain " 'has ceased (the Prior said to me) a strong excessive† elec- " 'tricity obtains, it is a sign that the weather will continue fair " 'for several days; if the electricity is but small, it is a sign " 'that such weather will not last so much as that whole day, " 'and that it will soon be cloudy again, or even will again " 'rain.'" The climate of this country is very different from that of Piedmont, where Beccaria and his friend made their observations, but their rule as to the "electricity of clearing weather" has been found frequently confirmed by the speaker. He therefore considered that, although it was still raining at the commencement of the meeting, the electrical indications they had seen gave fair promise‡ for the remainder of this evening, if not for a longer period. There can be no doubt but that electric indications, when sufficiently studied, will be found important additions to our means for prognosticating the weather; and the speaker hoped soon to see the atmospheric electrometer generally adopted as a useful and convenient weather-glass.

288. The speaker could not conclude without guarding himself against any imputation of having assumed the existence of two electric fluids or substances, because he had frequently spoken of the vitreous and resinous electricities. Dufay's very important discovery of two modes or qualities of electrification, led his followers too readily to admit his supposition of two distinct electric fluids. Franklin, Æpinus, and Cavendish,

---

* From Beccaria's first letter "On Terrestrial Atmospheric Electricity during Serene Weather."—*Garzegna di Mondovi*, May 16, 1775.

† *i.e.*, vitreous, or positive.

‡ At the conclusion of the meeting it was found that the rain had actually ceased. The weather continued fair during the remainder of the night, and three or four of the finest days of the season followed.

with a hypothesis of one electric fluid, opened the way for a
juster appeciation of the *unity* of nature in electric phenomena.
Beccaria, with his "electric atmospheres," somewhat vaguely
struggled to see deeper into the working of electric force, but
his views found little acceptance, and scarcely suggested inquiry or even meditation. The eighteenth century made a
school of science for itself, in which, for the not unnatural
dogma of the earlier schoolmen, "matter cannot act where it is
not," was substituted the most fantastic of paradoxes, *contact
does not exist.* Boscovich's theory was the consummation of
the eighteenth century school of physical science. This strange
idea took deep root, and from it grew up a barren tree, exhaust-
ing the soil and overshadowing the whole field of molecular
investigation, on which so much unavailing labour was spent
by the great mathematicians of the early part of our nineteenth
century. If Boscovich's theory no longer cumbers the ground,
it is because one true philosopher required more light for tracing lines of electric force.

289.   Mr Faraday's investigation of electrostatic induction
influences now every department of physical speculation, and
constitutes an era in science. If we can no longer regard
electric and magnetic fluids attracting or repelling at a distance
as realities, we may now also contemplate as a thing of the
past that belief in atoms and in vacuum, against which Leib-
nitz so earnestly contended in his memorable correspondence
with Dr Samuel Clarke.

290.   We now look on space as full. We know that light is
propagated like sound through pressure and motion. We know
that there is no substance of caloric—that inscrutably minute
motions cause the expansion which the thermometer marks,
and stimulate our sensation of heat—that fire is not laid up in
coal more than in this Leyden phial, or this weight: there is
potential fire in each. If electric force depends on a residual
surface action, a resultant of an inner tension experienced by
the insulating medium, we can conceive that electricity itself
is to be understood as not an accident, but an essence of matter.
Whatever electricity is, it seems quite certain that electricity
in motion IS *heat;* and that a certain alignment of axes of
revolution in this motion IS *magnetism.* Faraday's magneto-

optic experiment makes this not a hypothesis but a demonstrated conclusion*. Thus a rifle-bullet keeps its point foremost; Foucault's gyroscope finds the earth's axis of palpable rotation; and the magnetic needle shows that more subtle rotatory movement in matter of the earth, which we call terrestrial magnetism : all by one and the same dynamical action.

291. It is often asked, are we to fall back on facts and phenomena, and give up all idea of penetrating that mystery which hangs round the ultimate nature of matter? This is a question that must be answered by the metaphysician, and it does not belong to the domain of Natural Philosophy. But it does seem that the marvellous train of discovery, unparalleled in the history of experimental science, which the last years of the world has seen to emanate from experiments within these walls, must lead to a stage of knowledge, in which laws of inorganic nature will be understood in this sense—that one will be known as essentially connected with all, and in which unity of plan through an inexhaustibly varied execution, will be recognised as a universally manifested result of creative wisdom.

292. [Postscript, with diagram, communicated to the *Philosophical Magazine* in 1861; but now first published.]

Mr Balfour Stewart, Director of the Kew Meteorological Observatory, has, since the commencement of the present year (1861), brought into regular and satisfactory operation the self-recording atmospheric electrometer with water-dropping collector, described in the preceding abstract: a specimen of the results is exhibited in the accompanying photographic curves.

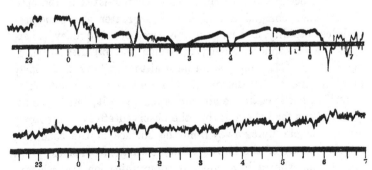

* See "Dynamical Illustrations of the Magnetic and the Helicoidal Rotatory Effects of Transparent Bodies on Polarized Light." By Prof. W. Thomson.—*Proceedings of the Royal Society*, June 12, 1856.

293. The diagram exhibits the variations of the electric force of the atmosphere, as photographically recorded by the divided ring electrometer at the Kew Observatory for two successive days, commencing on the 28th of April 1861. The prepared sensitive paper was made to move vertically at a uniform rate by means of clock-work, while a spot of light (the image of a portion of a gas-flame reflected from the mirror of the divided ring electrometer) moved horizontally across it according to the continually varying electric force of the atmosphere, and marked the curve photographically. The datum line, showing the position the spot of light would have if the electric force were zero, is produced by an image from the same source of light reflected from a fixed mirror attached to the case of the electrometer. The numbers indicate hours reckoned from noon as zero, up to 23. The same paper is, for the sake of economy, generally used to bear the record for two days.

Thus the distance of the spot of light from the datum line, on one side or other, indicates, and the photo-chemical action records, for each instant of time the electric potential, positive or negative, of the atmosphere at the point where the stream of water discharged from the insulated vessel breaks into drops.

ON ELECTRICAL "FREQUENCY."

[From *Report of British Association*, Aberdeen Meeting, 1859.]

294. Beccaria found that a conductor insulated in the open air becomes charged sometimes with greater and sometimes with less rapidity, and he gave the name of "frequency" to express the atmospheric quality on which the rapidity of charging depends. It might seem natural to attribute this quality to electrification of the air itself round the conductor, or to electrified particles in the air impinging upon it; but the author gave reasons for believing that the observed effects are entirely due to particles flying away from the surface of the conductor, in consequence of the impact of *non-electrified* particles against it. He had shown in a previous communication that when no electricity of separation (or, as it is more generally called, "frictional electricity," or "contact electricity") is called into

play, the tendency of particles continually flying off from a conductor is to destroy all electrification at the part of its surface from which they break away. Hence a conductor insulated in the open air, and exposed to mist or rain, with wind, will tend rapidly to the same electric potential as that of the air, beside that part of its surface from which there is the most frequent dropping, or flying away, of aqueous particles. The *rapid charging* indicated by the electrometer under cover, after putting it for an instant in connexion with the earth, is therefore, in reality, due to a *rapid discharging* of the exposed parts of the conductor. The author had been led to these views by remarking the extreme rapidity with which an electrometer, connected by a fine wire with a conductor insulated above the roof of his temporary electric observatory in the island of Arran, became charged, reaching its full indication in a few seconds, and sometimes in a fraction of a second, after being touched by the hand, during a gale of wind and rain. The conductor, a vertical cylinder about 10 inches long and 4 inches diameter, with its upper end flat and corner slightly rounded off, stood only 8 feet above the roof, or, in all, 20 feet above the ground, and was nearly surrounded by buildings rising to a higher level. Even with so moderate an exposure as this, sparks were frequently produced between an insulated and an uninsulated piece of metal, which may have been about $\frac{1}{40}$th of an inch apart, within the electrometer, and more than once a continuous line of fire was observed in the instrument during nearly a minute at a time, while rain was falling in torrents outside.

ON THE NECESSITY FOR INCESSANT RECORDING, AND FOR SIMULTANEOUS OBSERVATIONS IN DIFFERENT LOCALI-TIES, TO INVESTIGATE ATMOSPHERIC ELECTRICITY.

[From *Report of British Association*, Aberdeen Meeting, 1859.]

295. The necessity for incessantly recording the electric condition of the atmosphere was illustrated by reference to observations recently made by the author in the island of Arran, by which it appeared that even under a cloudless sky, without any

sensible wind, the negative electrification of the surface of the
earth, always found during serene weather, is constantly varying in degree. He had found it impossible, at any time, to
leave the electrometer without losing remarkable features of
the phenomenon. Beccaria, Professor of Natural Philosophy
in the University of Turin a century ago, used to retire to
Garzegna when his vacation commenced, and to make incessant observations on atmospheric electricity, night and day,
sleeping in the room with his electrometer in a lofty position,
from which he could watch the sky all round, limited by the
Alpine range on one side, and the great plain of Piedmont on
the other. Unless relays of observers can be got to follow his
example, and to take advantage of the more accurate instruments supplied by advanced electric science, a self-recording
apparatus must be applied to provide the data required for
obtaining knowledge in this most interesting field of nature.
The author pointed out certain simple and easily-executed
modifications of working electrometers (exhibited to the meeting), to render them self-recording. He also explained a new
collecting apparatus for atmospheric electricity, consisting of
an insulated vessel of water, discharging its contents in a
fine stream from a pointed tube. This stream carries away
electricity as long as any exists on its surface, where it breaks
into drops. The immediate object of this arrangement is to
maintain the whole insulated conductor, including the portion
of the electrometer connected with it and the connecting wire,
in the condition of no absolute charge ; that is to say, with as
much positive electricity on one side of a neutral line as of
negative on the other. Hence the position of the discharging
nozzle must be such, that the point where the stream breaks
into drops is in what would be the neutral line of the conductor, if first perfectly discharged under temporary cover, and
then exposed in its permanent open position, in which it will
become inductively electrified by the aërial electromotive force.
If the insulation is maintained in perfection, the dropping will
not be called on for any electrical effect, and sudden or slow
atmospheric changes will all instantaneously and perfectly induce their corresponding variations in the conductor, and give
their appropriate indications to the electrometer. The neces-

sary imperfection of the actual insulation, which tends to bring
the neutral line downwards or inwards, or the contrary effects
of aërial convection, which, when the insulation is good, generally preponderate, and which in some conditions of the atmosphere, especially during heavy wind and rain, are often very
large, are corrected by the tendency of the dropping to maintain the neutral line in the one definite position. The objects
to be attained by simultaneous observations in different localities
alluded to were—(1) to fix the constant for any observatory,
by which its observations are reduced to absolute measure of
electromotive force per foot of air; (2) to investigate the distribution of electricity in the air itself (whether on visible
clouds or in clear air) by a species of electrical trigonometry, of
which the general principles were slightly indicated. A portable electrometer, adapted for balloon and mountain observations, with a burning match, regulated by a spring so as to give
a cone of fire in the open air, in a definite position with reference to the instrument, was exhibited. It is easily carried,
with or without the aid of a shoulder-strap, and can be used
by the observer standing up, and simply holding the entire
apparatus in his hands, without a stand or rest of any kind.
Its indications distinguish positive from negative, and are reducible to absolute measure on the spot. The author gave the
result of a determination which he had made, with the assistance of Mr Joule, on the Links, a piece of level ground near
the sea, beside the city of Aberdeen, about 8 A.M. on the preceding day (September 14), under a cloudless sky, and with a
light north-west wind blowing, with the insulating stand of the
collecting part of the apparatus buried in the ground, and the
electrometer removed to a distance of 5 or 6 yards, and connected by a fine wire with the collecting conductor. The
height of the match was 3 feet above the ground, and the
observer at the electrometer lay on the ground to render the
electrical influence of his own body on the match insensible.
The result showed a difference of potentials between the earth
(negative) and the air (positive) at the match equal to that of
115 elements of Daniell's battery, and, therefore, at that time
and place, the aërial electromotive force per foot amounted to
that of thirty-eight Daniell's cells, or 1·2 cells per centimetre.

OBSERVATIONS ON ATMOSPHERIC ELECTRICITY.

[From the *Proceedings of the Literary and Philosophical Society of Manchester*, March, 1862.]

296. I find that atmospheric electricity is generally negative within doors, and almost always sensible to my divided ring reflecting electrometer. I use a spirit-lamp, on an insulated stand a few feet from walls, floor, or ceiling of my lecture room, and connect it by a fine wire with the insulated half ring of the electrometer. A decided negative effect is generally found, which shows a potential to be produced in the conductors connected with the flame, negative relatively to the earth by a difference amounting to several times the difference of potentials (or electromotive force) between two wires of one metal connected with the two plates of a single element of Daniell's. I have tested that the spirit-lamp gives no *idio-electric* effect amounting to so much as the effect of a single cell. The electric effect observed is therefore not due to thermal or chemical action in the flame. It cannot be due to contact electrifications of metallic or other bodies in conductive communication with the walls, floor, or ceiling, because the potentials of such must always fall short of the difference of potentials produced by a single cell. I have taken care to distinguish the observed natural effect from anything that can be produced by electrical operations for lecture or laboratory purposes. Thus I observe generally in the morning before any electrical operations have been performed, and find ordinarily results quite similar to those observed on the Monday mornings when the electrical machine has not been turned since the previous Friday. The effect, when there has been no artificial disturbance, has *always been found negative, except two or three times*, since the middle of November; but trustworthy observations have not been made on more than a quarter of the number of days.

297. A few turns of the electrical machine, with a spirit-lamp on its prime conductor, or a slightly charged Leyden phial, with its inside coating positive put in connexion with an insulated spirit-lamp, is enough to reverse the common negative indication. Another very striking way in which this may be done is to put a *negatively charged Leyden phial* below an insulated

XVI.] *Atmospheric Electricity.*

flame (a common gas-burner, for instance). The flame, becoming positively electrified by induction, keeps throwing off, by the dynamic power of its burning, portions of its own gaseous matter, and does not allow them to be electrically attracted down to the Leyden phial, but forces them to rise. These, on cooling, become, like common air, excellent non-conductors*, and, mixing with the air of the room, give a preponderance of positive influence to the testing insulated flame (that is to say, render the air potential positive at the place occupied by this flame).

298. Half an hour, or often much more, elapses after such an operation, before the natural negatively electrified air becomes again paramount in its influence on the testing flame.

299. That either positive or negative electricity may be carried, even through narrow passages, by air, I have tested by turning an electric machine, with a spirit-lamp on its prime conductor, for a short time in a room separated from the lecture room by an oblique passage about two yards long and then stopping the machine and extinguishing the lamp; so as to send a limited quantity of positive electricity into the air of that room. When the lecture-room window was kept open, and the door leading to the adjoining room shut, the testing spirit-lamp showed the natural negative. When the window was closed, and a small chink (an inch or less wide) opened of the door, the indication quickly became positive. If the door was then shut, and the window again opened, the natural effect was slowly recovered. A current of air, to feed the lecture-room fire, was found entering by either door or window when the other was shut. This alternate positive and negative electric ventilation may be repeated many times without renewing the positive electricity of the adjoining room by turning the machine afresh.

---

* I find that steam from a kettle boiling briskly on a common fire is an excellent insulator. I allow it to blow for a quarter of an hour or more against an insulated electrified conductor, without discovering that it has any effect on the retention of the charge. The electricity of the steam itself, in such circumstances, as is to be expected from Faraday's investigation, is not considerable. Common air loses nearly all its resisting power at some temperature between that of boiling water and red-hot iron, and conducts continuously (not, as I believe is generally supposed to be the case, by disruption) as glass does at some temperature below the boiling point, with so great ease as to discharge any common insulated conductor almost completely in a few seconds.

300. The out of doors air potential, as tested by a portable
electrometer in an open place, or even by a water dropping
nozzle outside, two or three feet from the walls of the lecture
room, was generally on these occasions positive, and the earth's
surface itself, therefore, of course, negative ;—the common fair
weather condition, which I am forced to conclude is due to a
paramount influence of positive electricity in higher regions of
the air, notwithstanding the negative electricity of the air in
the lower stratum near the earth's surface. On the two or three
occasions when the in-door atmospheric electricity was found
positive, and, therefore, the surface of the floor, walls, and ceiling negative, the potential outside was certainly positive,
and the earth's surface out of doors negative, as usual in fair
weather.

300′. Extract from letter addressed to General Sabine :—

"During my recent visit to Creuznach I became acquainted
with Mr Dellman of that place, who makes meteorological,
chiefly electrical, observations for the Prussian Government,
and I had opportunities of witnessing his method of electrical
observation. It consists in using a copper ball about 6 inches
diameter, to carry away an electrical effect from a position
about two yards above the roof of his house, depending simply
on the atmospheric potential' at the point to which the centre
of the ball is sent; and it is exactly the method of the 'carrier
ball' by which Faraday investigated the atmospheric potential
in the neighbourhood of a rubbed stick of shell-lac, and other
electrified bodies (*Experimental Researches*, Series XI. 1837).
The whole process only differs from Faraday's in not employing
the carrier ball directly, as the repeller in a Coulomb-electro-
meter, but putting it into communication with the conductor of
a separate electrometer of peculiar construction. The collecting
part of the apparatus is so simple and easily managed that an
amateur could, for a few shillings, set one up on his own house,
if at all s itable as regards roof and windows ; and, if provided
with a suitable electrometer, could make observations in atmospheric electricity with as much ease as thermometric or barometric observations. The electrometer used by Mr Dellman is
of his own construction (described in Poggendorff's *Annalen*,
1853, Vol. LXXXIX., also Vol. LXXXV.), and it appears to be very

satisfactory in its operation. It is, I believe, essentially more
accurate and sensitive than Peltier's, and it has a great advantage in affording a very easy and exact method for reducing its
indications to absolute measure. I was much struck with the
simplicity and excellence of Mr Dellman's whole system of
observation on atmospheric electricity; and it has occurred to
me that the Kew Committee might be disposed to adopt it, if
determined to carry out electrical observations. When I told
Mr Dellman that I intended to make a suggestion to this effect,
he at once offered to have an electrometer, if desired, made
under his own care. I wish also to suggest two other modes
of observing atmospheric electricity which have occurred to me,
as possessing each of them some advantages over any of the
systems hitherto followed. In one of these I propose to have
an uninsulated cylindrical iron funnel, about 7 inches diameter,
fixed to a height of two or three yards above the highest part
of the building, and a light moveable continuation (like the
telescope funnel of a steamer) of a yard and a half or two yards
more, which can be let down or pushed up at pleasure. Insulated by supports at the top of the fixed part of the funnel, I
would have a metal stem carrying a ball like Dellman's, standing to such a height that it can be covered by a hinged lid on
the top of the moveable joint of the funnel, when the latter is
pushed up; and a fine wire fixed to the lower end of the insulated stem, and hanging down, in the axis of the funnel to the
electrometer. When the apparatus is not in use, the moveable
joint would be kept at the highest, with its lid down, and the
ball uninsulated. To make an observation, the ball would be
insulated, the lid turned up rapidly, and the moveable joint
carrying it let down, an operation which could be effected in a
few seconds by a suitable mechanism. The electrometer would
immediately indicate an inductive electrification simply proportional to the atmospheric potential at the position occupied by
the centre of the ball, and would continue to indicate at each
instant the actual atmospheric potential, however variable, as
long as no sensible electrification or diselectrification has taken
place through imperfect insulation or convection by particles of
dust or currents of air (probably for a quarter or a half of an
hour, when care is taken to keep the insulation in good order).

This might be the best form of apparatus for making observations in the presence of thunder-clouds. But I think the best possible plan in most respects, if it turns out to be practicable, of which I can have little doubt, will be to use, instead of the ordinary fixed insulated conductor with a point, a fixed conductor of similar form, but hollow, and containing within itself an apparatus for making hydrogen, and blowing small soap-bubbles of that gas from a fine tube terminating as nearly as may be in a point, at a height of a few yards in the air. With this arrangement the insulation would only need to be good enough to make the loss of a charge by conduction very slow in comparison with convective loss by the bubbles; so that it would be easy to secure against any sensible error from defective insulation. If 100 or 200 bubbles, each $\frac{1}{10}$ inch in diameter, are blown from the top of the conductor per minute, the electrical potential in its interior will very rapidly follow variations of the atmospheric potential, and would be at any instant the same as the mean for the atmosphere during some period of a few minutes preceding. The action of a simple point is (as, I suppose, is generally admitted) essentially unsatisfactory, and as nearly as possible nugatory in its results. I am not aware how flame has been found to succeed, but I should think not well in the circumstances of atmospheric observations, in which it is essentially closed in a lantern; and I cannot see on any theoretical ground how its action in these circumstances can be *perfect*, like that of the soap-bubbles. I intend to make a trial of the practicability of blowing the bubbles; and if it proves satisfactory, there cannot be a doubt of the availability of the system for atmospheric observations."

[Addition, Feb. 1857.]—The author has now made various trials on the last-mentioned part of his proposal, and he has not succeeded in finding any practicable self-regulating apparatus for blowing bubbles and detaching them one by one from the tube. He has seen reason to doubt whether it will be possible to get bubbles so small as those proposed above, to rise at all; but he has not been led to believe that, if it is thought worth while to try, it will be found impracticable to construct a self-acting apparatus which will regularly blow and discharge separately, bubbles of considerably larger diameter, and so to secure the advantages mentioned, although with a proportionately larger consumption of the gas.

On the other hand, he finds that, by the aid of an extremely sensitive electrometer which he has recently constructed, he will be able, in all probability with great ease and at very small cost, to bring into practice the first of his two plans, constructed on a considerably smaller scale as regards height than proposed in the preceding statement.

ON SOME REMARKABLE EFFECTS OF LIGHTNING OBSERVED IN A FARM-HOUSE NEAR MONIEMAIL, CUPAR-FIFE.

(From *Proceedings of the Philosophical Society of Glasgow.*)

301. The following is an extract from a letter, addressed last autumn to me by Mr Leitch, minister of Moniemail parish:—

"MONIEMAIL MANSE, CUPAR-FIFE,
26*th August*, 1849.

" . . . We were visited on the 11th inst. with a violent thunder-storm, which did considerable damage to a farm-house in my immediate neighbourhood. I called shortly afterwards and brought away the wires and the paper which I enclose. . . .

"I have some difficulty in accounting for the appearance of the wires. You will observe that they have been partially fused, and when I got them first they adhered closely to one another. You will find that the flat sides exactly fit. They were both attached to one crank, and ran parallel to one another. The question is, how were they attracted so powerfully as to be compressed together? . . .

"You will observe that the paper is discoloured. This has been done, not by scorching, but by having some substance deposited on it. There was painted *wood* also discoloured, on which the stratum was much thicker. It could easily be rubbed off, when you saw the paint quite fresh beneath. . . .

"The farmer showed me a probang which hung on a nail.

The handle only was left. The rest, consisting of a twisted cane, had entirely disappeared. By minute examination I found a small fragment, which was not burnt, but broken off."

[The copper wires and the stained paper, enclosed with Mr Leitch's letter, were laid before the Society.]

The remarkable effects of lightning, described by Mr Leitch, are all extremely interesting. Those with reference to the copper wires are quite out of the common class of electrical phenomena; nothing of the kind having, so far as I am aware, been observed previously, either as resulting from natural discharges, or in experiments on electricity. It is not improbable that they are due to the electro-magnetic attraction which must have subsisted between the two wires during the discharge, it being a well-known fact that adjacent wires, with currents of electricity in similar directions along them, attract one another. It may certainly be doubted whether the inappreciably short ·time occupied by the electrical discharge could have been sufficient to allow the wires, after having been drawn into contact, to be pressed with sufficient force to make them adhere together, and to produce the remarkable impressions which they still retain. On the other hand, the electro-magnetic force must have been very considerable, since the currents in the wires were strong enough nearly to melt them, and since they appear to have been softened, if not partially fused; the flattening and remarkable impressions might readily have been produced by even a slight force subsisting after the wires came in contact.

The circumstances with reference to the probang, described by Mr Leitch, afford a remarkable illustration of the well-known fact, that an electrical discharge, when effected through the substance of a non-conducting (that is to say, a *powerfully resisting*) solid, shatters it, without producing any considerable elevation of its temperature; not leaving marks of combustion, if it be of an ordinary combustible material such as wood.

Dr Robert Thomson, at my request, kindly undertook to examine the paper removed from the wall of the farm-house, and enclosed with his letter to me by Mr Leitch; so as, if possible, by the application of chemical tests, to discover the staining substance deposited on its surface. Mr Leitch, in his letter, had suggested that it would be worth while to try whether this case is an example of the deposition of sulphur, which Fusinieri believed he had discovered in similar circumstances. Accordingly tests for sulphur were applied, but with entirely negative results. Stains presenting a similar appearance had been sometimes observed on paper in the neighbourhood of copper-wires through which powerful discharges in experiments with the hydro-electric machine had been passed; and from this it was suggested that the staining substance might have come from the bell-wires. Tests for copper were accordingly applied, and the results were most satisfactory. The front of the paper was scraped in different places, so as to remove some of the pigment in powder; and the powders from the stained, and from the not stained parts, were repeatedly examined. The presence of copper in the former was readily made manifest by the ordinary tests: in the latter, no traces of copper could be discovered. The back of the paper presented a green tint, having been torn from a wall which has probably been painted with Scheele's green; and matter scraped away from any part of the back was found to contain copper. Since, however, the stains in front were manifestly superficial, the discolouration being entirely removed by scraping, and since there was no appearance whatever of staining at the back of the paper, nor of any effect of the electrical discharge, it was impossible to attribute the stains to copper produced from the Scheele's green on the wall below the paper. Dr Thomson, therefore, considered the most probable explanation to be, that the stains of oxide of copper must have come from the bell-wire. To ascertain how far this explanation could be supported by the circumstances of the case, I wrote to Mr Leitch asking him for further particulars, especially with reference to this point, and I received the following answer:—

" MONIEMAIL, CUPAR-FIFE,
30th Nov. 1849.

" . . . . I received your letter to-day, and immediately called at Hall-hill, in the parish of Collessie, the farm-house which had been struck by the lightning. . . .

"I find that Dr Thomson's suggestion is fully borne out by the facts. I at first thought that the bell-wire did not run along the line of discolouration, but I now find that such was the case. . .

[From a drawing and explanation which Mr Leitch gives, it appears that the wire runs vertically along a corner of the room, from the floor, to about a yard from the ceiling, where it branches into two, connected with two cranks near one another, and close to the ceiling.]

"The efflorescence [the stains previously adverted to] was on each side of this perpendicular wire. In some places it extended more than a foot from the wire. The deposit seemed to vary in thickness according to the surface on which it was deposited. There was none on the plaster on the roof. It was thinnest upon the wall-paper, and thickest upon the wood facing of the door*. This last exhibited various colours. On the thickest part it appeared quite black; where there was only a slight film, it was green or yellow. . . .

"I may mention that the thunder-storm was that of the 11th of August last. It passed over most of Scotland, and has rarely been surpassed for terrific grandeur at least beyond the tropics. It commenced about nine o'clock P.M., and in the course of an hour it seemed to die away altogether. The peals became very faint, and the intervals between the flashes and the reports very great, when all at once a terrific crashing peal was heard, which did the damage. The storm ceased with this peal.

"The electricity must have been conducted along the lead on the ridge of the house, and have diverged into three streams; one down through the roof, and the two others along the roof to the chimneys. One of these appears to have struck a large stone out from the chimney, and to have been conducted down the chimney to the kitchen, where it left traces upon the floor. It had been washed over before I saw it, but still the traces were visible on the Arbroath flags. The stains were of a lighter

---

* These remarkable facts are probably connected with the conducting powers of the different surfaces. The plaster on the roof is not so good a conductor as the wall-paper, with its pigments; and the painted wood is probably a better conductor than either.—W. T.

tint than the stone, and the general appearance was as if a pail
of some light-coloured fluid had been dashed over the floor, so
as to produce various distinct streams. All along the course of
the discharge, and particularly in the neighbourhood of the bell-
wires, there were small holes in the wall about an inch deep,
like the marks that might be made by a finger in soft plaster.

"Most of the windows were shattered, and all the fragments
of glass were on the outside. I suppose this must be accounted
for by the expansion of the air within the house.

"The window-blind of the staircase, which was down at the
time, was riddled, as if with small shot. The diameter of the
space so riddled was about a foot. On minute examination I
found that the holes were not such as could readily be made
by a pointed instrument or a pellet. They were angular, the
cloth being torn along both the warp and the woof.

"The house was shattered from top to bottom. Two of the
serving-maids received a positive shock, but soon recovered.
A strong smell of what was supposed to be sulphur was perceived throughout the house, but particularly in the bed-room
in which the effects I described before took place."

XVII. SOUND PRODUCED BY THE DISCHARGE OF A CONDENSER.

[LETTER TO PROFESSOR TAIT.]

KILMICHAEL, BRODICK,
ISLE OF ARRAN, *Oct.* 10, 1863.

302. Yesterday evening, when engaged in measuring the
electrostatic capacities of some specimens of insulated wire
designed for submarine telegraph cables, I had occasion frequently to discharge, through a galvanometer coil, a condenser
consisting of two parallel plates of metal, separated by a space
of air about ·007 inch across, and charged to a difference
of potentials equal to that of about 800 Daniell's elements.
I remarked at an instant of discharge a sharp sound, with a
very slight prolonged resonance, which seemed to come from

the interior of the case containing the condenser, and which struck me as resembling a sound I had repeatedly heard before when the condenser had been overcharged and a spark passed across its air-space. But I ascertained that this sound was distinctly audible when there was no spark within the condenser, and the whole discharge took place fairly through the 2000 yards of fine wire, constituting the galvanometer coil. I arranged the circuit so that the place where the contact was made to produce the discharge was so far from my ear that the initiating spark was inaudible; but still I heard distinctly the same sound as before from within the condenser.

303. Using instead of the galvanometer coil either a short wire or my own body (as in taking a shock from a Leyden phial), I still heard the sound within the condenser. The shock was imperceptible except by a very faint prick on the finger in the place of the spark, and (the direct sound of the spark being barely, if at all, sensible) there was still a very audible sound, always of the same character, within the condenser, which I heard at the same instant as I felt the spark on my finger. Mr Macfarlane could hear it distinctly standing at a distance of several yards. We watched for light within the condenser, but could see none. I have since ascertained that suddenly charging the condenser out of one of the specimens of cable charged for the purpose produces the same sound within the condenser; also that it is produced by suddenly reversing the charge of the condenser.

304. Thus it is distinctly proved that a plate of air emits a sound on being suddenly subjected to electric force, or on experiencing a sudden change of electric force through it. This seems a most natural result when viewed in connexion with the new theory put forward by Faraday in his series regarding the part played by air or other dielectric in manifestations of electric force. It also tends to confirm the hypothesis I suggested to account for the remarkable observation made regarding lightning, when you told me of it about a year ago, and other similar observations which I believe have been reported, proving a sound to be heard at the instant of a flash of lightning in localities at considerable distances from any part of the line of discharge, and which by some have been supposed to de-

monstrate an error in the common theory of sound. I may
add that Mr Macfarlane tells me he believes he has heard, at
the instant of a flash of lightning, a sound as of a heavy body
striking the earth, and imagined at first that something close
to him had been struck, but heard the ordinary thunder at a
sensible time later.

XVIII. MEASUREMENT OF THE ELECTROSTATIC FORCE
PRODUCED BY A DANIELL'S BATTERY.

[*Proceedings Royal Society*, Feb. 23 and April 12, 1860, or *Phil. Mag.* 1860,
second half-year.]

305. In a paper "On Transient Electric Currents," published
in the *Philosophical Magazine* for June 1853, [Mathematical
and Physical Papers, Art. LXII.] I described a method for
measuring differences of electric potential in absolute electrostatic units, which seemed to me the best adapted for obtaining
accurate results.   The "absolute electrometer" which I exhibited to the British Association on the occasion of its
meeting at Glasgow in 1855, was constructed for the purpose
of putting this method into practice, and, as I then explained,
was adapted to reduce the indications of an electroscopic* or of
a torsion electrometer to absolute measure.

306. The want of sufficiently constant and accurate instruments of the latter class has long delayed my carrying out of
the plans then set forth. Efforts which I have made to produce
electrometers to fulfil certain conditions of sensibility, convenience, and constancy, for various objects, especially the
electrostatic measurement of galvanic forces, and of the differences of potential required to produce sparks in air, under
definite conditions, and the observation of natural atmospheric
electricity, have enabled me now to make a beginning of absolute determinations, which I hope to be able to carry out soon
in a much more accurate manner. In the meantime, I shall
give a slight description of the chief instruments and processes

* I have used the expression "electroscopic electrometer," to designate an
electrometer of which the indications are merely read off in each instance
by a single observation, without the necessity of applying any experimental
process of weighing, or of balancing by torsion, or of otherwise modifying the
conditions exhibited.

followed, and state the approximate results already obtained, as these may be made the foundation of various important estimates in several departments of electrical science

307.   The absolute electrometer alluded to above (compare § 358, below), consists of a plane metallic disc, insulated in a horizontal position, with a somewhat smaller plane metallic disc hung centrally over it, from one end of the beam of a balance. A metal case protects the suspended disc from currents of air, and from irregular electric influences, allowing a light vertical rod, rigidly connected with the disc at its lower end, and suspended from the balance above, to move up and down freely, through an aperture just wide enough not to touch it.   In the side of the case there is another aperture, through which projects an electrode rigidly connected with the lower insulated disc. The upper disc is kept in metallic communication with the case.

308.   In using this instrument to reduce the indications of an electroscopic or torsion electrometer to absolute electrostatic measure, the insulated part of the electrometer is kept in metallic communication with the insulated disc, while the cases enclosing the two instruments are also kept in metallic communication with one another.   A charge, either positive or negative, is communicated to the insulated part of the double apparatus.   The indication of the tested electrometer is read off, and at the same time the force required to keep the moveable disc at a stated distance from the fixed disc below it, is weighed by the balance.   This part of the operation is, as I anticipated, somewhat troublesome, in consequence of the instability of the equilibrium, but with a little care it may be managed with considerable accuracy.   The plan which I have hitherto followed, has been to limit the play of the arm of the balance to a very small arc, by means of firm stops suitably placed, thus allowing a range of motion to the upper disc through but a small part of its whole distance from the lower. A certain weight is put into the opposite scale of the balance, and the indications of the second electrometer are observed when the electric force is just sufficient to draw down the upper disc from resting in its upper position, and again when insufficient to keep it down with the beam pressed on its lower stop.   This operation is repeated at different distances,

and thus no considerable error depending on a want of parallelism between the discs could remain undetected. It may be
remarked that the upper disc is carefully balanced by means
of small weights attached to it, so as to make it hang as nearly
as possible parallel to the lower disc. The stem carrying it is
graduated to hundredths of an inch (·254 of a millimetre);
and by watching it through a telescope at a short distance, it
is easy to observe $\frac{1}{40}$ of a millimetre of its vertical motion.

309. I have recently applied this method to reduce to absolute electrostatic measure the indications of an electrometer
forming part of a portable apparatus for the observation of
atmospheric electricity. In this instrument (compare § 263)
a very light bar of aluminium attached at right angles to the
middle of a fine platinum wire, which is firmly stretched between the inside coatings of two Leyden phials, one occupying
an inverted position above the other, experiences and indicates
the electrical force which is the subject of measurement, and
which consists of repulsions in contrary directions on its two
ends, produced by two short bars of metal fixed on the two
sides of the top of a metal tube, supported by the inside coating of the lower phial.

310. The amount of the electrical force (or rather, as it should
be called in correct mechanical language, *couple*) is measured by
the angle through which the upper Leyden phial must be
turned round an axis coincident with the line of the wire, so
as to bring the index to a marked position. An independently
insulated metal case, bearing an electrode projecting outwards,
to which the body to be tested is applied, surrounds the index
and repelling bars, but leaves free apertures above and below,
for the wire to pass through it without touching it; and by
other apertures in its sides and top, it allows the motions of
the index to be observed, and the Leyden phials to be charged
or discharged at pleasure, by means of an electrode applied to
one of the fixed bars described above. When by means of such
an electrode the inside coatings of the Leyden phials are kept
connected with the earth, this electrometer becomes a plain
repulsion electrometer, on the same principle as Peltier's, with
the exception that the index, supported by a platinum wire
instead of on a pivot, is directed by elasticity of torsion instead

of by magnetism; and the electrical effect to be measured is produced by applying the electrified body to a conductor connected with a fixed metal case round the index and repelling bars, instead of with these conductors themselves.

311. This electrometer, being of suitable sensibility for direct comparison with the absolute electrometer according to the process described above, is not sufficiently sensitive to measure directly the electrostatic effect of any galvanic battery of fewer than two hundred cells with much accuracy. Not having at the time arrangements for working with a multiple battery of reliable character, I used a second torsion electrometer of a higher degree of sensibility as a medium for comparison, and determined the value of its indications by direct reference to a Daniell's battery of from six to twelve elements in good working order. This electrometer, in which a light aluminium index, suspended by means of a fine glass fibre, kept constantly electrified by means of a light platinum wire hanging down from it and dipping into some sulphuric acid in the bottom of a charged Leyden jar, exhibits the effects of electric force due to a difference of potentials between two halves of a metallic ring separately insulated in its neighbourhood, will be sufficiently described in another communication to the Royal Society. Slight descriptions of trial instruments of this kind have already been published in the *Transactions of the Pontifical Academy of Rome**, and in the second edition of Nichol's *Cyclopædia* (article Electricity, Atmospheric), 1860 (§§ 249, 266, above).

312. I hope soon to have another electrometer on the same general principle, but modified from those hitherto made, so as to be more convenient for accurate measurement in terms of constant units. In the meantime, I find that, by exercising sufficient care, I can obtain good measurements by means of the divided ring electrometer of the form described in Nichol's *Cyclopædia* (§ 263, above).

313. In the ordinary use of the portable electrometer, a considerable charge is communicated to the connected inside coatings of the Leyden phials, and the aluminium index is brought to an accurately marked position by torsion, while the insulated

---

* Accademia Pontificia dei Nuovi Lyncei, February 1857.

metal case surrounding it is kept connected with the earth.
The square root of the reading of the torsion-head thus obtained measures the potential, to which the inside coatings of
the phials have been electrified. If, now, the metal case
referred to is disconnected from the earth and put in connexion with a conductor whose potential is to be tested, the
square root of the altered reading of the torsion-head required
to bring the index to its marked position in the new circumstances measures similarly the difference between this last
potential and that of the inside coatings of the phials. Hence
the excess of the latter square root above the former expresses
in degree and in quality (positive or negative) the required
potential. This plan has not only the merit of indicating the
quality of the electricity to be tested, which is of great importance in atmospheric observation, but it also affords a much
higher degree of sensibility than the instrument has when used
as a plain repulsion electrometer; and, on account of this last-
mentioned advantage, it was adopted in the comparisons with
the divided ring electrometer. On the other hand, the portable
electrometer was used in its least sensitive state, that is to say,
with its Leyden phials connected with the earth, when the
comparisons with the absolute electrometer were made.

314. The general result of the weighings hitherto made, is
that when the discs of the absolute electrometer were at a distance of ·5080 of a centimetre, the number of degrees of torsion
in the portable electrometer was ·20924 times the number of
grammes' weight required to balance the attractive force; and
the number of degrees of torsion was ·4983 times the number
of grammes' weight found in other series of experiments in
which the distance between the discs was ·762 of a centimetre.
According to the law of inverse squares of the distances to
which the attraction between two parallel discs is subject when
a constant difference of potentials is maintained between them*,
the force at a distance of ·254 of a centimetre would have been
$\frac{1}{13\cdot46}$, according to the first of the preceding results, or, according to the second, $\frac{1}{13\cdot68}$ of the number of degrees of torsion.
The mean of these is $\frac{1}{13\cdot07}$, or ·0777; and we may consider this

---

* See § 11 of Elements of Mathematical Theory of Electricity appended to
the communication following this in the " Proceedings."

number as representing approximately the value in grammes' weight at ·254 of a centimetre distance between the discs of the absolute electrometer, corresponding to one degree of torsion of the portable electrometer. By comparing the indications of the portable electrometer with those of the divided ring electrometer, and by evaluating those of the latter in terms of the electromotive force of a Daniell's battery charged in the usual manner, I find that 284 times the square root of the number of degrees of torsion in the portable electrometer is approximately the number of cells of a Daniell's battery which would produce an electromotive force (or, which is the same thing, a difference of potentials) equal to that indicated. Hence the attraction between the discs of the portable electrometer, if at ·254 of a centimetre distance, and maintained at a difference of potentials amounting to that produced by 284 cells, is ·0777 of a gramme. The effect of 1000 cells would therefore be to give a force of ·965 of a gramme, since the force of attraction is proportional to the square of the difference of potentials between the discs. The diameter of the opposed circular areas between which the attraction observed took place, was 14·88 centimetres. Its area was therefore 174·0 square centimetres, and therefore the amount of attraction per square decimetre, according to the preceding estimate for ·254 of a centimetre distance and 1000 cells' difference of potential, is ·554 of a gramme. Hence, with an electromotive force or difference of potentials produced by 1000 cells of Daniell's battery, the force of attraction would be 3·57 grammes weight per square decimetre between discs separated to a distance of 1 millimetre. [The force in grammes weight is equal to $000357 \times n^2$, if the area of each of the opposed surfaces is equal to *a square* whose side is n times the distance between them, provided n be a large number.]

315.   This result differs very much from an estimate I have made according to Weber's comparison of electrostatic with electro-magnetic units and my theoretical estimate of 2,500,000 British electro-magnetic units for the electromotive force of a single element of Daniell's.   On the other hand, it agrees to a remarkable degree of accuracy with direct observations made for me, during my absence in Germany, by Mr Macfarlane, in

the months of June and July 1856, on the force of attraction produced by the direct application of a miniature Daniell's battery, of different numbers of elements, from 93 to 451, applied to the same absolute electrometer with its discs at 2006 of a centimetre asunder. These observations gave forces varying, on the whole, very closely according to the square of the number of cells used; and the mean result reduced according to this law to 1000 cells was 1·516 grammes. Reducing this to the distance of 1 millimetre, and dividing by 1·74, the area in square decimetres, we find 3·51 grammes per square decimetre at a distance of 1 millimetre.

316.   Although the experiments leading to this result were executed with great care by Mr Macfarlane, I delayed publishing it because of the great discrepance it presented from the estimate which I deduced from Weber's measurement, published while my preparations were in progress. I cannot doubt its general correctness now, when it is so decidedly confirmed by the electrometric experiments I have just described, which have been executed chiefly by Mr John Smith and Mr John Ferguson, working in my laboratory with much ability since the month of November. I am still unable to explain the discrepance, but it may possibly be owing to some miscalculation I have made in my deductions from Weber's result.

Glasgow College, *Jan.* 18, 1860.

[*Addition, April* 1870.—From experiments of the present date, performed by Mr William Leitch and Mr Dugald M'Kichan, with the new Absolute Electrometer (§ 364, below), it is deduced that with the difference of potentials produced by 1000 Daniell's cells in series, the force of attraction would be 5·7 grammes per square decimetre between discs separated to a distance of 1 millimetre, instead of 3·57 grammes as found in § 314. This new measurement, with Maxwell's correction of Weber's number, which diminishes it by about 8 per cent. (*Report of British Association for* 1869, page 438 :—Committee on Electrical Standards), seems to reduce to as nearly as may be nothing, the discrepance from my thermo-dynamic estimate of December 1851 (*Philosophical Magazine*) referred to in § 318, below. Calculating from it by § 339, we find 3·74 for the difference of potentials, or electromotive force in c. g. s. absolute electrostatic measure, produced by 1000 elements of Daniell's.]

POSTSCRIPT, *April* 12, 1860.

317. I have since found that I had inadvertently misinterpreted Weber's statement in the ratio of 2 to 1. I had always, as it appears to me most natural to do, regarded the transference of negative electricity in one direction, and of positive electricity in the other direction, as identical agencies, to which, in our ignorance as to the real nature of electricity, we may apply indiscriminately the one expression or the other, or a combination of the two. Hence I have always regarded a current of unit strength as a current in which the positive or vitreous electricity flows in one direction at the rate of a unit of electricity per unit of time; or the negative or resinous electricity in the other direction at the same rate; or (according to the infinitely improbable hypothesis of two electric fluids) the vitreous electricity flows in one direction at any rate less than a unit per second, and the resinous in the opposite direction at a rate equal to the remainder of the unit per second. I have only recently remarked that Weber's expressions are not only adapted to the hypothesis of two electric fluids, but that they also reckon as a current of unit strength, what I should have called a current of strength 2, namely, a flow of vitreous electricity in one direction at the rate of a unit of vitreous electricity per unit of time, and of the resinous electricity in the other direction simultaneously, at the rate of a unit of resinous electricity per unit of time.

318. Weber's result as to the relation between electrostatic and electro-magnetic units, when correctly interpreted, I now find would be in perfect accordance with my own results given above, if the electromotive force of a single element of the Daniell's battery used were 2,140,000 British electro-magnetic units instead of 2,500,000, as according to my thermo-dynamic estimate. This is as good an agreement as could be expected when the difficulties of the investigations, and the uncertainty which still exists as to the true measure of the

electromotive force of the Daniell's element are considered. It must indeed be remarked that the electromotive force of Daniell's battery varies by two or three or more per cent. with variations of the solutions used; that it varies also very sensibly with temperature; and that it seems also to be dependent, to some extent, on circumstances not hitherto elucidated. A thorough examination of the electromotive force of Daniell's and other forms of galvanic battery, is an object of high importance, which, it is to be hoped, will soon be attained. Until this has been done, at least for Daniell's battery, the results of the preceding paper may be regarded as having about as much accuracy as is desirable.

319. I may state, therefore, in conclusion, that the average electromotive force per cell of the Daniell's batteries which I have used, produces a difference of potentials amounting to 00296 [corrected to ·00374, April 1870,] in [c. g. s.] absolute electrostatic measure. This statement is perfectly equivalent to the following in more familiar terms :—

One thousand cells of Daniell's battery, with its two poles connected by wires with two parallel plates of metal 1 millimetre apart, and each a square decimetre in area, produces an electrical attraction equal to the weight of 3·57 [corrected to 5·7] grammes.

XIX.—MEASUREMENT OF THE ELECTROMOTIVE FORCE REQUIRED TO PRODUCE A SPARK IN AIR BETWEEN PARALLEL METAL PLATES AT DIFFERENT DISTANCES.

[*Proceedings Royal Society*, Feb. 23 and April 12, 1860, or *Phil. Mag.*, 1860, second half-year.]

320. THE electrometers used in this investigation were the absolute electrometer and the portable electrometer described in my last communication to the Royal Society, and the operations were executed by the same gentlemen, Mr Smith and Mr Ferguson. The conductors between which the sparks passed were two unvarnished plates of a condenser, of which one was moved by a micrometer screw, giving a motion of $\frac{1}{25}$ of an inch (about one millimetre) per turn, and having its head divided into 40 equal parts of circumference. The readings on the screw-head could be readily taken to tenth parts of a division, that is to say, to about $\frac{1}{400}$ of a millimetre on the distance to be measured. The point from which the spark would pass in successive trials being somewhat variable, and often near the edges of the discs, a thin flat piece of metal, made very slightly convex on its upper surface like an extremely flat watch-glass, was laid on the lower plate. It was then found that the spark always passed between the crown of this convex piece of metal and the flat upper plate. The curvature of the former was so small, that the physical circumstances of its own electrification near its crown, the opposite electrification of the opposed flat surface in the parts near the crown of the convex, and the electric pressure on or tension in the air between them could not, it was supposed, differ sensibly from those between two plane conducting surfaces at the same distance and maintained at the same difference of potentials.

321. The reading of the screw-head corresponding to the position of the moveable disc when touching the metal below, was always determined electrically by making a succession of sparks pass, and approaching the moveable disc gradually by the screw until all appearance of sparks ceased. Contact was thus produced without any force of pressure between the two bodies capable of sensibly distorting their supports.

With these arrangements several series of experiments were made, in which the differences of potentials producing sparks across different thicknesses of air were measured first by the absolute electrometer, and afterwards by the portable torsion electrometer. The following Tables exhibit the results hitherto obtained :—

322. TABLE I.—*December* 13, 1859. *Measurements by absolute electrometer of maximum electrostatic forces* across a stratum of air of different thicknesses.*

Area of each plate of absolute electrometer = 174 square centimetres. Distance between plates of absolute electrometer = ·508 of a centimetre.

Length of spark in inches. $s.$	Weight in grains required to balance in absolute electrometer. $w.$	Electromotive force in units of the electrometer. $\sqrt{w}.$	Electrostatic force, or electromotive force per inch of air, in temporary units. $\dfrac{\sqrt{w}}{s}.$
·007	6	2·4495	349·9
·0105	9	3·0000	285·7
·0115	10	3·1622	275·0
·014	13	3·6055	257·5
·017	16	4·0000	235·3
·018	19	4·3589	242·2
·024	30	5·4772	228·2
·0295	40	6·3245	214·4
·034	50	7·0710	208·0
·0385	60	7·7459	201·2
·041	70	8·3666	204·1
·0445	80	8·9442	201·0
·048	90	9·4868	197·6
·052	100	10·0000	192·3
·055	110	10·4880	190·7
·058	120	10·9544	188·9
·060	130	11·4017	190·0

323. These numbers demonstrate an unexpected and a very remarkable result,—that greater electromotive force per unit

---

* See § 331 below.

length of air is required to produce a spark at short distances
than at long. When it is considered that the absolute electrification of each of the opposed surfaces* depends simply on
the electromotive force per unit length of the space between
them, or, which is the same thing, the resultant electrostatic
force in the air occupying that space, it is difficult even to conjecture an explanation. Without attempting to explain it, we
are forced to recognise the fact that a thin stratum of air is
stronger than a thick one against the same disruptive tension
in the air, according to Faraday's view of its condition as transmitting electric force, or against the same lifting electric pressure from its bounding surfaces, according to the views of the
eighteenth century school, as represented by Poisson. The
same conclusion is established by a series of experiments with
the previously-described portable torsion electrometer substituted for the absolute electrometer, leading to results shown
in the following Table :—

324.   TABLE II.—*January* 17, 1860. *Measurements by portable
torsion electrometer of electromotive forces producing sparks
across a stratum of air of different thicknesses.*

Length of spark in inches. s.	Torsion in degrees required to balance in electrometer. θ.	Electromotive force in units of the electrometer. $\sqrt{\theta}$.	Electrostatic force, or electromotive force per inch of air, in temporary units. $\sqrt{\theta} \div s$.
·001	3	1·732	1732
·002	7	2·646	1323
·003	11	3·316	1105
·004	14	3·742	935
·005	18	4·243	849
·006	22	4·690	782
·007	27	5·196	742
·008	30	5·477	685
·009	33	5·744	638
·010	38	6·164	616
·011	43	6·557	596
·012	48·5	6·964	580
·013	54	7·348	565
·014	59	7·681	549
·015	66	8·124	542
·016	73	8·544	534
·017	79	8·888	523
·018	85	9·219	512

* See § 332 below.

325. The series of experiments here tabulated stops at the distance 18 thousandths of an inch, because it was found that the force in the electrometer corresponding to longer sparks than that, was too strong to be measured with certainty by the portable electrometer, whether from the elasticity of the platinum wire, or from the rigidity of its connexion with the aluminium index ·being liable to fail when more than 85° or 90° of torsion were applied. So far as it goes, it agrees remarkably well with the other experiments exhibited in Table I., as is shown by the following comparative Table, in which, along with results of actual observation extracted from Table II., are placed results deduced from Table I. by interpolation for the same lengths of spark :—

TABLE III.—*Experiments of December* 13, 1859, *and January* 17, 1860, *compared.*

Col. 1. ·Length of spark in inches. s.	Col. 2. Electromotive force per inch of air, Dec. 13, in temporary units of that day. $\frac{\sqrt{w}}{s}$.	Col. 3. Electromotive force per inch of air, Jan. 17, in temporary units of that day. $\frac{\sqrt{\theta}}{s}$.	Col. 4. Ratios of numbers in Col. 3 to numbers in Col. 2.
·007	349·3	742	2·13
·0105	285·7	606	2·12
·0115	275·0	588	2·14
·014	257·5	549	2·14
·017	235·3	523	2·22
·018	242·2	512	2·11
			Mean 2·14

The close agreement with one another of the numbers in Col. 4, derived from series differing so much as those in Cols. 2 and 3, and obtained by means of electrometers differing so much in construction, constitutes a very thorough confirmation of the remarkable result inferred above from the experiments of the first series, and shows that the law of variation of the electrostatic force in the air required to produce sparks of the different lengths, must be represented with some degree of accuracy by the numbers shown in the last column of either Table I. or Table III.

The following additional series of experiments were made on precisely the same plan as those of Table II :—

TABLE IV.—*January 21, 1860.   Measurements by portable torsion electrometer of electromotive forces producing sparks across a stratum of air of different thicknesses.*

Length of spark in inches. s.	Torsion in degrees required to balance in electrometer. θ.	Electromotive force in units of the electrometer. $\sqrt{\theta}$.	Electrostatic force, or electromotive force per inch of air, in temporary units. $\sqrt{\theta} \div s$.
·001	3·2	1·79	1790
·002	6·4	2·32	1160
·003	10·5	3·24	1080
·004	13·2	3·63	907
·005	14·2	3·77	754
·006	18·2	4·27	712
·007	21·7	4·66	666
·012	41·2	6·42	535
·013	46·7	6·83	525
·014	53·2	7·29	521
·015	57·2	7·56	504
·016	63·2	7·95	497
·017	68·2	8·26	486
·018	78·2	8·84	491

TABLE V.—*January 23, 1860.   Similar experiments repeated.*

s.	θ.	$\sqrt{\theta}$.	$\sqrt{\theta} \div s$.
·001	3·5	1·87	1870
·002	6·5	2·55	1275
·003	9·5	3·08	1027
·004	12·7	3·56	890
·005	15·5	3·94	788
·006	18·5	4·30	716
·007	23·0	4·80	686
·008	25 62	5·06	632
·009	30·5	5·52	613
·010	35·0	5·92	592
·011	39·5	6·28	571
·012	44·0	6·63	553
·013	50·0	7·07	544
·014	54·0	7·35	525
·015	59·0	7·68	512
·016	63·5	7·97	498
·017	69·5	8·34	490
·018	74·5	8·63	479

The difference between the numbers shown in these two Tables and in Table II. above, are probably due in part to true differences in the resistance of the air to electrical disruption; but variations in the electrometer, which was by no means of perfect construction, may have sensibly influenced the results, especially as regards the differences between those shown in Table II. and those shown in Tables IV. and V., which, agreeing on the whole closely with one another, fall considerably short of the former.

326. TABLE VI.—*Summary of results reduced to absolute measure.*

Col. 1. Length of spark in centimetres. s.	Col. 2. Electrostatic forces according to simple determinations of Dec. 13, 1859. $\frac{\sqrt{w}}{s} \times \cdot 508 \sqrt{\dfrac{981\cdot4 \times 8\pi}{174}}$* $= R$.	Col. 3. Electrostatic forces according to estimated average of various determinations. R.	Col. 4. Differences.	Col. 5. Pressures of electricity from either metallic surface balanced by air immediately before disruption, in grammes weight per square centimetre†. $\dfrac{R^2}{8\pi \times 981\cdot4}$.
·00254	...	527·7	...	11·290
·00508	...	367·8	...	5·484
·00762	...	314·4	...	4·007
·01016	...	267·6	...	2·903
·01270	...	234·0	...	2·220
·01524	...	216·1	...	1·893
·01778	211·4	208·2	+ 3·2	1·757
·02032	...	193·1	...	1·512
·02286	...	183·4	...	1·364
·02540	...	177·5	...	1·277
·02667	172·8	173·3	− 0·5	1·217
·02794	...	171·0	...	1·185
·02921	166·4	166·9	− 0·5	1·129
·03048	...	163·2	...	1·080
·03302	...	159·4	...	1·030
·03556	155·8	155·8	·0	·984
·03810	...	152·6	...	·944
·04064	...	149·9	...	·911
·04318	142·5	144·4	− 1·9	·845
·04572	146·7	145·7	+ 1·0	·860
·06096	142·5	...	...	·823
·07493	129·6	...	...	·681
·08636	126·0	...	...	·644
·09779	121·8	...	...	·601
·10414	123·7	...	...	·620
·11303	121·8	...	...	·601
·12192	119·5	...	...	·579
·13208	116·3	...	...	·548
·13970	115·4	...	...	·540
·14732	114·5	...	...	·531
·15240	114·9	...	...	·535

* Distance between discs of absolute electrometer = ·508 of a centimetre. Area of each = 174 square centimetres.

Force of gravity at Glasgow on unit mass = 981·4 dynamical units of force; that is to say, generates in one second a velocity of 981·4 centimetres per second.

† This is most directly obtained by finding the force between the discs of the absolute electrometer per square centimetre, and reducing, according to the inverse proportion of squares of distances, to what it would have been if the distance between them had been equal to the length of the spark.

APPENDIX (§§ 327-338).

327.   In order that the different expressions, " potential,"
" electromotive force," " electric force," or " electrostatic force,"
" pressure of electricity from a metallic surface balanced by air,"
used in the preceding statement, may be perfectly understood, I
add the following explanations and definitions belonging to the
ordinary elements of the mathematical theory of electricity:—

328.   *Measurement of quantities of electricity.*—The unit quantity of electricity is such a quantity, that, if collected in a point,
it will repel an equal quantity collected in a point at a unit
distance with a force equal to unity.

329   [In absolute measurements the unit distance is one
centimetre; and the unit force is that force which, acting on a
gramme of matter during a second of time, generates a velocity
of one centimetre per second.   The weight of a gramme at
Glasgow is 981·4 of these units of force.   The weight of a
gramme in any part of the earth's surface may be estimated
with about as much accuracy as it can be without a special
experiment to determine it for the particular locality, by the
following expression :—

In latitude λ, average weight of a gramme

$$= 978\!\cdot\!024 \times (1 + \cdot005133 \times \sin^2 \lambda) \text{ absolute kinetic units.}]$$

330.   *Electric density.*—This term was introduced by Coulomb
to designate the quantity of electricity per unit of area in any
part of the surface of a conductor.   He showed how to measure
it, though not in absolute measure, by his proof plane.

331.   *Resultant electric force at any point in an insulating fluid*
[compare § 65, above].—The resultant force at any point in air
or other insulating fluid in the neighbourhood of an electrified
body, is the force which a unit of electricity concentrated at
that point would experience if it exercised no influence on the
electric distributions in its neighbourhood.

332.   *Relation between electric density on the surface of a conductor, and electric force at points in the air close to it.*—According to a proposition of Coulomb's, requiring, however, correction,
and first correctly given by Laplace, the resultant force at any
point in the air close to the surface of a conductor is perpendi-

cular to the surface and equal to $4\pi\rho$, if ρ denotes the electric density of the surface in the neighbourhood (§ 87, Cor.).

333. *Electric pressure from the surface of a conductor balanced by air.*—A thin metallic shell or liquid film, as for instance a soap-bubble, if electrified, experiences a real mechanical force in a direction perpendicular to the surface outwards, equal in amount per unit of area to $2\pi\rho^2$, ρ denoting, as before, the electric density at the part of the surface considered (§ 88). This force may be called either a repulsion (as according to the views of the eighteenth century school) or an attraction effected by tension of air between the surface of the conductor and the conducting boundary of the air in which it is insulated, as it would probably be considered to be by Faraday ; but whatever may be the explanation of the *modus operandi* by which it is produced, it is a real mechanical force, and may be reckoned as in Col. 5 of the preceding Table, in grammes weight per square centimetre. In the case of the soap-bubble, for instance, its effect will be to cause a slight enlargement of the bubble on electrification with either vitreous or resinous electricity, and a corresponding collapse on being perfectly discharged. In every case we may regard it as constituting a deduction from the amount of air-pressure which the body experiences when unelectrified. The amount of this deduction being different in different parts according to the square of the electric density, its resultant action on the whole body disturbs its equilibrium, and constitutes in fact the resultant of the electric force experienced by the body.

334. *Collected formulæ of relation between electric density on the surface of a conductor, electric diminution of air-pressure upon it, and resultant force in the air close to the surface.*—Let, as before, ρ denote the first of these three elements, let p denote the second reckoned in units of force per unit of area, and let R denote the third. Then we have

$$R = 4\pi\rho,$$
$$p = 2\pi\rho^2 = \frac{1}{8\pi} R^2.$$

335. *Electric potential* [difference of potentials being what, after German usage, is still sometimes called "electromotive force." (*Addition, April* 1870.)]—The amount of work required

to move a unit of electricity against electric repulsion from any
one position to any other position, is equal to the excess of the
electric potential of the second position above the electric
potential of the first position.

Cor. 1. The electric potential at all points close to the surface
of an electrified metallic body has one value, since an electrified point, possessing so small a quantity of electricity as not
sensibly to influence the electrification of the metallic surface,
would, if held near the surface in any locality, experience a
force perpendicular to the surface in its neighbourhood.

Cor. 2. The electric potential throughout the interior of a
hollow metallic body, electrified in any way by external influence, or, if insulated, electrified either by influence or by communication of electricity to it, is constant, since there is no
electric force in the interior in such circumstances.

[It is easily shown by mathematical investigation, that the
electric force experienced by an electric point containing an
infinitely small quantity of electricity, when placed anywhere
in the neighbourhood of a hollow electrified metallic shell,
gradually diminishes to nothing if the electric point be moved
gradually from the exterior through a small aperture in the
shell into the interior. Hence the one value of the potential
close to the surface outside, mentioned in Cor. 1, is equal to
the constant value throughout the interior mentioned in Cor. 2.]

336. *Interpretation of measurement by electrometer.*—Every
kind of electrometer consists of a cage or case containing a moveable and a fixed conductor, of which one at least is insulated and
put in metallic communication, by what I shall call the principal electrode passing through an aperture in the case or cage,
with the conductor whose electricity is to be tested. In every
properly constructed electrometer, the electric force experienced by the moveable part in a given position cannot be
electrically influenced except by changing the difference of
potentials between the principal electrode and the uninsulated
conductor or conducting system in the electrometer. Even
the best of ordinary electrometers hitherto constructed do not
fulfil this condition, as the inner surface of the glass of which
the whole or part of the enclosing case is generally made, is
liable to become electrified, and inevitably does become so

when any very high electrification is designedly or accidentally introduced, even for a very short time; the consequence
of which is that the moving body will generally not return to
its zero position when the principal electrode is perfectly dis-
insulated. Faraday long ago showed how to obviate this radical defect by coating the interior of the glass case with a fine
network of tinfoil; and it seems strange that even at the present day electrometers for scientific research, as, for instance,
for the investigation of atmospheric electricity, should be constructed with so bad and obvious a defect uncured by so simple
and perfect a remedy. When it is desired to leave the interior
of the electrometer as much light as possible, and to allow it
to be clearly seen from any external position with as little
embarrassment as possible, a cage made like a bird's cage, with
an extremely fine wire on a metal frame, inside the glass shade
used to protect the instrument from currents of air, etc., may
be substituted with advantage for the tinfoil network lining of
the glass. It appears, therefore, that a properly constructed
electrometer is an instrument for measuring, by means of the
motions of a moveable conductor, the difference of potentials
of two conducting systems insulated from one another, of one
of which the case or cage of the apparatus forms part. It may
be remarked in passing, that it is sometimes convenient in
special researches to insulate the case or cage of the apparatus,
and allow it to acquire a potential differing from that of the
earth, and that then, as always, the subject of measurement is
the difference of potentials between the principal electrode and
the case or cage, while in the ordinary use of the instrument
the potential of the latter is the same as that of the earth.
Hence we may regard the electrometer merely as an instrument
for measuring differences of potential between two conducting
systems mutually insulated; and the object to be aimed at in
perfecting any kind of electrometer (more or less sensitive as it
may be, according to the subjects of investigation for which it
is to be used), is, *that accurate evaluations in absolute measure,
of differences of potential, may be immediately derivable from its
indications.*

337. *Relation between electrostatic force and variation of electric
potential.*—§ 335, otherwise stated, is equivalent to this:—The

average component electrostatic force in the straight line of
air between two points in the neighbourhood of an electrified
body is equal to their difference of potentials divided by their
distance. In other words, the rate of variation of electric
potential per unit of length in any direction is equal to the
component of the electrostatic force in that direction. Since
the average electrostatic force in the line joining two points at
which the values of the potential are equal is nothing, the
direction of the resultant electrostatic force at any point must
be perpendicular to the equipotential surface passing through
that point; or the lines of force (which are generally curves)
cut the series of equipotential surfaces at right angles. The
rate of variation of potential per unit of length along a line of
force is therefore equal to the electrostatic force at any point.

338. *Stratum of air between two parallel or nearly parallel
plane or curved metallic surfaces maintained at different potentials.*—Let a denote the distance between the metallic surfaces
on each side of the stratum of air at any part, and V the difference of potentials. It is easily shown that the resultant electrostatic force is sensibly constant through the whole distance,
from the one surface to the other; and being in a direction
sensibly perpendicular to each, it must (§ 337) be equal to $\dfrac{V}{a}$.
Hence (§ 332) the electric density on each of the opposed surfaces is equal to $\dfrac{V}{4\pi a}$. This is Green's theory of the Leyden
phial.

339. *Absolute Electrometer.*—As a particular case of § 338,
let the discs be plane and parallel: and let the distance between them be small in comparison with their diameters, or
with the distance of any part of either from any conductor
differing from it in potential. The electric density will be
uniform over the whole of each of the opposed surfaces and
equal to $\dfrac{V}{4\pi a}$, being positive on one and negative on the other;
and in all other parts of the surface of each the electrification
will be comparatively insensible. Hence the force of attraction
between them per unit of area (§§ 333 and 334) will be $\dfrac{V^2}{8\pi a^2}$;

if A denote the area of either of the opposed surfaces, the whole force of attraction between them is therefore $A \dfrac{V^2}{8\pi a^2}$.

Hence, if the observed force be equal to the weight of w grammes at Glasgow, we have

$$981\cdot4 \times w = A \frac{V^2}{8\pi a^2},$$

and therefore $\quad V = a \sqrt{\dfrac{981\cdot4 \times 8\pi \times w}{A}}$

ADDITION, DATED APRIL 12, 1860.

340. Experiments on precisely the same plan as those of Table I. December 13, have been repeated by the same two experimenters, with different distances from ·75 to 1·5 of a centimetre between the plates of the absolute electrometer, and results have been obtained confirming the general character of those shown in the preceding Tables.

The absolute evaluations derived from these later series must be more accurate than those deduced above from the single series of December 13, when the distance between the plates in the absolute electrometer was only ·5 of a centimetre. I therefore, by permission, add the following Table of absolute determinations:—

Length of spark in centimetres. $s.$	Electrostatic forces according to estimated average of determinations of February 15, 23, 28, and 29, and March 2. $R.$
·0086	267·1
·0127	257·0
·0127	262·2
·0190	224·2
·0281	200·6
·0408	151·5
·0563	144·1
·0584	139·6
·0688	140·8
·0904	134·9
·1056	132·1
·1325	131·0

These results, as well as those shown in the preceding Tables, demonstrate a much less rapid variation with distance, of the electrostatic force preceding a spark, at the greater than at the smaller distances. It seems most probable that at still greater distances the electrostatic force will be found to be sensibly constant, as it was certainly expected to be at all distances. The limiting value to which the results shown in the last Table seem to point must be something not much less than 130. This corresponds to a pressure of 68 grammes weight per square decimetre. We may therefore conclude that the ordinary atmospheric pressure of 103,200 grammes per square decimetre, is electrically relieved by the subtraction of not more than 68, on two very slightly convex metallic surfaces, before the air between them is cracked and a spark passes, provided the distance between them is not less than $\frac{1}{8}$ of a centimetre. By taking into account the result of my preceding communication to the Royal Society, we may also conclude that a Daniell's battery of 5510 elements can produce a spark between two slightly convex metallic surfaces at $\frac{1}{8}$ of a centimetre asunder in ordinary atmospheric air.

XX. ELECTROMETERS AND ELECTROSTATIC MEASUREMENTS.

[§ 340' from *British Association Report* of Glasgow 1855 Meeting, §§ 341—389 from *Report* of Dundee 1867 Meeting, being part of *Report of Committee on Standards of Electrical Resistance.*]

340'. IN this communication three instruments were described and exhibited to the Section: the first a standard electrometer, designed to measure, by a process of weighing the mutual attraction of two conducting discs, the difference of electrical potential between two bodies with which they are connected, an instrument which will be useful for determining the electromotive force of a galvanic battery in electrostatic measure, and for graduating electroscopic instruments so as to convert their scale indications into absolute measure; the second an electroscopic electrometer, which may be used for indicating electrical potentials in absolute measure, in ordinary experiments, and, probably with great advantage, in observations of atmospheric electricity; and the third, for which a scientific friend has suggested the name of Electroplatymeter, an instrument which may be applied either to measure the capacities of conducting surfaces for holding charges of electricity, or to determine the electric inductive capacities of insulating media.

341. An electrometer is an instrument for measuring differences of electric potential between two conductors through effects of electrostatic force, and is distinguished from the galvanometer, which, of whatever species, measures differences of electric potentials through electromagnetic effects of electric currents produced by them. When an electrometer merely indicates the existence of electric potential, without measuring its amount, it is commonly called an electroscope; but the name electrometer is properly applied when greater or less degrees of difference are indicated on any scale of reckoning,

if approximately constant, even during a single series of experiments. The first step towards accurate electrometry in every case is to deduce from the scale-readings, numbers which shall be in simple proportion to the difference of potentials to be determined. The next and last step is to assign the corresponding values in absolute electrostatic measure. Thus, when for any electrometer the first step has been taken, it remains only to determine the single constant coefficient by which the numbers, deduced from its indications as simply proportional to differences of potential, must be multiplied to give differences of potential in absolute electrostatic measure. This coefficient will be called, for brevity, the absolute coefficient of the instrument in question.

342. Thus, for example, the gold-leaf electrometer indicates differences of potential between the gold leaves and the solid walls enclosing the air-space in which they move. If this solid be of other than sufficiently perfect conducting material, of wood and glass, or of metal and glass, for instance, as in the instrument ordinarily made, it is quite imperfect and indefinite in its indications, and is not worthy of being even called an electroscope, as it may exhibit a divergence when the difference of potentials which the operator desires to discover is absolutely zero. It is interesting to remark (§ 336) that Faraday first remedied this defect by coating the interior of the glass case with tinfoil, cut away to leave apertures proper and sufficient to allow indications to be seen, but not enough to cause these indications to differ sensibly from what they would be if the conducting envelope were completely closed around it; and that not till a long time after did any other naturalist, mathematician, or instrument-maker seem to have noticed the defect, or even to have unconsciously remedied it.

343. Electrometers may be classified in genera and species according to the shape and kinematic relations of their parts; but as in plants and animals a perfect continuity of intermediate species has been imagined between the rudimentary plant and the most perfect animal, so in electrometers we may actually construct species having intermediate qualities continuous between the most widely different genera. But, notwithstanding, some such classification as the following is

convenient with reference to the several instruments commonly in use and now to be described :—

I.   Repulsion electrometers.
Pair of diverging straws as used by Beccaria, Volta, and others, last century.
Pair of diverging gold leaves (Bennet).
Peltier's electrometer.
Delmann's electrometer.
Old station-electrometer, described in lecture to the Royal Institution, May 1860 [§§ 274–275, above]; also in Nichol's *Cyclopædia*, article "Electricity, Atmospheric" (edition, 1860) [§ 263, above], and in Dr Everett's paper of 1867, "On Atmospheric Electricity" (*Philosophical Transactions*).

II.   Symmetrical electrometers.
Bohnenberger's electrometer.
Divided-ring electrometers.

III.   Attracted disc electrometers.
Absolute electrometer.
Long-range electrometer.
Portable electrometer.
Spring-standard electrometer.

344.   Class I. is sufficiently illustrated by the examples referred to ; and it is not necessary to explain any of these instruments minutely at present, as they are, for the present at all events, superseded by the divided-ring electrometer and electrometers of the third class.

There are at present only two known species of the second class; but it is intended to include all electrometers in which a symmetrical field of electric force is constituted by two symmetrical fixed conductors at different electric potentials, and in which the indication of the force is produced by means of an electrified body moveable symmetrically in either direction from a middle position in this field. This definition is obviously fulfilled by Bohnenberger's well-known instrument*.

---

* A single gold leaf hanging between the upper ends of two equal and similar dry piles standing vertically on a horizontal plate of metal, one with its positive and the other with its negative pole up.

345.   My first published description of a divided-ring electrometer is to be found in the *Memoirs of the Roman Academy of
Sciences** for February 1857; but since that time I have made
great improvements in the instrument—first, by applying a
light mirror to indicate deflections of the moving body; next,
by substituting for two half rings four quadrants, and consequently for au electrified body projecting on one side only of
the axis, an electrified body projecting symmetrically on the
two sides, and moveable round an axis; and lastly, by various
mechanical improvements, and by the addition of a simple
gauge to test the electrification of the moveable body, and of
a replenisher to raise this electrification to any desired degree.

346.   In the accompanying drawings, Plate I. fig. 1 represents the front elevation of the instrument, of which the chief
bulk consists of a jar of white glass (flint) supported on three
legs by a brass mounting, cemented round the outside of its
mouth, which is closed by a plate of stout sheet-brass, with
a lantern-shaped cover standing over a wide aperture in its
centre.   For brevity, in what follows these three parts will be
called the jar, the main cover, and the lantern.

Fig. 5 represents the quadrants as seen from above; they
are shown in elevation at *a* and *b*, fig. 1, and in section at *c* and
d, fig. 2.   They consist of four quarters of a flat circular box
of brass, with circular apertures in the centres of its top and
bottom.   Their position in the instrument is shown in figs.
1, 2, and 6.   Each of the four quadrants is supported on a
glass stem passing downwards through a slot in the main cover
of the jar, from a brass mounting on the outside of it, and
admits of being drawn outwards for a space of about 1 centimetre ($\frac{2}{5}$ of an inch) from the positions they occupy when the
instrument is in use, which are approximately those shown in
the drawings.   Three of them are secured in their proper positions by nuts (*e, e, e*) on the outside of the chief flat lid of the
jar shown in fig. 4.   The upper end of the stem, carrying the
fourth, is attached to a brass piece (*f*, fig. 6) resting on three
short legs on the upper side of the main cover, two of these
legs being guided by a straight V-groove at (*g*) to give them

---

* Accademia Pontificia dei Nuovi Lincei.

freedom to move in a straight line inwards or outwards, and to prevent any other motion. This brass piece is pressed outwards and downwards by a properly arranged spring (*h*), and is kept from sliding out by a micrometer-screw (*i*) turning in a fixed nut. This simple kinematic arrangement gives great steadiness to the fourth quadrant when the screw is turned inwards or outwards, and then left in any position; and at the same time produces but little friction against the sliding in either direction. The opposite quadrants are connected in two pairs by wires, as shown in fig. 5; and two stout vertical wires (*l, m*), called the chief electrodes, passing through holes in the roof of the lantern, are firmly supported by long perforated vulcanite columns passing through those holes, and serve to connect the pairs of quadrants with the external conductors whose difference of potentials is to be tested. Springs (*n, o*) at the lower ends of these columns, shown in figs. 1 and 2, maintain metallic contact between the chief electrodes and the upper sides of two contiguous quadrants (*a* and *b*) when the lantern is set down in its proper position, but allow the lantern to be removed, carrying the chief electrodes with it, and to be replaced at pleasure without disturbing the quadrants. The lantern also carries an insulated charging-rod (*p*), or temporary electrode, for charging the inner coating of the jar (§ 351) to a small degree, to be increased by the replenisher (§ 352), or, it may be, for making special experiments in which the potential of the interior coating of the jar is to be measured by a separate electrometer, or kept at any stated amount of difference from that of the outer coating. When not in use this temporary electrode is secured in a position in which it is disconnected from the inner coating.

347. The main cover supports a glass column (*q*, fig. 2) projecting vertically upwards through its central aperture, to the upper end of which is attached a brass piece (*r*), which bears above it a fixed attracting disc (*s*), to be described later (§ 353); and projecting down from it a fixed plate bearing the silk-fibre suspension of the mirror (*t*), needle (*u*), etc., seen in figs. 1 and 2, and fixed guard tubes (*v, w*), to be described presently. To the main cover also is attached the circular level (fig. 6), which is adjusted to indicate the position of the

instrument in which the quadrants are level, and the guard-tubes just mentioned vertical. Its lower surface which rests on the cover is slightly rounded, like a convex lens, so as to admit of a slight further adjustment (see end of § 348, *Addition*) by varying the relative pressure of the three screws by which it is fastened down to the cover.

348. The moveable conductor of the instrument consists of a stiff platinum wire (*x*), about 8 centimetres (3½ inches) long, with the needle rigidly attached in a plane perpendicular to it, and connected with sulphuric acid in the bottom of the jar by a fine platinum wire hanging down from its lower end and kept stretched by a platinum weight under the level of the liquid. The upper end of the stiff platinum wire is supported by a single silk-fibre so that it hangs down vertically. The mirror is attached to it just below its upper end. Thus the mirror, the needle, and the stiff platinum stem constitute a rigid body having very perfect freedom to move round a vertical axis (the line of the bearing fibre), and yet practically prevented from any other motion in the regular use of the instrument by the weight of its own mass and that of the loose piece of platinum hanging from it below the surface of the liquid in the jar. A very small magnet is attached to the needle, which, by strong magnets fixed outside the jar, is directed to one position, about which it oscillates after it is turned through any angle round the vertical axis, and then left to itself. The external magnets are so placed that when there is magnetic equilibrium the needle is in the symmetrical position shown in figs. 5 and 6 with reference to the quadrants*.

[*Addition, April* 1870.—The success of the experiments referred to in the footnote has led to the adoption of the bifilar suspension in all the Quadrant Electrometers now made. It is represented in the margin. The stiff platinum wire which carries the mirror and needle has a cross piece at its upper end, to which are attached the lower ends of the two suspending silk fibres ; the other ends being wound upon the two pins *c, d*, which may be turned in their sockets by a square-pointed key, to

---

* Recently I have made experiments on a bifilar suspension with a view to superseding the magnetic adjustment, which promise well.

equalize the tensions of the fibres, and make the needle hang midway between the upper and under surfaces of the quadrants. The pins _c_, _d_, are pivoted in blocks carried by springs _e_, _f_, to allow them to be shifted horizontally when adjusting the position of the points of suspension. The screws _a_, _b_, which traverse these blocks, have their points bearing against the fixed plate behind, so that when _a_ or _b_ is turned in the direction of the hands of a watch, the neighbouring point of suspension is brought forward, and conversely. The needle may thus be made to turn through an angle, till it lies in the symmetrical position represented in fig. 5, Plate I., when all electrical disturbance has been guarded against by connecting the quadrants with the inside and outside of the jar. The conical pin _h_ passes between the two springs and screws into the plate behind; by screwing it inwards the points of

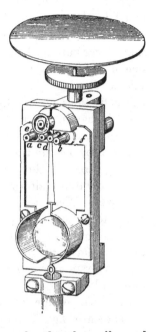

suspension are made to recede from each other laterally, and the sensibility of the needle to a deflecting couple is diminished, and conversely.

The method employed to test the symmetry of the suspension is suggested by the consideration that if the tension be equally distributed between the two fibres, the sensibility of the needle to the same deflecting couple will be less than if the whole or the greater part of the weight were supported by one fibre; also, the sensibility being a minimum, a small deviation from the conditions which make it so will produce the least change of sensibility, by the known property of a maximum or minimum. To test whether these conditions are attained, raise first one side of the instrument a little (one turn of the foot-screw on that side is usually sufficient), and then produce an equal deviation in the opposite direction from the position marked by the attached level (§ 347); and in each position of the instrument observe the deflection of the image on the scale produced by some constant difference of potentials, as that between the two poles of a Daniell's cell. This deflection ought to be very nearly equal in the three positions, but exactly equal in the two disturbed positions, and somewhat greater in these than in the middle or level position. When the instrument is far out of adjustment, the deviation will be greater in one of the disturbed positions and less in the other than in the middle position. When it is but slightly out of adjustment, the deflections in the disturbed positions may both somewhat exceed that in the middle position, but to different degrees. An approximation to symmetry thus far at least should be obtained by merely turning the pins (*c, d*) in their sockets as already directed, through the minutest angles sensible to the operator, without altering the adjustment of the spirit-level on the cover. When that has been done, the level on the cover ought to be adjusted (§ 347) by successive trials to indicate the position of the instrument such that when equally disturbed from it in opposite directions, the deflections obtained are equally in excess of the deflection obtained in the indicated position.]

349. The needle (*u*) is of thin sheet aluminium cut to the shape seen in figs. 5 and 6; the very thinnest sheet that gives the requisite stiffness being chosen. Its area is $4\frac{1}{8}$ square centimetres, and weight ·07 of a gramme. If the four quadrants are in a perfectly symmetrical position round it, and if they are kept at one electric potential by a metallic arc connecting the chief electrodes outside, the needle may be strongly electrified without being disturbed from its position of magnetic equilibrium; but if it is electrified, and if the external electrodes be disconnected, and any difference of potentials established between them, the needle will clearly experience a couple turning it round its vertical axis, its two ends being driven from the positive quadrants towards the negative, if it is itself positively electrified. It is kept positive rather than negative in the ordinary use of the instrument, because I find that when a conductor with sharp edges or points is surrounded by another presenting everywhere a smooth surface, a much greater difference of potentials can be established between

them, without producing disruptive discharge, if the points and edges are positive than if they are negative.

350.   The mirror (t) serves to indicate, by reflecting a ray of light from a lamp, small angular motions of the needle round the vertical axis.   It is a very light, concave, silvered glass mirror, being of only 8 millimetres ($\frac{1}{3}$ of an inch) diameter, and 22 milligrammes ($\frac{1}{3}$ of a grain) weight.   I had for many years experienced great difficulty in getting suitable mirrors for my form of mirror galvanometer; but they are now supplied in very great perfection by Mr Becker, of Messrs Elliott Brothers, London.   [*Addition, May* 1870.—I have not succeeded in getting more of these light ground concave mirrors giving good images, after a few supplied by Mr Becker at the time when the report was written.   The lightest ground mirrors that Mr Becker can guarantee to give good images, weigh $\frac{1}{22}$ of a gramme ($\frac{7}{10}$ of a grain).   These answer well enough for the electrometers, because the aluminium needle weighing $\frac{1}{14}$ of a gramme ($1\frac{1}{10}$ grain), and being of much greater linear dimensions, its moment of inertia is not largely increased by the addition of a mirror of that weight; and they are preferred for this purpose to the exquisite light mirrors supplied by Mr White, as being stronger and less liable to warp in being mounted.   But for galvanometers, and especially telegraph-signal galvanometers, it is important that the mirrors be the very lightest possible.   The only mirrors suitable for this purpose which I can now obtain are supplied by Mr White. They give very perfect images, and weigh $\frac{1}{50}$ of a gramme ($\frac{3}{10}$ of a grain) without the magnets, and $\frac{1}{30}$ of a gramme with the magnets attached.   Mr White produces them by cutting out and silvering a large number of circles of the thinnest microscope glass, attaching the magnets (four on the back of each mirror), and finally testing for the image.   Out of fifty tried, about ten or fifteen are generally found satisfactory. A mirror may give a good image before the magnets are attached, and become warped out of shape and give a bad image after the magnets have been cemented to it.]   The focus for parallel rays is about 50 centimetres (20 inches) from the mirror, and thus the rays of the lamp placed at a distance of 1 metre (or 40 inches) are brought to a focus at

the same distance. The lamp is usually placed close behind the vertical screen a little below or above the normal line of the mirror, and the image is thrown on a graduated scale extending horizontally above or below the aperture in the screen through which the lamp sends its light. When the mirror is at its magnetic zero position, the lamp is so placed that its image is, as nearly as may be, in a vertical plane with itself, and not more than an inch above or below its level, so that there is as little obliquity as possible in the reflection, and the line traversed by the image on the screen during the deflection is, as nearly as may be, straight. The distance of the lamp and screen from the mirror is adjusted so as to give as perfect an image as possible of a fine wire which is stretched vertically in the plane of the screen across the aperture through which the lamp shines on the mirror; and with Mr Becker's mirrors, as with Mr White's selected galvanometer mirrors, I find it easy to read the horizontal motions of the dark image to an accuracy of the tenth of a millimetre. In the ordinary use of the instrument a white paper screen, printed from a copper-plate, divided to fortieths of an inch, is employed, and the readings are commonly taken to about a quarter of a scale-division; but with a little practice they may, when so much accuracy is desired, be read with considerable accuracy to the tenth of a scale-division. Formerly a slit in front of the lamp was used, but the wire giving a dark line in the middle of the image of the flame is a very great improvement, first introduced by Dr. Everett (in consequence of a suggestion made by Professor P. G. Tait) in his experiments on the elasticity of solids made in the Natural Philosophy Laboratory of Glasgow University*.

351. The charge of the needle remains sensibly constant from hour to hour, and even from day to day, in virtue of the arrangement by which it is kept in communication with sulphuric acid in the bottom of the jar, the outside of the

---

* A Drummond light placed about 70 centimetres from the mirror gives an image, on a screen about 3 metres distant, brilliant enough for lecture-illustrations, and with sufficient definition to allow accurate readings of the positions on a scale marked by the image of a fine vertical wire in front of the light.

jar being coated with tinfoil and connected with the earth, so that it is in reality a Leyden jar.   The whole outside of the jar, even where not coated with tinfoil, is in the ordinary use of the instrument, especially in our moist climate, kept virtually at one potential through conduction along its surface.   This potential is generally, by connecting wires or metal pieces, kept the same as that of the brass legs and framework of the instrument.   To prevent disturbance in case of strongly electrified bodies being brought into the neighbourhood of the instrument, a wire is either wrapped round the jar from top to bottom, or a cage or network of wire, or any convenient metal case, is placed round it; but this ought to be easily removed or opened at any time to permit the interior to be seen.   When the instrument is left to itself from day to day in ordinary use, the needle, connected with the inner coating of the jar as just described, loses, of course, unless replenished, something of its charge; but not in general more than $\frac{1}{2}$ per cent. per day, when the jar is of flint-glass made in Glasgow.   On trying similar jars of green glass I found that they lost their charge more rapidly per hour than the white glass jars per month.   I have occasionally, but very rarely, found white glass jars to be as defective as those green ones, and it is possible that the defect I found in the green jars may have been an accident to the jars tested, and not an essential property of that kind of glass.

352.   I have recently made the very useful addition of a replenisher to restore electricity to the jar from time to time when required.   It consists of (1) a turning vertical shaft of vulcanite bearing two metal pieces called carriers (b, b, figs. 17 and 18); (2) two springs (d, d, figs. 16 and 18), connected by a metallic arc, making contact with the carriers once every half turn of the shaft, and therefore called connectors; and (3) two inductors (a, a) with receiving springs (c, c) attached to them, which make contact with the carriers once every half turn, shortly before the connecting contacts are made.   The inductors (a, a, figs. 16 and 18) are pieces of sheet metal bent into circular cylindrical shapes of about 120° each; they are placed so as to deviate in the manner shown in the drawing from parts of a cylindrical surface coaxial with the turning-shaft, leaving gaps of about 60° on each side.   The diameter of

T. E.   18

this cylindrical surface is about 15 millimeters (about ⅗ of an inch). The carriers (*b*, *b*, figs. 17 and 18) are also of sheet metal bent to cylindrical surfaces, but not exactly circular cylinders; and are so placed on the bearing vulcanite shaft that each is rubbed by the contact springs over a very short space, about 1 millimeter beyond its foremost edge, when turned in the proper direction for replenishing. The receiving springs (*c*, *c*, figs. 17 and 18) make their contacts with each carrier immediately after it has got fairly under cover, as it were, of the inductor. Each carrier subtends an angle of about 60° at the axis of the turning-shaft. The connecting contacts are completed just before the carriers commence emerging from being under cover of the inductors. The carriers may be said to be under cover of the inductors when they are within the angle of 120° subtended by the inductors on each side of the axis. One of the inductors is in metallic communication with the outside coating of the jar, the other with the inside. Figs. 16, 17, and 18 illustrate sufficiently the shape of carriers and the succession of the contacts. The arrow-head indicates the direction to turn for replenishing. When it is desired to diminish the charge, the replenisher is turned backwards. A small charge having been given to the jar from an independent source, the replenisher when turned forwards increases the difference of potentials between the two inductors and therefore between the two coatings of the jar connected with them by a constant percentage per half turn, unless it is raised to so high a degree as to break down the air-insulation by disruptive discharge. The electric action is explained simply thus:—The carriers, when connected by the connecting springs, receive opposite charges by induction, of which they deposit large proportions the next time they touch receiving springs. Thus, for example, if the jar be charged positively, the carrier emerging from the inductor connected with the inner coating carries a negative charge round to the receiving spring connected with the outside coating, while the other carrier, emerging from the inductor connected with the outside coating, carries a positive charge round to the receiving spring connected with the inside coating. If the carriers are not sufficiently well under cover of the inductors during both the receiving contacts and the connecting contacts to render the charges which they acquire by induction during the connecting contacts greater than that which they carry away with them from the receiving contacts, the rotation, even in the proper direction for replenishing, does not increase, but, on the contrary, diminishes the charge of the jar. The deviations of the inductors from the circular cylinder, referred to above, have been adopted to give greater security against this failure. A steel pivot fixed to the top of the vulcanite shaft, and passing through the main cover, carries a small milled head (*y*, fig. 1) above, on the outside, which is spun rapidly round in either direction by the finger, and thus in less than a minute a small charge in the jar may be doubled. The diminution of the charge, when the instrument is left to itself for twenty-four hours, is sometimes imperceptible; but when any loss is discovered to have taken place, even if to the extent of 10 per cent., a few moments' use of the replenisher suffices to restore it, and to adjust it with minute accuracy to the required degree by aid of the gauge to be described presently. The principle of the "replenisher" is identical with that of the "doubler" of Bennet. In the essentials of its construction it is the same as Varley's improved form of Nicholson's "revolving doubler."

353. The gauge consists of an electrometer of Class III. The moveable attracted disc is a square portion of a piece of very thin sheet aluminium of the shape shown at *a* in fig. 4. It is supported on a stretched platinum wire passing through two holes in the sheet, and over a very small projecting ridge of bent sheet aluminium placed in the manner shown in the magnified drawing, fig. 3. The ends of this wire are passed through holes in curved springs, shown in fig. 4, and are bent round them so as to give a secure attachment without solder, and without touching the straight stretched part of the wire. The ends of the platinum wire (*β, β*) are attached by cement to the springs, merely to prevent them from becoming loose, care being taken that the cement does not prevent metallic contact between some part of the platinum wire and one or both of the brass springs. I have constantly found fine platinum wire rendered brittle by ordinary solder applied to it. The use of these springs is to keep the platinum wire stretched

18—2

with an approximately constant tension from year to year, and at various temperatures. Their fixed ends are attached to round pins, which are held with their axes in a line with the fibre by friction, in bearings forming parts of two adjustable brass pieces (γ, γ) indicated in fig. 4; these pieces are adjusted once for all to stretch the wire with sufficient force, and to keep the square attracted disc in its proper position. The round pins bearing the stretching springs are turned through very small angles by pressing on the projecting springs with the finger. They are set so as to give a proper amount of torsion tending to tilt the attracted disc (α) upwards, and the long end of the aluminium lever (δ), of which it forms a part, downwards. The downward motion of the long end is limited by a properly placed stop. Another stop (ϵ) above limits the upward motion, which takes place under the influence of electrification in the use of the instrument. A very fine opaque black hair (that of a small black-and-tan terrier I have found much superior to any hitherto tried) is stretched across the forked portion of the sheet aluminium in which the long arm of the lever terminates. Looked at horizontally from the outside of the instrument it is seen, as shown in fig. 7, Plate L, against a white background, marked with two very fine black circles. These sight-plates in the instruments, as now made by Mr White, are of the same material as the ordinary enamel watch-dials, with black figures on a white ground. The white space between the two circles should be a very little less than the breadth of the hair. The sight-plate is set to be as near the hair as it can be without impeding its motion in any part of its range; it is slightly convex forwards, and is so placed that the hair is nearer to it when in the middle between the black circles than when in any other part of its range. It is thus made very easy, even without optical aid, to avoid any considerable error of parallax in estimating the position of the hair relatively to the two black circles. By a simple plano-convex lens (ϕ, fig. 2), with the convex side turned inwards, it is easy, in the ordinary use of the instrument, to distinguish a motion up or down of the hair amounting to $\frac{1}{5000}$ of an inch. With a little care I have ascertained, Dr Joule assisting, that a motion of no more than $\frac{1}{50.000}$ of an inch from one definite central position can be securely tested without the aid of other magnifying power than that given by the simple lens. The lens during use is in a fixed position relatively to the framework bearing the needle, but it may be drawn out or pushed in to suit the focus of each observer. To give great magnification, it ought to be drawn out so far that the hair and sight-plate behind may be but little nearer to the lens than its principal focus, and the observer's eye ought to be at a very considerable distance from the instrument, no less than 20 centimetres (8 inches) to get good magnification; and a short-sighted person should use his ordinary concave eye-lens close to his eye. The reason for turning the convexity of the small plano-convex lens inwards is, that with such a lens so placed, if the eye of the observer is too high or too low, the hair seems to him curved upwards or downwards, and he is thus guided to keep his eye on a level sufficiently constant to do away with all sensible effects of parallax on the position of the hair relatively to the black circles. The framework carrying the stretched platinum wire and moveable attracted disc is above the brass roof of the lantern, in which a square aperture is cut to allow the square portion constituting the short arm of the aluminium balance to be attracted downwards by the fixed attracting disc (§ 347), to be presently described. A side view of the attracting plate, the brass roof of the lantern, the aluminium balance, the sight-plate, the hair, and the plano-convex lens is given in section (fig. 2); also a glass upper roof to protect the gauge and the interior of the instrument below from dust and disturbance by currents of air, to which, without this upper roof, it would be exposed, through the small vacant space around the moveable aluminium square. The fixed attracting disc is borne by a vertical screw screwing into the upper brass mounting (z, fig. 2) (§ 347), connected with the inner coating of the Leyden jar through the guard tubes, etc., and is secured in any position by the "jam nut," shown in the drawing at z, fig. 2. This disc (s) is circular, and about 38 millimetres ($1\frac{1}{2}$ inch) in diameter, and is placed horizontally with its centre under the centre of the square aperture in the roof of the lantern. Its distance from the lower surface of the roof and of the moveable attracted disc may be from $2\frac{1}{2}$ to 5 millimetres (from $\frac{1}{10}$ to $\frac{1}{5}$ of an inch), and is to be adjusted, along with the amount of torsion in the platinum wire bearing the aluminium balance-arm, so as to give the proper sensibility to the gauge. The sensibility is increased by diminishing the distance from the attracting to the attracted plate, and increasing the amount of torsion. Or, again, the degree of the potential indicated by it when the hair is in the sighted position is increased by increasing the distance between the plates, or by increasing the amount of torsion. If the electrification of the needle is too great, its proper position of equilibrium becomes unstable; or before this there is sometimes a liability to discharge by a spark across some of the air-spaces. The instrument works extremely well with the needle charged but little less than to give rise to one or both of these faults, and I adjust the gauge accordingly.

354. The strength of the fixed steel directing magnets is to be adjusted to give the desired amount of deflection with any stated difference of potentials maintained between the two chief electrodes, when the jar is charged to the degree which brings the hair of the gauge to its sighted position. In the instruments already made, the deflection* by a single cell of Daniell's amounts to about 100 scale-divisions (of $\frac{1}{40}$ of an inch each and at a distance of 40 inches), if the magnetic directive force is such as to give a period of vibration equal to about 1·5 seconds, when the jar is discharged and the four quadrants are connected with one another and with the inner coating of the jar. Lower degrees of sensibility may be attained better by increasing the magnetic directing force than by diminishing the charge of the jar. Thus, for instance, when it is to be used for measuring and photographically recording the potential of atmospheric electricity at the point where the stream of the water-dropping collector† breaks into drops, the magnetic directing force may be made from 10 to 100 times greater than that just described. When this is to be done it may be convenient to attach a somewhat more powerful magnetic needle than that which has been made in the most recent instruments where a high degree of sensibility has been provided for. But it

---

* That is to say, the number of scale-divisions over which the luminous image moves when the chief electrodes are disconnected from one another and put in metallic connexion with the two plates of a Daniell's battery.

† See Royal Institution Lecture, May 18, 1860 (§§ 278, 279, above), or Nichol's *Cyclopædia*, article "Electricity, Atmospheric" (Edition 1860) (§ 262, above).

is to be remarked that in general the directing-force of the external steel magnets cannot be too strong, as the stronger it is the less is the disturbance produced by magnetic bodies moving in the neighbourhood of the instrument*. In laboratory work, where numerous magnetic experiments are being performed in the immediate neighbourhood, and in telegraph factories where there is constant disturbance by large moving masses of iron, the artificial magnetic field of the electrometer ought to be made very strong. To allow this, and yet leave sufficient sensibility to the instrument, the suspended magnetic needle has been made smaller and smaller, until it is now reduced to two small pieces of steel side by side, 6 millimetres ($\frac{1}{4}$ of an inch) long. For a meteorological observatory all that is necessary is, that the directing magnetic force may be so great that the greatest disturbance experienced in magnetic storms shall not sensibly deflect the luminous image.

355. The sensibility of the gauge should be so adjusted that a variation in the charge of the jar, producing an easily perceived change in the position of the hair, shall produce no sensible change in the deflection of the luminous image produced by the greatest difference of potentials between the quadrants, which is to be measured in the use of the instrument. I believe the instruments already made, when adjusted to fulfil these conditions, may be trusted to measure the difference of potentials produced by a single cell of Daniell's to an accuracy of a quarter per cent. It must be remembered that the constancy of value of the unit of each instrument depends not only on the constancy of the potential indicated by the gauge, but also on the constancy of the magnetic force in the field traversed by the suspended magnet, and on the constancy of the magnetic moment of the latter. As each of these may be expected to decrease gradually from year to year (although very slowly after the first few hours or weeks), rigorous methods must be adopted to take such variations into account, if the instrument is to be trusted as giving accurately comparable indications at all times. The only method hitherto provided

---

* All embarrassment from this source will be done away with if the bifilar plan be adopted (see § 348, *Addition*).

for this most important object consists in the observation of the deflection produced by a measured motion of one of the quadrants by the micrometer screw (*i*) when the four quadrants are put in metallic communication with one another through the principal electrodes; the jar being brought to one constant potential by aid of the gauge, and therefore the force producing the deflection being constant. The amount of the deflection will show whether or not the force of the magnetic field has changed, and will render it easy at any time to adjust the strength of the magnets, if necessary, to secure this constancy. But to attain this object by these means, the three quadrants not moved by the micrometer screw must be clamped by their fixing-screws so that they may be always in the same position.

356. The absolute constancy of the gauge cannot be altogether relied upon. It certainly changes to a sensible degree with temperature; and in different instruments, to very different degrees, and even in different directions, as will be seen (§ 377) in connexion with the description of the portable electrometer to be given later. But this temperature variation does not amount in ordinary cases probably to as much as one per cent ; and it is probable that after a year or two any continued secular variation of the platinum torsion spring will be quite insensible. It is to be remarked, however, that secular experiments on the elasticity of metals are wanting, and ought at least to be commenced in our generation. In the meantime it will be desirable, both on account of the temperature variation and of the possible secular variation in the couple of torsion, to check the gauge by accurate measurements of the time of oscillation of the needle with its appurtenances. The moment of inertia of this rigid body, except in so far as it may be influenced by oxidation of the metal, of which I have as yet discovered no signs, may be regarded as constant, and therefore the amount of the directing couple due to the magnets may be determined with great accuracy by finding the period of an oscillation when the four quadrants are put in connexion through the charging rod with the metal mounting bearing the guard plates, etc. I have not as yet put into practice any of the obvious methods, founded on the general principle of coincidences used in pendulum observations, for determining the period of the oscillation; but

although not more than twenty or thirty complete oscillations can be counted, it seems certain that with a little trouble the period of one of them may be easily determined to an accuracy of about $\frac{1}{10}$ per cent.

357. [*Addition, May* 1870.—The most direct and obvious method of using the Quadrant Electrometer is to connect the two chief electrodes, with the two bodies whose difference of potentials is to be measured, and one of them with the case of the instrument. With the instruments made at the present date, a difference of potentials equal to that of the opposite poles of a single Daniell's cell gives, when measured in this manner, a deflection of the image over about 60 scale-divisions, more or less according to the distance at which the points of suspension of the silk fibres have been adjusted (§ 348, *Addition*). The difference of potentials due to six cells in series would thus deflect the image to the extremity of the scale, and be the greatest difference of potentials that could be measured by the electrometer, if these were the only connexions available for measurements. A second and much lower grade of sensibility is obtained by simply raising, so as to disconnect from the quadrant beneath it, the electrode connected with the case.

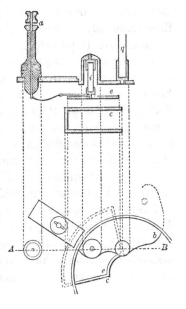

This being done, it requires a battery of about 10 or 15 cells to produce the deflection previously produced by a single cell. Several still lower grades of sensibility have been provided for in the instruments recently made, by the addition of an induction-plate, insulated directly over one of the quadrants behind the mirror. The sketch in the margin represents a vertical section through the induction-plate (*e*), insulating glass stem (*i*) by which it is supported, its electrode (*a*), the quadrant (*c*), and main glass stem (*q*). The line *AB* in the horizontal plan below is the line of section, passing through the centres of the electrode and insulating stem of the induction-plate, and that of the main glass stem, which are in one straight line. The plan represents that part of the main cover as seen from above, when the lantern and upper works are removed. The plate (*b*) which supports the main stem (*q*) has been enlarged to bear also the insulating support (*i*) of the induction-plate. The outline of the induction-plate falls within that of the quadrant beneath it by ·16 of a centimetre ($\frac{1}{16}$ of an inch) all round. It is distant ·48 of a centimetre ($\frac{3}{16}$ of an inch) from the upper surface of the quadrant. The dimensions in the figure are half full size.

With an electrometer fitted with the induction-plate, the usual connexions for the first or direct method of measurement are the same as above mentioned. The electrode of the induction-plate may be connected with that of the quadrant beneath it, or with the case, or it may be insulated, without sensibly affecting the indications of the instrument. For the second grade of sensibility the induction-plate is connected with the case, and the difference of potentials to be measured is established between it and the distant pair of quadrants, the nearer pair being insulated by raising their electrode. To free the latter from the induced charge which they commonly receive by the act of raising their electrode, a disinsulator is provided, consisting of a light arm or spring which may be turned so as to make contact with the quadrant by means of a small milled head projecting above the cover. For a certain lower grade the arrangement is the same, except that the distant pair of quadrants, instead of the induction-plate, is connected with the cover, and the difference of potentials to be measured is established between the cover and the induction-plate. With this arrangement the deflections measure about five times the difference of potentials producing the same deflections by the second grade.

The connexions may be further varied so as to produce other degrees of sensibility giving indications perfectly trustworthy and available for comparative measurements. The different methods of forming the connexions, with or without an inductor, are indicated in the following table, where *R* means the electrode of the pair of quadrants marked RR' in the figure, L that of the pair LL', and I that of the induction-plate; C is the conductor led from one of the bodies experimented upon, O the conductor led from the other and connected to the outer metallic case of the instrument, which may be insulated from the table if necessary by placing a small block or cake of clean paraffin under each of the three feet on which the instrument stands; (R) or (L) means that the electrode of RR' or LL' is to be raised so as to be disconnected from its pair of quadrants. Thus in the grade of diminished power or sensibility standing first in the table on the right, the electrode L is raised, one conductor is connected with R; I and the other with the case of the instrument. The grade standing last in the table, in which L and R are both raised, is the least sensitive of all. In each of these methods the correctness of the indications has been verified by measurements taken simultaneously with the Standard Electrometer (§ 379), the measured difference of potentials being that of the earth and of a Leyden jar fitted with a replenisher, by means of which its potential was varied so as to make the deflected image stand at all points between the extremity of the scale and the zero position. The working of the replenisher being suspended at intervals to allow an accurate reading to be taken of the position of the image and the indication of the Standard Electrometer, the subsistence of a correct proportion between the deflection and the measurement obtained from the Standard Electrometer was verified at all points of the range.

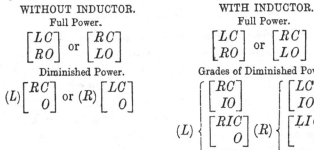

WITHOUT INDUCTOR.
Full Power.
$$\begin{bmatrix} LC \\ RO \end{bmatrix} \text{ or } \begin{bmatrix} RC \\ LO \end{bmatrix}$$
Diminished Power.
$$(L)\begin{bmatrix} RC \\ O \end{bmatrix} \text{ or } (R)\begin{bmatrix} LC \\ O \end{bmatrix}$$

WITH INDUCTOR.
Full Power.
$$\begin{bmatrix} LC \\ RO \end{bmatrix} \text{ or } \begin{bmatrix} RC \\ LO \end{bmatrix}$$
Grades of Diminished Power.
$$(L)\begin{cases} \begin{bmatrix} RC \\ IO \end{bmatrix} \\ \begin{bmatrix} RIC \\ O \end{bmatrix} \\ \begin{bmatrix} IC \\ RO \end{bmatrix} \end{cases} (R)\begin{cases} \begin{bmatrix} LC \\ IO \end{bmatrix} \\ \begin{bmatrix} LIC \\ O \end{bmatrix} \\ \begin{bmatrix} IC \\ LO \end{bmatrix} \end{cases}$$
$$(RL)\begin{bmatrix} IC \\ O \end{bmatrix}$$

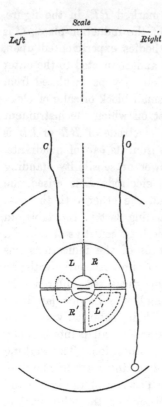

The facility afforded by the number of these arrangements for varying the sensibility of the instrument even to a moderate or slight degree without altering the adjustment of the fibres, will be found useful in some kinds of observations. For instance, if it be desired to observe the fluctuations of a varying potential, a degree of sensibility which throws the deflected image nearly to the extremity of the scale will cause the fluctuations to be twice as sensible and accurately read as if the deflection were only half as much, as they will bear the same proportion to the whole deflection in the two cases.

It is intended in future to make the induction-plate smaller and more distant from the quadrant, in order to diminish the inductive effect and permit of the measurement of from 100 to 5000 cells by the least sensitive method. In some electrometers also the first two grades of sensibility may be considered sufficient, and the induction-plate dispensed with.]

ABSOLUTE ELECTROMETER.

358.   The absolute electrometer (fig. 11, Plate II.) and the other instruments of Class III. are founded on a method of experimenting introduced by Sir William Snow Harris, and described in his first paper "On the Elementary Laws of Electricity*," thirty-four years ago.   In these experiments a conductor, hung from one arm of a balance and kept in metallic communication with the earth, is attracted by a fixed insulated conductor, which is electrified, and, for the sake of keeping its electric potential constant, is connected with the

---

inner coating of a Leyden battery. The first result which he announced is, that, when other circumstances remain the same, the attraction varies with the square of the quantity of electricity with which the insulated body is charged and is independent of the unopposed parts. " It is readily seen " that, in the case of Mr Harris's experiments, it will be " so slight on the unopposed portions that it could not be " perceived without experiments of a very refined nature, such " as might be made by the proof plane of Coulomb, which is, " in fact, with a slight modification, the instrument employed " by Mr Faraday in the investigation. Now to the degree of " approximation to which the electrification of the unopposed " parts may be neglected, the laws observed by Mr Harris when " the opposed surfaces are plane may be readily deduced from " the mathematical theory. Thus let v be the potential in the " interior of A, the charged body, a quantity which will depend " solely on the state of the interior coating of the battery with " which, in Mr Harris's experiments, A is connected, and will " therefore be sensibly constant for different positions of A " relative to the uninsulated opposed body B. Let a be the " distance between the plane opposed faces of A and B, and " let S be the area of the opposed parts of these faces, which " will in general be the area of the smaller, if they be unequal. " When the distance a is so small that we may entirely neglect " the intensity on all the unopposed parts of the bodies, it is " readily shown, from the mathematical theory, that (since the " difference of the potentials at the surfaces of A and B is v) " the intensity of the electricity produced by induction at any " point of the portion of the surface of B which is opposed to

" A is $\dfrac{v}{4\pi a}$, the intensity at any point which is not so situated

" being insensible. Hence the attraction on any small element " ω, of the portion S of the surface of B, will be in a direction

" perpendicular to the plane and equal to $2\pi \left(\dfrac{v}{4\pi a}\right)^2 \omega$*. Hence " the whole attraction on B is

$$\frac{v^2 S}{8\pi a^2}.$$

---

* See *Mathematical Journal*, vol. III. p. 275 (VII. above, §§ 146, 147).

"This formula expresses all the laws stated by Mr Harris
"as results of his experiments in the case when the opposed
surfaces are plane*"

359   After many trials to make an absolute electrometer
founded on the repulsion between two electrified spherical
conductors for which I had given a convenient mathematical
formula in § 4 of the paper just quoted (§ 30, above), it occurred
to me to take advantage of the fact noticed by Harris, but easily
seen as an immediate consequence of Green's mathematical
theory, that the mutual attraction between two conductors used
as in his experiments is but little influenced by the form of the
unopposed parts; and in 1853, in a paper "On Transient Electric
Currents†," I described a method for measuring differences of
electric potential in absolute electrostatic measure founded on
that idea.   The "absolute electrometer," which I exhibited to
the British Association at its Glasgow Meeting in 1855, was constructed for the purpose of putting these methods into practice.
This instrument consists of a plane metal disc insulated in a
fixed horizontal position with a somewhat smaller fixed metal
disc hung centrally over it, from one end of the beam of a
balance.   In two papers‡ entitled "Measurement of Electrostatic Force produced by a Battery," and "Measurement of the
Electromotive Force required to produce a Spark in Air between
Parallel Metal Plates at Different Distances," published in the
Proceedings of the Royal Society for February 1860, I described
applications of this electrometer, in which, for the first time I
believe, absolute electrostatic measurements were made.   The
calculations of differences of potential in absolute measure were
made according to the formula quoted above (§ 358) from my
old paper on "The Elementary Laws of Statical Electricity."

360.   This formula is rigorous only if the distance between
the discs is infinitely small in comparison with their diameters;
and therefore, in my earliest attempt to make absolute electrostatic measurements, I used very small distances.   I found

---

* "On the Elementary Laws of Statical Electricity," *Cambridge and Dublin
Mathematical Journal*, 1846; and *Philosophical Magazine*, July, 1854 (II. above,
§ 27).

† *Philosophical Magazine*, June, 1853.

‡ XVIII. and XIX. above, §§ 310—340.

great difficulty in securing that the distance should be nearly enough equal between different parts of the plates, and in measuring its absolute amount with sufficient accuracy; and found besides serious inconveniences in respect of sensibility and electric range: later I made a great improvement in the instrument by making only a small central area of one of the discs moveable. Thus the electric part of the instrument becomes two large parallel plates with a circular aperture in one of them, nearly filled up by a light circular disc supported properly to admit of its electrical attraction towards the other being accurately measured in absolute units of force. The disc and the perforated plate surrounding it will be called, for brevity, the disc and the guard-plate. The faces of these two next the other plate must be as nearly as possible in one plane when the disc is precisely in the position for measuring the electric force upon it, which, for brevity, will be called its sighted position. The space between the disc and the inner edge of its guard-ring must be a very small part of the diameter of the aperture, and must be very small in comparison with the distance between the plates; but the diameter of the disc may be greater than, equal to, or less than the distance between the plates.

361. Mathematical theory shows that the electric attraction experienced by the disc is the same as that experienced by a certain part of one of two infinite planes at the same distance, with the same difference of electric potentials, this area being very approximately the mean between the area of the aperture and the area of the disc, and that the approximation is very good, even should the distance between the plates be as much as a fourth or fifth, and the diameter of the disc as much as three-fourths of the diameter of the smaller of the two plates. This conclusion will be readily assented to when we consider that* the resultant electric force at any point in the air between the two plates is equal numerically to the rate of conduction of heat per unit area across the corresponding space in the following thermal analogue. Let a solid of uniform thermal conductivity replace all the air between and around the plates; and in

---

* "On the Uniform Conduction of Heat through Solid Bodies, and its connexion with the Mathematical Theory of Electricity," *Cambridge Mathematical Journal*, Feb. 1842; and *Philosophical Magazine*, July, 1854 (I. above, §§ 1—6).

place of the plates let there be hollow spaces in this solid. Let these hollow spaces be kept at two uniform temperatures, differing by a number of degrees equal numerically to the difference of potentials in the electric system, the space corresponding to the disc and guard-ring being at one temperature, and that corresponding to the opposite plate at the other temperature; and let the thermal conductivity of the solid be unity. If we attempt to draw the isothermal surfaces between the hollow corresponding to the continuous plate on the one side, and that corresponding to the disc and guard-ring on the other, we see immediately that they must be very nearly plane, from very near the disc all the way across to the corresponding central portion of the opposite plate, but that there will be a convexity towards the annular space between the disc and guard-ring.

362.   Thus we see that the resultant electric force will, to a very close approximation, be equal to $\dfrac{V}{D}$ for all points of the air between the plates at distances from the outer bounding edges exceeding two or three times the distance between the plates, and at distances from the interstice between the guard-ring and disc not less than the breadth of this interstice. Hence. if ρ denote the electric density of any point of the plate or disc far enough from the edges, we have

$$\rho = \frac{V}{4\pi D}.$$

But the outward force experienced by the surface of the electrified conductor per unit of area at any point is $2\pi\rho^2$, and therefore if F denote the force experienced by any area A of the fixed plate, no part of which comes near its edge, we have

$$F = \frac{V^2 A}{8\pi D^2},$$

which will clearly be equal to the attraction experienced by the moveable disc, if A be the mean area defined above. This gives $V = D\sqrt{\dfrac{8\pi F}{A}}$, the formula by which difference of potentials in absolute electrostatic measure is calculated from the result of a measurement of the force F, which, it must be remembered, is to be expressed in kinetic units. Thus if W be the mass in grammes to which the weight is equal, we have

$$F = g W,$$

where g is the force of gravity in centimetres per second per second.

The difficulty which, in first applying this method about twelve years ago, I found in measuring accurately the distance D between the plates and in avoiding error from their not being rigorously parallel, I now elude by measuring only *differences* of distance, and deducing the desired results from the difference of the corresponding differences of potentials. Thus let V' be the difference of potentials between the plates required to give the same force F; when the difference of potentials is V' instead of V, we have

$$V' - V = (D' - D) \sqrt{\frac{8\pi F}{A}}.$$

363. The plan of proceeding which I now use is as follows:
—Each plate (fig. 11, Plate II.) is insulated; one of them, the continuous one, for instance, is kept at a potential differing from the earth by a fixed amount tested by aid of a separate idiostatic* electrometer†; the other plate (the guard-ring and moveable disc in metallic communication with one another) is alternately connected with the earth and with the body whose potential is to be measured. The lower plate is moved up or down by a micrometer screw until the moveable disc balances in a definite position, indicated by the hair (with background of white with black dots) seen through a lens, as shown in fig. 11. Before and after commencing each series of electrical experiments, a known weight is placed on the disc, and a small wire rider on the lever from which the disc hangs is adjusted to bring the hair to its sighted position when there is no electric force. This last condition is secured by putting the two plates

---

* See § 385, below.
† [A Leyden jar with an idiostatic gauge and replenisher fitted to the cover by which it is closed has been found very suitable for this purpose. The gauge can be adjusted to a higher degree of sensibility than is attainable in an electrometer for general purposes, as the Standard or the Portable Electrometer, and the micrometer movements and graduations of these electrometers are not required.—*May*, 1870.]

in metallic communication with one another. For the electric experiments the weight is removed, so that when the hair is in the sighted position the electric attraction on the moveable disc is equal to the force of gravity on the weight. The electric connexions suitable in using this instrument for determining in absolute electrostatic measure the difference of potentials maintained by a galvanic battery between its two electrodes are indicated in fig. 11. No details as to the case for preventing disturbance by currents of air, and for maintaining a dry atmosphere, by aid of pumice impregnated with strong sulphuric acid, are shown, because they are by no means convenient in the instrument at present in use, which has undergone so many transformations that scarcely any part of the original structure remains. I hope soon to construct a compact instrument convenient for general use. The amount of force which is constant in each series of experiments may be varied from one series to another by changing the position of the small wire rider on the lever.

The electric system here described is heterostatic (§ 385 below), there being an independent electrification besides that whose difference of potential is to be measured.

New Absolute Electrometer.

[§ 364...367 *added May*, 1870.]

364. Plate III. is a sketch in perspective of this instrument, one-third of the full size. As in the Absolute Electrometer just described, the electric system is heterostatic; with this addition, that the potential of the auxiliary charge is tested and maintained, not by a separate electrometer and electric machine, but by an idiostatic arrangement forming part of the instrument itself. This consists of a Leyden jar, forming the case of the instrument; a gauge; and a replenisher. The Leyden jar is a white (flint) glass cylinder, coated inside and outside with tinfoil to nearly the height of the circular plate (A); apertures being left to admit the requisite light to the interior and allow the indications of the vertical scale (r) and divided circle (t) to be read. A brass mounting is cemented round the upper rim of the jar, to which is screwed the cover

of stout sheet-brass (*C*), which closes the jar at the top.   By another brass mounting cemented round its lower rim, the jar is fastened down to the cast-iron sole-plate (*D*) which closes its lower end.   The sole-plate is supported on three legs similar to those shown in fig. 13, Plate II.   The cover (*C*) supports the replenisher (*E*), and the aluminium balance-lever of the idiostatic gauge, which are identical in construction with those described in §§ 352, 353, but on a larger scale.   The air inside is kept dry by aid of pumice soaked with strong sulphuric acid, contained in glass vessels placed in the bottom of the jar.

The moveable disc or balance (*c*) hangs in a circular aperture in the plate (*A*), which rests on three fixed supports ($z, z,$,) cemented to the interior surface of the jar, and in metallic connexion with the inside coating; the manner of support is that of the hole, slot, and plane, described in § 380, (2), below. This perforated plate or guard-plate supports on a brass pillar the attracting plate (*F*) of the idiostatic gauge, which thus tests the potential of the guard-plate, balance, and inside coating.   This potential is kept constant during any series of experiments by using the replenisher according to the indications of the gauge, which is made extremely sensitive by a proper adjustment of the distance from the attracting plate (*F*) to the balance-lever and of the torsion by which the electrical attraction is balanced (see end of § 353).   The replenisher has metallic contact with the guard-plate through the spring (*e*). The jar is charged by an insulated charging-rod let down for the occasion through a hole in the cover.

365.   The balance (*c*) is a light aluminium disc, about 46 millimetres in diameter, strengthened by an elevated rim and radial ribs on its upper surface, but having its lower surface plane and smooth.   It nearly fills the aperture in the guard-plate, sufficient clearance being left ('75 of a millimetre all round) to allow it to move up and down without risk of friction.   It is supported by three delicate steel springs, each of which consists of two parts; the upper end of the upper part is attached to the lower extremity of a vertical insulating stem (*i*) directly above the centre of the disc, where the corresponding end of the lower part is fixed.   The opposite ends,

19—2

which project considerably beyond the circumference of the disc, are riveted together. One of these springs (*s*) is shown in the figure. Their general form may be compared to that of coach-springs. The point of attachment of their upper parts is moved vertically by a kinematic arrangement precisely the same as that employed in the Portable Electrometer (§ 369). The insulating stem (*i*) is attached to a brass tube (*a*), which slides up and down in V guides by the action of a micrometer screw. This micrometer screw is worked by means of the milled head (*m*) projecting above the cover (*C*); the guides for the tube (*a*) and index (*x*) which moves up and down with the tube, are similar to those represented more fully in fig. 10, Plate II., and are rigidly attached to a strong brass plate (*b*) lying across the mouth of the jar below the cover, and resting upon the flange of the brass mounting, to which it is fastened by screws. The plate (*b*) is so adjusted that the balance may hang concentric with the perforation in the guard-plate. The tube (*a*) is similar in construction to that represented in fig. 8, Plate II., and described in § 369, below. The micrometer screw carries a horizontal circular disc (*d*) graduated by 100 equal angular divisions. An aperture is left in the cover through which its indications can be read off by reference to a fixed mark on the sloping edge of the aperture. This, together with the scale (*f*), each division of which corresponds to one full turn of the micrometer screw, measures the vertical distance through which the tube (*a*) and the points of attachment of the springs are moved.

Metallic communication between the balance and the guard-plate is maintained by a light spiral wire attached to the pillar (*g*) and to the upper support of the springs, which is a brass piece cemented to the insulating stem. An arm, not seen in the figure, projects from the guard-plate over the disc so that its extremity is between the centre of the disc and the upper end, bent horizontally, of an upright fixed to the disc; thus serving as a stop to confine the motion of the disc between certain limits. A very fine opaque black hair (§ 353) is stretched between two small uprights (one of which is seen in the figure) standing in the centre of the disc. An achromatic convex lens (*h*), fixed on the guard-plate, stands opposite, and produces an image of the hair in the conjugate focus, which is just over the outer edge of the guard-plate. The two opposed screw-points (k) are adjusted to touch each side of the image thus thrown by the lens, which, on the principle of the astronomical telescope, is observed through an eye-lens (l), attached outside of the jar to the upper brass mounting. By this arrangement the error of parallax in observing the position of the hair relatively to the two points is avoided; the position of the eye may be varied in any direction without causing any change in the apparent relative position of the hair (image) and points. In adjusting these different parts, it is arranged that when the image of the hair is exactly between the two points, or in what is called the sighted position, the under surfaces of the balance and guard-plate may be as nearly as possible in one horizontal plane.

The balance and springs are protected, in the use of the instrument, from disturbing electrical forces, by a brass cover in two halves (y, y), one of which is represented displaced in the figure, to show the interior arrangements. The two halves, when placed together, form a circular box, with an aperture in front in which the lens (h) stands, and another aperture behind to admit light from the sky or from a lamp placed outside of the jar in the line of the hair, lens, and points.

366. The electrical part of the instrument is completed by the continuous attracting plate (B), under and parallel to the guard-plate and spring-balance. This is a stiff circular brass plate with parts cut out to allow it to move freely past the fixed supports (z, z, .) of the guard-plate. An electrode (n) projecting through a hole in the sole-plate from an insulating stem (p) is kept in metallic communication by a spiral wire with an arm projecting from the centre of the continuous plate. The plate (B) is supported by a brass pillar (q), from which it is insulated by a short glass stem. It is moved vertically by the micrometer screw (w) (step $\frac{1}{50}$ of an inch); and this motion is measured by a vertical scale (r) and horizontal graduated circle (t) attached to the screw. The screw projects below the sole-plate, and is worked by the milled head (u), the nut (v) being fixed in the centre of the sole-plate. The pillar (q) moves in

V or ring guides, and rests upon the upper end of the screw in the manner represented in fig. 14, Plate II.

367.   Before this instrument is available for absolute electrostatic measurements, the force required to move the balance through any fixed vertical distance (the point of suspension being unmoved) must be known.   This is ascertained by weighings conducted in the following manner:—The cover (C) is removed, and all electrical force upon the balance is guarded against by putting the electrode (n) in metallic communication with the guard-plate.   The balance is then brought, by turning the micrometer circle (d), to the sighted position; and the reading on the scale (f) and graduated circle (d) is noted.   A known weight is then distributed symmetrically over the disc ($\frac{6}{10}$ of a gramme has been used hitherto), which displaces it below the sighted position.   It is now raised to the sighted position by turning the disc (d), and the altered micrometer reading is noted.   The difference between the two readings measures the distance through which the given weight displaces the balance in opposition to the tension of the springs; and conversely, when the balance has been displaced through the same distance by electrical attraction between it and the continuous plate below it, this known weight is the measure of the force exerted upon it.   It has been thus found by repeated weighings, that a weight of $\frac{6}{10}$ of a gramme displaces the balance through a distance corresponding to two full turns of the micrometer screw and a fraction of one division of the circle, in the instrument belonging to the Laboratory of the Glasgow University. This distance having been ascertained with all possible care and at different temperatures, in view of the possible effect of temperature on the elasticity of the springs, the plan of proceeding to absolute electrostatic measurements is as follows, the weights being removed and covers (y, y, C) replaced.

All electrical influence having been removed by a wire led from the electrode (n) through the hole in the cover (C) to the guard-plate, the balance is brought to the sighted position. Starting from this point, it is raised by the micrometer screw through any distance which has been ascertained to correspond to a known weight, *e.g.* the distance just mentioned.   This corresponds exactly to the removal of the weight (§ 363) in the use of the Absolute Electrometer already described. The jar is then charged, and the potential is kept constant during the experiments by using the replenisher according to the indications of the gauge, which, as already said, has been made extremely sensitive for the purpose. The attracting plate (B) is connected by its electrode (n) alternately with the outside coating of the jar (which may be either connected with the earth or insulated) and with the body the difference of whose potential from that of the outside coating is to be measured. In each case the balance is brought to the sighted position by moving the plate (B) up or down by the micrometer screw (w), and the reading on the vertical scale (r) and graduated circle (t) is noted. The difference of the two readings gives the difference of the two distances between balance and attracting plate, from which the difference of potentials is deduced by the formula at the end of § 362. In measuring the difference of potentials between the poles of a voltaic battery, it is found very convenient to connect the poles, through a Steinheil (or double Bavarian) key, either with the outer coating of the jar (or earth), the other with the insulated electrode (n). The reading being taken and the key reversed, the difference of readings, it is evident, measures a difference of potentials double that of the poles of the battery. Two observers are convenient, one to watch the gauge and use the replenisher accordingly, the other to take the readings.

PORTABLE ELECTROMETER.

368. In the ordinary use of the portable electrometer (figs. 8, 9, and 10, Plate II.), the electric system is heterostatic and quite similar to that of the absolute electrometer, when used in the manner described above in § 363. But the balance is not adapted for absolute measure of the amount of force of attraction experienced by the moveable disc; on the contrary, it is precisely the same as that described for the gauge of the quadrant electrometer in § 353 above, only turned upside down.

Thus, in the portable instrument, the square disc (f) forming part of the lever of thin sheet aluminium is attracted *upwards* by a solid circular disc of sheet-brass (g), thick enough for stiffness. Every part of the aluminium lever except this square portion is protected from electric attraction by a fixed brass plate $(h\ h)$ with a square hole in it, as nearly as may be stopped by the square part of the sheet aluminium destined to experience the electric attraction, all other parts of the aluminium balance-lever being below this guard-plate. The aluminium lever $(i\ k)$, as shown in figs. 8 and 10, is shaped so that when the hair (l) at the end of its long arm is in its sighted position, the upper surfaces of the fixed guard-plate (h) and moveable aluminium square (f) are as nearly as may be in one plane. The mode of suspension is precisely the same as that described (§ 353) for the gauge of the quadrant electrometer. In the portable instrument, careful attention is given by the maker to balance the aluminium lever by adding to it small masses of shellac or other convenient substance, so that its centre of gravity may be in the line of its platinum-wire axis, or, more properly speaking, in such a position that the instrument shall give, when electrified, the same "earth-readings" when held in any positions, either upright, or inclined, or inverted (§ 375 below). Thus the condition of equilibrium of the balance, when the hair is in its sighted position, is that the moment of electric attraction round the axis of suspension shall be equal to the moment of the couple of torsion, the latter being as constant as the properties of the matter concerned (platinum wire, brass stretching-springs, etc.) will allow.

369. The guard-plate carrying, by the platinum-wire suspension, the aluminium balance, is attached to the bottom of a small glass Leyden jar $(m\ m)$, and is in permanent metallic communication with its inside coating of tinfoil. The outside tinfoil coating of this jar is in permanent metallic communication with the outside brass protecting case. The upper open mouth of this case is closed by a lid or roof, which bears on its inner side a firm frame projecting downwards. This frame has two V notches, in which a stout brass tube (o) slides, kept in the Vs by a properly placed spring (p) [(*May* 1870) better two springs, one pressing directly towards each V], giving it

freedom to slide up and down in one definite line *. Firmly fixed
in the upper end of this tube is a nut (*a*, fig. 8), which is made
to move up and down by a micrometer screw. The lower end
of the shaft of this screw has attached to it a convex piece of
polished steel (*b*, fig. 8), which is pressed upon a horizontal
agate plate rigidly attached to the framework above mentioned
by a stiff brass piece projecting into the interior of the brass
tube through a slot long enough to allow the requisite range
of motion. This arrangement will be readily understood from
the accompanying drawings. It has been designed upon obvious
geometrical principles, which have been hitherto neglected, so
far as I know, in all micrometer screw mechanisms, whether
for astronomical instruments or other purposes. The screw-
shaft is turned by a milled head, fixed to it at the top outside of
the roof of the instrument; and the angles through which it is
turned are read on a circle divided into one hundred equal parts
of the circumference (or $3°\cdot6$ each) by reference to a fixed mark
on the roof of the instrument. The hole in the roof through
which the screw-shaft passes is wide enough to allow the shaft
to turn without touching it, and the lower edge of the graduated circle turning with the screw is everywhere very near the
upper side of the roof, but must not touch it at any point. A
second nut (*c*, fig. 8) above the effective nut fits easily, but
somewhat accurately, in the hollow brass tube, and is prevented
from turning round in the tube by a proper projection and slot.
Thus the screw is rendered sufficiently steady, with reference
to the sliding tube; that is to say, its axis is prevented from
any but excessively small deviations from the axis of the
sliding tube and fixed guides; and when the nut is kept from
being turned round its proper axis, it forms along with the
sliding tube virtually a rigid body. A carefully arranged

---

* In consequence of suggestions by Mr Jenkin, it is probable that the spring
may be done away with, and the Vs replaced by rings approximately fitting
round the tube, but leaving it quite free to fall down by its own weight. In
consequence of the symmetrical position of the convex end of the screw over
the centre of the attracted disc, slight lateral motions of the tube produce no
sensible effect on the electric attraction. [*May*, 1870.—Various trials both on
the portable and stationary instruments have but very partially fulfilled this
anticipation; and have confirmed the practical value of the Vs. The con-
structional advantages of the rings and geometrical merits of the Vs are easily
combined.]

spiral spring presses the two nuts asunder, and so causes the upper side of the thread of the screw-shaft always to press against the under side of the thread of the effective nut, thus doing away with what is technically called in mechanics "lost time." In turning the micrometer screw, the operator presses its head gently downwards with his finger, to secure that its lower end bears firmly upon the agate plate. It would be the reverse of an improvement to introduce a spring attached to the roof of the instrument outside to press the screw-head downwards, inasmuch as however smooth the top of. the screw-shaft might be made, and however smooth the spring pressing it down, there would still be a very injurious friction impeding the proper settlement of the sliding tube into its Vs. A stiff fork (q) stretching over the graduated circle is firmly attached to the roof outside, to prevent the screw from being lifted up by more than a very small space; about $\frac{1}{20}$ of an inch at most. In using the instrument, the observer should occasionally pull up the screw-head and press it down again, and give it small horizontal motions, to make sure that when it is being used it is pressed in properly to its Vs and down upon the agate-plate. A long arm (d, figs. 8 and 10) (or two arms one above the other), firmly attached to the sliding-tube, carries an index which moves up and down with it. Two fixed guiding-cheeks on each side of this index prevent the tube from being carried round too far in either direction when the screw is turned: one of these cheeks is graduated so that each division is equal in length to the step of the micrometer screw; this enables the operator to ascertain the number of times he has turned the screw. These two cheeks must never simultaneously press upon the sliding-pointer; on the contrary, they must leave it a slight amount of lateral freedom to move. If this does not amount to 36 of a degree, the amount of "lost time" produced by it will not exceed $\frac{1}{10}$ of a division of the micrometer circle, and will not produce any sensible error in the use of the instrument. A glass rod cemented to the lower end of the tube prolongs its axis downwards, and bears the continuous attracting-plate of the electrometer at its lower end.

The object aimed at in the mechanism just described is to prevent the nut and other parts rigidly connected with it from any other motion than parallel to one definite line, and to leave it freedom to move in this line, unimpeded by any other friction than that which is indispensable in the arrangement for keeping the sliding tube in its Vs.

370. If the inner tinfoil covering of the Leyden jar were completed up to the guard-plate bearing the aluminium balance-lever, the long arm of this lever being in the interior of a hollow conductor would experience no electric influence, and no force from the electrification of the Leyden jar, or from separate electrification of the upper attracting plate, or, more strictly speaking, the electric density and consequent electric force on the long arm of the lever would be absolutely insensible to the most refined test we could apply, because of the smallness of the gap between the moveable aluminium square and the boundary of the square aperture in the guard-plate. But to see the hair on the long end of the lever, and the white background with black dots behind it, a not inconsiderable portion of the glass under the guard-plate must be cleared of tinfoil outside and inside. Thus the electric potential of the inner coating of the Leyden jar will not be continued quite uniformly over the inner surface of the bared portion of the glass, and a disturbance affecting chiefly the most sensitive part of the lever will be introduced. To diminish this as much as possible without inconveniently impeding vision, a double screen of thin wire fencing, in metallic communication with the inner tinfoil coating and the guard-plate, is introduced between the end of the lever and the glass through which it is observed.

371. A very light spiral spring (r) connects the upper attracting plate with a brass piece supported upon a fixed vertical glass column projecting downwards from the roof of the instrument. This brass piece bears a stout wire (s), called the main electrode, projecting vertically upwards along the axis of a brass tube open at each end, fixed in an aperture in the roof so as to project above and below, as shown in fig. 9.

372. The top of the main electrode bears a brass sliding piece (t), which, when raised a little, serves for umbrella and wind-guard without disturbing the insulation; and when pressed down closes the aperture and puts the electrode in metallic

connexion with the roof of the instrument. When the instrument is to be used for atmospheric electricity (unless at a fixed station), a steel wire, about 20 centimetres long, is placed in the hole on the top of the sliding brass piece just mentioned, and is thus held in the vertical position. A burning match is attached to its upper end, which has the effect of bringing the potential of the chief electrode and upper attracting plate, etc., all to the potential of the air at the point where the match burns*. The instrument is either held in the observer's hand, or it is placed upon a fixed support, and care taken that its outer brass case is in connexion with the earth. When the difference of potentials between two conductors is to be tested, one of these is connected with the brass case of the instrument, and the other with the chief electrode, the umbrella being kept up. If both of these conductors must be kept insulated from the earth, the brass case of the electrometer must be put on an insulating stand, and the micrometer screw turned by an insulating handle.

373. A lead cup (*e e*, fig. 8), supported by metal pillars from the roof and carrying pieces of pumice-stone, held in their place by India-rubber bands, completes the instrument. The inner surface of the glass must be clean, and particles of dust, minute shreds or fibres, etc., removed as carefully as possible, especially from the lower surface of the upper attracting-plate, and the upper surface of the guard-plate and aluminium square facing it from below. The pumice is prepared by moistening it with a few drops of strong pure sulphuric acid. Ordinary sulphuric acid of commerce should be boiled with sulphate of ammonia to free it from volatile acid vapours, and to strengthen it sufficiently by removing water if the acid be not of the strongest. There should not be so much acid applied to the pumice as to make it have the appearance of being moist, but there must be enough to maintain a sufficiently dry atmosphere within the instrument for very perfect insulation of the Leyden jar, which I find does not in general lose more of its charge

---

* See Nichol's *Cyclopædia*, article "Electricity, Atmospheric," 2nd edition, 1860 (§ 266, above); or "Royal Institution Lecture on Atmospheric Electricity," May, 1860 (§§ 277, 278, above).

than five per cent. per week, when the pumice is properly impregnated with acid. Thus there is no tendency of the liquid to drop out of the pumice; and the pumice being properly secured by the India-rubber bands, the instrument may be thrown about with any force, short of that which might break the glass jar or either of the glass stems, without doing any damage; but to insure this hardiness the sheet aluminium of which the bal-· ance is made must be *very thin*. After several weeks' use the pumice may begin to look moist, and even slight traces of moisture may be seen on the outside of the lead cup, in consequence of watery vapour attracted by the sulphuric acid from the atmosphere; but the pumice should then be taken out and dried. At all events this must be done in good time, before enough of liquid has collected to give any tendency to drop. In all climates in which I have hitherto tested the instrument, I have found the pumice effective for insulation and safe in keeping all the liquid to itself for two months. But Mr Becker having reported to me that many instruments have been returned to him in a ruinous condition from drops of sulphuric acid having become scattered through their metal work, I now cause to be engraved conspicuously on the outer case of the instrument "PUMICE DANGEROUS, IF NOT DRIED ONCE A MONTH;" also a frame carrying a card, on which the dates of drying are inscribed, to be placed in a convenient position on the roof of the instrument.

374. To prepare the instrument for use, the inner coating of the Leyden jar must be charged through a charging rod, insulated in a vulcanite or glass tube, and let down for the occasion through a hole in the roof of the instrument, by aid of a small electrophorus, which generally accompanies the instrument, or by an electrical machine. I generally prefer to give a negative charge to the inner coating, as I have not found any physical reason, such as that mentioned in § 349 above, to prefer a positive charge to a negative charge; and the negative charge gives increased readings of the micrometer, in the ordinary use of the instrument, to correspond to positive charges of the principal electrode, as will be presently explained. Before commencing to charge the jar, the upper attracting-plate should be moved to nearly the highest position of its range by the micrometer

screw, otherwise too strong a force of electric attraction may be put upon the aluminium square; and besides, the jar will discharge itself between the upper plate and the extreme edge of the aluminium square, when it is pulled very much above the level of the guard-plate by the electric attraction. I have not found any injury or change of electric value of the scale-divisions to arise from any such rough usage; but still, to guard against such a possibility, I propose to add to the guard-plate checks to prevent the corners of the aluminium from rising much, if at all, above its level, and to conduct the discharge and protect the aluminium and platinum from the shock, in case of the upper plate being brought too near the lower. When the instrument is being charged, or when it is out of use at any time, the umbrella should always be kept down; but it must be raised to insulate the principal electrode, of course, before proceeding to apply this to a body whose difference of potential from a body connected with the case of the instrument is to be measured.

375.   In using the instrument the umbrella must very frequently be lowered, or metallic communication established in any other convenient way between the chief electrode and the outer brass case, the micrometer screw turned until the hair takes its sighted position, and the reading taken, the hundreds being read on the interior vertical scale, and the units (or single divisions of the circle) on the graduated circle above. The number thus found is called the earth-reading. It measures the distance from an arbitrary zero position to the position in which the upper attracting-plate must be placed to give the amount of electric force on the aluminium square which balances the lever in its sighted position. A constant added to the earth-reading, or subtracted from it, gives (§ 341) a number simply proportional to the difference of potentials between the upper and lower plate; that is to say, between the two coatings of the Leyden jar. The vertical scale and micrometer circle are numbered, so that increased distances between the plates give increased readings; and the zero reading should correspond as nearly as may be to zero distance between them; although in the instruments hitherto made no pains have been taken to secure this condition, even somewhat approximately.

If it is desired to know the constant, an electrical experiment must be made to determine it, which is done with ease; but this is not necessary for the ordinary use of the instrument, which is as follows:—

376. First, an earth-reading is taken, then the upper electrode is insulated by raising the umbrella, or otherwise breaking connexion between the principal electrode and the outer metal case of the instrument. The principal electrode and the outer case are then connected with the two bodies whose difference of potential is to be determined, and the micrometer screw is turned until the hair is brought to its sighted position. The reading of hundreds on the vertical scale and units on the circle is then taken. Lastly, the principal electrode is again connected with the case of the instrument and another earth-reading is taken. If the second earth-reading differs from the first, the observer must estimate the most probable earth-reading for the moment when the hair was in its sighted position, with the upper plate and the metal case in connexion with the two bodies whose difference of potential is to be measured. The estimated earth-reading is to be subtracted from the reading taken in connexion with the bodies to be tested. This difference measures (§ 362) the required difference of potentials between them in units of the instrument. The value of the unit of the instrument ought to be known in absolute electrostatic measure; and the difference of reading found in any experiment is to be multiplied by this, which is called (§ 341) the absolute coefficient of the instrument, to give the required difference of potentials in absolute measure. It so happens that, in the portable electrometers of the kind now described which have been hitherto constructed, the absolute coefficient is somewhere about ·01, so that one turn of the screw, or one hundred divisions of the circle, corresponds to somewhere about one electrostatic unit, with a gramme for the unit of mass, a centimetre for the unit of distance, and a second for the unit of time; but the different instruments differ from one another by as much as ten or twenty per cent. in their absolute coefficients. In all of these I have found between three and four Daniell's cells to correspond to the unit division; that is to say, between three hundred and four hundred cells to a full

turn of the screw. With great care, the observer may measure small differences of potentials by this instrument to the tenth part of a division (or to about half a Daniell's cell). With a very moderate amount of practice and care, an error of as much as half a division may be avoided in each reading.

377. But there are imperfections in the instrument itself which make it difficult or impossible to secure very minute accuracy, especially in measurements through wide ranges.

(1) In the first place, I am not sure that the end of the needle carrying the hair is protected sufficiently by the wire fences (§ 370) from electric disturbance to provide against any error from this source, which possibly introduces serious irregularities.

(2) In the second place, the capacity of the jar in the small portable instrument is not sufficient to secure that the potential of its inner coating shall not differ sensibly with the different distances to which the upper plate is brought, to balance the aluminium lever with the hair in its sighted position. But on this point it is to be remarked that the electric density on the upper surface of the guard-plate is in its central parts always the same when the hair is in its sighted position; and it is therefore only the comparatively small difference of the quantity of electricity on this surface, towards the rim, corresponding to different distances of the attracted plate, that causes difference of potential in the inner coating of the jar. But if the upper attracting-plate be kept for several minutes at any distance, differing by a few turns of the screw, from that which brings the hair to its sighted position, the electricity creeps along the inner unconnected surface of the glass so as to diminish the charge of the inner metallic coating, or increase it, according as the distance is too great or too small. If then quickly the screw be turned and the earth-reading taken, it is found smaller or greater, as the case may be, than previously; but after a few minutes more it returns to its previous value very approximately. Error from this source may be practically avoided by taking care never to allow the hair to remain for more than a few minutes far from its sighted position; never so far, for instance, as above the centre of the upper, or below the centre of the lower spot.

(3) A third source of error arises from change of temperature influencing the indications. In most of the instruments hitherto made I have found that the warmth of the hand produces in a few minutes a very notable augmentation of the earth-reading (as it were an increased charge in the jar); but in the last instrument which I have tested (White, No. 18) I find the reverse effect, the earth-reading becoming smaller as the instrument is warmed, or larger when it is cooled. I have ascertained that these changes are not due to changes in the electric capacities of the Leyden jars; and I have found that the change, if any, of specific inductive capacity of glass by change of temperature is excessively small, in comparison to what would be required to account for the temperature errors of these instruments, which probably must be due to thermo-elastic properties of the platinum wire, or of the stretching-springs, or of the aluminium balance-lever, or to a combination of the effects depending on such properties; but I have endeavoured in vain, for several years, and made many experiments, to discover the precise cause. It surely will be found, and means invented for remedying the error, now when I have an instrument in which the error is in the opposite direction to that of most of the other instruments. It is of course much greater in some instruments than in others: in some it is so great that the earth-reading is varied by as much as twenty divisions by the warmth of the hand in the course of five or ten minutes after commencing to use the instrument, if it has been· previously for some time in a cold place. Its influence may be eliminated, not quite rigorously, but nearly enough so for most practical purposes, by frequently taking earth-readings (§ 375) and proceeding according to the directions of § 376.

(4) A fourth fault in the portable electrometer is, that the diameter of the guard-plate and upper attracting disc, which ought to be infinite, are not sufficiently great, in proportion to the greatest distance between them, to render the scale quite uniform in its electric value throughout. A careful observer will, however, remedy the greater part of the error due to this defect, by measuring experimentally the relative (or absolute) values of the scale-division in different parts of the range. There will, however, remain uncorrected some irregularity, due

to influence of the distribution of electricity over the uncoated inner surface, in the instruments as hitherto made, in all of which the inner surface of the jar is coated with tinfoil only below the guard-plate, so that the upper surface of the guard-plate may be seen clearly, in order that the observer may always see that all is in order about the aluminium square and aperture round it; and particularly that there are no injurious shreds or minute fibres. But the irregular influence of the electrification of the uncoated glass, if found sensible, will be rendered insensible by continuing the tinfoil coating an inch above the upper surface of the guard-plate.

378. All faults, except the temperature error, depend on the smallness of the instrument; and if the observer chooses to regard as portable an instrument of thirty centimetres (or a foot) diameter, with all other dimensions, and all details of construction, the same as those of the instrument described above, he may have a portable electrometer practically free from three of the four faults described. It is scarcely to be expected that a small instrument ($12\frac{1}{2}$ centimetres high, and $8\frac{1}{2}$ centimetres in diameter) which may be carried about in the pocket can be free from such errors. But they are so far remedied as to be probably not perceptible, in the large stationary instrument which I now proceed to describe.

STANDARD ELECTROMETER.

379. This instrument (figs. 12, 13, and 14, Plate II.) differs from the portable electrometer only in dimensions, and in certain mechanical details, which are arranged to give greater accuracy by taking advantage of freedom from the exigencies of a small portable instrument. It is at present called the standard electrometer, in anticipation of either remedying, or of learning to perfectly allow for, the temperature error, and of finding by secular experiments on the elasticity of metals, that their properties used in the instrument are satisfactory as regards the permanence from year to year, and from century to century, of the electric value of its reading. It is an instrument capable of being applied with great ease to very accurate

measurements of differences of potential, in terms of its own unit. The value of the unit for each such standard instrument ought, of course, to be determined with the greatest possible accuracy in absolute measure; and until confidence can be felt as to its secular constancy, determinations should frequently be made by aid of the absolute electrometer.

380. The Leyden jar of the standard electrometer consists of a large thin white-glass shade coated inside and outside to within 6 centimetres of its lip, and placed over the instrument as an ordinary glass shade, to protect against dust, currents of air, and change of atmosphere. It may be removed at pleasure from the cast-iron sole of the instrument, and then the interior works are seen, consisting of—

(1) A continuous disc of brass supported on a glass stem, in prolongation of a stout brass rod or tube sliding vertically in Vs, in which it is kept by a spring [better by two springs (§ 369)], and resting with its lower flat end on the upper end of a micrometer screw shaft, shown in fig. 13, where the screw, graduated circle, and stout brass rod are as seen in the instrument; the manner in which the lower end of the rod or tube is constructed to keep the round upper end of the screw-shaft in position is shown in section in fig. 14.

(2) Resting on three glass columns, a guard-plate with a square aperture in its centre, and carrying on its upper side the stretching springs and thin platinum wire suspension of an aluminium balance-lever, shaped like those of the gauge (§ 353) and the portable (§ 368) already described, but somewhat larger. The tops of the three glass columns are rounded; a round hole and a short slot in line with this hole are cut in the guard-plate, and receive the rounded ends of two of the columns, which are somewhat longer than the third. The flat smooth lower surface of the guard-plate rests simply on the top of the third glass column. The diameter of the round hole and the breadth of the slot in the guard-plate may be about $\frac{1}{\sqrt{2}}$ of the diameter of curvature of the upper hemispherical rounded ends of the glass columns, so that the bearing portions of the rounded ends in the round hole and in the slot respectively may be inclined somewhere about 45° to the plane of the

plate. This well-known but too often neglected geometrical arrangement gives perfect steadiness to the supported plate, without putting any transverse strain upon the supporting glass columns, such as was almost inevitable, and caused the breakage of many glass stems, before the mental inertia opposing deviations from the ordinary instrument-maker's plan (of screwing the guard-plate to brass mountings cemented to the tops of the glass columns) was overcome. It has also the advantage of allowing the guard-plate to be lifted off and replaced in a moment.

(3) Principal electrode projecting downwards through a hole in the sole of the instrument, and rigidly supported from above by a brass mounting cemented to the top of a thick vertical glass column, connected by a light spiral spring with the lower attracting plate moved up and down by the micrometer screw. The aperture round the principal electrode may be ordinarily stopped by a perforated column of well paraffined vulcanite projecting some distance above and below the aperture, which I find to insulate extremely well, even in the smoky, dusty, and acidulated atmosphere of Glasgow. When an extremely perfect insulation of the principal electrode and connected attracting plate is required, the vulcanite stopper surrounding it may be withdrawn from the aperture, so that the only communication between the electrode and the case of the instrument may be along the two glass columns in the artificially dried interior atmosphere of the case; but from day to day, when the instrument is out of use, the aperture round the principal electrode should be kept carefully stopped, if not by a vulcanite insulator by a perforated cork; (although I find but little loss of insulation, either along the inner glass surface of the Leyden jar or along the three glass columns, when this precaution is neglected).

(4) Temporary charging-rod enclosed in and supported by a vertical insulating column of paraffined vulcanite, or a glass tube well varnished outside and thickly paraffined inside. This insulating column bearing the charging-rod is turned round till a horizontal spring projecting from its upper end touches the inner coating of the jar, when this is to be charged from an independent source, or when, for any other experimental

XX.] *On Electrometers and Electrostatic Measurements.* 309

reason, it is to be put in connexion with a conductor outside the case of the instrument.

(5) A small replenisher of the kind described for the quadrant electrometer (§ 352), but with much wider air-spaces to prevent discharge by sparks.

(6) A large glass or lead dish to hold as large masses of pumice as may be, which are to be kept sufficiently impregnated with strong sulphuric acid.

381. A considerable portion of the jar above the guard-plate is left uncoated to allow the observer to see easily the hair and white background with black dots; also several other smaller parts of the glass above the guard-plate are left uncoated to admit light to allow a small circular level on the upper side of the guard-plate to be seen. The long arm of the aluminium balance-lever is very thoroughly guarded by double cages and fences of wire (§ 370), so that it can experience no sensible influence from electric disturbing forces when the covering jar is put in position and electric connexion is established between its inner coating and the guard-plate by projecting flexible wires or slips of metal.

382. The aluminium square plate is somewhat larger, and the platinum bearing wire somewhat longer in this instrument than in the portable electrometer, to render it sensible to smaller differences of potential. The step of the screw is the same as in the portable ($\frac{1}{50}$ of an inch), and one division ($\frac{1}{100}$ of the circumference of the screw-head) corresponds to a difference of potentials which, roughly speaking, is equal to about that of a single cell of Daniell's. The effective range of the instrument is about sixty turns of the screw, and therefore about 6000 cells of Daniell's. That of the portable electrometer is about 15 turns of the screw (equivalent to about 5000 cells). Neither of these instruments has sufficient range to measure the potential to which Leyden jars are charged in ordinary electric experiments, or those reached by the prime conductor of a powerful electric machine. The stationary instrument with its long screw and its large plates now described, would go far towards meeting this want if its aluminium lever and platinum suspension were made on the same scale as those of the portable electrometer; but for an instrument never wanted to

directly measure differences of potentials of less than two or three thousand cells, the heterostatic (§ 385) principle is in general not useful, and therefore I have constructed the following very simple idiostatic (§ 385) instrument, which is adapted to measure with considerable accuracy differences of potential from 4000 cells upwards, to about 80,000 cells.

LONG-RANGE ELECTROMETER.

383.   In this (fig. 15, Plate VI.) the continuous attracting-plate is above, and the guard-plate with aluminium balance below, as in the portable electrometer; but, as in the standard stationary electrometer, the upper plate is fixed and the lower plate is moved up and down by a micrometer-screw. The mechanism of the screw and slide has all the simplicity and consequent accuracy of that of the standard electrometer. In the only long-range instrument yet constructed the step of the screw is the same as that of the others ($\frac{1}{50}$ of an inch). In future instruments it would be well either to have a longer step or to have a simple mechanism (which can be easily added) to give a quick motion; as in the use of the present instrument, the turning of the screw required for great changes of the potential measured is very tedious. The guard-plate projects by more than an inch all round beyond the rim of the upper attracting-plate; partly to obviate the necessity of giving it a thick rim, which would be required to prevent brushes and sparks from originating in it, if it had only the same diameter as the continuous plate above, and partly to guard the observer from receiving a spark or shock in measuring the potential of an electric machine or of a Leyden battery, and to prevent his hair from being attracted to the upper plate. Thus the guard-plate is allowed to be no thicker than suffices for stiffness, and this allows the observer to see the hair at the end of the aluminium balance-lever without the lever being made of a dynamically disadvantageous shape, as would be necessary if the guard-plate were thick, or had a thick rim added to it. No glass case is required for this instrument. The smallness of the needle and the greatness of the electric force acting on

it are such that I find in practice no disturbance to any inconvenient degree by ordinary currents of air; although it and all these attracted disc instruments show the influence of sudden change of barometric pressure, such as that produced by opening or shutting a door. If not kept under a glass shade when out of use, the lower surface of the upper attracting-plate, and the lower surface of the guard-plate and attracted aluminium square, should be carefully dusted by a dry cool hand. Generally speaking, none of the vital electric organs of an electrometer should be touched by a cloth, as this is almost sure to leave shreds fatal to their healthy action.

[(Addition, 1870) I intend to cover the whole instrument with a glass shade, well varnished over a large space round an aperture in its top, into which an insulated electrode for the· upper plate will be cemented: because with the instrument open as it is at present great difficulty has been experienced in measuring high tension on account of dust and shreds which impair the insulation.]

384. The effective range of this instrument is about 200 turns of the screw. Rather greater force of torsion is given than in the portable electrometer, and a rather smaller attracted disc may be used, so that upwards of four cells may be the electric value of one division. The instrument in its present state measures nearly but not quite the highest potential I can ordinarily produce in the conductor of a good Winter's electric machine, which sometimes gives sparks and brushes a foot long.

385. The classification of electrometers given above is founded on the shape and kinematic relations of their chief organic parts; but it will be remarked that another principle of classification is presented by the different electric systems used in them, which may be divided into two classes :—

I. Idiostatic, that in which the whole electric force depends on the electrification which is itself the subject of the test.

II. Heterostatic, in which, besides the electrification to be tested, another electrification maintained independently of it is taken advantage of.

Thus, for example, the long-range electrometer (§§ 383, 384) is simply idiostatic, and is not adapted for heterostatic use; but each of them may be used idiostatically. The absolute electrometer was at first simply idiostatic (§§ 358-362); more recently it has been used heterostatically, and is about to acquire (§ 363) special organs adapted for heterostatic use; as yet, however, no species of the absolute electrometer promising permanence has come into existence.   [See §§ 364-367 describing a heterostatic absolute electrometer of a species which (Jan. 1871) promises to be permanent.]

386.   It is instructive to trace the origin of various heterostatic species of electrometers by natural selection.   A body hanging, or otherwise symmetrically balanced, in the middle of a symmetrical field of force, but free to move in one direction or the other in a line tangential to a line of force, moves in one direction or the opposite when electrified positively or negatively.   Bohnenberger's arrangement of this kind has a convenient and approximately constant field of force; and his instrument was chosen in preference to others which may have been equally sensitive, but were less convenient and constant, and it became a permanent species.

387.   Bennet's gold-leaf electroscope, constructed with care to secure good insulation, electrified sufficiently to produce a moderate divergence, has been often used to test, by aid of this electrification, the quality of the electrification of an electrified body brought into the neighbourhood of its upper projecting electrode, causing, if its electricity is of the same sign as that of the gold leaves, increase of divergence; if of the opposite sign, diminution.   By connecting the upper electrode with the inner coating of a Leyden jar with internal artificially dried atmosphere, the charge of the gold leaves may be made to last with little loss from day to day; and by insulating Faraday's metal cage (§ 342) round the gold leaves, and alternately connecting it with the earth and with a conductor whose difference of potentials from the earth is to be tested, an increase or a diminution of divergence is observed according as this difference is negative or positive, the gold leaves being positive. Hence (through Peltier's and Delmann's forms) the heterostatic stationary and portable repulsion electrometers, described (§§ 274-277, 263 above) in the Royal Institution Lecture on "Atmospheric Electricity," and in Nichol's *Cyclopædia,* article "Electricity, Atmospheric," already referred to, of which

one species still survives in King's College, Nova Scotia, and in the Natural Philosophy Classroom of Edinburgh University. The same form of the heterostatic principle applied to Snow Harris's attracted disc electrometer gave the portable and standard electrometers described above.

388. A modification of Bohnenberger's electroscope, in which the two knobs on the two sides of the hanging gold leaf became transformed into halves of a circular cylinder, with its axis horizontal and the gold leaf hung on a wire insulated in a position coinciding with its axis; producing a species designed for telegraphic purposes, but which did not acquire permanence by natural selection, and is only known to exist in one fossil specimen. In this instrument the wire 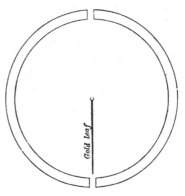 bearing the gold leaf was connected with a charged Leyden jar, and the semi-cylinders with the bodies whose difference of potential was to be tested. But various modifications of the divided-cylinder or divided-ring class with the axis vertical and plane of motion horizontal have done some practical work, and one species, the new quadrant electrometer (§ 346), promises to become permanent.

389. The heterostatic principle in one form or other is essential to distinguish between positive and negative. As remarked above (§ 387), the original type of this use of it is to be found in the old system of testing the quality of the charge taken by the diverging straws or gold leaves of the electroscopes used for the observation of atmospheric electricity; which was done by bringing a piece of rubbed sealing-wax into the neighbourhood, and observing whether this caused increase or diminution of the divergence. A doubt which still exists as to the sign (§ 252) of the atmospheric electricity observed by Professor Piazzi Smyth on the Peak of Teneriffe, is owing to the imperfection of this way of applying the principle. It is, indeed, to be doubted in any one instance whether it is not vitreous

electricity that the rubbed sealing-wax acquires. And, again (§ 342), it is not certain that the glass case enclosing the gold leaves, especially if very clean and surrounded by a very dry natural atmosphere, screens them sufficiently from direct influence of the piece of sealing-wax to make sure that the divergence due to vitreous electricity could not be increased by the presence of the resinously electrified sealing-wax if held nearer the gold leaves than the upper projecting stem.

390.   The heterostatic principle has a very great advantage as regards sensibility over any simple idiostatic arrangement, inasmuch as, for infinitely small differences of potential to be measured, the force is as the squares of the differences in any idiostatic arrangement, but is simply proportional to the differences in every heterostatic arrangement.

NEW APPARATUS FOR OBSERVING ATMOSPHERIC ELECTRICITY†.

[*Proceedings Literary and Philosophical Society of Manchester*, March 8, 1859.]

391. Dr Joule read an extract from a letter he had some time ago received from Professor W. Thomson.—"I have had an apparatus for Atmospheric Electricity put up on the roof of my lecture-room, and got a good trial of it yesterday, which proved most satisfactory. It consists of a hollow conductor supported by a glass rod attached to its own roof, with an internal atmosphere kept dry by sulphuric acid : the lower end of the glass rod is attached to the top of an iron bar, by which the hollow conductor is held about two feet above the inclined roof of the building. A can, open at the top, slides up and down on the iron bar which passes through a hole in the centre of its bottom, and, being supported by a tube with pulleys, etc. below, can easily be raised or lowered at pleasure. A wire attached to the insulated conductor passes through a wide hole in the bottom of the can, and is held by a suitable insulated support inside the building, so that it may be led away to an electrometer below. To make an observation, the wire is connected with the earth, while the can is up, and envelopes the conductor—its position when the instrument is not in use. The earth connexion is then broken, and the can is drawn down about eighteen inches. Immediately the electrometer shows a large effect (from five to fifteen degrees on my divided ring electrometer, in the state it chanced to be in, requiring more than one hundred degrees of torsion to bring it back to zero, in the few observations I made). When the surface of the earth is (as usual when the sky is cloud-

---

* The two articles constituting this chapter were accidentally omitted from Chapter XVI.

† It was with the insulated conductor of an apparatus of this kind afterwards set up in the island of Arran that the observations described in § 294 were made.

less) negative, the electrometer shows positive electricity. But when a negative cloud (natural, or of smoke) passes over, the indication is negative. The insulation is so good that the changes may be observed for a quarter of an hour or more, and when the can is put up the electrometer comes sensibly to zero again, showing scarcely any sensible change when the earth connection is made, before making a new start."

Dr Joule stated that he had recently witnessed experiments with Professor Thomson's new Atmospheric Electrometer, the merit of which consisted in its extreme sensitiveness, and the facility with which accurate observations could be made with it.

NOTES ON ATMOSPHERIC ELECTRICITY*.

[From the *Philosophical Magazine*, Fourth Series, Nov. 1860.]

392. Two water-dropping collectors for atmospheric electricity were prepared, and placed, one at a window of the Natural Philosophy Lecture-room, and the other at a window of the College Tower of the University of Glasgow. A divided ring-electrometer was used at the last-mentioned station; an electrometer adapted for absolute measurement, nearly in the form now constructed as an ordinary house electrometer, was used in the lecture-room. Four students of the Natural Philosophy Class, Messrs Lorimer, Lyon, M'Kerrow, and Wilson, after having persevered in preliminary experiments and arrangements from the month of November, devoted themselves with much ardour and constancy during February, March, and April to the work of observation. During periods of observation, at various times of day, early and late, measurements were completed and recorded every quarter minute or every half minute,—the continual variations of the phænomenon rendering solitary observations almost nugatory. During several hours each day, simultaneous observation was carried on on this plan at the two stations. A comparison of the

---

results manifested often great discordance, and never complete agreement. It was thus ascertained that electrification of the air, if not of solid particles in the air (which have no claim to exclusive consideration in this respect), between the two stations and round them, at distances from them not very great in comparison with their mutual distance, was largely operative in the observed phænomena. It was generally found that after the indications had been negative for some time at both stations, the transition to positive took place earlier by several minutes at the tower station (upper) than at the lecture-room (lower). Sometimes during several minutes, preceded and followed by positive indications, there were negative indications at the lower, while there were only positive at the upper. In these cases the circumambient air must have contained negative (or resinous) electricity. A horizontal stratum of air several hundred feet thick overhead, containing as much positive electricity per cubic foot as there must have been of negative per cubic foot of the air about the College buildings on those occasions, would produce electrical manifestations at the earth's surface similar in character and amount to those ordinarily observed during fair weather.

393. Beccaria has remarked on the rare occurrence of negative atmospheric indications during fair weather, of which he can only record six during a period of fifteen years of very persevering observation by himself and the Prior Ceca. On some, if not all, of those occasions there was a squally and variable wind, changing about rapidly between N.E. and N.W. On several days of unbroken fair weather in April and May of the present year the atmospheric indication was negative during short periods, and on each occasion there was a sudden change of wind, generally from N.E. to N.W., W., or S.W. For instance, on the 3rd of May, after a warm, sunny, and very dry day, with a gentle N.E. breeze, and slight easterly haze in the air, I found, about 8.30 P.M., the expected positive atmospheric indication. After dark (nearly an hour later) it was so calm that I was able to carry an unprotected candle into the open air and make an observation with my portable electrometer. To my surprise I found a somewhat strong negative indication, which I observed for several minutes. Although there was no sensible wind in

the locality where I stood*, I perceived by the line of smoke from a high chimney at some distance that there was a decided breeze from W. or S.W. A little later a gentle S.W. wind set in all round, and with the aid of a lantern I found strong positive indications, which continued as long as I observed. During all this time the sky was cloudy, or nearly so. That reversed electric indications should often be observed about the time of a change of wind may be explained, with a considerable degree of probability, thus :—

394. The lower air up to some height above the earth must in general be more or less electrified with the same kind of electricity as that of the earth's surface; and, since this reaches a high degree of intensity on every tree-top and pointed vegetable fibre, it must therefore cause always more or less of the phænomenon which becomes conspicuous as the "light of Castor and Pollux" known to the ancients, or the "fire of St. Elmo" described by modern sailors in the Mediterranean, and which consists of a flow of electricity, of the kind possessed by the earth, into the air. Hence in fair weather the lower air must be negative, although the atmospheric potential, even close to the earth's surface, is still generally positive. But if a considerable area of this lower stratum is carried upwards into a column over any locality by wind blowing inwards from different directions, its effect may for a time predominate, and give rise to a negative potential in the air, and a positive electrification of the earth's surface.

395. If this explanation is correct, a whirlwind (such as is often experienced on a small scale in hot weather) must diminish, and may reverse, the ordinary positive indication.

396. Since the beginning of the present month I have had two or three opportunities of observing electrical indications with my portable electrometer during day thunder-storms. I commenced the observation on each occasion after having heard thunder, and I perceived frequent impulses on the needle which caused it to vibrate, indicating sudden changes of electric potential at the place where I stood. I could connect the larger of these impulses with thunder heard some time later,

---

* About five miles south of Glasgow.

XXI.] *Atmospheric Electricity.* 319

with about the same degree of certainty as the brighter flashes of lightning during a thunder-storm by night are usually recognised as distinctly connected with distinct peals of thunder. By counting time I estimated the distance of the discharge not nearer on any occasion than about four or five miles. There were besides many smaller impulses; and most frequently I observed several of these between one of the larger and the thunder with which I connected it. The frequency of these smaller disturbances, which sometimes kept the needle in a constant state of flickering, often prevented me from identifying the thunder in connexion with any particular one of the impulses I had observed. They demonstrated countless discharges, smaller or more distant than those that give rise to audible thunder. On none of these occasions have I seen any lightning. The absolute potential at the position of the burning match was sometimes positive and sometimes negative; and the sudden change demonstrated by the impulses on the needle were, so far as I could judge, as often augmentations of positive or diminutions of negative, as diminutions of positive or augmentations of negative. This afternoon, for instance (Thursday, June 28), I heard several peals of thunder, and found the usual abrupt changes indicated by the electrometer. For several minutes the absolute potential was small positive, with two or three abrupt changes to somewhat strong positive, falling back to weak positive, and gathering again to a discharge. This was precisely what the same instrument would have shown anywhere within a few yards of an electrical machine turned slowly so as to cause a slow succession of sparks from its prime conductor to a conductor connected with the earth.

397. I have repeatedly observed the electric potential in the neighbourhood of a locomotive engine at work on a railway, sometimes by holding the portable electrometer out at a window of one of the carriages of a train, sometimes by using it while standing on the engine itself, and sometimes while standing on the ground beside the line. I have thus obtained consistent results, to the effect that the steam from the funnel was *always negative*, and the steam from the safety-valve always positive. I have observed *extremely strong* effects of each class from

carriages even far removed from the engine. I have found
strong negative indications in the air after an engine had disappeared round a curve, and its cloud of steam had dissolved
out of sight.

398. In almost all parts of a large manufactory, with steam-
pipes passing through them for various heating purposes, I
have found decided indications of positive electricity. In
most of these localities there was some slight escape of high-
pressure steam, which appeared to be the origin of the positive
indications.

399. These phænomena seem in accordance with Faraday's
observations on the electricity of steam, which showed high-
pressure steam escaping into the air to be in general positive,
but negative when it carried globules of oil along with it.

[*Proceedings Literary and Philosophical Society of Manchester*, Jan. 21, 1862.]

THE following extract of a letter from Professor W. Thomson, LL.D., etc., to the President, was read:—

400. "About two years ago I wrote to you that a metal bar, insulated so as to be moveable about an axis perpendicular to the plane of a metal ring made up half of copper and half of zinc, the two halves being soldered together, turns from the zinc towards the copper when vitreously electrified, and from the copper towards the zinc when resinously electrified. [See diagram of § 270 (4).]

"If the copper half and the zinc half of the ring are insulated from one another, and if they are connected by means of wires with two pieces of one metal maintained at any stated difference of potential by proper apparatus for dividing the electro-motive force of the two plates of a Daniell's element into 100 parts, from 60 to 70 of those parts are required to reduce the zinc half ring and the copper half ring to such a state that the moveable bar remains at rest whether it is electrified vitreously or resinously.

"If the copper half ring is oxidized by heat, the amount of electro-motive force then required to neutralize the two halves is much increased. If, after oxidizing the copper one day by heat, I leave the apparatus till the next day, the effect is generally diminished, though something of it still remains. After again heating the copper by laying it for some time on a red-hot iron heater and allowing it to cool, I found the effect almost exactly 100 parts. I have no doubt that by making the coat of oxide very complete and thick enough, and by cleaning the zinc perfectly, I shall be able to get considerably above the electro-motive force of a single Daniell's element. I remembered perfectly what you told me a long time ago about heating the coppers of a battery and getting a strong effect, for some time equal to that of the Daniell's cell, when I tried the effect of oxidizing the copper plate by heat.

"I believe there are also electrical effects of heat itself; so that if one half of a ring of one metal is hot and the other is cold, the needle will show a difference according as it is charged positively or negatively.

"For nearly two years I have felt quite sure that the proper explanation of voltaic action in the common voltaic arrangement is very near Volta's, which fell into discredit because Volta or his followers neglected the principle of conservation of force. I now think it quite certain that two metals dipped in one electrolytic liquid will (when polarization is done away with) reduce two dry pieces of the same metals, when connected each to each by metallic arcs, to the same potential.

"There cannot be a doubt that the whole thing is simply chemical action at a distance. Zinc and copper connected by a metallic arc attract one another from any distance. So do platinum plates coated with oxygen and hydrogen respectively. I can now tell the amount of the force, and calculate how great a proportion of chemical affinity is used up electrolytically, before two such discs come within $\frac{1}{1000}$th of an inch of one another, or any less distance down to a limit within which molecular heterogeneousness becomes sensible. This, of course, will give a definite limit for the sizes of atoms, or rather, as I do not believe in atoms, for the dimensions of molecular structures." [In an article on the "Size of Atoms" published in "Nature" for March 31, 1870, it has been shown, by the principle of reckoning here proposed, that "plates of copper "and zinc of a three-hundred-millionth of a centimetre thick, "placed close together alternately, form a near approximation "to a chemical combination, if indeed such thin plates could "be made without splitting atoms."]

ON A SELF-ACTING APPARATUS FOR MULTIPLYING AND MAIN-
TAINING ELECTRIC CHARGES, WITH APPLICATIONS TO ILLUS-
TRATE THE VOLTAIC THEORY.

[From the *Proceedings of the Royal Society* for June 20, 1867.]

401. In explaining the water-dropping collector for atmospheric electricity, in a lecture in the Royal Institution in 1860
(§ 285, above), I pointed out how, by disinsulating the water jar
and collecting the drops in an insulated vessel, a self-acting electric condenser is obtained. If, owing to electrified bodies in the
neighbourhood, the potential in the air round the place where the
stream breaks into drops is positive, the drops fall away negatively electrified; or *vice versa*, if the potential is negative, the
drops fall away positively electrified. The stream of water
descending does not in any way detract from the charges of
the electrified bodies to which its electric action is due, provided always these bodies are kept properly insulated; but by
the dynamical energy of fluid-motion, and work performed by
gravity upon the descending drops, electricity may be unceas-
ingly produced on the same principle as by the electrophorus.
But, as in the electrophorus there was no provision except good
insulation for maintaining the charge of the electrified body
or bodies from which the induction originates, this want is
supplied by the following reciprocal arrangement, in which the
body charged by the drops of water is made the inductor for
another stream, the drops from which in their turn keep up
the charge of the inductor of the first.

402. To stems connected with the inside coatings of two
Leyden phials are connected metal pieces, which, to avoid circumlocution, I shall call inductors and receivers. Each stem
bears an inductor and a receiver, the inductor of the first jar being

vertically over the receiver of the second jar, and *vice versa.*
Each inductor consists of a vertical metal cylinder (fig. 1) open
at each end.   Each receiver consists of a vertical metal cylinder

Fig. 1.

open at each end, but partially stopped in its middle
by a small funnel (fig. 1), with its narrow mouth
pointing downwards, and situated a little above the
middle of the cylinder.   Two fine vertical streams
of uninsulated water are arranged to break into.
drops, one as near as may be to the centre of each
inductor.   The drops fall along the remainder of the
axis of the inductor, and thence downwards, along the
upper part of the axis of the receiver of the other jar,
until they meet the funnel.   The water re-forms into
drops at the fine mouth of the funnel, which fall along
the lower part of the axis of the receiver and are
carried off by a proper drain below the apparatus.
Suppose now a small positive charge of electricity be
given to the first jar.   Its inductor electrifies negatively each drop of water breaking away in its centre
from the continuous uninsulated water above ; all
these drops give up their electricity to the second jar,
when they meet the funnel in its receiver.   The drops

a Water Jet.
b Inductor.
c Receiver.

falling away from the lower fine mouth of the funnel carry away
excessively little electricity, however highly the jar may be
charged ; because the place where they break away is, as it were,
in the interior of a conductor, and therefore has nearly zero electrification.   The negative electrification thus produced in the
second jar acts, through its inductor, on the receiver of the first
jar, to augment the positive electrification of the first jar, and
causes the negative electrification of the second jar to go on
more rapidly, and so on.   The dynamical value of the electrifications thus produced is drawn from the energy of the descending water, and is very approximately equal to the integral work
done by gravity against electric force on the drops, in their path
from the point where they break away from the uninsulated
water above, to contact with the funnel of the receiver below.
In the first part of this course each drop will be assisted downwards by electric repulsion from the inductively electrified
water and tube above it ; but below a certain point of its course

the resultant electric force upon it will be upwards, and, according to the ordinary way of viewing the composition of electric forces, may be regarded as being at first chiefly upward repulsion of the receiver diminished by downward repulsion from the water and tube, and latterly the sum of upward repulsion of the receiver and upward attraction of the inductor. The potential method gives the integral amount, being the excess of work done *against* electric force, above work performed *by* electric force on each drop in its whole path. It is of course equal to m V, if m denote the quantity of electricity carried by each drop, as it breaks from the continuous water above, and V the potential of the inner coating of the jar bearing the receiver, the potential of the uninsulated water being taken as zero. The practical limit to the charges acquired is when one of them is so strong as to cause sparks to pass across some of the separating air-spaces, or to throw the drops of water out of their proper course and cause them to fall outside the receiver through which they ought to pass. It is curious, after com-

Fig. 2.

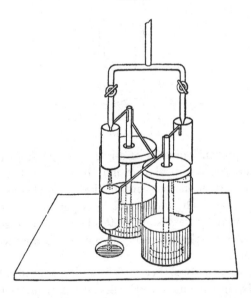

mencing with no electricity except a feeble charge in one of the jars, only discoverable by a delicate electrometer, to see in the course of a few minutes a somewhat rapid succession of sparks pass in some part of the apparatus, or to see the drops of water scattered about over the lips of one or both the receivers.

403.   The Leyden jars represented in the sketch (fig. 2) are open-mouthed jars of ordinary flint glass, which, when very dry, I generally find to insulate electricity with wonderful perfection. The inside coatings consist of strong liquid sulphuric acid, and heavy lead tripods with vertical stems projecting upwards above the level of the acid, which, by arms projecting horizontally above the lip of the jar, bear the inductors and receivers, as shown in fig. 2.   Lids of gutta percha or sheet metal close the mouth of each jar, except a small air-space of from $\frac{1}{8}$ to $\frac{1}{4}$ of an inch round the projecting stems.   If a tube (fig. 3) be added

Fig. 3.

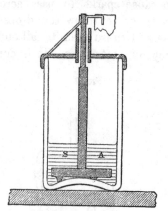

SA Sulphuric Acid.

to the lid to prevent currents of air from circulating into the interior of the jar, the insulation may be so good that the loss may be no more than one per cent. of the whole charge in three or four days.   Two such jars may be kept permanently charged from year to year by very slow water-dropping arrangements, a drop from each nozzle once every two or three minutes being quite sufficient.

404.   The mathematical theory of the action, appended below*, is particularly simple, but nevertheless curiously interesting.

---

* Let c, c' be the capacities of the two jars, l, l' their rates of loss per unit

405.   The reciprocal electrostatic arrangement now described presents an interesting analogy to the self-sustaining electro-magnetic system recently brought before the Royal Society by Mr C. W. Siemens and Professor Wheatstone, and mathematically investigated by Professor Clerk Maxwell.   Indeed it was from the fundamental principle of this electromagnetic system that the reciprocal part of the electrostatic arrangement occurred to me recently.   The particular form of self-acting electrophorus condenser now described, I first constructed many years ago. I may take this opportunity of describing an application of it to illustrate a very important fundamental part of electric theory,   I hope soon to communicate to the Royal Society a description of some other experiments which I made seven years ago on the same subject, and which I hope now to be able to prosecute further.

406.   Using only a single inductor and a single receiver, as shown in fig. 1, let the inductor be put in metallic communication with a metal vessel or cistern whence the water flows; and let the receiver be put in communication with a delicate electro-scope or electrometer.   If the lining of the cistern and the inner metallic surface of the inductor be different metals, an electric effect is generally found to accumulate in the receiver and electrometer.   Thus, for instance, if the inner surface of the

---

potential of charge, per unit of time, and D, D′ the values of the water-droppers influenced by them. Let $+v$ and $-v'$ be their potentials at time t; v and v' being of one sign in the ordinary use of the apparatus described in the text. The action is expressed by the following equations:—

$$c\frac{dv}{dt} = D'v' - lv; \qquad c'\frac{dv'}{dt} = Dv - l'v'.$$

If c, D, l, c', D′, l' were all constant, the solution of these equations would be, for the case of commencing with the first jar charged to potential 1, and the second zero,

$$v = \frac{(c'\rho + l')\epsilon\rho t - (c'\sigma + l')\,\epsilon\sigma t}{c'\,(\rho - \sigma)}, \qquad v' = D\frac{\epsilon\rho t - \epsilon\sigma t}{c'(\rho - \sigma)}.$$

with the corresponding symmetrical expression for the case in which the second jar is charged, and the first at zero, in the beginning; the roots of the quadratic

$$(cx + l)\,(c'x + l') - DD' = 0$$

being denoted by ρ and σ. When $ll' >$ DD′, both roots are negative; and the electrification comes to zero in time, whatever may be the initial charges. But when $ll' <$ DD′, one root is positive and the other negative, and ultimately the charges augment in proportion to $\epsilon\rho t$ if ρ be the positive root.

inductor be dry polished zinc, and the vessel of water above be
copper, the receiver acquires a continually increasing charge
of negative electricity. There is little or no effect, either positive or negative, if the inductor present a surface of polished
copper to the drops where they break from the continuous water
above: but if the copper surface be oxidized by the heat of a lamp,
until, instead of a bright metallic surface of copper, it presents
a slate-coloured surface of oxide of copper to the drops, these
become positively electrified, as is proved by a
continually increasing positive charge exhibited by the electrometer. When the inner
surface of the inductor is of bright metallic
colour, either zinc or copper, there seems to be
little difference in the effect whether it be wet
with water or quite dry; also I have not found
a considerable difference produced by lining
the inner surface of the inductor with moist or
dry paper. Copper filings falling from a copper
funnel and breaking away from contact in the
middle of a zinc inductor, in metallic communication with a copper funnel, as shown in fig. 4,
produce a rapidly increasing negative charge in
a small insulated can catching them below.

Fig. 4.

a Copper Filings.
b Inductor—Zinc.
c Receiver.

  The quadrant divided-ring electrometer* indicating, by the image of a lamp on a scale, angular motions of
a small concave mirror ($\frac{1}{8}$ of a grain in weight) such as I use
in galvanometers, is very convenient for exhibiting these results.
Its sensibility is such that it gives a deflection of 100 scale-
divisions ($\frac{1}{40}$ of an inch each) on either side of zero, as the
effect of a single cell of Daniell's; the focusing, by small concave mirrors supplied to me by Mr Becker, being so good that
a deflection can easily be read with accuracy to a quarter of a
scale-division. By adopting Peltier's method of a small magnetic needle attached to the electric moveable body (or "needle"),
and by using fixed steel magnets outside the instrument to give
directing force (instead of the glass-fibre suspension of the

---

* See Nichol's *Encyclopædia*, 1860, article "Electricity, Atmospheric;"
or *Proceedings of the Royal Institution*, May 1860, Lecture on Atmospheric
Electricity [§§ 249...293, above].

divided-ring electrometers described in the articles referred to), and by giving a measurable motion by means of a micrometer screw to one of the quadrants, I have a few weeks ago succeeded in making this instrument into an independent electrometer, instead of a mere electroscope, or an electrometer in virtue of a separate gauge electrometer, as in the Kew recording atmospheric electrometer, described in the Royal Institution lecture.

407. Reverting to the arrangement described above of a copper vessel of water discharging water in drops from a nozzle through an inductor of zinc in metallic connection with the copper, let the receiver be connected with a second inductor, this inductor insulated; and let a second nozzle, from an uninsulated stream of water, discharge drops through it to a second receiver. Let this second receiver be connected with a third inductor used to electrify a third stream of water to be caught in a third receiver, and so on. We thus have an ascending scale of electrophorus action analogous to the beautiful mechanical electric multiplier of Mr. C. F. Varley, with which, by purely electrostatic induction, he obtained a rapid succession of sparks from an ordinary single voltaic element. This result is easily obtained by the self-acting arrangement now described, with the important modification in the voltaic element according to which no chemical action is called into play, and work done by gravity is substituted for work done by the combination of chemical elements.

ON A UNIFORM ELECTRIC CURRENT ACCUMULATOR.

[From the *Philosophical Magazine*, January 1868.]

408. CONCEIVE a closed circuit, $C\,T\,A\,B\,C$, according to the following description:—One portion of it, $T\,A$, tangential to a circular disc of conducting material and somewhat longer than the radius; the continuation, $A\,B$, at right angles to this in the plane of the wheel, of a length equal to the radius; and the completion of the circuit by a fork, $B\,C$, extending to an axle bearing the wheel. If all of the wheel were cut away except a portion, $C\,T$, from the axle to the point of contact at the circumference, the circuit would form a simple rectangle, $C\,T\,A\,B$, except the bifurcation of the side $B\,C$. Let a

fixed magnet be placed so as to give lines of force perpendicular to the wheel, in the parts of it between C the centre
and T the point of the circumference touched by the fixed

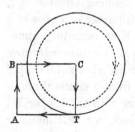

conductor; and let power be applied
to cause the wheel to rotate in the
direction towards A. According to
Faraday's well-known discovery, a
current is induced in the circuit in
such a direction that the mutual
electromagnetic action between it and
the fixed magnet resists the motion of
the wheel. Now the mutual elec-
tromagnetic force between the portions A B and C T of
the circuit is repulsive, according to the well-known elementary law of Ampère, and therefore resists the actual motion
of the wheel; hence, if the magnet be removed, there will
still be electromagnetic induction tending to maintain the
current. Let us suppose the velocity of the wheel to have been
at first no greater than that practically attained in ordinary experiments with Barlow's electromagnetic disc. As the magnet
is gradually withdrawn let the velocity be gradually increased
so as to keep the strength of the current constant, and, when
the magnet is quite away, to maintain the current solely by
electromagnetic induction between the fixed and moveable portions of the circuit. If, when the magnet is away, the wheel
be forced to rotate faster than the limiting velocity of our previous supposition, the current will be augmented according to
the law of compound interest, and would go on thus increasing
without limit were it not that the resistance of the circuit would
become greater in virtue of the elevation of temperature produced by the current. The velocity of rotation which gives by
induction an electromotive force exactly equal to that required
to maintain the current, is clearly independent of the strength
of the current. The mathematical determination of it becomes
complicated by the necessity of taking into account the diffusion
of the current through portions of the disc not in a straight line
between C and T; but it is very simple and easy if we prevent
this diffusion by cutting the wheel into an infinite number of
infinitely thin spokes, a great number of which are to be simul-

taneously in contact with the fixed conductor at T. The linear velocity of the circumference of the wheel in the limiting case bears to the velocity which measures, in absolute measure, the resistance of the circuit, a ratio (determinable by the solution of the mathematical problem) which depends on the proportions of the rectangle $CTAB$, and is independent of its absolute dimensions.

409. Lastly, suppose the wheel to be kept rotating at any constant velocity, whether above or below the velocity determined by the preceding considerations; and suppose the current to be temporarily excited in any way (for instance, by bringing a magnet into the neighbourhood and then withdrawing it); the strength of this current will diminish towards zero or will increase towards infinity, according as the velocity is below or above the critical velocity. The diminution or augmentation would follow the compound interest law if the resistance in the circuit remained constant. The conclusion presents us with this wonderful result : that if we commence with absolutely no electric current and give the wheel any velocity of rotation exceeding the critical velocity, the electric equilibrium is unstable : an infinitesimal current in either direction would augment until, by heating the circuit, the electric resistance becomes increased to such an extent that the electromotive force of induction just suffices to keep the current constant.

410. It will be difficult, perhaps impossible, to realize this result in practice, because of the great velocity required, and the difficulty of maintaining good frictional contact at the circumference, without enormous friction, and consequently frictional generation of heat.

411. The electromagnetic augmentation and maintenance of a current discovered by Siemens, and put in practice by him, with the aid of soft iron, and proved by Maxwell to be theoretically possible without soft iron, suggested the subject of this communication to the author, and led him to endeavour to arrive at a similar result with only a single circuit, and no making and breaking of contacts; and it is only these characteristics that constitute the peculiarity of the arrangement which he now describes.

ON VOLTA-CONVECTION BY FLAME.

[From the *Philosophical Magazine*, January 1868.]

412. In Nichol's *Cyclopædia*, article "Electricity, Atmospheric" (2d edition), and in the *Proceedings of the Royal Institution* May 1860 (Lecture on Atmospheric Electricity), [§§ 249...293, above] the author had pointed out that the effect of the flame of an insulated lamp is to reduce the lamp and other conducting material connected with it to the same potential as that of the air in the neighbourhood of the flame, and that the effect of a fine jet of water from an insulated vessel is to bring the vessel and other conducting material connected with it to the same potential as that of the air at the point where the jet breaks into drops. In a recent communication to the Royal Society "On a Self-acting Apparatus for Multiplying and Maintaining Electric Charges, with applications to illustrate the Voltaic Theory," [§§ 401...407, above,] an experiment was described in which a water-dropping apparatus was employed to prove the difference of potential in the air, in the neighbourhood of bright metallic surfaces of zinc and copper metallically connected with one another, which is to be expected from Volta's discovery of contact-electricity. In the present communication a similar experiment is described, in which the flame of a spirit-lamp is used instead of a jet of water breaking into drops.

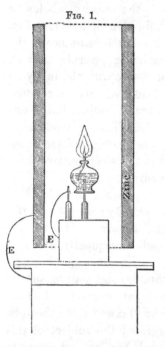

Fig. 1.

413. A spirit-lamp is placed on an insulated stand connected with a very delicate electrometer. Copper and zinc cylinders, in metallic connection with the metal case of the electrometer, are alternately held vertically in such a position that the

flame burns nearly in the centre of the cylinder, which is open
at both ends. If the electrometer reading, with the copper
cylinder surrounding the flame, is called zero, the reading
observed with the zinc cylinder surrounding the flame indicates
positive electrification of the insulated stand bearing the lamp.

414. It is to be remarked that the differential method here
followed eliminates the ambiguity involved in what is meant by
the potential of a conducting system composed partly of flame,
partly of alcohol, and partly of metal. In a merely illustrative
experiment, which the author has already made, the amount of
difference made by substituting the zinc cylinder for the copper
cylinder round the flame was rather more than half the difference of potential maintained by a single cell of Daniell's. Thus,
when the sensibility of the quadrant divided-ring electrometer
(§ 406) was such that a single cell of Daniell's gave a deflection
of 79 scale-divisions, the difference of the reading when the zinc
cylinder was substituted for the copper cylinder round the insulated lamp was 39 scale-divisions. From other experiments
on contact-electricity made seven years ago by the author, and
agreeing with results which have been published by Hankel, it
appears that the difference of potentials in the air in the neighbourhood of bright metallic surfaces of zinc and copper in
metallic connexion with one another is about three-quarters of
that of a single cell of Daniell's. It is quite certain that the
difference produced in the metal connected with the insulated
lamp would be exactly equal to the true contact difference of
the metals, if the interior surfaces of the metal cylinders were
perfectly metallic (free from oxidation or any other tarnishing,
such as by sulphur, iodine, or any other body); provided the
distance of the inner surface of the cylinder from the flame were
everywhere sufficient to prevent conduction by heated air between them, and provided the length of the cylinder were
infinite (or, practically, anything more than three or four times
its diameter).

415. The author hopes before long to be able to publish a
complete account of his old experiments on contact-electricity,
of which a slight notice appeared in the Proceedings of the
Literary and Philosophical Society of Manchester [§ 400,
above].

ON ELECTRIC MACHINES FOUNDED ON INDUCTION AND CONVECTION.

[From the *Philosophical Magazine*, January 1868.]

416. To facilitate the application of an instrument, which I have recently patented, for recording the signals of the Atlantic Cable, a small electric machine running easily enough to be driven by the wheelwork of an ordinary Morse instrument was desired; and I have therefore designed a combination of the electrophorus principle with the system of reciprocal induction explained in [§§ 401...407] a recent communication to the Royal Society (*Proceedings*, June 1867), which may be briefly described as follows :—

417. A wheel of vulcanite, with a large number of pieces of metal (called carriers, for brevity) attached to its rim, is kept rotating rapidly round a fixed axis. The carriers are very lightly touched at opposite ends of a diameter by two fixed tangent springs. One of these springs (the earth-spring) is connected with the earth, and the other (the receiver-spring) with an insulated piece of metal called the receiver, which is analogous to the "prime conductor" of an ordinary electric machine. The point of contact of the earth-spring with the carriers is exposed to the influence of an electrified body (generally an insulated piece of metal) called the inductor. When this is negatively electrified, each carrier comes away from contact with the earth-spring, carrying positive electricity, which it gives up, through the receiver-spring, to the receiver. The receiver and inductor are each hollowed out to a proper shape, and are properly placed to surround, each as nearly as may be, the point of contact of the corresponding spring.

418. The inductor, for the good working of the machine, should be kept electrified to a constant potential. This is effected by an adjunct called the replenisher, which may be applied to the main wheel, but which, for a large instrument, ought to be worked by a much smaller carrier-wheel, attached either to the same or to another turning-shaft.

419. The replenisher consists chiefly of two properly shaped pieces of metal called inductors, which are fixed in the neighbourhood of a carrier-wheel, such as that described above, and four

fixed springs touching the carriers at the ends of two diameters. Two of these springs (called receiver-springs) are connected respectively with the inductors; and the other two (called connecting springs) are insulated and connected with one another (one of the inductors is generally connected with the earth, and the other insulated). They are so situated that they are touched by the carriers on emerging from the inductors, and shortly after

Fig. 1.

Section.   Elevation.

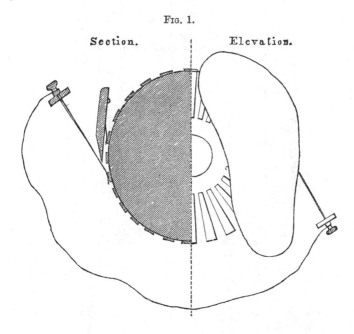

the contacts with the receiver-springs. If any difference of potential between the inductors is given to begin with, the action of the carriers, as is easily seen, increases it according to the compound-interest law as long as the insulation is perfect. Practically, in a few seconds after the machine is started running, bright flashes and sparks begin to fly about in various parts of the apparatus, even although the inductors and connectors have been kept for days as carefully discharged as possible. Forty elements of a dry pile (zinc, copper, paper), applied with one pole to one of the inductors, and the other for a moment to the connecting springs and the other inductor, may be used to determine, or to suddenly reverse, the character (vitreous or resinous) of the electrification of the insulated inductor. The only instrument yet made is a very small one (with carrier-wheel 2 inches in diameter), constructed for the Atlantic

FIG. 2.

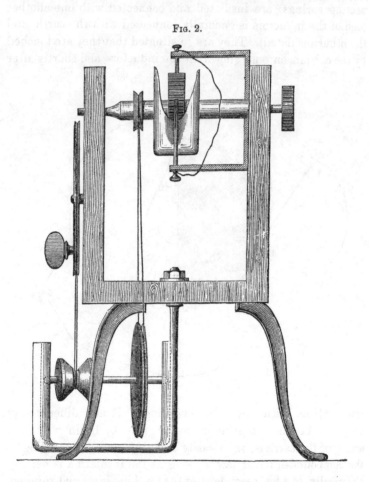

Telegraph application; but its action has been so startlingly successful that good effect may be expected from larger machines on the same plan.

420. When this instrument is used to replenish the charge of the inductor in the constant electric machine, described above, one of its own inductors is connected with the earth, and the other with the inductor to be replenished. When accurate constancy

is desired, a gauge-electroscope is applied to break and make contact between the connector-springs of the replenisher when the potential to be maintained rises above or falls below a certain limit.

421. Several useful applications of the replenisher for scientific observation were shown by the author at the recent meeting of the British Association (Dundee),—among others, to keep up the charge in the Leyden jar for the divided-ring mirror-electrometer, especially when this instrument is used for recording atmospheric electricity. A small replenisher, attached to the instrument within the jar, is worked by a little milled head on the outside, a few turns of which will suffice to replenish the loss of twenty-four hours.

POSTSCRIPT, *Nov.* 23, 1867.

422. As has been stated, this machine was planned originally for recording the signals of the Atlantic Cable. The small "replenisher" represented in the diagrams has proved perfectly suitable for this purpose. The first experiments on the method for recording signals which I recently patented were made more than a year ago by aid of an ordinary plate-glass machine worked by hand. This day the small "replenisher" has been connected with the wheelwork drawing the Morse paper on which signals are recorded, and, with only the ordinary driving-weight as moving power, has proved quite successful.

423. The scientific applications indicated when the communication was made to the British Association have been tested within the last few weeks, and especially to-day, with the assistance of Professor Tait. The small replenisher is now made as part of each quadrant electrometer. It is permanently placed in the interior of the glass Leyden jar ; and a few turns by the finger applied to a milled head on the outside of the lid are found sufficient to replenish the loss of twenty-four hours. A small instrument has also been made and tested for putting in practice the plan of equalizing potentials, described verbally in the communication to the British Association, which consisted in a mechanical arrangement to produce effects of the same character as those of the water-dropping system, described several

years ago at the Royal Institution*. The instrument is represented in the annexed sketch (fig. 3). *AT* and *A'T'* are two
springs touching a circular row of small brass pegs† insulated
from one another in a vulcanite disc. These springs are insulated, one or both, and are connected with the two electrodes of

FIG. 3.

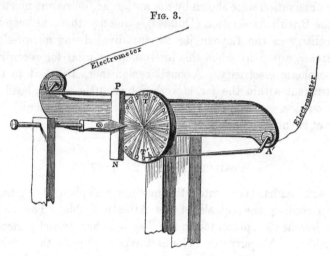

the electrometer—or one of them with the insulated part of the
electrometer, and the other with the metal enclosing the case
when there is only one insulated electrode. One application is
to test the "pyro-electricity" of crystals; thus a crystal of tourmaline, *PN*, by means of a metal arm holding its middle, is supported symmetrically with reference to the disc in a position
parallel to the line *TT'*, and joining the lines of contact of the
springs. When warmed (as is conveniently done by a metal
plate at a considerable distance from it), it gives by ordinary
tests, as is well known, indications of positive electrification to-

---

* Lecture on Atmospheric Electricity, *Proceedings of the Royal Institution*,
May 1860. See also Nichol's *Cyclopædia*, article "Electricity, Atmospheric"
[§§ 249...293].

† [I now find a smaller number of larger discs to be preferable, as considerable disturbances are produced by the numerous breakings of contact unless
the two springs are in precisely the same condition as to quality and clean-
ness of metal surface. Thin stiff platinum pins attached to the discs, and
very fine platinum springs touching them as they pass, will probably give
good and steady results if the springs are kept very clean. The smallest
quantity of the paraffin (with which, as usual in electric instruments, the
vulcanite is coated), if getting on either spring, would probably produce im-
mense disturbance.—December 23, 1867.]

wards the one end P, and of negative electrification towards the
other end N. The wheel in the arrangement now described is
kept turning at a rapid rate; and the effect of the carrier is to
produce in the springs TA, $T'A'$ the same potentials, approximately, as those which would exist in the air at the points T, T'
if the wheel and springs were removed. The springs being
connected with the electrodes of the divided-ring quadrant
electrometer, the spot of light is deflected to the right, let us
say. After continuing the application of heat for some time
the hot plate is removed, and a little later the spot of light goes
to zero and passes to the left, remaining there for a long time,
and indicating a difference of potentials between the springs, in
the direction $A'T'$ positive and AT negative. The electrometer
being of such sensibility as to give a deflection of about 100
scale-divisions to the right or left when tested by a single galvanic cell, and having a range of 300 scale-divisions on each
side, it is necessary to place the tourmaline at a distance of
several inches from the disc to keep the amount of the deflection within the limits of the scale.

424. Another application of this instrument is for the
experimental investigation of the voltaic theory, according
to the general principle described [§ 406] in the communication to the Royal Society already referred to.* In it two
inductors are placed as represented in fig. 4. The inner

FIG. 4.

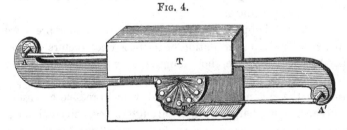

surface of each of these is of smooth brass; and one of them
is lined wholly, or partially, with sheet zinc, copper, silver,
or other metal to be tested. Thus, to experiment upon
the contact difference of potentials between zinc and copper,

---

* *Proceedings of the Royal Society*, May 1867.

one of the inductors is wholly lined with sheet zinc or with sheet copper, and the two inductors are placed in metallic communication with one another. The springs are each in metallic communication with the electrodes of the quadrant mirror electrometer, and the wheel is kept turning. The spot of light is observed to take positions differing, according as the lining is zinc or copper, by $72\frac{1}{2}$ per cent. of the difference produced by disconnecting the two inductors from one another and connecting them with the two plates of a single Daniell's cell, when either the zinc or the copper lining is left in one of them. These differences are very approximately in simple proportion to the differences of potentials between the pairs of the opposite quadrants of the electrometer in the different cases. The difference between the effects of zinc and of copper in this arrangement is of course in the direction corresponding to the positive electrification of the quadrants connected with the spring whose point of contact is exposed to the zinc-lined inducing surface. It must be remembered, however, as is to be expected from Hankel's observations, that the difference measured will be much affected by a slight degree of tarnishing by oxidation, or otherwise, of the inner surface of either inductor. When the copper surface is brought to a slate-colour by oxidation under the influence of heat, the contact difference between it and polished zinc amounts sometimes, as I found in experiments made seven years ago [§ 400, above], to 125, that of a single cell of Daniell's being called 100.

425. A useful application of the little instrument represented in fig. 4 is for testing insulation of insulated conductors of small capacity, as for instance, short lengths (2 or 3 feet) of submarine cable, when the electrometer used is such that its direct application to the conductor to be tested would produce a sensible disturbance in its charge, whether through the capacity of the electrometer being too great, or from inductive effects due to motion of the moveable part, or parts, especially if the electrometer is "heterostatic" [§ 385]. In this application one of the inductors is kept in connection with a metal plate in the water surrounding the specimen of cable to be tested; and the other is connected with the specimen, or is successively connected with the different specimens under examination. The springs are

connected with the two electrodes of the electrometer as usual.
The small constant capacity of the insulated inductor, and the
practically perfect insulation which may with ease be secured
for the single glass or vulcanite stem bearing it, are such that
the application of the testing apparatus to the body to be tested
produces either no sensible change, or a small change which
can be easily allowed for. It will be seen that the small metal
pegs carried away by the turning-wheel from the point of the
insulated spring, in the arrangement last described, correspond
precisely to the drops of water breaking away from the nozzle
in the water-dropping collector for atmospheric electricity.

426. A form bearing the same relation to that represented
in the drawings that a glass-cylinder electric machine bears to a
plate-glass machine of the ordinary kind will be more easily
made, and will probably be found preferable, when the dimensions
are not so great as to render it cumbrous. In it, it is proposed
to make the carrier-wheel nearly after the pattern of a mouse-
mill, with discs of vulcanite instead of wood for its ends.
The inductor and receiver of the rotatory electrophorus or
the two inductor-receivers of the replenisher, may, when this
pattern is adopted, be mere tangent planes; but it will probably
be found better to bend them somewhat to a curved cylindrical
shape not differing very much from tangent planes. When,
however, great intensity is desired, the best pattern will probably be had by substituting for the carrier-wheel an endless
rope ladder, as it were, with cross bars of metal and longitudinal
cords of silk or other flexible insulating material. This, by an
action analogous to that of the chain-pump, will be made to
move with great rapidity, carrying electricity from a properly
placed inductor to a properly shaped and properly placed receiver at a distance from the inductor which may be as much
as several feet.

ON THE RECIPROCAL ELECTROPHORUS.

[From the *Philosophical Magazine*, April 1868.]

427. Having been informed by Mr. Fleeming Jenkin that he
had heard from Mr. Clerk Maxwell that the instrument which I
described under the name "Replenisher," in the *Philosophical*

Magazine for January 1868, was founded on precisely the same principle as an instrument "for generating electricity" which had been patented some years ago by Mr. C. F. Varley, I was surprised; for I remembered his inductive machine which had been so much admired at the Exhibition of 1862, and which certainly did not contain the peculiar principle of the "Replenisher." But I took the earliest opportunity of looking into Mr. Varley's patent (1860), and found, as was to be expected, that Mr. Maxwell was perfectly right. In that patent Mr. Varley describes an instrument agreeing in almost every detail with the general description of the "Replenisher" which I gave in the article of the *Philosophical Magazine* already referred to. The only essential difference is that no contacts are made in Mr. Varley's instrument, but, instead, the carriers pass, each at four points of its circular path, within such short distances of four metallic pieces that when a sufficient intensity of charge has been reached, sparks pass across the air-intervals. Hence to give a commencement of action to Mr. Varley's instrument, one of the inductors must be charged from an independent source to a considerable potential (that of several thousand cells for instance), to make sure that sparks will pass between the carriers and the metal piece (corresponding to one of my connecting springs) which it passes under the influence of that inductor. In my "Replenisher," however well discharged it may be to begin with, electrification enough is reached after a few seconds (on the compound interest principle, with an infinitesimal capital to begin with) to produce sparks and flashes in various parts of the instrument. In Mr. Varley's instrument, what corresponds to my connector is described as being connected with the ground; and the effect is to produce positive and negative electrification of the two inductors. In this respect it agrees with the self-acting apparatus for multiplying and maintaining electric charges, described in a communication to the Royal Society last May.* From this arrangement I passed to the "Replenisher" by using a wheel with carriers as a substitute for the water-droppers, and arranging that the connectors might be insulated and one of the inductors con-

---

* *Proceedings of the Royal Society*, 1867; or, *Phil. Mag.*, November 1867.

nected with the earth, which, of course, may be done in Mr.
Varley's instrument, and which renders it identical with mine,
with the exception of the difference of spring-contacts instead
of sparks. This difference is essential for some of the applications of the "Replenisher," which I described, and have found
very useful, especially the small internal replenisher, for replenishing, when needed, the charges of the Leyden jar of my
heterostatic electrometers. But the reciprocal-electrophorus
principle, which seemed to me a novelty in the communication
to the Royal Society and in the *Philosophical Magazine* article
of last January referred to, had, as I now find, been invented and
published by Mr. Varley long before, in his patent of 1860,
when it was, I believe, really new to science.

428. POSTSCRIPT.—GLASGOW COLLEGE, *March* 20, 1868.—In
looking further into Mr. Varley's patent, I find that he describes
an arrangement for making spring-contacts instead of the narrow
air-spaces for sparks,—and that he uses the spring-contacts to
enable him to commence with a very small difference of potentials, and to magnify on the compound interest principle. He
even states that he can commence with such a difference of
potentials as can be produced by a single thermo-electric element,
and by the use of his inductive instrument can multiply this in
a measured proportion until he reaches a difference of potentials
measurable by an ordinary electrometer. Thus it appears that his
anticipation of all that I have done in my "Replenisher" is even
more complete than I supposed when writing the preceding.

429. SECOND POSTSCRIPT (1870).—On having had my attention called to Nicholson's "Revolving Doubler," I find in it the
same compound interest principle of electrophoric action. It
seems certain that the discovery is Nicholson's, and about one
hundred years old. Holtz's now celebrated electric machine,
which is closely analogous in principle to Varley's of 1860, is, I
believe, a descendant of Nicholson's. Its great power depends
on the abolition by Holtz of metallic carriers, and of metallic
make-and-break contacts. Its inductive principle is identical
with that of Varley's earlier and my own later invention. It
differs from Varley's and mine in leaving the inductors to themselves, and using the current in the "connecting" arc (§ 419),
which, when sparks are to be produced, is broken.

XXIV.—A MATHEMATICAL THEORY OF MAGNETISM.

[Abstract from the *Proceedings of the Royal Society*, June 1849.]

430. THE theory of magnetism was first mathematically treated in a complete form by Poisson. Brief sketches of his theory, with some simplifications, have been given by Green and Murphy in their works on Electricity and Magnetism. In all these writings a hypothesis of two magnetic fluids has been adopted, and strictly adhered to throughout. No physical evidence can be adduced in support of such a hypothesis; but on the contrary, recent discoveries, especially in electromagnetism, render it extremely improbable. Hence it is of importance that all reasoning with reference to magnetism should be conducted without assuming the existence of those hypothetical fluids.

431. The writer of the present paper endeavours to show that a complete mathematical theory of magnetism may be established upon the sole foundation of facts generally known, and Coulomb's special experimental researches. The positive parts of this theory agree with those of Poisson's mathematical theory, and consequently the elementary mathematical formulæ coincide with those which have been previously given by Poisson.

The paper at present laid before the Royal Society is restricted to the elements of the mathematical theory, exclusively of those parts in which the phenomena of magnetic induction are considered.

The author hopes to have the honour of laying before the Society a continuation, containing some original mathematical investigations on magnetic distributions, and a theory of induction, in ferromagnetic or diamagnetic substances.

[*Transactions of the Royal Society* for June 1849, and June 1850.]

Introduction.

432. THE existence of magnetism is recognised by certain phenomena of force which are attributed to it as their cause. Other physical effects are found to be produced by the same agency; as in the operation of magnetism with reference to polarized light, recently discovered by Mr Faraday; but we must still regard magnetic force as the characteristic of magnetism, and, however interesting such other phenomena may be in themselves, however essential a knowledge of them may be for enabling us to arrive at any satisfactory ideas regarding the physical nature of magnetism, and its connexion with the general properties of matter, we must still consider the investigation of the laws, according to which the development and the action of magnetic force are regulated, to be the primary object of a Mathematical Theory in this branch of Natural Philosophy.

433. Magnetic bodies, when put near one another, in general exert very sensible mutual forces; but a body which is not magnetic can experience no force in virtue of the magnetism of bodies in its neighbourhood. It may indeed be observed that a body, *M*, will exert a force upon another body *A*; and again, on a third body *B*; although when *A* and *B* are both removed to a considerable distance from *M*, no mutual action can be discovered between themselves; but in all such cases *A* and *B* are, when in the neighbourhood of *M*, temporarily magnetic; and when both are under the influence of *M* at the same time, they are found to act upon one another with a mutual force. All these phenomena are investigated in the mathematical theory of magnetism, which, therefore, comprehends two distinct kinds of magnetic action—the mutual forces exercised between bodies possessing magnetism, and the magnetization induced in other bodies through the influence of magnets. The First Part of this paper is confined to the more *descriptive* and *positive* details of the subject, with reference to the former class of phenomena. After a sufficient foundation has been laid in it, by the mathematical exposition of the distribution of magnetism in bodies, and by the determination and expression

of the general laws of magnetic force, a Second Part will be devoted to the theory of magnetization by influence, or magnetic induction.

FIRST PART.—ON MAGNETS, AND THE MUTUAL FORCES BETWEEN MAGNETS.

CHAPTER I.—*Preliminary Definitions and Explanations.*

434. A magnet is a substance which intrinsically possesses magnetic properties.

A piece of loadstone, a piece of magnetized steel, a galvanic circuit, are examples of the varieties of natural and artificial magnets at present known; but a piece of soft iron, or a piece of bismuth temporarily magnetized by induction, cannot, in unqualified terms, be called a magnet.

A galvanic circuit is frequently, for the sake of distinction, called an "electro-magnet;" but, according to the preceding definition of a magnet, the simple term, without qualification, may be applied to such an arrangement. On the other hand, a piece of apparatus consisting of a galvanic coil, with a soft iron core, although often called simply "an electro-magnet," is in reality a complex arrangement involving an electro-magnet (which is intrinsically magnetic as long as the electric current is sustained) and a body transiently magnetized by induction.

435. In the following analysis of magnets, the magnetism of every magnetic substance considered will be regarded as absolutely permanent under all circumstances. This condition is not rigorously fulfilled either for magnetized steel or for loadstone, as the magnetism of any such substance is always liable to modification by induction, and may therefore be affected either by bringing another magnet into its neighbourhood, or by breaking the mass itself and separating the fragments. When, however, we consider the magnetism of any fragment taken from a steel or loadstone magnet, the hypothesis will be that it retains without any alteration the magnetic state which it actually had in its position in the body. The general theory of the distribution of magnetism founded upon conceptions of this kind, will be independent of the truth or falseness of any such hypothesis which may be made for the sake of conveni-

ence in studying the subject; but of course any actual experiments in illustration of the analysis or synthesis of a magnet would be affected by a want of *rigidity* in the magnetism of the matter operated on.   For such illustrations electro-magnets [without iron or other magnetic substance] are extremely appropriate, as in them, except during the motion by which any alteration in their form or arrangement is effected, no appreciable inductive action can exist.

436.   In selecting from the known phenomena of magnetism those elementary facts which are to serve for the foundation of the theory, all complex actions depending on the irregularities of the bodies made use of should be excluded.   Thus if we were to attempt an experimental investigation of the action between two amorphous fragments of loadstone, or between two pieces of steel magnetized by ordinary processes, we should probably fail to recognise the simple laws on which the actions resulting from such complicated circumstances depend; and we must look for a simpler case of magnetic action before we can make an analysis which may lead to the establishment of the fundamental principles of the theory.   Much complication will be avoided if we take a case in which the irregularities of one, at least, of the bodies do not affect the phenomena to be considered.   Now, the earth, as was first shown by Gilbert, is a magnet; and its dimensions are so great that there is no sensible variation in its action on different parts of any ordinary magnet upon which we can experiment, and consequently, in the circumstances, no complicacy depending on the actual distribution of terrestrial magnetism.   We may therefore, with advantage, commence by examining the action which the earth produces upon a magnet of any kind at its surface.

437.   At a very early period in the history of magnetic discovery the remarkable property of "pointing north and south" was observed to be possessed by fragments of loadstone and magnetized steel needles.   To form a clear conception of this phenomenon, we must consider the total action produced by the earth upon a magnet of any kind, and endeavour to distinguish between the effects of gravitation which the earth exerts upon the body in virtue of its weight, and those which result from the magnetic agency.

438. In the first place, it is to be remarked that the magnetic agency of the earth gives rise to no resultant force of sensible magnitude, upon any magnet with reference to which we can perform experiments [that is to say, small enough to be a subject for laboratory experiments], as is proved by the following observed facts:—

(1.) A magnet placed in any manner, and allowed to move with perfect freedom in any horizontal direction (by being floated, for example, on the surface of a liquid), experiences no action which tends to set its centre of gravity in motion, and there is therefore no [directly observable] horizontal force upon the body.

(2.) The magnetism of a body may be altered in any way, without affecting its weight as indicated by a balance. Hence there can be no [directly observable] vertical force upon it depending on its magnetism.

439. It follows that any magnetic action which the earth can exert upon a magnet [of dimensions suitable for laboratory experiments] must be [sensibly] a couple. To ascertain the manner in which this action takes place, let us conceive a magnet to be supported by its centre of gravity* and left perfectly free to turn round this point, so that, without any constraint being exerted which could balance the magnetic action, the body may be in circumstances the same as if it were without weight. The magnetic action of the earth upon the magnet gives rise to the following phenomena:—

(1.) The magnet does not remain in equilibrium in every position in which it may be brought to rest, as it would do did it experience no action but that of gravitation.

(2.) If the magnet be placed in a position of equilibrium there is a certain axis (which, for the present, we may conceive to be found by trial), such that if the magnet be turned round it, through any angle, and be brought to rest, it will remain in equilibrium.

---

* The ordinary process for finding experimentally the centre of gravity of a body fails when there is any magnetic action to interfere with the effects of gravitaton. It is, however, for our present purpose, sufficient to know that the centre of gravity exists; that is, that there is a point such that the vertical line of the resultant action of gravity passes through it, in whatever position the body be held. If it were of any consequence, a process somewhat complicated by the magnetic action, for actually determining, by experiment, the centre of gravity of a magnet might be indicated, and thus the experimental treatment of the subject in the text would be completed.

(3.) If the magnet be turned through 180°, about an axis perpendicular to this, it will again be in a position of equilibrium.

(4.) Any motion of the magnet whatever, which is not of either of the kinds just described, nor compounded of the two, will bring it into a position in which it will not be in equilibrium.

(5.) The directing couple experienced by the magnet in any position depends solely on the angle of inclination of the axis described in (*1*.) to the line along which it lies when the magnet is in equilibrium; being independent of the position of the plane of this angle, and of the different positions into which the magnet is brought by turning it round that axis.

440. From these observations we draw the conclusion that a magnet always experiences a directing couple from the earth unless a certain axis belonging to it is placed in a determinate position. This line of the magnet is called its magnetic axis.*

441. The direction towards which the magnetic axis of the magnet tends in virtue of the earth's action, is called "the line of dip," or "the direction of the total terrestrial magnetic force," at the locality of the observation.

442. No further explanation regarding phenomena which depend on terrestrial magnetism is required in the present chapter; but, as the facts have been stated in part, it may be right to complete the statement, as far as regards the action experienced by a magnet of any kind when held in different positions in a given locality, by mentioning the following conclusions, deduced in a very obvious manner from the general laws of magnetic action stated below, and verified fully by experiment :—

If a magnet be held with its magnetic axis inclined at any angle to the line of dip, it will experience a couple, the moment of which is proportional to the sine of the angle of inclination, acting in a plane containing the magnetic axis and the line of dip. The position of equilibrium towards which this couple tends to bring the magnetic axis is stable, and if the direction of the magnetic axis be reversed, the magnet may be left balanced, but it will be in unstable equilibrium.

---

* Any line in the body parallel to this might, with as good reason, be called a magnetic axis, but when we conceive the magnet to be supported by its centre of gravity, the magnetic axis is naturally taken as a line through this point. [See addition to § 444.]

443. The directive tendency observed in magnetic bodies being found to depend on their geographical position, and to be related, in some degree, to the terrestrial poles, received the name of *polarity*, probably on account of a false hypothesis of forces exercised by the pole-star* or by the earth's poles upon certain points of the loadstone or needle, thence called the "poles of the magnet." The terms "polarity" and "poles" are still retained, but the use of them, which has very generally been made, is nearly as vague as the ideas from which they had their origin. Thus, when the magnet is an elongated mass, its ends are called poles if its magnetic axis be in the direction of its length; no definite points, such as those in which the surface of the body is cut by the magnetic axis, being precisely indicated by the term as it is generally used. If, however, the body be symmetrical about its magnetic axis, and symmetrically magnetized, whether elongated in that direction or not, the poles might be definitely the *ends of the magnetic axis* (or the points in which the surface is cut by it), unless the magnet be annular and not cut by its magnetic axis (a ring electro-magnet, for instance), in which case the ordinary conception of *poles* fails. Notwithstanding this vagueness, however, the terms poles and polarity are extremely convenient, and, with the following explanations, they will frequently be made use of in this paper:—

444. Let O be any point in a magnet, and let NOS be a straight line parallel to the line defined above as the magnetic axis through the centre of gravity. If the point O, however it has been chosen, be called the centre of the magnet, the line NS, terminated either at the surface, on each side, or in any arbitrary manner, is called the magnetic axis, and the ends N, S, of the magnetic axis are called the poles of the magnet.†

---

* In the poem of Guiot de Provence (quoted in Whewell's *History of the Inductive Sciences*, vol. ii. p. 46), a needle is described as being magnetized and placed in or on a straw (floating on water it is to be presumed)—

> "Puis se torne la pointe toute
> Contre l'estoile sans doute."

† A definition of poles at variance with this is adopted in some special cases, especially in that of the earth considered as a great magnet, but the manner in which the term will be used in this paper will be such as to produce no confusion on this account.

[*Addition*, 1871.—Later, § 494, a proper central axis, to be called *the* magnetic axis, and a point in it which may be called the magnetic centre, will be defined according to purely magnetic conditions.]

445. That pole (marked N) which points, on the whole, from the north, and, in northern latitudes, upwards, is called the north pole, and the other (S), which points from the south, is called the south pole.

446. The *sides* of the body towards its north pole and south pole are said to possess "northern polarity" and "southern polarity" respectively, an expression obviously founded on the idea that the surface of a magnet may in general be contemplated as a locus of poles.

447. If a magnetic body be broken up into any number of fragments, each morsel is found to be a complete magnet, presenting in itself all the phenomena of poles and polarity. This property is generally contemplated when, in modern writings on physical subjects, polarity is mentioned as a property belonging to a solid body ; and a corresponding idea is involved in the term when it is applied with reference to the electric state which Mr Faraday discovered to be induced in non-conductors of electricity ("dielectrics") when subjected to the influence of electrified bodies.* However different are the physical circumstances of magnetic and electric polarity, it appears that the positive laws of the phenomena are the same,† and therefore the mathematical theories are identical. Either subject might be taken as an example of a very important branch of physical mathematics, which might be called "A Mathematical Theory of Polar Forces."

448. Although we have seen that any magnet, in general, experiences from the earth an action subject to certain very simple laws, yet the actual distribution of the magnetism which it possesses may be extremely irregular. We may certainly conceive that if the magnetized substance be a regular crystal of magnetic iron ore, the magnetism is distri-

---

* Faraday's *Experimental Researches in Electricity*, Eleventh Series.
† See a paper "On the Elementary Laws of Statical Electricity," published in the *Cambridge and Dublin Mathematical Journal* (vol. i.) in December 1845.

buted through it according to some simple law; but by taking an amorphous and heterogeneous fragment of ore presenting magnetic properties, by magnetizing in any way an irregular mass of steel, by connecting any number of morsels of magnetic matter so as to make up a complex magnet, or by bending a galvanic wire into any form, we may obtain magnets in which the magnetic property is distributed in any arbitrary manner, however irregular. Excluding for the present the last-mentioned case, let us endeavour to form a conception of the distribution of magnetism in actually magnetized matter, such as steel or loadstone, and to lay down the principles according to which it may in any instance be mathematically expressed.

449. In general we may consider a magnet as composed of matter which is magnetized throughout, since, in general, it is found that any fragment cut out of a magnetic mass is itself a magnet possessing properties entirely similar to those which have been described as possessed by any magnet whatever. It may be, however, that a small portion cut out of a certain position in a magnet, may present no magnetic phenomena; and if we cut equal and similar portions from different positions, we may find them to possess magnetic properties differing to any extent both in intensity and in the directions of their magnetic axes.

450. If we find that equal and similar portions, cut in parallel directions, from any different positions in a given magnetic mass, possess equal and similar magnetic properties, the mass is said to be uniformly magnetized.

451. In general, however, the intensity of magnetization must be supposed to vary from one part to another, and the magnetic axes of the different parts to be not parallel to one another. Hence, to lay down determinately a specification of the distribution of magnetism through a magnet of any kind, we must be able to express the *intensity* and the *direction* of magnetization at each point. Before attempting to define a standard for the numerical expression of intensity of magnetization, it will be convenient to examine the elementary laws upon which the phenomena of magnetic force depend, since it is by these effects that the nature and energy of the *magnetism* to which they are due must be estimated.

CHAPTER II.— *On the Laws of Magnetic Force, and on the Distribution of Magnetism in Magnetized Matter*

452. The object of the elementary magnetic researches of Coulomb was the determination of the mutual action between two infinitely thin, uniformly and longitudinally magnetized bars. The magnets which he used were in strictness neither uniformly nor longitudinally magnetized, such a state being unattainable by any actual process of magnetization; but, as the bars were very thin cylindrical steel wires, and were symmetrically magnetized, the resultant actions were sensibly the same as if they were in reality infinitely thin, and longitudinally magnetized; and from experiments which he made, it appears that the intensity of the magnetization must have been very nearly constant from the middle of each of the bars to within a short distance from either end, where a gradual decrease of intensity is sensible*.

453. These circumstances having been attended to, Coulomb was able to deduce from his experiments the true laws of the phenomena, and arrived at the following conclusions :—

(1) If two thin uniformly and longitudinally magnetized bars be held near one another, an action is exerted between them. which consists of four distinct forces, along the four lines joining their extremities.

(2) The forces between like ends of the two bars are repulsive†.

(3) The forces between unlike ends are attractive.

(4) If the bars be held so that the four distances between their extremities, two and two, are equal, the four forces between them will be equal.

(5) If the relative positions of the bars be altered, each force will vary inversely as the square of the mutual distance of the poles between which it acts.

---

* See note on § 469, below.

† Hence we see the propriety of the terms *north* and *south* applied to the opposite polarities of a magnet, as explained above. Thus we designate the polarity, or the imaginary magnetic matter of the northern and southern magnetic hemispheres of the earth, as northern and southern respectively; and since the poles of ordinary magnets which are repelled by the earth's northern or southern polarity must be *similar*, these also are called northern or southern, as the case may be.

454. To establish a standard for estimating the *strength* of a magnet, let us conceive two infinitely thin bars to be placed so that either end of one may be at unit of distance from an end of the other. Then, if the bars be equally magnetized, each uniformly and longitudinally, to such a degree that the force between those ends shall be unity, the strength of each bar-magnet is unity*.

455. If any number, m, of such unit bars, of equal length, be put with like ends together, so as to constitute a single complex bar, the strength of the magnet so formed is denoted by m.

If there be any number of thin bar-magnets of equal length, and each of them of such a strength that q of them, with like ends together, would constitute a unit-bar; and if p of those bars be put with like ends together, the strength of the complex magnet so formed will be $\frac{p}{q}$.

456. If a single infinitely thin bar be magnetized to such a degree that in the same positions it would produce the same effects as a complex bar of any strength m (an integer or fraction), the strength of this magnet is denoted by m.

457. If two complex bar-magnets, of the kind described above, be put near one another, each bar of one will act on each bar of the other with the same forces as if all the other bars were removed. Hence, if the distance between the two poles be unity, and if the strengths of the bars be respectively m and m' (whether these numbers be integral or fractional), the force between those poles will be mm'. If, now, the relative position of the magnets be altered, so that the distance between two poles may be f, the force between them will, according to Coulomb's law, be

$$\frac{mm'}{f^2}.$$

---

* The Royal Society, in its Instructions for making observations on Terrestrial Magnetism, adopts one foot as the unit of length; and that force which, if acting on a grain of matter, would in one second of time generate one foot per second of velocity, as the unit of force; which is consequently very nearly $\frac{1}{32.2}$ of the weight, in any part of Great Britain or Ireland, of one grain. [*Note*, 1871. — The British Association's Committee on Electric Measurement have recently adopted the centimetre as unit of length, and the gramme as unit of mass, instead of the foot and grain.]

According to the definition given above of the strength of a simple bar-magnet, it follows that the same expression gives the force between two poles of any thin uniformly and longitudinally magnetized bars, of strengths m and m'.

458. The *magnetic moment* of an infinitely thin, uniformly and longitudinally magnetized bar, is the product of its length into its strength.

459. If any number of equally strong, uniformly and longitudinally magnetized rectangular bars of equal infinitely small sections, be put together with like ends towards the same parts, a complex uniformly magnetized solid of any form may be produced. The *magnetic moment* of such a magnet is equal to the sum of the magnetic moments of the bars of which it is composed.

460. The magnetic moment of any continuous solid, uniformly magnetized in parallel lines, is equal to the sum of the magnetic moments of all the thin uniformly and longitudinally magnetized bars into which it may be divided.

It follows that the magnetic moment of any part of a uniformly magnetized mass is proportional to its volume.

461. The *intensity of magnetization* of a uniformly magnetized solid is the magnetic moment of a unit of its volume.

It follows that the magnetic moment of a uniformly magnetized solid, of any form and dimensions, is equal to the product of its volume into the intensity of its magnetization.

462. If a body be magnetized in any arbitrary regular or irregular manner, a portion may be taken in any position, so small in all its dimensions that the distribution of magnetism through it will be sensibly uniform. The quotient obtained by dividing the magnetic moment of such a portion, in any position P, by its volume, is the *intensity of magnetization* of the substance at the point P; and a line through P parallel to its lines of magnetization, is the *direction of magnetization*, at P.

CHAPTER III.—*On the Imaginary Magnetic Matter by means of which the Polarity of a Magnetized Body may be represented.*

463. It will very often be convenient to refer the phenomena of magnetic force to attractions or repulsions mutually exerted

between portions of an imaginary magnetic matter, which, as we shall see, may be conceived to represent the polarity of a magnet of any kind. This imaginary substance possesses none of the primary qualities of ordinary matter, and it would be wrong to call it either a solid, or the "magnetic fluid" or "fluids"; but, without making any hypothesis whatever, we may call it "magnetic matter," on the understanding that it possesses only the property of attracting or repelling magnets, or other portions of "matter" of its own kind, according to certain determinate laws, which may be stated as follows :—

(1) There are two kinds of imaginary magnetic matter, northern and southern, to represent respectively the northern and southern magnetic polarities of the earth, or the similar polarities of any magnet whatever.

(2) Like portions of magnetic matter repel, and unlike portions attract, mutually.

(3) Any two small portions of magnetic matter exert a mutual force which varies inversely as the square of the distance between them.

(4) Two units of magnetic matter, at a unit of distance from one another, exert a unit of force, mutually.

464. If quantities of magnetic matter be measured numerically in such units, and if the positive or negative sign be prefixed to denote the *species* of matter, whether *northern* (which, by convention, we may call *positive*) or *southern*, all the preceding laws are expressed in the following proposition :—

If quantities m *and* m', *of magnetic matter be concentrated respectively at points at a distance,* f, *from one another, they will repel with a force algebraically equal to*

$$\frac{mm'}{f^2}.$$

465. It appears from the explanations given above that the circumstances of a uniformly magnetized needle may be represented if we imagine equal quantities of northern and southern magnetic matter to be concentrated at its two poles, the numerical measure of these equal quantities being the same as that of the "strength" of the magnet.

The mutual action between two needles would thus be

reduced to forces of attraction and repulsion between the portions of magnetic matter by which their poles are represented.

466. Any magnetic mass whatever may, as we have seen, be regarded as composed of infinitely small bar-magnets put together in such a way as to produce the distribution of magnetism which it actually possesses; and hence, by substituting imaginary magnetic matter for the poles of these magnets, we obtain a distribution of equal quantities of northern and southern magnetic matter through the magnetized substance, by which its actual magnetic condition may be represented. The distribution of this matter becomes very much simplified, from the circumstance that we have in general unlike poles of the elementary magnets in contact, by which the opposite kinds of magnetic matter are partially (or in a class of cases *wholly**) destroyed through the interior of the body. The determination of the resulting distribution of magnetic matter, which represents in the simplest possible manner the polarity of any given magnet, is of much interest, and even importance, in the theory of magnetism, and we may therefore make this an object of investigation, before going further.

467. Let it be required to find the distribution of imaginary magnetic matter to represent the polarity of any number of uniformly magnetized needles, $S_1 N_1$, $S_2 N_2$,...$S_n N_n$, of strengths $\mu_1, \mu_2, ... \mu_n$ respectively, when they are placed together, end to end (not necessarily in the same straight line).

If A denote the position occupied by S_1 when the bars are in their places; if N_1 and S_2 are placed in contact at K_1; N_2 and S_3, at K_2; and so on until we have the last magnet, with its end S_n, in contact with N_{n-1}, at K_{n-1}, and its other end, N_n, free, at a point B; we shall have to imagine

μ_1 units of southern magnetic matter to be placed at A;

μ_1 units of northern, and μ_2 units of southern matter at K_1;

μ_2 units of northern, and μ_3 of southern matter at K_2;

. . . .

μ_{n-1} units of northern, and μ_n of southern matter at K_{n-1};
and lastly,

μ_n units of northern matter at B.

---

* In all cases when the distribution is "solenoidal." See below, Chap. v. § 499; communicated to the Royal Society, June 20, 1850.

Hence the final distribution of magnetic matter is as follows:—

$$-\mu_1 \quad\ldots\ldots\ldots\ldots \text{ at } A$$
$$\mu_1 - \mu_2 \quad\ldots\ldots\ldots\ldots K_1$$
$$\mu_2 - \mu_3 \quad\ldots\ldots\ldots\ldots K_2$$
$$\cdots\cdots\cdots\cdots\cdots\cdots$$
$$\cdots\cdots\cdots\cdots\cdots\cdots$$
$$\mu_{n-1} - \mu_n \ldots\ldots\ldots\ldots K_{n-1}$$

and $\qquad \mu_n \quad\ldots\ldots\ldots\ldots B.$

468. The complex magnet $AK_1K_2\ldots K_{n-1}B$ consists of a number of parts, each of which is uniformly and longitudinally magnetized, and it will act in the same way as a simple bar of the same length, similarly magnetized; and hence the magnetic matter which represents a bar-magnet AB of this kind is concentrated in a series of points, at the ends of the whole bar, and at all the places where there is a variation in the strength* of its magnetization.

469. If the length of each part through which the strength of the magnetism is constant, be diminished without limit, and if the entire number of the parts be increased indefinitely, a straight or curved infinitely thin bar may be conceived to be produced, which shall possess a distribution of longitudinal magnetism varying continuously from one end to the other according to any arbitrary law. If the strength of the magnetism at any point P of this bar be denoted by μ, and if $[\mu]$ and (μ) denote the values of μ at the points A and B, the investigation of § 467, with the elementary principles and notation of the differential calculus, leads at once to the determination of the ultimate distribution of magnetic matter by which such a bar-magnet may be represented. Thus if AP be denoted by s; μ will be a function of s, which may be supposed to be known, and its differential coefficient will express the continuous distribution of magnetic matter which replaces the group of material points at K_1, K_2, etc.; so that the entire distribution of polarity in the bar and at its ends will be as follows:—in

---

* This expression is equivalent to *the product of the intensity of magnetization into the section of the bar;* and by retaining it we are enabled to include cases in which the bar is not of uniform section.

any infinitely small length, σ, of the bar, a quantity of matter equal to

$$-\frac{d\mu}{ds}\ \sigma,$$

and, besides, terminal accumulations, of quantities

$$-\ [\mu]\ \text{at}\ A,$$

and $\qquad\qquad (\mu)\ \text{at}\ B.$

It follows that if, through any part of the length of a bar, the strength of the magnetism is constant, there will be no magnetic matter to be distributed through this portion of the magnet; but if the strength of the magnetism varies, then, according as it diminishes or increases from the north to the south pole of any small portion, there will be a distribution of northern or southern magnetic matter to represent the polarity which results from this variation.

Corresponding inferences may be made conversely, with reference to the distribution of magnetism, when the distribution of the imaginary magnetic matter is known. Thus Coulomb found that his long thin cylindrical bar-magnets acted upon one another as if each had a symmetrical distribution of the two kinds of magnetic matter, northern within a limited space from one end, and southern within a limited space from the other, the intermediate space (constituting generally the greater part of the bar) being unoccupied; from which we infer that no variation in the magnetism was sensible through the middle part of the bar, but that, through a limited space on each side, the intensity of the magnetization must have decreased gradually towards the ends*.

---

* This circumstance was alluded to above, in § 452. Interesting views on the subject of the distribution of magnetism in bar-magnets are obtained by taking arbitrary examples to illustrate the investigation of the text. Thus we may either consider a uniform bar variably magnetized, or a thin bar of varying thickness, cut from a uniformly magnetized substance; and, according to the arbitrary data assumed, various remarkable results may be obtained. We shall see afterwards that any such data, however arbitrary, may be actually produced in electro-magnets, and we have therefore the means of illustrating the subject experimentally, in as complete a manner as can be conceived, although from the practical *non-rigidity* of the magnetism of magnetized substances, ordinary steel or loadstone magnets would not afford such satisfactory illustrations of *arbitrary* cases as might be desired. The distribution of longitudinal magnetism in steel needles actually magnetized in different ways, and especially " magnetized to saturation," has been the

470.   The distribution of magnetic matter which represents the polarity of a uniformly magnetized body of any form, may be immediately determined if we imagine it divided into infinitely thin bars, in the directions of its lines of magnetization; for each of these bars will be uniformly and longitudinally magnetized, and therefore there will be no distribution of matter except at their ends.   Now the bars are all terminated on each side by the surface of the body, and consequently the whole magnetic effect is represented by a certain superficial distribution of northern and southern magnetic matter.   It only remains to determine the actual form of this distribution; but, for the sake of simplicity in expression, it will be convenient to state previously the following definition, borrowed from Coulomb's writings on electricity :—

471.   If any kind of matter be distributed over a surface, the *superficial density* at any point is the quotient obtained by dividing the quantity of matter on an infinitely small element of the surface in the neighbourhood of that point, by the area of the element.

472.   To determine the superficial density at any point in the case at present under consideration, let ω be the area of the perpendicular section of an infinitely thin uniform bar of the solid, with one end at that point.   Then, if i be the intensity of magnetization of the solid, $i\omega$ will be, as may be readily shown, the "strength" of the bar-magnet.   Hence at the two ends of the bar we must suppose to be placed quantities of northern and southern imaginary magnetic matter each equal to $i\omega$.   In the distribution over the surface of the given magnet, these quantities of matter must be imagined to be spread over the oblique ends of the bar.   Now if θ denote the inclination of the bar to a normal to the surface through one end, the area of that end will be $\dfrac{\omega}{\cos \theta}$, and therefore in that part of the surface we have a quantity of matter equal to $i\omega$ spread over an area $\dfrac{\omega}{\cos \theta}$.   Hence the superficial density is

$$i \cos \theta.$$

object of interesting experimental and theoretical investigations by Coulomb, Biot, Green, and Riess.

This expression gives the superficial density at any point, P, of the surface, and its algebraic sign indicates the kind of matter, provided the angle denoted by θ be taken between the external part of the normal, and a line drawn from P in the same direction as that of the motion of a point carried from the south pole to the north pole, of a portion close to P, of the infinitely thin bar-magnet which we have been considering.

473. Let it be required, in the last place, to determine the entire distribution of magnetic matter necessary to represent the polarity of any given magnet.

We may conceive the whole magnetized mass to be divided into infinitely small parallelepipeds by planes parallel to three planes of rectangular co-ordinates. Let α, β, γ denote the three edges of one of these parallelepipeds having its centre at a point P (x, y, z). Let i denote the given intensity, and l, m, n the given direction cosines of the magnetization at P. It will follow from the preceding investigation that the polarity of this infinitely small uniformly magnetized parallelepiped may be represented by imaginary magnetic matter distributed over its six faces in such a manner that the density will be uniform over each face, and that the quantities of matter on the six faces will be as follows :—

$-il.\beta\gamma$, and $il.\beta\gamma$; on the two faces parallel to YOZ ;

$-im.\gamma\alpha$, and $im.\gamma\alpha$; on the two faces parallel to ZOX ;

$-in.\alpha\beta$, and $in.\alpha\beta$; on the two faces parallel to XOY.

Now if we consider adjacent parallelepipeds of equal dimensions, touching the six faces of the one we have been considering, we should find from each of them a second distribution of magnetic matter, to be placed upon that one of those six faces which it touches. Thus if we consider the first face $\beta\gamma$, or that of which the distance from YOZ is $x - \frac{1}{2}\alpha$; we shall have a second distribution upon it derived from a parallelepiped, the co-ordinates of the centre of which are $x - \alpha, y, z$; and the quantity of matter in this second distribution will be

$$\left\{ il + \frac{d\,(il)}{dx}\,(-\alpha) \right\} \beta\gamma.$$

This, added to that which was found above, gives

$$\frac{d\,(il)}{dx}(-\alpha)\,.\,\beta\gamma, \text{ or } -\frac{d\,(il)}{dx}\,.\,\alpha\beta\gamma$$

for the total amount of matter upon this face. Again, the quantity in the second distribution on the other face, $\beta\gamma$, is equal to

$$-\left\{il+\frac{d\,(il)}{dx}\,.\,\alpha\right\}\beta\gamma,$$

and therefore the total amount of matter on this face will be

$$-\frac{d\,(il)}{dx}\,.\,\alpha\beta\gamma.$$

By determining in a similar way the final quantities of matter on the other faces of the parallelepiped, we find that the total amount of matter to be distributed over its surface is

$$-2\left\{\frac{d\,(il)}{dx}+\frac{d\,(im)}{dy}+\frac{d\,(in)}{dz}\right\}\alpha\beta\gamma.$$

Now as the parallelepipeds into which we imagined the whole mass divided are infinitely small, we may substitute a continuous distribution of matter through them, in place of the superficial distributions on their faces which have been determined; and in making this substitution, the quantity of matter which we must suppose to be spread through the interior of any one of them must be half the total quantity on its surface, since each of its faces is common to it and another parallelepiped. Hence the quantity of matter to be distributed through the parallelepiped $\alpha\beta\gamma$ is equal to

$$-\left\{\frac{d\,(il)}{dx}+\frac{d\,(im)}{dy}+\frac{d\,(in)}{dz}\right\}\alpha\beta\gamma.$$

Besides this continuous distribution through the interior of the magnet, there must be a superficial distribution to represent the un-neutralized polarity at its surface. If ρ denote the density of this distribution at any point; $[l]$, $[m]$, $[n]$ the direction-cosines, and $[i]$ the intensity of the magnetization of the solid close to it; and λ, μ, ν the direction-cosines of a normal to the surface, we shall have, as in the case of the uniformly magnetized solid previously considered,

$$\rho = [i]\cos\theta = [il]\,.\,\lambda + [im]\,.\,\mu + [in]\,.\,\nu \;\;........(1).$$

If according to the usual definition of " density," k denote the

density of the magnetic matter at P, in the continuous distribution through the interior, the expression found above for the
quantity of matter in the element α, β, γ, leads to the formula

$$k = - \left\{ \frac{d\,(il)}{dx} + \frac{d\,(im)}{dy} + \frac{d\,(in)}{dz} \right\} \dots\dots\dots (2).$$

These two equations express respectively the superficial distribution, and the continuous distribution through the solid, of
the magnetic matter which entirely represents the polarity of
the given magnet. The fact that the quantity of northern
matter is equal to the quantity of southern in the entire distribution, is readily verified by showing from these formulæ, as
may readily be done by integration, that the total quantity of
matter is algebraically equal to nothing.

474. If there be an abrupt change in the intensity or direction of the magnetization from one part of the magnetized substance to another, a slight modification in the formulæ given
above will be convenient. Thus we may take a case differing
very little from a given case, but which, instead of presenting
finite differences in the intensity or direction of magnetization
on the two sides of any surface in the substance of the magnet,
has merely very sudden continuous changes in the values of
those elements: we may conceive the distribution to be made
more and more nearly the same as the given distribution, with
its abrupt transitions, and we may determine the limit towards
which the value of the expression (2) approximates, and thus,
although according to the ordinary rules of the differential
calculus this formula fails in the limiting case, we may still
derive the true result from it. It is very easily shown in this
way, that, besides the continuous distribution given by the
expression (2) applied to all points of the substance for which
it does not fail, there will be a superficial distribution of magnetic matter on any surface of discontinuity; and that the
density of this superficial distribution will be the difference
between the products of the intensity of magnetization into
the cosine of the inclination of its direction to the normal, on
the two sides of the surface.

475. This result, obtained by the interpretation of formula
(2) in the extreme case, might have been obtained directly
from the original investigation, by taking into account the

abrupt variation of the magnetization at the surface of discontinuity, as (§ 472) we did the abrupt termination of the magnetized substance at the boundary of the magnet, and representing the un-neutralized polarity which results, by a superficial distribution of magnetic matter.

CHAPTER IV.—*Determination of the Mutual Actions between any Given Portions of Magnetized Matter.*

476.   The synthetical part of the theory of magnetism has for its ultimate object the determination of the total action between two magnets, when the distribution of magnetism in each is given.   The principles according to which the data of such a problem may be specified have been already laid down (§§ 459...62), and we have seen that, with sufficient data in any case, Coulomb's laws of magnetic force are sufficient to enable us to apply ordinary statical principles to the solution of the problem.   Hence the elements of this part of the theory may be regarded as complete, and we may proceed to the mathematical treatment of the subject.

477.   The investigations of the preceding chapter, which show us how we may conventionally represent any given magnet, in its agency upon other bodies, by an imaginary magnetic matter distributed on its surface and through its interior; enable us to reduce the problem of finding the action between any two magnets, to the known problem of determining the resultant of the attractions or repulsions exerted between the particles of two groups of matter, according to the law of force which is met with so universally in natural phenomena.   The direct formulæ applicable for this object are so readily obtained by means of the elementary principles of statics, and so well known, that it is unnecessary to cite them here, and we may regard equations (1) and (2) of the preceding chapter (§ 473) as sufficient for indicating the manner in which the details of the problem may be worked out in any particular case.   The expression for the "potential," and other formulæ of importance in Laplace's method of treating this subject, are given below (§ 482), as derived from the results expressed in equations (1) and (2).

478. The preceding solution of the problem, although extremely simple and often convenient, must be regarded as very artificial, since in it the resultant action is found by the composition of mutual actions between the particles of an imaginary magnetic matter, which are not the same as the real mutual actions between the different parts of the magnets themselves, although the resultant action between the entire groups of matter is necessarily the same as the real resultant action between the entire magnets. Hence it is very desirable to investigate another solution, of a less artificial form, in which the required resultant action may be obtained by compounding the real actions between the different parts into which we may conceive the magnets to be divided. The remainder of the chapter, after some preliminary explanations and definitions, will be devoted to this object.

479. The "resultant magnetic force at any point" is an expression which will very frequently be employed in what follows, and it is therefore of importance that its signification should be clearly defined. For this purpose, let us consider separately the cases of an external point in the neighbourhood of a magnet, and a point in space which is actually occupied by magnetic matter.

(1) The resultant force at a point in space, void of magnetized matter, is the force that the north pole of a unit-bar (or a positive unit of imaginary magnetic matter), if placed at this point, would experience.

(2) The resultant force at a point situated in space occupied by magnetized matter, is an expression the signification of which is somewhat arbitrary. If we conceive the magnetic substance to be removed from an infinitely small space round the point, the preceding definition would be applicable; since, if we imagine a very small bar-magnet to be placed in a definite position in this space, the force upon either end would be determinate. The circumstances of this case are made clear by considering the distribution of imaginary magnetic matter required to represent the given magnet, without the small portion we have conceived to be removed from its interior; which will differ from the distribution that represents the entire given magnet, in wanting the small portion of the

continuous interior distribution corresponding to the removed portion, and in having instead a superficial distribution on the small internal surface bounding the hollow space. If we consider the portion removed to be infinitely small, the want of the small portion of the *solid* magnetic [imaginary] matter will produce no finite effect upon any point; but the superficial distribution at the boundary of the hollow space will produce a finite force upon any magnetic point within it. Hence the resultant force upon the given point round which the space was conceived to be hollowed, may be regarded as compounded of two forces, one due to the polarity of the complete magnet, and the other to the superficial polarity left free by the removal of the magnetized substance*. The former component is the force meant by the expression "the resultant force at a point within a magnetic substance," when employed in the present paper†.

480. The conventional language and ideas with reference to the imaginary magnetic matter, explained above (§§ 463...75), enable us to give the following simple statement of the definition, including both the cases which we have been considering.

---

* If the portion removed be spherical and infinitely small, it may be proved that the force at any point within it, resulting from the free polarity of the solid at the surface bounding the hollow space, is in the direction of the lines of magnetization of the substance round it, and is equal to $\frac{4\pi i}{3}$.

This theorem (due to Poisson) will be demonstrated at the commencement of the Theory of Magnetic Induction, because we shall have to consider the "magnetizing force" upon any small portion of an inductively magnetized substance as the actual resultant force that would exist within the hollow space that would be left if the portion considered were removed, and the magnetism of the remainder constrained to remain unaltered.

† If we imagine a magnet to be divided into two parts by any plane passing through the line of magnetization at any internal point, *P*, and if we imagine the two parts to be separated by an infinitely small interval, and a unit north pole to be placed between them at *P*, the force which this pole would experience is "the resultant force at a point, *P*, of the magnetic substance." This is the most direct definition of the expression that could have been given, and it agrees with the definition I have actually adopted; but I have preferred the explanation and statement in the text, as being practically more simple, and more directly connected with the various investigations in which the expression will be employed.

[*Note added June* 15, 1850.—Some subsequent investigations on the comparison of common magnets and electro-magnets have altered my opinion, that the definition in the text is to be preferred; and I now believe the definition in the note to present the subject in the simplest possible manner, and in that which, for the applications to be made in the continuation of this Essay, is most convenient on the whole.]

The resultant magnetic force at any point, whether in the neighbourhood of a magnet or in its interior, is the force that a unit of northern magnetic matter would experience if it were placed at that point, and if all the magnetized substance were replaced by the corresponding distribution of imaginary magnetic matter.

481. The determination of the resultant force at any point is, as we shall see, much facilitated by means of a method first introduced by Laplace in the mathematical treatment of the theory of attraction, and developed to a very remarkable extent by Green in his " Essay on the Application of Mathematical Analysis to the Theories of Electricity and Magnetism" (Nottingham, 1828), and in his other writings on the same and on allied subjects in the *Cambridge Philosophical Transactions,* and in the *Transactions of the Royal Society of Edinburgh.* Laplace's fundamental theorem is so well known that it is unnecessary to demonstrate it here ; but for the sake of reference, the following enunciation of it is given. The term " potential," defined in connexion with it, was first introduced by Green in his Essay (1828). It was at a later date introduced independently by Gauss, and is now in very general use.

Theorem (Laplace).—The resultant force produced by a body, or a group of attracting or repelling particles, upon a unit particle placed at any point P, is such that the difference between the values of a certain function, at any two points p and p' infinitely near P, divided by the distance pp', is equal to its component in the direction of the line joining p and p'.

Definition (Green).—This function, which, for a given mass, has a determinate value at any point P, of space, is called the potential of the mass, at the point P.

It follows from Laplace's general demonstration, that, when the law of force is that of the inverse square of the distance, the potential is found by dividing the quantity of matter in any infinitely small part of the mass, by its distance from P, and adding all the quotients so obtained.

482. The same demonstration is applicable to prove, in virtue of Coulomb's fundamental laws of magnetic force, the same theorem with reference to any kind of magnet that can

be conceived to be composed of uniformly magnetized bars, either finite or infinitely small, put together in any way, that is, of any magnet other than an electro-magnet; and the investigation, in the preceding chapter, of the resulting distribution of magnetic matter that may be imagined as representing in the simplest possible way the polarity of such a magnet, enables us to determine at once, from equations (1) and (2) of § 473, its potential at any point. Thus if V denote the potential at a point P, whose co-ordinates are ξ, η, ζ, and if dS denote an element of the surface of the magnet, situated at a point whose co-ordinates are $[x]$, $[y]$, $[z]$, we have, by the proposition enunciated at the end of § 480,—

$$V=\iint\frac{[il]\lambda+[im]\mu+[in]\nu}{[\Delta]}\,dS-\iiint\frac{\dfrac{d\,(il)}{dx}+\dfrac{d\,(im)}{dy}+\dfrac{d\,(in)}{dz}}{\Delta}\,dxdydz...(3),$$

where Δ and $[\Delta]$ are respectively the distances of the points x, y, z and $[x, y, z]$ from the point P, and are given by the equations
$$\Delta^2 = (\xi-x)^2 + (\eta-y)^2 + (\zeta-z)^2$$
$$[\Delta]^2 = (\xi-[x])^2 + (\eta-[y])^2 + (\zeta-[z])^2.$$
The double and triple integrals in the first and second terms of this expression are to be taken respectively over the whole surface bounding the magnet, and throughout the entire magnetized substance. Since, as is easily shown, the value of that portion of the triple integral in the second member which corresponds to an infinitely small portion of the solid containing (ξ, η, ζ), when this point is internal, is infinitely small, it follows that the magnetic force at any internal point, as defined in § 479, is derivable from a potential expressed by equation (3).

483.   The expressions for the resultant force at any point, and its direction, may be immediately obtained when the potential function has been determined, by the rules of the differential calculus. Thus, if V has been determined in terms of the rectangular co-ordinates, ξ, η, ζ, of the point P, the three components, X, Y, Z, of the resultant force on this point will be given, in virtue of Laplace's fundamental theorem enunciated in § 481, by the formulæ,

$$X=-\frac{dV}{d\xi},\quad Y=-\frac{dV}{d\eta},\quad Z=-\frac{dV}{d\zeta}\dots\dots\dots(4),$$

where the negative signs are introduced, because the potential is estimated in such a way that it diminishes in the direction along which a north pole is urged. If we take the expression (3) for V, and actually differentiate with reference to ξ, η, ζ under the integral signs, we obtain expressions for X, Y, and Z which agree with the expressions that might have been obtained directly, by means of the first principles of statics (see § 477), and thus the theorem is verified. Such a verification, extended so as to be applicable to a body acting according to any law of force, constitutes virtually the ordinary demonstration of the theorem.

484. The formulæ of the preceding paragraphs are applicable to the determination of the potential and the resultant force, at any point, whether within the magnetized substance or not, according to the general definition of § 480. The case of a point in the magnetized substance, according to the conventional second definition of § 479, cannot present itself in problems with reference to the mutual action between two actual magnets. This case being therefore excluded, we may proceed to the investigations indicated in § 478.

485. In the method which is now to be followed, the magnetized substances considered must be conceived to be divided into an infinite number of infinitely small parts, and the actual magnetism of each part will be taken into account, whether in determining the potential of the magnet at a given external point, or in investigating the mutual action between two magnets. In the first place, let us determine the potential due to an infinitely small element of magnetized substance, and for this purpose we may commence by considering an infinitely thin, uniformly magnetized bar of finite length. If m denote the strength of the bar, and if N and S be its north and south poles respectively, its potential at any point, P will be according to §§ 465 and 481,
$$\frac{m}{NP} - \frac{m}{SP}.$$

Let Δ denote the distance of the point of bisection of the bar from P, and θ the angle between this line and the direction of the bar measured from its centre towards its north pole. Then, if a be the length of the bar, the expression for the potential becomes

$$m \left\{ \frac{1}{(\Delta^2 - a\Delta \cos\theta + \frac{1}{4}a^2)^{\frac{1}{2}}} - \frac{1}{(\Delta^2 + a\Delta \cos\theta + \frac{1}{4}a^2)^{\frac{1}{2}}} \right\}.$$

By expanding this in ascending powers of a, and neglecting all the terms after the first, we find for the potential of an infinitely small bar-magnet,

$$\frac{ma \cos\theta}{\Delta^2}$$

If now we suppose any number of such bar-magnets to be put together so as to constitute a mass magnetized in parallel lines, infinitely small in all its dimensions, the values of θ and Δ, and consequently the value of $\frac{\cos\theta}{\Delta^2}$, will be infinitely nearly the same for all of them, and the product of this into the sum of the values of ma for all the bar-magnets will express the potential of the entire mass. Hence, if the total magnetic moment be denoted by μ, the potential will be equal to

$$\frac{\mu \cos\theta}{\Delta^2}.$$

Now if we conceive the bars to have been arranged so as to constitute a uniformly magnetized mass, occupying a volume ϕ, we should have (§ 461), for the intensity of magnetization, $i = \frac{\mu}{\phi}$. Hence if ϕ denote the volume of an infinitely small element of uniformly magnetized matter, and i the intensity of its magnetization, the potential which it produces at any point P, at a finite distance from it, will be

$$\frac{i\phi \cdot \cos\theta}{\Delta^2},$$

where Δ denotes the distance of P from any point, E, within the element, and θ the angle between EP and a line drawn through E, in the direction of magnetization of the element, *towards the side of it which has northern polarity.*

486.　Let us now suppose the element E to be a part of a magnet of finite dimensions, of which it is required to determine the total potential at an external point, P. Let ξ, η, ζ be the co-ordinates of P, referred to a system of rectangular axes, and let x, y, z be those of E. We shall have

$$\Delta^2 = (\xi - x)^2 + (\eta - y)^2 + (\zeta - z)^2;$$

and, if l, m, n denote the direction-cosines of the magnetization

at E, $$\cos\theta = l\frac{\xi-x}{\Delta} + m\frac{\eta-y}{\Delta} + n\frac{\zeta-z}{\Delta}$$

Hence the expression for the potential of the element E becomes

$$\frac{i\phi\{l(\xi-x) + m(\eta-y) + n(\zeta-z)\}}{\{(\xi-x)^2 + (\eta-y)^2 + (\zeta-z)^2\}^{\frac{3}{2}}}.$$

Now the potential of a whole is equal to the sum of the potentials of all its parts, and hence, if we take $\phi = dx\,dy\,dz$, we have, by the integral calculus, the expression

$$V = \iiint \frac{il.(\xi-x) + im.(\eta-y) + in.(\zeta-z)}{\{(\xi-x)^2 + (\eta-y)^2 + (\zeta-z)^2\}^{\frac{3}{2}}} dx\,dy\,dz \ldots(5),$$

for the potential at the point P, due to the entire magnet.*

[§§ 487...494 *added September,* 1871.]

[487. The expansion of this in ascending powers of $\frac{\xi}{r}, \frac{\eta}{r}, \frac{\zeta}{r}$,

where $$r = \sqrt{(\xi^2 + \eta^2 + \zeta^2)},$$

is necessarily convergent for all space outside the least spherical surface with the origin of co-ordinates for centre, enclosing the whole magnet.   To find it, we have first to expand

$$\frac{il(\xi-x) + im(\eta-y) + in(\zeta-z)}{\{(\xi-x)^2 + (\eta-y)^2 + (\zeta-z)^2\}^{\frac{3}{2}}}$$

by Taylor's Theorem, in a series of ascending powers of x, y, z, which is necessarily convergent or divergent according as $\sqrt{(x^2 + y^2 + z^2)}$ is less or greater than $\sqrt{(\xi^2 + \eta^2 + \zeta^2)}$. Thus, for the part of V depending on il, we find

$$\Sigma\Sigma\Sigma(-1)^{s+t+u} \frac{\dfrac{d^s}{d\xi^s}\dfrac{d^t}{d\eta^t}\dfrac{d^u}{d\zeta^u}\dfrac{\xi}{r^3}}{1.2...s.1.2...t.1.2...u} \iiint il\,dx\,dy\,dz\,x^s y^t z^u \ldots(6),$$

where $\Sigma\Sigma\Sigma$ denotes summation from 0 to ∞ relatively to integers s, t, u. Hence, remarking that $\dfrac{\xi}{r^3} = -\dfrac{d}{d\xi}\dfrac{1}{r}$, and putting

---

* From the form of definition given in the second footnote on § 479, for the magnetic force at an internal point, it may be shown that the expression (5), as well as the expression (3), is applicable to the potential at any point, whether internal or external. The same thing may be shown by proving, as may easily be done, that the investigation of § 487 does not fail or become nugatory when (ξ, η, ζ) is included in the limits of integration.

$$\iiint \left(s\frac{il}{x} + t\frac{im}{y} + u\frac{in}{z} \right) x^s y^t z^u dx dy dz = [s,\, t,\, u] \;\ldots\ldots(7)$$

subject to the exception that terms of the first member involving x^{-1}, or y^{-1}, or z^{-1}, are to be omitted, we have

$$V = \Sigma\Sigma\Sigma\, (-1)^{s+t+u}\, [s,\, t,\, u]\, \frac{\dfrac{d^s}{d\xi^s}\dfrac{d^t}{d\eta^t}\dfrac{d^u}{d\zeta^u}\dfrac{1}{r}}{1.2\ldots s \,.\, 1.2\ldots t \,.\, 1.2\ldots u}\ldots(8).$$

Each term of this expansion is a solid harmonic function of $\xi,\, \eta,\, \zeta$ [Thomson and Tait's *Natural Philosophy*, App. B. (*b*), and (*g*), (14) (15) (21)].

488.. Neglecting all terms of higher orders than the second, and putting x, y, z for $\xi,\, \eta,\, \zeta$, we have, as an approximate expression for the potential at a very distant point $(x,\, y,\, z)$,

$$V = \frac{Lx + My + Nz}{(x^2 + y^2 + z^2)^{\frac{3}{2}}}$$
$$+ \frac{A\,(2x^2 - y^2 - z^2) + B\,(2y^2 - z^2 - x^2) + C\,(2z^2 - x^2 - y^2) + 3\,(ayz + bzx + cxy)}{(x^2 + y^2 + z^2)^{\frac{5}{2}}}\ldots(9),$$

where $L, M, N, A, B, C, a, b, c$ are constants (depending on the magnetism of the magnet, and the position relatively to it of the axis of co-ordinates) given by the equations

$$L = \iiint il\,dx dy dz, \quad M = \iiint im\,dx dy dz, \quad N = \iiint in\,dx dy dz \ldots\ldots\ldots\ldots\ldots\ldots(10),$$
$$A = \iiint il x\,dx dy dz, \quad B = \iiint im y\,dx dy dz, \quad C = \iiint in z\,dx dy dz$$
$$a = \iiint (imz + iny)\,dx dy dz, \quad b = \iiint (inx + ilz)\,dx dy dz, \quad c = \iiint (ily + imx)\,dx dy dz \left.\right\}(11).$$

489. If we put

$$K = \sqrt{(L^2 + M^2 + N^2)}\ldots\ldots\ldots\ldots(12),$$

and

$$\cos\theta = \frac{L}{K}\frac{x}{r} + \frac{M}{K}\frac{y}{r} + \frac{N}{K}\frac{z}{r}\ldots\ldots\ldots\ldots(13),$$

in the first term of (9) it becomes

$$\frac{K\cos\theta}{r^2}\ldots\ldots\ldots\ldots\ldots\ldots(14),$$

which is the first approximate expression for the potential of the magnet at a very distant point, and agrees with the rigorous expression (§ 485) for the potential of an infinitely small uniformly magnetized magnet at the origin of co-ordinates, having its magnetic moment equal to K, and its direction of magnetization specified by the direction-cosines

$$\frac{L}{K},\ \frac{M}{K},\ \frac{N}{K}\ldots\ldots\ldots\ldots\ldots(15).$$

Hence K, given by (12) and (10), is defined as *the magnetic moment* of the given magnet; and the direction (15) is readily proved to fulfil the condition stated in §§ 439, 440 as the definition of a magnetic axis, determinate in direction but (§ 444) left till now indeterminate as to its position in the magnet. It is to be remarked that the values of L, M, N given by (10) are independent of the position of the origin of co-ordinates, and depend only on the positions of the co-ordinate axes relatively to the magnet.

490. Let now the axes of co-ordinates be turned to bring one of the three into parallelism with the direction of the magnetic axis (15). Calling this OX, and using the same symbols, x, y, z, l, m, n, for co-ordinates and direction-cosines relatively to the new axes, we have, instead of (9) and (10),

$$V = \frac{Kx}{(x^2+y^2+z^2)^{\frac{3}{2}}}$$
$$+ \frac{A(2x^2 - y^2 - z^2) + B(2y^2 - z^2 - x^2) + C(2z^2 - x^2 - y^2) + 2(ayz + bzx + cxy)}{(x^2+y^2+z^2)^{\frac{5}{2}}} \quad ...(16),$$

$$K = \iiint il\,dxdydz\,;\quad \iiint im\,dxdydz = 0\,;\quad \iiint in\,dxdydz = 0 ...(17),$$
with equations (11) unchanged.

491. Secondly, let the axis of x be transferred from OX to the parallel line through any point for which

$$z = \frac{b}{K},\quad y = \frac{c}{K} ....(18).$$

The values of the integrals for the new axes corresponding to b and c are each zero, as is readily seen from (11) and (17). Hence, altering the notation y, z to correspond to the new axes, we have

$$V = \frac{Kx}{(x^2+y^2+z^2)^{\frac{3}{2}}} + \frac{A(2x^2 - y^2 - z^2) + B(2y^2 - z^2 - x^2) + C(2z^2 - x^2 - y^2) + 2ayz}{(x^2+y^2+z^2)^{\frac{5}{2}}} \quad (19),$$

with $\iiint (inx + ilz)\,dxdydz = 0\,;\quad \iiint (ily + imx)\,dxdydz = 0$ (20),

and (11) in other respects unchanged. Now for

$$2y^2 - z^2 - x^2, \text{ and } 2z^2 - x^2 - y^2,$$

we may write

$$-\tfrac{1}{2}(2x^2 - y^2 - z^2) + \tfrac{3}{2}(y^2 - z^2), \text{ and } -\tfrac{1}{2}(2x^2 - y^2 - z^2) - \tfrac{3}{2}(y^2 - z^2) \right\} (20),$$

a transformation which, simple as it is, has an important significance in "spherical harmonics."   Hence if we put

$$a = \tfrac{1}{2}\iiint(2ilx - imy - inz)\,dxdydz, \text{ and } \beta = \tfrac{3}{2}\iiint(imy - inz)\,dxdydz \ (21),$$

(19) becomes

$$V = \frac{Kx}{(x^2 + y^2 + z^2)^{\frac{3}{2}}} + \frac{\alpha\,(2x^2 - y^2 - z^2) + \beta\,(y^2 - z^2) + 2ayz}{(x^2 + y^2 + z^2)^{\frac{5}{2}}} \quad (22).$$

492. Thirdly, shift the origin from O to the point

$$x = \frac{\alpha}{K} \dots\dots\dots\dots\dots\dots\dots\dots(23)$$

in OX; that is to say, for x substitute $x + \dfrac{\alpha}{K}$. By (21) and (17) we have

$\iiint (2ilx - imy - inz)\,dxdydz = 0$; $\beta = \frac{3}{2}\iiint (imy - inz)\,dxdydz$ (24);

and (22) becomes

$$V = \frac{Kx}{(x^2 + y^2 + z^2)^{\frac{3}{2}}} + \frac{\beta\,(y^2 - z^2) + 2ayz}{(x^2 + y^2 + z^2)^{\frac{5}{2}}} \dots\dots\dots(25).$$

493. Lastly, turn the axes OY, OZ, round OX through an angle equal to

$$\tfrac{1}{2}\tan^{-1}\frac{\beta}{\alpha} \dots\dots\dots\dots\dots\dots\dots..(26).$$

Relatively to OX, OY, OZ in this final position we have (17) and (24) unchanged, and

$\iiint (imz + iny)\,dxdydz = 0$, $\iiint (inx + ilz)\,dxdydz = 0$, $\iiint (ily + imx)\,dxdydz = 0$ (27);

and (25) becomes reduced to

$$V = \frac{Kx}{(x^2 + y^2 + z^2)^{\frac{3}{2}}} + \frac{(\alpha^2 + \beta^2)^{\frac{1}{2}}\,(y^2 - z^2)}{(x^2 + y^2 + z^2)^{\frac{5}{2}}} \dots\dots\dots(28).$$

494. This is the simplest expression to the second degree of approximation for the distant potential of a magnet having any irregular distribution of magnetism. The axis determined by § 489 (15) and § 491 (18) is *the magnetic axis*, and the point in it determined by § 492 (23) is *the magnetic centre*, of which definitions were promised in the addition to § 444.]

495. The expression (5) of § 486 is susceptible of a very remarkable modification, by integration by parts. Thus we may divide the second member into three terms, of which the following is one:

$$\iiint \frac{il\,.\,(\xi - x)\,dx}{\{(\xi - x)^2 + (\eta - y)^2 + (\zeta - z)^2\}^{\frac{3}{2}}}\,dydz.$$

Integrating here by parts, with reference to x, we obtain

$$\left[\iint \frac{il\,.\,dydz}{\Delta}\right] - \iiint \frac{\dfrac{d\,(il)}{dx}}{\Delta}\,dxdydz,$$

where the brackets enclosing the double integral denote that the variables in it must belong to some point of the surface. If λ, μ, ν denote the direction-cosines of a normal to the surface at any point $[\xi, \eta, \zeta]$, and dS an element of the surface, we may take $dy\,dz = \lambda \,.\, dS$, and hence the double integral is reduced to

$$\iint \frac{[il]\,\lambda\,.\,dS}{[\Delta]}\,;$$

and, as we readily see by tracing the limits of the first integral with reference to x, for all possible values of y and z this double integral must be extended over the entire surface of the magnet. By treating in a similar manner the other two terms of the preceding expression for V, we obtain, finally,

$$V = \iint \frac{[il]\lambda + [im]\mu + [in]\nu}{[\Delta]}\,dS - \iiint \frac{\dfrac{d\,(il)}{dx} + \dfrac{d\,(im)}{dy} + \dfrac{d\,(in)}{dz}}{\Delta}\,dx\,dy\,dz.$$

The second member of this equation is the expression for the potential of a certain complex distribution of matter, consisting of a superficial distribution and a continuous internal distribution. The superficial density of the distribution on the surface, and the density of the continuous distribution at any internal point, are expressed respectively by $[il]\,\lambda + [im]\,\mu + [in]\,\nu$, and

$$-\left\{\frac{d\,(il)}{dx} + \frac{d\,(im)}{dy} + \frac{d\,(in)}{dz}\right\}.$$

Hence we infer that the action of the complete magnet upon any external point is the same as would be produced by a certain distribution of imaginary magnetic matter, determinable by means of these expressions, when the actual distribution of magnetism in the magnet is given.* The demonstration of the same theorem, given above (§ 473), illustrates in a very interesting manner the process of integration by parts applied to a triple integral.

496. The mutual action of any two magnets, considered as the resultant of the mutual actions between the infinitely small elements into which we may conceive them to be divided, consists of a force and a couple of which the components will be expressed by means of six triple integrals. Simpler expres-

---

* This very remarkable theorem is due to Poisson, and the demonstration, as it has been just given in the text, is to be found in his first memoir on Magnetism. The demonstration which I have given in § 473 may be regarded as exhibiting, by the theory of polarity, the physical principles expressed in the analytical formulæ.

sions for the same results may be obtained by employing a
notation for subsidiary results derived from triple integration
with reference to one of the bodies, in the following manner :—

497. Let us in the first place determine the action exerted
by a given magnet upon an infinitely thin uniformly and longitudinally magnetized bar, placed in a given position in its
neighbourhood.

We may suppose the rectangular co-ordinates, ξ, η, ζ, of the
north pole, and ξ', η', ζ' of the south pole of the bar to be given,
and hence the components X, Y, Z and X', Y', Z', of the resultant forces at those points due to the other given magnet
may be regarded as known. Then, if β denote the "strength"
of the bar-magnet, the components of the forces on its two
poles will be respectively

$$\beta X, \beta Y, \beta Z, \text{ on the point } (\xi, \eta, \zeta),$$

and $\quad -\beta X', -\beta Y', -\beta Z', \text{ on the point } (\xi', \eta', \zeta').$

The resultant action due to this system of forces may be determined by means of the elementary principles of statics. Thus
if we conceive the forces to be transferred to the middle of the
bar by the introduction of couples, the system will be reduced
to a force, on this point, whose components are

$$\beta(X-X'), \ \beta(Y-Y'), \ \beta(Z-Z'),$$

and a couple, whose components are

$$\{\beta(Z+Z')\tfrac{1}{2}(\eta-\eta') - \beta(Y+Y').\tfrac{1}{2}(\zeta-\zeta')\},$$
$$\{\beta(X+X').\tfrac{1}{2}(\zeta-\zeta') - \beta(Z+Z').\tfrac{1}{2}(\xi-\xi')\},$$
$$\{\beta(Y+Y').\tfrac{1}{2}(\xi-\xi') - \beta(X+X').\tfrac{1}{2}(\eta-\eta')\}.$$

498. Let l, m, n denote the direction-cosines of a line drawn
along the bar, from its middle towards its north pole, and if a
be the length of the bar, we shall have

$$\xi - \xi' = al, \quad \eta - \eta' = am, \quad \zeta - \zeta' = an.$$

Hence, if the bar be infinitely short, and if x, y, z denote the
co-ordinates of its middle point, we have

$$X - X' = \frac{dX}{dx}.al + \frac{dX}{dy}.am + \frac{dX}{dz}.an,$$

$$Y - Y' = \frac{dY}{dx}.al + \frac{dY}{dy}.am + \frac{dY}{dz}.an,$$

and $\quad Z - Z' = \frac{dZ}{dx}.al + \frac{dZ}{dy}.am + \frac{dZ}{dz}.an.$

Multiplying each member of these equations by β, we obtain the expressions for the components of the force in this case ; and the expressions for the components of the couples are found in their simpler forms, by substituting for $\xi - \xi'$, etc., their values given above ; and, on account of the infinitely small factor which each term contains, taking $2X$, $2Y$, and $2Z$, in place of $X + X'$, $Y + Y'$, and $Z + Z'$.

499.   Let us now suppose an infinite number of such infinitely small bar-magnets to be put together so as to constitute a mass, infinitely small in all its dimensions, uniformly magnetized in the direction (l, m, n) to such an intensity that its magnetic moment is μ.   We infer, from the preceding investigation, that the total action on this body, when placed at the point x, y, z, will be composed of a force whose components are

$$\mu \left(\frac{dX}{dx} l + \frac{dX}{dy} m + \frac{dX}{dz} n \right),$$

$$\mu \left(\frac{dY}{dx} l + \frac{dY}{dy} m + \frac{dY}{dz} n \right),$$

$$\mu \left(\frac{dZ}{dx} l + \frac{dZ}{dy} m + \frac{dZ}{dz} n \right),$$

acting at the centre of gravity of the solid supposed homogeneous ; and a couple of which the components are

$$\mu (Zm - Yn),$$
$$\mu (Xn - Zl),$$
$$\mu (Yl - Xm).$$

500.   The preceding investigation enables us, by means of the integral calculus, to determine the total mutual action between any two given magnets.   For, if we take X, Y, Z to denote the components of the resultant force due to one of the magnets, at any point (x, y, z) of the other, and if i denote the intensity and (l, m, n) the direction of magnetization of the substance of the second magnet at this point, we may take $\mu = i . dxdydz$ in the expressions which were obtained, and they will then express the action which one of the magnets exerts upon an element $dxdydz$ of the other.   To determine the total resultant action, we may transfer all the forces to the origin of co-ordinates, by introducing additional couples ; and, by the usual process, we find, for the mutual action between the two magnets,

a force in a line through this point, and a couple, of which the components, F, G, H, and L, M, N, are given by the equations

$$F = \iiint \left(il \frac{dX}{dx} + im \frac{dX}{dy} + in \frac{dX}{dz} \right) dxdydz$$

$$G = \iiint \left(il \frac{dY}{dx} + im \frac{dY}{dy} + in \frac{dY}{dz} \right) dxdydz$$

$$H = \iiint \left(il \frac{dZ}{dx} + im \frac{dZ}{dy} + in \frac{dZ}{dz} \right) dxdydz$$

$$L = \iiint \left\{ imZ - inY + y \left(il \frac{dZ}{dx} + im \frac{dZ}{dy} + in \frac{dZ}{dz} \right) \right.$$
$$\left. - z \left(il \frac{dY}{dx} + im \frac{dY}{dy} + in \frac{dY}{dz} \right) \right\} dxdydz$$

$$M = \iiint \left\{ inX - ilZ + z \left(il \frac{dX}{dx} + im \frac{dX}{dy} + in \frac{dX}{dz} \right) \right.$$
$$\left. - x \left(il \frac{dZ}{dx} + im \frac{dZ}{dy} + in \frac{dZ}{dz} \right) \right\} dxdydz$$

$$N = \iiint \left\{ ilY - imX + x \left(il \frac{dY}{dx} + im \frac{dY}{dy} + in \frac{dY}{dz} \right) \right.$$
$$\left. - y \left(il \frac{dX}{dx} + im \frac{dX}{dy} + in \frac{dX}{dz} \right) \right\} dxdydz$$

501. If, in the second members of these equations, we employ for X, Y, Z respectively their values obtained, as indicated in equations (4) of § 483, by the differentiation of the expression (5) for V in § 486, we obtain expressions for F, G, H, L, M, N, which may readily be put under symmetrical forms with reference to the two magnets, exhibiting the parts of those quantities depending on the mutual action between an element of one of the magnets, and an element of the other. Again, expressions exhibiting the mutual action between any element of the imaginary magnetic matter of one magnet, and any element of the imaginary magnetic matter of the other, may be found by first modifying by integration by parts, as in § 495, from the expressions which we have actually obtained for F, G, H, L, M, N; and then substituting for X, Y, and Z their values obtained by the differentiation of the expression (3) of § 482, for V.

It is unnecessary here to do more than indicate how such other formulæ may be derived from those given above; for whenever it may be required, there can be no difficulty in applying the principles which have been established in this paper to obtain any desired form of expression for the mutual action between two given magnets.

§§ 502 and 503.*—*On the Expression of Mutual Action between two Magnets by means of the Differential Coefficients of a Function of their relative Position.*

502. By a simple application of the theory of the potential, it may be shown that the amount of mechanical work spent or gained in any motion of a permanent magnet, effected under the action of another permanent magnet in a fixed position, depends solely on the initial and final positions, and not at all upon the positions successively occupied by the magnet in passing from one to the other. Hence the amount of work requisite to bring a given magnet from being infinitely distant from all magnetic bodies into a certain position in the neighbourhood of a given fixed magnet, depends solely upon the distributions of magnetism in the two magnets, and on the relative position which they have acquired. Denoting this amount by Q, we may consider Q as a function of co-ordinates which fix the relative position of the two magnets; and the variation which Q experiences when this is altered in any way will be the amount of work spent or lost, as the case may be, in effecting the alteration. This enables us to express completely the mutual action between the two magnets, by means of differential coefficients of Q, in the following manner :—

If we suppose one of the magnets to remain fixed during the alterations of relative position conceived to take place, the quantity Q will be a function of the linear and angular co-ordinates by which the variable position of the other is expressed. Without specifying any particular system of co-ordinates to be adopted, we may denote by $d_\xi Q$ the augmenta-

* Communicated June 20, 1850.

tion of Q when the moveable magnet is pushed through an infinitely small space $d\xi$ in any given direction, and by $d_\phi Q$ the augmentation of Q when it is turned round any given axis, through an infinitely small angle $d\phi$. Then, if F denote the force upon the magnet in the direction of $d\xi$, and L the moment round the fixed axis of all the forces acting upon it (or the component, round the fixed axis, of the resultant couple obtained when all the forces on the different parts of the magnet are transferred to any point on this axis), we shall have

$$- Fd\xi = d_\xi Q, \text{ and } - Ld\phi = d_\phi Q,$$

since a force equal to $-F$ is overcome through the space $d\xi$ in the first case, and a couple, of which the moment is equal to $-L$, is overcome through an angle $d\phi$ in the second case of motion. Hence we have

$$F = - \frac{d_\xi Q}{d\xi}$$

$$L = - \frac{d_\phi Q}{d\phi}.$$

503. It only remains to show how the function Q may be determined when the distributions of magnetism in the two magnets and the relative positions of the bodies are given. For this purpose, let us consider points P and P', in the two magnets respectively, and let their co-ordinates with reference to three fixed rectangular axes be denoted by x, y, z and x', y', z'; let also the intensity of magnetization at P be denoted by i, and its direction-cosines by l, m, n; and let the corresponding quantities, with reference to P', be denoted by i', l', m', n'. Then it may be demonstrated without difficulty that

$$Q = \iiiiii dx\,dy\,dz\,dx'\,dy'\,dz'\,ii'\left\{ ll'\frac{d^2\frac{1}{\Delta}}{dx\,dx'} + lm'\frac{d^2\frac{1}{\Delta}}{dx\,dy'} + ln'\frac{d^2\frac{1}{\Delta}}{dx\,dz'} \right.$$
$$+ ml'\frac{d^2\frac{1}{\Delta}}{dy\,dx'} + mm'\frac{d^2\frac{1}{\Delta}}{dy\,dy'} + mn'\frac{d^2\frac{1}{\Delta}}{dy\,dz'}$$
$$\left. + nl'\frac{d^2\frac{1}{\Delta}}{dz\,dx'} + nm'\frac{d^2\frac{1}{\Delta}}{dz\,dy'} + nn'\frac{d^2\frac{1}{\Delta}}{dz\,dz'} \right\} \quad (1),$$

where, for brevity, Δ is taken to denote $\{(x - x')^2 + (y - y')^2 +$ $(z - z')^2\}^{\frac{1}{2}}$, and the differentiations upon $\frac{1}{\Delta}$ are merely indicated.

Now, by any of the ordinary formulæ for the transformation of co-ordinates, the values of x, y, z, and x', y', z', may be expressed in terms of co-ordinates of the point P with reference to axes fixed in the magnet to which it belongs, of the co-ordinates of the point P with reference to axes fixed in the other, and of the co-ordinates adopted to express the relative position of the two magnets : and so the preceding expression for Q may be transformed into an expression involving explicitly the relative co-ordinates, and containing the co-ordinates of the points P and P' in the two bodies only as variables in integrations, the limits of which, depending only on the forms and dimensions of the two bodies, are absolutely constant. Thus Q is obtained as a function of the relative co-ordinates of the bodies, and the solution of the problem is complete.

There is no difficulty in working out the result by this method, so as actually to obtain either the expressions of § 500, or the expressions indicated in § 501, although the process is somewhat long. [*Addition, Dec.* 11, 1871.—If in the formula for Q we suppose the integration with respect to x', y', z' to be performed, we have

$$Q = -\int_{-\infty}^{\infty} \int_{-\infty}^{\infty} \int_{-\infty}^{\infty} dx\,dy\,dz\, (\alpha \mathfrak{X}' + \beta \mathfrak{Y}' + \gamma \mathfrak{Z}') \dots\dots (2)$$

where α, β, γ are put for il, im, in; and $\mathfrak{X}'$, $\mathfrak{Y}'$, $\mathfrak{Z}'$ denote the components of the force at (x, y, z) due to the second magnet, to be taken according to the definition of § 480 when (x, y, z) is in the magnetized substance of this magnet. For simplicity, without loss of generality, suppose α, β, γ to vary continuously from finite values in the magnet to zero in space void of magnetized substance : and, putting

$$\mathfrak{X}' = -\frac{d\mathfrak{V}'}{dx}, \quad \mathfrak{Y}' = -\frac{d\mathfrak{V}'}{dy}, \quad \mathfrak{Z}' = -\frac{d\mathfrak{V}'}{dz} \quad \dots\dots\dots (3),$$

integrate by parts in the usual manner (§ 495). Thus

$$Q = -\iiint dx\,dy\,dz \left(\frac{d\alpha}{dx} + \frac{d\beta}{dy} + \frac{d\gamma}{dz}\right) \mathfrak{V}'.$$

But [§ 474 (2) and Poisson's Theorem]

$$\frac{d\alpha}{dx} + \frac{d\beta}{dy} + \frac{d\gamma}{dz} = -\frac{1}{4\pi}\left(\frac{d\mathfrak{X}}{dx} + \frac{d\mathfrak{Y}}{dy} + \frac{d\mathfrak{Z}}{dz}\right) \quad\ldots\ldots\ldots\ldots(4).$$

Hence, by a reverse integration by parts,

$$Q = \frac{1}{4\pi}\int_{-\infty}^{\infty}\int_{-\infty}^{\infty}\int_{-\infty}^{\infty} dx\,dy\,dz\,(\mathfrak{X}\mathfrak{X}' + \mathfrak{Y}\mathfrak{Y}' + \mathfrak{Z}\mathfrak{Z}') \quad\ldots\ldots(5).$$

This is a very important result, as we shall see in Chapter VII. Compare § 561.]

The method just explained for expressing the mutual action between two magnets in terms of a function of their relative position, has been added to this chapter rather for the sake of completing the mathematical theory of the division of the subject to which it is devoted, than for its practical usefulness in actual problems regarding magnetic force, for which the most convenient solutions may generally be obtained by some of the more synthetical methods explained in the preceding parts of the chapter.   There is, however, a far more important application of the principles upon which this last method is founded which remains to be made.   The mechanical value of a distribution of magnetism, although it has not, I believe, been noticed in any writings hitherto published on the mathematical theory of magnetism, is a subject of investigation of great interest, and, as I hope on a later occasion* to have an opportunity of showing, of much consequence, on account of its maximum and minimum problems, which lead to demonstrations of important theorems in the solutions of inverse problems regarding magnetic distribution.

CHAPTER V.—*On Solenoidal and Lamellar Distributions of Magnetism.*†

504.   In the course of some researches upon inverse problems regarding distributions of magnetism, and upon the comparison

---

* [Chap. VII....X. below; Dec. 1871.]
† Communicated to the Royal Society June 20, 1850.

of electro-magnets and common magnets, I have found it extremely convenient to make use of definite terms to express certain distributions of magnetism and forms of magnetized matter possessing remarkable properties. The use of such terms will be of still greater consequence in describing the results of these researches, and therefore, before proceeding to do so, I shall give definitions of the terms which I have adopted, and explain briefly the principal properties of the magnetic distributions to which they are applied. The remainder of this chapter will be devoted to three new methods of analysing the expressions for the resultant force of a magnet at any point, suggested by the consideration of these special forms of magnetic distribution. A Mathematical Theory of Electro-Magnets, and Inverse Problems regarding magnetic distributions, are the subjects of papers which I hope to be able to lay before the Royal Society on a subsequent occasion. [They are published for the first time in this volume : Chaps. VI....X.]

505. *Definitions and explanations regarding Magnetic Solenoids.*

(1) A magnetic solenoid* is an infinitely thin bar of any form, longitudinally magnetized with an intensity varying inversely as the area of the normal section in different parts.

The constant product of the intensity of magnetization into the area of the normal section, is called the magnetic strength, or sometimes simply the strength of the solenoid. Hence the magnetic moment of any straight portion, or of an infinitely small portion of a curved solenoid, is equal to the product of the magnetic strength into the length of the portion.

(2) A number of magnetic solenoids of different lengths may be put together so as to constitute what is, as far as regards magnetic action, equivalent to a single infinitely thin bar of any form, longitudinally magnetized with an intensity varying

---

* This term (from σωλήν, *a tube*) is suggested by the term "electro-dynamic solenoid" applied by Ampère to a certain tube-like arrangement of galvanic circuits which produces precisely the same external magnetic effect as is produced by ordinary magnetism distributed in the manner defined in the text. The especial appropriateness of the term to the magnetic distribution is manifest from the relation indicated in the footnote on § 513 below, between the intensity and direction of magnetization in a solenoid, and the velocity and direction of motion of a liquid flowing through a tube of constant or varying section.

arbitrarily from one end of the bar to the other. Hence such a magnet may be called a complex magnetic solenoid.

The magnetic strength of a complex solenoid is not uniform, but varies from one part to another.

(3) An infinitely thin closed ring, magnetized in the manner described in (1), is called a closed magnetic solenoid.

506. *Definitions and explanations regarding Magnetic Shells.*

(1) A magnetic shell is an infinitely thin sheet of any form, normally magnetized with an intensity varying inversely as the thickness in different parts.

The constant product of the intensity of magnetization into the thickness is called the magnetic strength, or sometimes simply the strength of the shell. Hence the magnetic moment of any plane portion, or of an infinitely small portion of a curved magnetic shell, is equal to the product of the magnetic strength into the area of the portion.

(2) A number of magnetic shells of different areas may be put together so as to constitute what is, as far as regards magnetic action, equivalent to a single infinitely thin sheet of any form, normally magnetized with an intensity varying arbitrarily over the whole sheet. Hence such a magnet may be called a complex magnetic shell.

The magnetic strength of a complex shell is not uniform, but varies from one part to another.

(3) An infinitely thin sheet, of which the two sides are closed surfaces, is called a closed magnetic shell.

507. *Solenoidal and Lamellar Distributions of Magnetism.*— If a finite magnet of any form be capable of division into an infinite number of solenoids which are either closed or have their ends in the bounding surface, the distribution of magnetism in it is said to be solenoidal, and the substance is said to be solenoidally magnetized.

If a finite magnet of any form be capable of division into an infinite number of magnetic shells which are either closed or have their edges in the bounding surface, the distribution of magnetism in it is said to be lamellar,* and the substance is said to be lamellarly magnetized.

---

* The term *lamellar*, adopted for want of a better, is preferred to "laminated"; since this might be objected to as rather meaning composed of plane

508. *Complex Lamellar Distributions of Magnetism.*—If a
finite magnet of any form be capable of division into an infinite
number of complex magnetic shells, it is said to possess a complex lamellar distribution of magnetism.

509. *Complex Solenoidal Distributions of Magnetism.*—Since,
by cutting it along lines of magnetization, every magnet of finite
dimensions may be divided into an infinite number of longitudinally magnetized infinitely thin bars or rings, any distribution of magnetism which is not solenoidal might be called a
complex solenoidal distribution ; but no advantage is obtained
by the use of this expression, which is only alluded to here,
on account of the analogy with the subject of the preceding
definition.

510. PROP.—*The action of a magnetic solenoid is the same as
if a quantity of positive or northern imaginary magnetic matter
numerically equal to its magnetic strength were placed at one end,
and an equal absolute quantity of negative or southern matter at
the other end.*

The truth of this proposition follows at once from the investigation of Chap. III. §§ 467, 468, 469.

Cor. 1.—The action of a magnetic solenoid is independent of
its form, and depends solely on its strength and the positions
of its extremities.

Cor. 2.—A closed solenoid exerts no action on any other
magnet.

Cor. 3.—The "resultant force" (defined in Chap. IV. § 480)
at any point in the substance of a closed magnetic solenoid
vanishes.

511. PROP—*If* i *be the intensity of magnetization, and* ω *the
area of the normal section at any point* P, *at a distance* s *from one
extremity of a complex solenoid, and if* [iω] *and* {iω} *denote the
values of the product of these quantities at the extremity from
which* s *is measured, and at the other extremity respectively ; the
magnetic action will be the same as if there were a distribution of
imaginary magnetic matter, through the length of the bar of which
the quantity is an infinitely small portion* ds, *of the length at the*

---

plates, than *composed of shells* whether plane or curve, and is besides too much
associated with a mechanical structure such as that of slate or mica, to be a
convenient term for the magnetic distributions defined in the text.

point P, *would be* $-\dfrac{d\,(i\omega)}{ds}\,ds$, *and accumulations of quantities*
equal to $-[i\omega]$ *and* $\{i\omega\}$ *respectively at the two extremities.*

The truth of this proposition follows immediately from the conclusions of Chap. III. § 469.

512. PROP.—*The potential of a magnetic shell at any point is equal to the solid angle which it subtends at that point multiplied by its magnetic strength*.*

Let dS denote the area of an infinitely small element of the shell, Δ the distance of this element from the point P, at which the potential is considered, and θ the angle between this line, and a normal to the shell drawn through the north polar side of dS. Then if λ denote the magnetic strength of the shell, the magnetic moment of the element dS will be λdS, and (§ 485) the potential due to it at P will be

$$\frac{\lambda dS\,.\,\cos\theta}{\Delta^2}\,.$$

Now $\dfrac{dS\,.\,\cos\theta}{\Delta^2}$ is the solid angle subtended at P by the element dS, and therefore the potential due to any infinitely small element, is equal to the product of its magnetic strength into the solid angle which its area subtends at P. But the potential due to the whole is equal to the sum of the potentials due to the parts, and the strength is the same for all the parts. Hence the potential due to the whole shell is equal to the product of its strength into the sum of the solid angles which all its parts, or the solid angle which the whole, subtends at P.

Cor. 1.—The expression $\dfrac{dS\,.\,\cos\theta}{\Delta^2}$, which occurs in the preceding demonstration, being positive or negative according as θ is acute or obtuse, it appears that the solid angle subtended by different parts of the shell at P must be considered as positive or negative according as their north polar or their south polar sides are towards this point.

---

*. This theorem is due to Gauss (see his paper "On the General Theory of Terrestrial Magnetism," § 38 ; of which a translation is published in Taylor's *Scientific Memoirs*, vol. II.). Ampère's well-known theorem, referred to by Gauss, that a closed galvanic circuit produces the same magnetic effect as a magnetic shell of any form having the circuit for its edge, implies obviously the truth of the first part of Cor. 2 below.

Cor. 2.—The potential at any point due to a magnetic shell is independent of the form of the shell itself, and depends solely on its bounding line or edge, subject to an ambiguity, the nature of which is made clear by the following statement:—

If two shells of equal magnetic strength, λ, have a common boundary, and if the north polar side of one, and the south polar side of the other be towards the enclosed space, the potentials due to them at any external point will be equal; and the potential at any point in the enclosed space, due to that one of which the northern polarity is on the inside, will exceed the potential due to the other by the constant $4\pi\lambda$.

Cor. 3.—Of two points infinitely near one another on the two sides of a magnetic shell, but not infinitely near its edge, the potential at that one which is on the north polar side exceeds the potential at the other by the constant $4\pi\lambda$.

Cor. 4.—The potential of a closed magnetic shell of strength λ, with its northern polarity on the inside, is $4\pi\lambda$, for all points in the enclosed space, and 0 for all external points; and for points in the magnetized substance it varies continuously from the inside, where it is $4\pi\lambda$ to the outside, where it is 0.

Cor. 5.—A closed magnetic shell exerts no force on any other magnet.

Cor. 6.—The "resultant force" as defined at §§ 479, 480 [polar definition], is equal to $\dfrac{4\pi\lambda}{\tau}$, at any point in the substance of a closed magnetic shell, if τ be the thickness, or to $4\pi i$, if i be the intensity of magnetization of the shell in the neighbourhood of the point, and is in the direction of a normal drawn from the point through the south polar side of the shell. [The "resultant force" as defined below in § 517, by the electromagnetic definition, is zero at any point in the substance of a closed magnetic shell, or of a lamellar distribution consisting of closed shells.]

Cor. 7.—If the intensity of magnetization of an open shell be finite, the resultant force at any external point not infinitely near the edge is infinitely small; but the force at any point in the substance not infinitely near the edge is finite, and is equal to $4\pi i$, if i be the intensity of the magnetization in the neighbourhood of the point, and is in the direction of a normal through the south polar side.

513. PROP.—*A distribution of magnetism expressed by* $\{(\alpha,\ \beta,\ \gamma)$ *at* $(x,\ y,\ z)\}$* *is solenoidal if, and is not solenoidal unless,* $\dfrac{d\alpha}{dx} + \dfrac{d\beta}{dy} + \dfrac{d\gamma}{dz} = 0.$

The condition that a given distribution of magnetism, in a substance of finite dimensions, may be solenoidal, is readily deduced from the investigations of § 473, by means of the propositions of §§ 510 and 511. For, if the distribution of magnetism be solenoidal, the imaginary magnetic matter by which the polarity of the whole magnet may be represented will be situated at the ends of the solenoids, according to § 510, and therefore (§ 507) will be spread over the bounding surface. On the other hand, if the distribution be not solenoidal, that is, if the magnet be divisible into solenoids, of which some, if not all, are complex; there will, according to § 511, be an internal distribution of imaginary magnetic matter in the representation of the polarity of the whole magnet. Hence it follows from § 473 that if α, β, γ denote the components of the intensity of magnetization at any internal point $(x,\ y,\ z)$, the equation

$$\frac{d\alpha}{dx} + \frac{d\beta}{dy} + \frac{d\gamma}{dz} = 0 \dots\dots\dots\dots\dots\dots\dots(\text{I.})$$

expresses that the distribution of magnetism is solenoidal†.

---

* Where α, β, γ, which may be called the components, parallel to the axes of co-ordinates, of the magnetization at $(x,\ y,\ z)$, denote respectively the products of the intensity into the direction cosines of the magnetization.

† The analogy between the circumstances of this expression and those of the cinematical condition expressed by "the equation of continuity" to which the motion of a homogeneous incompressible fluid is subject, is so obvious that it is scarcely necessary to point it out. When an incompressible fluid flows through a tube of variable infinitely small section, the velocity (or rather the mean velocity) in any part is inversely proportional to the area of the section. Hence the intensity and direction of magnetization, in a solenoid, according to the definition, are subject to the same law as the mean fluid velocity in a tube with an incompressible fluid flowing through it. Again, if any finite portion of a mass of incompressible fluid in motion be at any instant divided into an infinite number of solenoids (that is, tube-like parts), by following the lines of motion, the velocity in any one of these parts will, at different points of it, be inversely proportional to the area of its section. Hence the intensity and direction of magnetization in a solenoidal distribution of magnetism, according to the definition, are subject to the same condition as the fluid-velocity and its direction, at any point in an incompressible fluid in motion. It may be remarked, that by making an investigation on the plan of § 473 to express merely the condition that there may be no internal distribution of imaginary magnetic

514. PROP.—*A distribution of magnetism $\{(\alpha,\beta,\gamma)$ at $(x,y,z)\}$ is lamellar if, and is not lamellar unless, $\alpha dx + \beta dy + \gamma dz$ is the differential of a function of three independent variables.*

Let ψ be a variable which has a certain value for each of the series of surfaces, by which the magnet may be divided into magnetic shells; so that, if ψ be considered as a function of x, y, z, any one of these surfaces will be represented by the equation $$\psi(x,y,z) = \Pi \ldots\ldots\ldots\ldots(a);$$ and the entire series will be obtained by giving the parameter Π, successively a series of values each greater than that which precedes it by an infinitely small amount. According to the definition of a magnetic shell (§ 506), the lines of magnetization must cut these surfaces orthogonally; and hence, since α, β, γ denote quantities proportional to the direction cosines of the magnetization at any point, we must have

$$\frac{\alpha}{\dfrac{d\psi}{dx}} = \frac{\beta}{\dfrac{d\psi}{dy}} = \frac{\gamma}{\dfrac{d\psi}{dz}} \ldots\ldots\ldots\ldots(b).$$

Let us consider the magnetic shell between two of the consecutive surfaces corresponding to values of the parameter of which the infinitely small difference is ϖ. The thickness of this shell at any point (x, y, z) will be

$$\frac{\varpi}{\left(\dfrac{d\psi^2}{dx^2} + \dfrac{d\psi^2}{dy^2} + \dfrac{d\psi^2}{dz^2}\right)^{\frac{1}{2}}}$$

Now the product of the intensity of magnetization, into the thickness of the shell, must be constant for all points of the

---

matter, the equation $\frac{d\alpha}{dx} + \frac{d\beta}{dy} + \frac{d\gamma}{dz} = 0$ is obtained in a manner precisely similar to a mode of investigating the equation of continuity for an incompressible fluid, now well known, which is given in Duhamel's *Cours de Mécanique*, and in the *Cambridge and Dublin Mathematical Journal*, vol. II. p. 282. The following very remarkable proposition is an immediate consequence of the proposition that "a closed solenoid exerts no action on any other magnet" (§ 510, Cor. 2 above), in virtue of the analogy here indicated.

"If a closed vessel, of any internal shape, be completely filled with an incompressible fluid, the fluid set into any possible state of motion, and the vessel held at rest; and if a solid mass of steel of the same shape as the space within the vessel be magnetized at each point with an intensity proportional and in a direction corresponding to the velocity and direction of the motion at the corresponding point of the fluid at any instant; the magnet thus formed will exercise no force on any external magnet."

same shell; and hence, since ϖ is constant, and since α, β, γ denote quantities such that $(\alpha^2 + \beta^2 + \gamma^2)^{\frac{1}{2}}$ is the intensity of magnetization at any point, we must have

$$\frac{(\alpha^2 + \beta^2 + \gamma^2)^{\frac{1}{2}}}{\left(\dfrac{d\psi^2}{dx^2} + \dfrac{d\psi^2}{dy^2} + \dfrac{d\psi^2}{dz^2}\right)^{\frac{1}{2}}} = F(\psi) \ldots\ldots\ldots\ldots(c),$$

where $F(\psi)$ denotes a quantity which is constant when ψ is constant. This equation, and the two equations (b), express all the conditions required to make the given distribution lamellar. By combining them we obtain the following three, which are equivalent to them :—

$$\alpha = F(\psi)\frac{d\psi}{dx}, \quad \beta = F(\psi)\frac{d\psi}{dy}, \quad \gamma = F(\psi)\frac{d\psi}{dz} ;$$

and hence, if $\int F(\psi)d\psi$ be denoted by ϕ, we have

$$\alpha = \frac{d\phi}{dx}, \quad \beta = \frac{d\phi}{dy}, \quad \gamma = \frac{d\phi}{dz} \ldots\ldots\ldots\ldots\ldots(\text{II.}),$$

where ϕ is some function of x, y, and z. Hence the condition that a magnetic distribution (α, β, γ) may be lamellar, is simply that $\alpha dx + \beta dy + \gamma dz$ must be the differential of a function of three independent variables. The equations to express this are obtained in their simplest forms by eliminating the arbitrary function ϕ by differentiation; and are of course

$$\left.\begin{array}{l} \dfrac{d\beta}{dz} - \dfrac{d\gamma}{dy} = 0 \\[2mm] \dfrac{d\gamma}{dx} - \dfrac{d\alpha}{dz} = 0 \\[2mm] \dfrac{d\alpha}{dy} - \dfrac{d\beta}{dx} = 0 \end{array}\right\} \ldots\ldots\ldots\ldots\ldots\ldots(\text{III.}).$$

Cor.—It follows from the first part of the preceding investigation that equations (b) express that the distribution, if not lamellar, is complex-lamellar. By eliminating the arbitrary function ψ from those equations (which merely express that $\alpha dx + \beta dy + \gamma dz$ is integrable by a factor), we obtain the well-known equation

$$\alpha\left(\frac{d\beta}{dz} - \frac{d\gamma}{dy}\right) + \beta\left(\frac{d\gamma}{dx} - \frac{d\alpha}{dz}\right) + \gamma\left(\frac{d\alpha}{dy} - \frac{d\beta}{dx}\right) = 0 \quad (\text{IV.}),$$

as the simplest expression of the condition that α, β, γ must

satisfy, in order that the distribution which they represent may
be complex-lamellar; and we also conclude that if this equation be satisfied the distribution must be complex-lamellar,
unless each term of the first number vanishes by equations
(III.) being satisfied, in which case the distribution is, as we
have seen, lamellar.

515. The resultant force at any point external to a lamellarly-
magnetized magnet will, according to § 512 (Cors. 2 and 4),
depend solely upon the edges of the shells into which it may be
divided by surfaces perpendicular to the lines of magnetization
(or the bands into which those surfaces cut the bounding
surface), and not at all on the forms of these shells, within the
bounding surface, nor upon any closed shells of which part of
the magnet may consist; and the resultant force at any
internal point may (§ 512, Cors. 2, 4, and 7) be obtained by
compounding a force depending solely on those edges, with a
force in the direction contrary to that of the magnetization
of the substance at the point, and equal to the product of 4π
into the intensity of the magnetization. For either an external
or an internal point, the resultant force may be expressed by
means of a potential, according to § 480; and the value of this
potential may be obtained by means of the theorems of § 512,
in the following manner:—

Let us suppose all the open shells, that is to say all the
shells cut by the bounding surface of the given magnet,
to be removed, and a series of shells having the same edges,
and the same magnetic strengths, and coinciding with the
bounding surface, substituted for them; and, for the sake
of definiteness, let us suppose each of these shells to have its
north polar side outwards, and to occupy a part of the surface
for which the value of ϕ is greater than at its edge. The whole
surface will thus be occupied by a series of superimposed
magnetic shells, constituting a complex magnetic shell which
will produce a potential at any external point the same as that
due to the whole of the given magnet; and, at any internal
point a potential, which, together with the potential due to
the closed shells round it, if there are any, and (§ 512, Cor. 2)
together with the product of 4π into the sum of the strengths
of any open shells having it between them and their superficial

substitutes, will be the potential due to the whole of the given magnet at this point.

Now if $d\phi$ denote the difference between the values of ϕ at two consecutive surfaces of the series, by which we may conceive the whole magnet to be divided into shells, it follows, from the investigation of § 514, that the magnetic strength of the shell is equal to $d\phi$. Hence if A denote the least value of ϕ at any part of the bounding surface, and ϕ be supposed to correspond to a point in the surface, the strength of the complex magnetic shell, found by adding the strengths of all of the imagined series of shells superimposed at this point, will be $\phi - A$; and if P be an internal point, and the value of ϕ at it be denoted by (ϕ), the sum of the strengths of all the shells between that which passes through P and that which corresponds to A, will be $(\phi) - A$, from which it may be demonstrated*, that, whether (ϕ) be $>$ or $< A$, and whatever be the nature of the shells, whether all open or some open and some closed, the quantity to be added to the potential due to the imagined complex shell coinciding with the surface of the magnet to find the actual potential at P, is $4\pi \{(\phi) - A\}$. Now, from what we have seen above, it follows that the potential at any point P, due to an element, dS, of this complex shell is $\dfrac{\phi \cdot \cos \theta \, dS}{\Delta^2}$: if θ denote the angle which an external normal, or a normal through the north polar side of dS, makes with a line drawn from dS to P; and Δ the length of this line. Hence the total potential at P, due to the whole complex shell, is equal to

$$\iint \frac{\{\phi - A\} \cos \theta \, dS}{\Delta^2},$$

in which the integration includes the whole bounding surface of the magnet. Hence, if V denote the potential at P, we have the following expression, according as P is external or internal,—

$$V = \iint \frac{\{\phi - A\} \cos \theta \, dS}{\Delta^2},$$

or

$$V = \iint \frac{\{\phi - A\} \cos \theta \, dS}{\Delta^2} + 4\pi \{(\phi) - A\}.$$

---

* See second footnote on § 479 above, and Cors. 2, 3, § 515 below.

These expressions may be simplified if we remark that, for any external point,

$$\iint \frac{\cos\theta\, dS}{\Delta^2} = 0,$$

and that, for any internal point,

$$\iint \frac{\cos\theta\, dS}{\Delta^2} = -4\pi$$

(since θ is the angle between the line Δ and the *external* normal through dS). We thus obtain, for an external point,

$$V = \iint \frac{\phi\cdot\cos\theta\, dS}{\Delta^2};$$

and for an internal point,

$$V = \iint \frac{\phi\cdot\cos\theta\, dS}{\Delta^2} + 4\pi\,(\phi),$$

$$\left.\right\} \quad \ldots\ldots\ldots (V.).$$

Cor. 1.—The potentials at two points infinitely near one another, even if one be in the magnetized substance and the other be external, differ infinitely little; for the value of

$$\iint \frac{\phi\cdot\cos\theta\, dS}{\Delta^2},$$

at a point infinitely near the surface and within it, is found by adding $-4\pi(\phi)$ to the value of the same expression at an external point infinitely near the former.

Cor. 2.—If the value of

$$\iint \frac{\phi\cdot\cos\theta\, dS}{\Delta^2}$$

be denoted by $-Q$ for any internal point, x, y, z; and if (α), (β), (γ) denote the components of the intensity of magnetization, and X, Y, Z the components of the resultant magnetic force at this point (that is, according to the definition in the second foot-note on § 479, the force at a point in an infinitely small crevass tangential to the lines of magnetization at x, y, z), we have

$$X = -\frac{dV}{dx} = \frac{dQ}{dx} - 4\pi\,(\alpha)$$
$$Y = -\frac{dV}{dy} = \frac{dQ}{dy} - 4\pi\,(\beta)$$
$$Z = -\frac{dV}{dz} = \frac{dQ}{dz} - 4\pi\,(\gamma)$$

$$\left.\right\} \quad \ldots\ldots\ldots\ldots (VI.).$$

The resultant of the partial components, $-4\pi(\alpha)$, $-4\pi(\beta)$, $-4\pi(\gamma)$, is a force equal to $4\pi(i)$ acting in a direction contrary to that of magnetization, and this, compounded with the resultant of

$$\frac{dQ}{dx}, \ \frac{dQ}{dy}, \ \frac{dQ}{dz},$$

which depends solely on the edges of the shells, gives the total resultant force at the internal point. We thus see precisely how the statements made at the commencement of § 515 are fulfilled.

Cor. 3.—It is obvious, by the preceding investigation, that

$$\frac{dQ}{dx}, \ \frac{dQ}{dy}, \ \frac{dQ}{dz}$$

are the components of the force at a point in an infinitely small crevass perpendicular to the lines of magnetization at x, y, z.

516. An analytical demonstration of these expressions may be obtained by a partial integration of the general expression for the potential in the case of a lamellar distribution, in the following manner :—

In equation (5) of § 486, which, as was remarked in the footnote, expresses the potential for any point, whether internal or external, let $\frac{d\phi}{dx}$, $\frac{d\phi}{dy}$, and $\frac{d\phi}{dz}$ be substituted in place of il, im, and in respectively; and, for the sake of brevity, let

$$\{(\xi - x)^2 + (\eta - y)^2 + (\zeta - z)^2\}^{\frac{1}{2}}$$

be denoted by Δ: then observing that $\dfrac{\xi - x}{\Delta^3} = \dfrac{d\frac{1}{\Delta}}{dx}$, and so for the similar terms; we have

$$V = \iint\left(\frac{d\phi}{dx}\frac{d\frac{1}{\Delta}}{dx} + \frac{d\phi}{dy}\frac{d\frac{1}{\Delta}}{dy} + \frac{d\phi}{dz}\frac{d\frac{1}{\Delta}}{dz}\right) dxdydz.....(a).$$

Dividing the second member into three terms, integrating the first by parts commencing with the factor $\frac{d\phi}{dx} dx$, and so for the other terms; we obtain

$$\left.\begin{array}{l} V = \left[\iint \phi\left(\dfrac{d\frac{1}{\Delta}}{dx} dydz + \dfrac{d\frac{1}{\Delta}}{dy} dzdx + \dfrac{d\frac{1}{\Delta}}{dz} dxdy\right)\right] \\[4mm] \qquad - \iiint \phi\left(\dfrac{d^2\frac{1}{\Delta}}{dx^2} + \dfrac{d^2\frac{1}{\Delta}}{dy^2} + \dfrac{d^2\frac{1}{\Delta}}{dz^2}\right)dxdydz \end{array}\right\} \quad (b),$$

where the brackets which enclose the double integral denote

that it has reference to the surface of the body. Now, for any set of values of x, y, z, for which $\frac{1}{\Delta}$ is finite, we have, as is well known,

$$\frac{d^2\frac{1}{\Delta}}{dx^2} + \frac{d^2\frac{1}{\Delta}}{dy^2} + \frac{d^2\frac{1}{\Delta}}{dx^2} = 0 \ldots\ldots\ldots\ldots\ldots(c);$$

and consequently, if the point ξ, η, ζ is not in the space included by the triple integral in the expression for V, each element of this integral, and therefore also the whole, vanishes. In the contrary case, the simultaneous values $x = \xi$, $y = \eta$, and $z = \zeta$ will be included in the limits of integration, and, as these values make $\frac{1}{\Delta}$ infinitely great, the equation (c) will fail for one element of the integral, although it still holds for all elements corresponding to points at a finite distance from (ξ, η, ζ). Hence, if (ϕ) denote the value assumed by the function ϕ at this point, we have

$$\iiint \phi \left(\frac{d^2\frac{1}{\Delta}}{dx^2} + \frac{d^2\frac{1}{\Delta}}{dy^2} + \frac{d^2\frac{1}{\Delta}}{dz^2} \right) dxdydz = (\phi)\iiint \left(\frac{d^2\frac{1}{\Delta}}{dx^2} + \frac{d^2\frac{1}{\Delta}}{dy^2} + \frac{d^2\frac{1}{\Delta}}{dz^2} \right) dxdydz,$$

where the limits of integration may correspond to any surface whatever which completely surrounds the point (ξ, η, ζ). Now it is easily proved (as is well known) that the value of

$$\iiint \left(\frac{d^2\frac{1}{\Delta}}{dx^2} + \frac{d^2\frac{1}{\Delta}}{dy^2} + \frac{d^2\frac{1}{\Delta}}{dx^2} \right) dxdydz$$

is -4π, when (ξ, η, ζ) is included in the limits of integration; and therefore the value of the triple integral, in the expression for V, is $-4\pi(\phi)$. Hence, according as the point (ξ, η, ζ) is external or internal with reference to the magnet, the potential at it is given by the expressions

$$(1) \quad V = \left[\iint \phi \cdot \left(\frac{d\frac{1}{\Delta}}{dx} dydz + \frac{d\frac{1}{\Delta}}{dy} dzdx + \frac{d\frac{1}{\Delta}}{dz} dxdy \right) \right]$$

or

$$(2)^* \quad V = \left[\iint \phi \cdot \left(\frac{d\frac{1}{\Delta}}{dx} dydz + \frac{d\frac{1}{\Delta}}{dy} dzdx + \frac{d\frac{1}{\Delta}}{dz} dxdy \right) \right] + 4\pi(\phi)$$

$$\left. \right\} \text{(VII.).}$$

---

* It may be proved that the force derived from a potential having the same

These agree with the expressions obtained above in § 515; the same double integral with reference to the surface being here expressed symmetrically by means of rectangular co-ordinates.

517.   The value of ϕ at any point in the surface of the magnet, which, as appears from the preceding investigations, is all that is necessary for determining the potential due to a lamellar magnet at any point not contained in the magnetized substance, may, according to well-known principles, be determined by integration, if the tangential component of the magnetization at every point of the magnet infinitely near its surface be given. It appears therefore that, if it be known that a magnet is lamellarly magnetized throughout its interior, it is sufficient to know the tangential component of its magnetization at every point infinitely near the surface, or to have enough of data for determining it, without any further specification regarding the interior distribution than that it is lamellar, to enable us to determine completely its external magnetic action. This conclusion is analogous to a conclusion which may be drawn, for the case of a solenoidal distribution, from the expression obtained in § 482, for the potential of a magnet of any kind. For, from this expression, we have, according to § 513, the following in the case of a solenoidal distribution:

$$V = \left[\iint \frac{(l\alpha + m\beta + n\gamma)\, dS}{\Delta} \right] \dots\dots\dots\text{(VIII.)};$$

from which we conclude, that without further data regarding the interior distribution than that it is solenoidal, it is sufficient to know the normal component of the magnetization at every point infinitely near the surface to enable us to determine the external magnetic action. Yet, although analogous conclusions are thus drawn from these two formulæ, the formulæ themselves are not analogous, as the former (that of § 482) is applicable to all distributions, whether solenoidal or not, and shows precisely how the resultant magnetic action will in general depend on the interior distribution besides the normal

---

expression (VII.) (1) as for external points, is, for any internal point, the force at a point within an infinitely small crevass perpendicular to the lines of magnetization; as it is easily shown that the differential coefficients of $4\pi(\phi)$ are the rectangular components of the force at such a point due [§ 7 (5)] to the free contrary polarities on the two sides of the crevass.

magnetization near the surface, according to the deviation from being solenoidal which it presents; while the formulæ of § 515 merely express a fact with reference to lamellar distributions, and being only applicable to lamellar distributions, do not indicate the effect of a deviation from being lamellar, in a distribution of general form. Certain considerations regarding the comparison between common magnets and electromagnets, suggested by Ampère's theorem that the magnetic action of a closed galvanic circuit is the same as that of a "magnetic shell" (as defined in § 506) of any figure having its edge coincident with the circuit, led me to a synthetical investigation [§ 554 below] of a distribution of galvanism through the interior and at the surface of a magnet magnetized in any arbitrary manner, from which I deduced formulæ for the resultant force at any external or internal point, giving the desired indication regarding effect of a deviation from being lamellar, on expressions which, for lamellar distributions, depend solely on the tangential component of magnetization at points infinitely near the surface. These galvanic elements throughout the body, from the action of which the resultant force at any external point is compounded, produce effects which are not separately expressible by means of a potential, and therefore, although of course when the three components X, Y, Z of the total resultant force have been obtained, they will be found to be such that $Xdx + Ydy + Zdz$ is a complete differential, the separate infinitely small elements of which these forces are compounded by integration with reference to the elements of the magnet, do not separately satisfy such a condition. Hence the investigation does not lead to an expression for the potential; but by means of it the following expressions for the three components of the force at any external point, or at a point within any infinitely small crevass perpendicular to the lines of magnetization, have been obtained*:—

---

* The expression $Xdx + Ydy + Zdz$ will not be a complete differential for internal points unless the distribution of magnetism be lamellar, since, for any internal point, X, Y, Z differ from the rectangular components of the "resultant force," as defined in § 479, by the quantities $4\pi a$, $4\pi\beta$, $4\pi\gamma$, respectively, and since (§ 483) the "resultant force," for all points, whether internal or external, is derivable from a potential. (See Postscript to § 517.)

$$X = \iiint dxdydz \left\{ \frac{\eta - y}{\Delta^3} \left(\frac{d\alpha}{dy} - \frac{d\beta}{dx} \right) - \frac{\zeta - z}{\Delta^3} \left(\frac{d\gamma}{dx} - \frac{d\alpha}{dz} \right) \right\}$$
$$\qquad - \left[\iint \left\{ \frac{\eta - y}{\Delta^3} (m\alpha - l\beta) - \frac{\zeta - z}{\Delta^3} (l\gamma - n\alpha) \right\} dS \right]$$

$$Y = \iiint dxdydz \left\{ \frac{\zeta - z}{\Delta^3} \left(\frac{d\beta}{dz} - \frac{d\gamma}{dy} \right) - \frac{\xi - x}{\Delta^3} \left(\frac{d\alpha}{dy} - \frac{d\beta}{dx} \right) \right\}$$
$$\qquad - \left[\iint \left\{ \frac{\zeta - z}{\Delta^3} (n\beta - m\gamma) - \frac{\xi - x}{\Delta^3} (m\alpha - l\beta) \right\} dS \right] \qquad \text{(IX.).}$$

$$Z = \iiint dxdydz \left\{ \frac{\xi - x}{\Delta^3} \left(\frac{d\gamma}{dx} - \frac{d\alpha}{dz} \right) - \frac{\eta - y}{\Delta^3} \left(\frac{d\beta}{dz} - \frac{d\gamma}{dy} \right) \right\}$$
$$\qquad - \left[\iint \left\{ \frac{\xi - x}{\Delta^3} (l\gamma - n\alpha) - \frac{\eta - y}{\Delta^3} (n\beta - m\gamma) \right\} dS \right]$$

[*Postscript to* § 517, *Nov.* 17, 1871.—These expressions, to be proved in § 518 for external points, may be taken as *a* definition for "resultant force" at points in the magnetized substance. They are simplified by putting

$$\frac{d\gamma}{dy} - \frac{d\beta}{dz} = u, \quad \frac{d\alpha}{dz} - \frac{d\gamma}{dx} = v, \quad \frac{d\beta}{dx} - \frac{d\alpha}{dy} = w \right\} \dots\dots(a),$$
and $\qquad n\beta - m\gamma = U, \ l\gamma - n\alpha = V, \ m\alpha - l\beta = W$

which, with x', y', z' substituted for x, y, z; u', v', w' for u, v, w; and x, y, z for ξ, η, ζ; reduces them to

$$X = \iiint dx'dy'dz' \frac{(z-z')v' - (y-y')w'}{\Delta^3} + \left[\iint \frac{(z-z')V' - (y-y')W'}{\Delta^3} dS \right] \ (b)$$

with the symmetrical forms for Y and Z. Now observe that

$$\iiint dx'dy'dz' \frac{(y-y')w'}{\Delta^3}$$

is the y-component of the resultant force at (x, y, z) due to a distribution of imaginary matter through the magnet and over its surface, having w for density at any interior point (x, y, z), and W for surface density at $[x, y, z]$; and for the other terms of (*b*), etc., consider corresponding distributions (v, V), and (u, U); and therefore instead of (*b*), etc., write

$$X = \frac{dN}{dy} - \frac{dM}{dz}, \quad Y = \frac{dL}{dz} - \frac{dN}{dx}, \quad Z = \frac{dM}{dx} - \frac{dL}{dy} \ \dots\dots(c)$$

denoting * by

---

* This notation has been introduced to agree with that used by Helmholtz in corresponding formulæ with reference to Vortex Motion. It is to be remarked

L the potential of distribution (u, U)

M „   „   „ (v, V) $\Bigg\}$ ...... (d)

N „   „   „ (w, W)

so that $L = \iiint \dfrac{u' dx' dy' dz'}{\Delta} + \left[\iint \dfrac{U' dS}{\Delta} \right]$, $M =$ etc., $N =$ etc. (e);

and, by Poisson's theorem,

$$\nabla^2 L = -4\pi u, \quad \nabla^2 M = -4\pi v, \quad \nabla^2 N = -4\pi w \ldots\ldots\ldots(f),$$

where, as is now usual, $\dfrac{d^2}{dx^2} + \dfrac{d^2}{dy^2} + \dfrac{d^2}{dz^2}$ is denoted by ∇^2. The second members of (f) vanish for all points external to the magnet, because there $u = 0$, $v = 0$, $w = 0$. Now for simplicity suppose the magnetization to diminish gradually, not abruptly, to zero at the boundary of the magnet. The second terms of the expressions (d) for L, M, N will disappear, and by differentiations and summation we have

$$\frac{dL}{dx} + \frac{dM}{dy} + \frac{dN}{dz} = \iiint \frac{\left(\dfrac{du'}{dx'} + \dfrac{dv'}{dy'} + \dfrac{dw'}{dz'} \right) dx' dy' dz'}{D}.$$

But (a) show that $\dfrac{du}{dx} + \dfrac{dv}{dy} + \dfrac{dw}{dz} = 0$ ....(g);

and therefore $\dfrac{dL}{dx} + \dfrac{dM}{dy} + \dfrac{dN}{dz} = 0$ ....(h).

However quick the gradation from finite values of u, v, w within the magnet, to zero through external space, this equation holds, and therefore it holds in the limit, when the magnetization comes to an end abruptly at the boundary. To prove (h) directly from the expressions (e), with the surface-terms included, will be found a good exercise for the student.

From (c) by differentiations, and application of (f) and (h), we find

$$\frac{dX}{dx} + \frac{dY}{dy} + \frac{dZ}{dz} = 0 \ldots\ldots\ldots\ldots\ldots(k)$$

$\dfrac{dZ}{dy} - \dfrac{dY}{dz} = 4\pi u, \quad \dfrac{dX}{dz} - \dfrac{dZ}{dx} = 4\pi v, \quad \dfrac{dY}{dx} - \dfrac{dX}{dy} = 4\pi w,$ $\Bigg\}$

or in virtue of (a) (l)

$\dfrac{dZ}{dy} - \dfrac{dY}{dz} = 4\pi\left(\dfrac{d\gamma}{dy} - \dfrac{d\beta}{dz}\right), \dfrac{dX}{dz} - \dfrac{dZ}{dx} = 4\pi\left(\dfrac{d\alpha}{dz} - \dfrac{d\gamma}{dx}\right), \dfrac{dY}{dx} - \dfrac{dX}{dy} = 4\pi\left(\dfrac{d\beta}{dx} - \dfrac{d\alpha}{dy}\right)$

---

that the quantities u, v, w, U, V, W thus introduced fulfil the equations (1) and (2) of § 539. They represent the components of the internal and superficial distributions of electric currents, in the electro-magnetic representative (§ 554) of the given polar magnet.

The corresponding properties of $\mathfrak{X}$, $\mathfrak{Y}$, $\mathfrak{Z}$, if these denote the components of the "resultant force" as defined in §§ 479, 480, are [see § 473 (2) and § 483]

$$\frac{d\mathfrak{X}}{dx} + \frac{d\mathfrak{Y}}{dy} + \frac{d\mathfrak{Z}}{dz} = -4\pi \left(\frac{d\alpha}{dx} + \frac{d\beta}{dy} + \frac{d\gamma}{dz} \right) \ldots\ldots\ldots(m),$$

$$\frac{d\mathfrak{Z}}{dy} - \frac{d\mathfrak{Y}}{dz} = 0, \quad \frac{d\mathfrak{X}}{dz} - \frac{d\mathfrak{Z}}{dx} = 0, \quad \frac{d\mathfrak{Y}}{dx} - \frac{d\mathfrak{X}}{dy} = 0 \ldots\ldots\ldots(n).$$

These equations, as well as (k) and (l), hold through all space, the values of α, β, γ being zero in every part of space not containing magnetized matter. Some if not all of the differential coefficients appearing in them become infinite when the magnetization varies abruptly from one side to the other of any surface, but the interpretation presents no difficulty. Taking for instance the case when the magnetization, finite up to the boundary of the magnet, comes to an end abruptly there, let $X_{,}$ and $X_{,,}$ denote the values of X at points infinitely near one another outside and inside the boundary; and similarly for Y, Z, $\mathfrak{X}$, $\mathfrak{Y}$, $\mathfrak{Z}$. We have by § 7 (5), § 517 (c) and (e), and § 473 (1),

$$X_{,} - X_{,,} = 4\pi\,(nV - mW), \quad Y_{,} - Y_{,,} = 4\pi\,(lW - nU), \quad Z_{,} - Z_{,,} = 4\pi\,(mU - lV) \;\; (o)$$

and $\quad \mathfrak{X}_{,} - \mathfrak{X}_{,,} = 4\pi\rho l, \quad \mathfrak{Y}_{,} - \mathfrak{Y}_{,,} = 4\pi\rho m, \quad \mathfrak{Z}_{,} - \mathfrak{Z}_{,,} = 4\pi\rho n \left.\right\}$
where $\qquad\qquad\qquad \rho = lu + mv + nw \qquad\qquad\qquad (p).$

By (a) we have

$$nV - mW = l\,(m\beta + n\gamma) - (m^2 + n^2)\,\alpha = l\,(l\alpha + m\beta + n\gamma) - \alpha.$$

Hence, with the notation of (p), (o) becomes

$$X_{,} - X_{,,} = 4\pi\,(l\rho - \alpha), \quad Y_{,} - Y_{,,} = 4\pi\,(m\rho - \beta), \quad Z_{,} - Z_{,,} = 4\pi\,(n\rho - \gamma) \;\; (q).$$

In a foot-note to § 517 above it was stated that the values of X, Y, Z differ from what in this postscript I call $\mathfrak{X}$, $\mathfrak{Y}$, $\mathfrak{Z}$ by quantities equal to $4\pi\alpha$, $4\pi\beta$, $4\pi\gamma$, respectively; a statement which is no doubt to be proved directly by carefully examining the meaning of the integrals of § 518 for internal points. We may now verify it by taking the difference between (k) and (m), and the differences between (l) and (n). If in these we put

$$X - \mathfrak{X} - 4\pi\alpha = P, \quad Y - \mathfrak{Y} - 4\pi\beta = Q, \quad Z - \mathfrak{Z} - 4\pi\gamma = R,$$

they give $\qquad\qquad \dfrac{dP}{dx} + \dfrac{dQ}{dy} + \dfrac{dR}{dz} = 0$

$$\frac{dR}{dy} - \frac{dQ}{dz} = 0, \ \frac{dP}{dz} - \frac{dR}{dx} = 0, \ \frac{dQ}{dx} - \frac{dP}{dy} = 0.$$

These last three equations show that

$$P = \frac{d\psi}{dx}, \ \ Q = \frac{d\psi}{dy}, \ \ R = \frac{d\psi}{dz}$$

where ψ if not zero is a function of x, y, z: and the first then becomes

$$\frac{d^2\psi}{dx^2} + \frac{d^2\psi}{dy^2} + \frac{d^2\psi}{dz^2} = 0.$$

This equation must hold through all space when there is no abrupt variation of magnetization; and, as ψ must vanish at an infinite distance from the magnet in any direction, we must (§ 206 above) therefore (whether there are abrupt variations or not) have $\psi = 0$. The proof may be illustrated for abrupt variations, by taking the differences of equations (q) and (p), which show that

$$(X - \mathfrak{X} - 4\pi\alpha)_{,} - (X - \mathfrak{X} - 4\pi\alpha)_{,,} = 0; \ (Y \text{ etc., } Z \text{ etc.});$$

or

$$P_{,} - P_{,,} = 0, \ Q_{,} - Q_{,,} = 0, \ R_{,} - R_{,,} = 0;$$

which prove that $\psi_{,} - \psi_{,,} = 0,$

the suffixed accents denoting values for infinitely near points on the two sides of the surface of abrupt change.

We conclude that through all space

$$X = \mathfrak{X} + 4\pi\alpha, \ Y = \mathfrak{Y} + 4\pi\beta, \ Z = \mathfrak{Z} + 4\pi\gamma \dots\dots(r);$$

which, for space unoccupied by magnetized matter, give (what we knew before)

$$X = \mathfrak{X}, \ \ Y = \mathfrak{Y}, \ \ Z = \mathfrak{Z}.$$

For space within the magnet, it was shown in § 479 that the force $(\mathfrak{X}, \mathfrak{Y}, \mathfrak{Z})$ is the resultant force experienced by a unit pole in a crevass tangential to the lines of magnetization. From this, and (r), it follows that, as was asserted in § 517, the force (X, Y, Z) is the resultant force experienced by a unit pole in a crevass perpendicular to the lines of magnetization. Of these two definitions of "resultant force" for space within a magnet, the former, as suitable to a *polar magnet* (§ 549), will sometimes be called the "polar definition," and the latter, as suitable for an electromagnet, the "electromagnetic definition," for the sake of brevity.]

518. The investigation by which I originally obtained the expressions (IX. of § 517) is, with reference to galvanism, precisely analogous to the investigation in § 473 with reference to imaginary magnetic matter. It cannot be given without explanations regarding the elements of electro-magnetism which would exceed the limits of the present communication[*]; but when I had once discovered the formulæ I had no difficulty in working out the subjoined analytical demonstration of them for the case of an external point, which is precisely analogous to Poisson's original investigation (given in § 495 above) of the formula of § 482.

Equations (3) and (4) of §§ 482 and 483 lead to expressions for the components of the resultant force at any point in the neighbourhood of a magnet. Taking X only (since the expressions for the three components are symmetrical), we have

$$X = -\frac{d}{d\xi}\iiint dx\,dy\,dz\left\{\alpha\frac{d\frac{1}{\Delta}}{dx}+\beta\frac{d\frac{1}{\Delta}}{dy}+\gamma\frac{d\frac{1}{\Delta}}{dz}\right\}.$$

Now if the factor of $dx\,dy\,dz$ in the second member of this equation be differentiated with reference to ξ, an expression is obtained which does not become infinitely great for any values of x, y, z included within the limits of integration, since the point (ξ, η, ζ) is considered to be external in the present investigation. Hence the differentiation with reference to ξ may be performed under the integral sign; and, since

$$\frac{d\frac{1}{\Delta}}{d\xi} = -\frac{d\frac{1}{\Delta}}{dx},$$

we thus obtain

$$X = \iiint dx\,dy\,dz\left\{\alpha\frac{d^2\frac{1}{\Delta}}{dx^2}+\beta\frac{d^2\frac{1}{\Delta}}{dx\,dy}+\gamma\frac{d^2\frac{1}{\Delta}}{dx\,dz}\right\}[†].$$

---

[*] [*Note, Nov.* 1871.—It is given in § 554, below.]

[†] If the point (ξ, η, ζ) be either within the magnet or infinitely near it, the factor of $dx\,dy\,dz$ in this integral is infinitely great for values of (x, y, z) included

Now, for all points included within the limits of integration, we have, from Laplace's well-known equation,

$$\frac{d^2 \frac{1}{\Delta}}{dx^2} = -\left(\frac{d^2 \frac{1}{\Delta}}{dy^2} + \frac{d^2 \frac{1}{\Delta}}{dz^2} \right),$$

and therefore

$$X = \iiint dx\,dy\,dz \left\{ -\alpha \left(\frac{d^2 \frac{1}{\Delta}}{dy^2} + \frac{d^2 \frac{1}{\Delta}}{dz^2} \right) + \beta \frac{d^2 \frac{1}{\Delta}}{dx\,dy} + \gamma \frac{d^2 \frac{1}{\Delta}}{dx\,dz} \right\}$$

Dividing the second member into four terms, and applying an obvious process of integration by parts, we deduce

$$X = \Bigg[\iint \left\{ -\alpha \frac{d \frac{1}{\Delta}}{dy} dx\,dz - \alpha \frac{d \frac{1}{\Delta}}{dz} dx\,dy + \beta \frac{d \frac{1}{\Delta}}{dy} dy\,dz + \gamma \frac{d \frac{1}{\Delta}}{dz} dy\,dz \right\}$$

$$+ \iiint dx\,dy\,dz \left\{ \frac{d\alpha}{dy} \frac{d \frac{1}{\Delta}}{dy} + \frac{d\alpha}{dz} \frac{d \frac{1}{\Delta}}{dz} - \frac{d\beta}{dx} \frac{d \frac{1}{\Delta}}{dy} - \frac{d\gamma}{dx} \frac{d \frac{1}{\Delta}}{dz} \right\} \Bigg]$$

Modifying the double integral by assuming, in its different terms,	$dy\,dz = l\,dS$; $dz\,dx = m\,dS$; $dx\,dy = n\,dS$,

and altering the order of all the terms, we obtain

---

within the limits of integration; and it may be demonstrated that the value of a part of the integral corresponding to any infinitely small portion of the magnet infinitely near the point (ξ, η, ζ) is in general finite, and that it depends on the form of this portion, on its position with reference to the line of magnetization through (ξ, η, ζ), and on the proportions of the distances of its different parts from this point. It follows that if the point (ξ, η, ζ) be internal, and if a portion of the magnet round it be omitted from the integral, the value of the integral will be affected by the form of the omitted portion, however small its dimensions may be, and consequently the complete integral has no determinate value if the point (ξ, η, ζ) be internal. Hence although, as we have seen above (§§ 482, 483),

$$-\frac{d}{d\xi} \iiint dx\,dy\,dz \left\{ \alpha \frac{d \frac{1}{\Delta}}{dx} + \beta \frac{d \frac{1}{\Delta}}{dy} + \gamma \frac{d \frac{1}{\Delta}}{dz} \right\}$$

has in all cases a determinate value, which, by the definition (§ 479), is called the component parallel to OX of the resultant force at (ξ, η, ζ), the expression

$$\iiint dx\,dy\,dz \frac{d}{d\xi} \left\{ \alpha \frac{d \frac{1}{\Delta}}{dx} + \beta \frac{d \frac{1}{\Delta}}{dy} + \gamma \frac{d \frac{1}{\Delta}}{dz} \right\}$$

has no meaning when (ξ, η, ζ) is in the substance of the magnet.

$$X = \iiint dxdydz \left\{ \frac{d\frac{1}{\Delta}}{dy} \left(\frac{d\alpha}{dy} - \frac{d\beta}{dx} \right) - \frac{d\frac{1}{\Delta}}{dz} \left(\frac{d\gamma}{dx} - \frac{d\alpha}{dz} \right) \right\}$$

$$- \left[\iint \left\{ \frac{d\frac{1}{\Delta}}{dy}(m\alpha - l\beta) - \frac{d\frac{1}{\Delta}}{dz}(l\gamma - n\alpha) \right\} dS \right].$$

This expression, when the indicated differentiations are actually performed upon $\frac{1}{\Delta}$, becomes identical with the expression for X at the end of § 517, and the formulæ which it was required to prove are therefore established.

519. The triple integrals in these expressions vanish in the case of a lamellar distribution, in virtue of the equations (III.) of § 514; and we have simply

$$\left. \begin{array}{l} X = - \left[\iint \left\{ \dfrac{d\frac{1}{\Delta}}{dy}(m\alpha - l\beta) - \dfrac{d\frac{1}{\Delta}}{dz}(l\gamma - n\alpha) \right\} dS \right] \\[3mm] Y = - \left[\iint \left\{ \dfrac{d\frac{1}{\Delta}}{dz}(n\beta - m\gamma) - \dfrac{d\frac{1}{\Delta}}{dx}(m\alpha - l\beta) \right\} dS \right] \\[3mm] Z = - \left[\iint \left\{ \dfrac{d\frac{1}{\Delta}}{dx}(l\gamma - n\alpha) - \dfrac{d\frac{1}{\Delta}}{dy}(n\beta - m\gamma) \right\} dS \right] \end{array} \right\} \dots \text{(X.)}$$

To interpret these expressions, let us assume, for brevity,

$$U = n\beta - m\gamma; \quad V = l\gamma - n\alpha; \quad W = m\alpha - l\beta \dots \text{(XI.)}$$

From these we deduce

$$\left. \begin{array}{l} mW - nV = \alpha - l\,(l\alpha + m\beta + n\gamma) = \alpha_{,} \\ nU - lW = \beta - m\,(l\alpha + m\beta + n\gamma) = \beta_{,} \\ lV - mU = \gamma - n\,(l\alpha + m\beta + n\gamma) = \gamma_{,} \end{array} \right\} \text{(XII.)};$$

where $\alpha_{,}$, $\beta_{,}$, $\gamma_{,}$ denote the rectangular components of the tangential component of the magnetization at a point infinitely near the surface. Conversely, from these equations we deduce

$$U = n\beta_{,} - m\gamma_{,}; \quad V = l\gamma_{,} - n\alpha_{,}; \quad W = m\alpha_{,} - l\beta_{,} \dots \text{(XIII.)}$$

Now the direct data required for obtaining the values of X, Y, and Z, by means of formulæ (X.), are simply the values of U, V, W at all points of its surface. Equations (XII.) show that with these data the values of $\alpha_{,}$, $\beta_{,}$, $\gamma_{,}$ may be calculated; and again, equations (XIII.) show conversely that if $\alpha_{,}$, $\beta_{,}$, $\gamma_{,}$

be given the required data for the problem may be immediately deduced. We infer that the necessary and sufficient data for determining the resultant force of a lamellar magnet, at any external point, by means of formulæ (X.), are equivalent to a specification of the direction and magnitude of the tangential component of the intensity of magnetization at every point infinitely near the surface of the magnet; and we conclude, as we did in § 517 from a very different process of reasoning, that besides these data, nothing but that it is lamellar throughout need be known of the interior distribution.

520. The close analogy which exists between solenoidal and lamellar distributions of magnetism having led me to the new formulæ which have just been given, it occurred to me that a formula (or formulæ, if it were necessary here to separate the cases of internal and external points), for solenoidal distributions analogous to the formulæ (VII.) of § 516 for lamellar distributions might be discovered. Taking an analytical view of the problem (the synthetical view, although itself much more obvious, not showing any very obvious way of arriving at a formula of the desired kind), I observed that the formula $\iint \frac{\phi \cdot \cos \theta dS}{\Delta^2}$ is deduced from the general expression for the potential by a partial integration performed upon factors involving α, β, γ, and depending on the integrability of the function $\alpha dx + \beta dy + \gamma dz$, insured by the equations

$$\frac{d\beta}{dz} - \frac{d\gamma}{dy} = 0, \ \frac{d\gamma}{dx} - \frac{d\alpha}{dz} = 0, \ \frac{d\alpha}{dy} - \frac{d\beta}{dx} = 0,$$

for a lamellar distribution ; and I endeavoured to find a corresponding mode of treatment for solenoidal distributions, to consist of a partial integration, commencing still with factors involving α, β, γ, but depending now upon the single equation

$$\frac{d\alpha}{dx} + \frac{d\beta}{dy} + \frac{d\gamma}{dz} = 0 \dots\dots\dots\dots\dots(a),$$

instead of the three equations required in the former process. After some fruitless attempts to connect this equation with the integrability of some function of two independent variables, I fell upon the following investigation, which exactly answered my expectations :—

521. In virtue of the preceding equation (a), we may assume

$$a = \frac{dH}{dy} - \frac{dG}{dz}, \quad \beta = \frac{dF}{dz} - \frac{dH}{dx}, \quad \gamma = \frac{dG}{dx} - \frac{dF}{dy} \dots\text{(XIV.)},$$

where F, G, H are three functions to a certain extent arbitrary. These functions I have since found, have for their most general expressions

$$\left. \begin{aligned} F &= \tfrac{1}{3} \iint dydz \left(\frac{d\beta}{dy} - \frac{d\gamma}{dz} \right) + \frac{d\psi}{dx} \\ G &= \tfrac{1}{3} \iint dzdx \left(\frac{d\gamma}{dz} - \frac{da}{dx} \right) + \frac{d\psi}{dy} \\ H &= \tfrac{1}{3} \iint dxdy \left(\frac{da}{dx} - \frac{d\beta}{dy} \right) + \frac{d\psi}{dz} \end{aligned} \right\} \dots\dots\text{(XV.)};$$

where ψ denotes an absolutely arbitrary function; and the indicated integrations are indefinite, with the arbitraries which they introduce subject to the equations (XIV.).

The demonstration of these equations follows immediately from the results obtained by differentiating the three equations (XIV.) with reference to x, y, and z respectively. The simplest final forms for F, G, and H are the following, which are deduced from the preceding by integration:—

$$\left. \begin{aligned} F &= \tfrac{1}{3} \int (\beta dz - \gamma dy) + \frac{d\psi}{dx} \\ G &= \tfrac{1}{3} \int (\gamma dx - a dz) + \frac{d\psi}{dy} \\ H &= \tfrac{1}{3} \int (a dy - \beta dx) + \frac{d\psi}{dz} \end{aligned} \right\} \dots\dots\text{(XVI.)}.$$

Making substitutions according to the formulæ (XIV.) for a, β, γ in the general expression for the potential, we have

$$V = \iiint dxdydz \left\{ \left(\frac{dH}{dy} - \frac{dG}{dz} \right) \frac{d\frac{1}{\Delta}}{dx} + \left(\frac{dF}{dz} - \frac{dH}{dx} \right) \frac{d\frac{1}{\Delta}}{dy} + \left(\frac{dG}{dx} - \frac{dF}{dy} \right) \frac{d\frac{1}{\Delta}}{dz} \right\}.$$

Dividing the second member into six terms, and integrating each by parts, commencing upon the factors such as $\frac{dH}{dy} dy$, we obtain an expression, with a triple integral involving six terms which destroy one another two and two because of properties such as

$$\frac{d}{dy}\frac{d\frac{1}{\Delta}}{dx} = \frac{d}{dx}\frac{d\frac{1}{\Delta}}{dy};$$

and besides, a double integral, which may be reduced in the usual manner to a form involving dS, an element of the surface. We thus obtain, finally,

$$V = \left[\iint\left\{(mH - nG)\frac{d\frac{1}{\Delta}}{dx} + (nF - lH)\frac{d\frac{1}{\Delta}}{dy} + (lG - mF)\frac{d\frac{1}{\Delta}}{dz}\right\}dS\right]$$

$$\text{(XVII.).}$$

522. The second member of this equation expresses the potential of a certain distribution of magnetism in an infinitely thin sheet coinciding with the surface of the body; the total magnetic moment of the magnetism in the area dS being

$$\{(mH - nG)^2 + (nF - lH)^2 + (lG - mF)^2\}^{\frac{1}{2}}\, dS,$$

and its direction cosines proportional to

$$mH - nG,\quad nF - lH,\quad lG - mF.$$

Now we have identically,

$$l(mH - nG) + m(nF - lH) + n(lG - mF) = 0;$$

and hence the direction of this imaginary magnetization at every point of the surface is perpendicular to the normal. It follows that we have found a distribution of tangential magnetism in an infinitely thin sheet coinciding with the bounding surface which produces the same potential at any point, internal or external, as the given solenoidal magnet. [It is remarkable that the *imaginary tangential* magnetization thus found 'depends (§ 523) upon the *normal* component of the actual magnetization infinitely near the surface; so that, besides this normal component, nothing need be known of the actual magnetization except that it is solenoidal. Compare conclusion of § 519.]

523. The conclusion of § 522 may be arrived at synthetically in a very obvious manner, by taking into account the property of a solenoid stated in § 510, according to which it appears that any two solenoids of equal strength, with the same ends, produce the same force at any point whether in the magnetized substance of either, or not. For it follows from

this, that when a magnet is divisible into solenoids with their ends on its surface, we may by joining the two ends of each solenoid by any arbitrary curve on this surface, and laying a solenoid of equal strength along this curve, obtain a series of solenoids, constituting by their superposition, a tangential distribution of magnetism in an infinitely thin sheet coinciding with the bounding surface, which produces the same resultant force at any internal or external point as the given magnet. It is not, however, easy to deduce from this synthesis a formula involving the requisite arbitrary functions to express a superficial distribution satisfying the existing conditions in the most general manner. The analytical investigation given above, supplies, in reality, a complete solution of this problem.

It may be remarked that the sole condition which F, G and H considered as functions of the co-ordinates, x, y, z, of some point in the surface of the magnet, and therefore functions of two independent variables, must satisfy in order that (XVII.) may express correctly the potential at any point, is—

$$l\left(\frac{dH}{dy}-\frac{dG}{dz}\right)+m\left(\frac{dF}{dz}-\frac{dH}{dx}\right)+n\left(\frac{dG}{dx}-\frac{dF}{dy}\right)=l\alpha+m\beta+n\gamma \text{(XVIII.)},$$

x, y, and z of course being supposed to satisfy the equation to the surface ; and it may be proved, by a demonstration independent of the investigation which has been given, that the second member of (XVII.) has the same value for any functions F, G, H whatever, which are subject to this relation.

[*Postscript, Dec.* 7, 1871, *and Jan.* 6, 1872.—Inasmuch as the second member of (XVIII.) is (§ 473 (1)), the surface density of the imaginary magnetic matter, representing the polarity of the given solenoidal magnet, we may eliminate the idea of magnetization, and so arrive at the following remarkable theorem :—

Let ρ be the density at any point of a superficial distribution of matter on a surface S, which may be either a closed surface or an open shell, there being as much negative matter as positive in the whole distribution, and let F, G, H be any three quantities such that

$$l\left(\frac{dH}{dy}-\frac{dG}{dz}\right)+m\left(\frac{dF}{dz}-\frac{dH}{dx}\right)+n\left(\frac{dG}{dx}-\frac{dF}{dy}\right)=\rho\ldots \text{(XIX.)};$$

the potential of this distribution, that is to say $\iint \frac{\rho dS}{\Delta}$, is correctly expressed by the formula (XVII.). When S is a closed surface this expression holds for the space within S, as well as for external space. From the remark with reference to (XVII.) and (XVIII.) at the conclusion of the section in the original now numbered § 523, it appears that the values of F, G, H given by (XVI.), although expressing the most general solution of (XIV.), are not the most general expressions for functions F, G, H to satisfy (XVII.); and that instead of F, G, H in (XVII.) we may substitute

$$F + F', \quad G + G', \quad H + H'$$

where F, G, H are given by (XVI.) and F', G', H' are any three functions of x, y, z, which, over the whole surface S, satisfy the equation

$$l\left(\frac{dH'}{dy} - \frac{dG'}{dz}\right) + m\left(\frac{dF'}{dz} - \frac{dH'}{dx}\right) + n\left(\frac{dG'}{dx} - \frac{dF'}{dy}\right) = 0 \ldots (\text{XX.})$$

The surface distribution of tangential magnetization specified by F', G', H' in accordance with the explanations of § 522, consists of closed solenoids lying on the surface S.]

CHAPTER VI.*—*On Electromagnets.*

524. Oersted's discovery of the mutual forces between magnets and conductors containing electric currents gave rise to the science of electromagnetism. It was soon found that there are also mutual forces between different conductors and between different parts of the same conductor conveying electric currents: and various very remarkable electro-magenetic phenomena were observed by different experimenters, of which the most remarkable are the continuous rotations of portions of conductors round magnets and of magnets round conductors,

---

* [*Note, October* 1871.—This chapter was written twenty-two years ago, and has lain in manuscript ever since, because I had not succeeded in finding time to write a sequel on inverse problems. It is now printed from the original manuscript with only a few verbal alterations, and it will be followed in this volume (Chap. IX.) by the long-projected article on inverse problems, of which something was communicated to the British Association at its Oxford Meeting of 1847, but not published except in the very short abstract contained in the Report of that meeting.]

discovered by Faraday to result, in certain circumstances, from
their mutual actions. The laws to which all these actions are
subject were first completely investigated by Ampère. His
experiments are the foundation, and the conclusions which he
deduces from them constitute the elements, of the Mathematical
Theory of Electromagnetism. As a complete and satisfactory
account of these researches is to be found in Ampère's
original papers*, and a succinct exposition of the mathematical part of the investigations, in Murphy's *Treatise on Electricity*, the results will be considered as fully established, and
those of them which are required in the present essay will be
quoted.

525.   Let P and P' be points in two conductors, of which the
lateral dimensions are very small compared with the distance
PP'; let σ and σ' be the length of infinitely small elements
of these conductors, with their centres at the points P and P'
respectively, and terminated by planes perpendicular to the
directions of the conductors; let PP' be denoted by r; let θ
and θ' denote the angles at which the directions of the conductors at P and P' are inclined to the line PP'; and let ϕ be
the angle between two planes each passing through PP', and
respectively containing the directions of the conductors. Thus,
if there be electrical currents in the two conductors, they will
mutually act and react with a system of force which is the
same as would result from mutual forces, in lines joining all
the infinitely small arcs σ of the one, and σ' of the other, given
in amount (attractions reckoned positive and repulsions negative) by the following formula:—

$$\frac{\gamma\sigma\,\gamma'\sigma'}{r^2}\,(2\sin\theta\sin\theta'\cos\phi - \cos\theta\cos\theta')\dagger$$

---

* "Sur la Théorie Mathématique des phénomènes électro-dynamiques."
Collection of six "Mémoires" of dates 4th and 20th December 1820, 10th June
1822, 22nd December 1823, 12th September and 21st November 1825. Published
in the *Mémoires* of the French Academy, 1827.

† [*Note, Oct.* 1871.—In the original manuscript the formula stands

$$\frac{\gamma\sigma.\gamma'\sigma'}{r^2}\,(\sin\theta\sin\theta'\cos\phi - \tfrac{1}{2}\cos\theta\cos\theta').$$

I have doubled its second member to avoid the inconvenient distinction between
"electro-dynamic" and "electro-magnetic" units to which in its original form,
(the form in which Weber used it in his system of absolute units,) it leads. See
below, § 531.]

where γ and γ' denote quantities invariable in value for the same two conductors, with the same electrical currents flowing through them. These quantities (γ, γ') are the numerical measures of the strengths of the currents.

526. Again, let N be the north pole of an infinitely thin uniformly and longitudinally magnetized bar, and S its south pole: let P be a point in a conductor, and let σ and γ denote the same as before, with reference to this conductor. Let NP and SP be denoted by Δ and Δ' respectively; and let the angles between NP and σ and between NSP and σ be denoted by ϕ and ϕ' respectively. There will be such a mutual action between the magnet and the galvanic arc σ that each will experience a force, the resultant of two forces through P perpendicular respectively to the planes of NP and σ, and of SP and σ, given in amount by the following expressions respectively:—

$$\frac{m \cdot \gamma \sigma}{\Delta^2} \sin \phi, \quad \text{and} \frac{m \cdot \gamma \sigma}{\Delta'^2} \sin \phi'.$$

The directions of these forces, upon the element, if the direction of the current be from east to west, and if N and S be each north of P, will be;—the former obliquely or directly downwards, and the latter,—upwards. [For mnemonic principle see below, § 547.] The magnet will be acted upon as if a point in the position of P, rigidly connected with it, experienced two forces equal and opposite to the forces of which the action on σ is compounded.

527. Let (x, y, z) denote the middle point of the element σ, and (x', y', z') the middle point of the element σ', according to ordinary rectangular co-ordinates. Let also l, m, n be the direction cosines of the former element, and l', m', n' those of the latter; quantities which will be all positive when the current in each element is in a similar direction to that of a point moving from the origin towards the space between the positive parts of the co-ordinate planes. The expression for the force between the elements in terms of these data, will be

$$\frac{\gamma \sigma \cdot \gamma' \sigma' \{2(ll'+mm'+nn')[(x-x')^2+(y-y')^2+(z-z')^2]-3[l(x-x')+m(y-y')+n(z-z')][l'(x-x')+m'(y-y')+n'(z-z')]\}}{\{(x-x')^2+(y-y')^2+(z-z')^2\}^2}$$

528. Again, if ξ, η, ζ denote the co-ordinates of a unit north

pole, and x, y, z those of an infinitely small element σ of an electric current of strength γ, in a direction (l, m, n), the mutual action will be

$$\frac{\gamma\sigma.\sin\phi}{(\xi-x)^2+(\eta-y)^2+(\zeta-z)^2}, \text{ or } \frac{\gamma\sigma.\sin\phi}{\Delta^2},$$

and will be in a line of which the direction cosines are

$$\frac{m(\zeta-z)-n(\eta-y)}{\Delta\sin\phi}, \frac{n(\xi-x)-l(\zeta-z)}{\Delta\sin\phi}, \frac{l(\eta-y)-m(\xi-x)}{\Delta\sin\phi}.$$

529. Hence the components of the force experienced by the element of electric current are given in magnitude and direction by the following expressions:—

$$\frac{\gamma\sigma[m(\zeta-z)-n(\eta-y)]}{\Delta^3}, \frac{\gamma\sigma[n(\xi-x)-l(\zeta-z)]}{\Delta^3}, \frac{\gamma\sigma[l(\eta-y)-m(\xi-x)]}{\Delta^3}.$$

If the axes of co-ordinates be so chosen that when OX is from south to north, and OY from east to west, OZ will be vertically upwards, these expressions will be applicable, as far as regards signs, to the direction of the action which the electric arc experiences; and it would be necessary to change the sign of each, to make them applicable to the direction of the force upon the pole.

530. These expressions, since they involve l, m, n only linearly, show that a galvanic arc σ, of strength γ, in the direction l, m, n, produces the same effect either upon another arc, or upon a magnet, as three arcs parallel to the axes of co-ordinates, each of the same strength, γ, and of lengths respectively equal to σl, σm, σn.

531. The factor γ being taken as the numerical measure of the strength of the current in the circuit of which σ is an arc, the unit of strength for an electric current may be defined in the following manner:—

If a galvanic current, in a conductor of infinitely small section, be such that the mutual action between any infinitely small arc of it, and a unit magnetic pole held in a direction perpendicular to the length of the arc, at a unit of distance, is numerically equal to σ the infinitely small length of the element, the strength of the current is unity.

Or in the following manner :—

If a galvanic current in a conductor of infinitely small section be such that the action between two infinitely small portions of it in line with one another and at a distance unity from one another, is numerically equal to the product of the length of the elements, the strength of the current is unity.

532. If what is called "an electric current" be in reality the transference of matter along the conductor in which it exists, the "strength of the current" numerically measured in the manner which has been explained, will depend upon the quantity of this matter transmitted in a given time; and a unit of time may be chosen, according to the unit of electrical quantity which is adopted, so that the quantity γ, measured as above explained by the electro-magnetic action of the conductor, may be numerically the quantity of electricity which flows across any section of it in a unit of time.

533. In a continuous current, this quantity is of course the same for every section; and, as it is impossible that a continuous stream of electricity can emanate from one body, and be discharged into another, the current must be *re-entering*, or every continuous current must form what is termed "a closed circuit." It is found by experiment that whatever be the dimensions or material of the different parts of the conductor along which the current flows, provided always the dimensions of the section be small compared with the distances through which the electro-magnetic action is observed, the quantity γ has the same value for all parts of it; and even in the places where the electro-motive force operates, as has been shown by Faraday, as in the liquid of any ordinary galvanic battery, or in a conductor in motion in the neighbourhood of a magnet, the electro-magnetic effects are observable and probably to exactly the same degree; so that it would probably be found that a galvanic circuit consisting of a battery of small cells arranged in a circular arc, and a wire completing the circuit by joining the poles, would produce the same electro-magnetic effects at all points symmetrically situated with reference to the circle, irrespectively of the part of the circuit, whether the cells or the wire, provided always that the distances considered be great compared with either the dimensions of a section of the

wire, or of any of the cells made by planes perpendicular to the plane of the circle, through its centre.

534. *Hypothesis of Matter flowing.*—In the theory of electro-magnetism it is quite unnecessary to adopt any such hypothesis as this, however probable or improbable it may be as an ulterior theory; and all that we could introduce as depending upon it is that, for a linear circuit of varying section or material, the quantity γ is the same throughout the circuit, and that all finite circuits possessing continuous currents are necessarily closed; two facts which cannot be assumed *a priori*, but which are in reality established by satisfactory experimental evidence.

535. *Division of Electromagnets into three Classes—Linear, Superficial, and Solid.*—If all the dimensions of any section of the conductor along which the current is communicated be infinitely small, the complete circuit constitutes what will be called a linear electromagnet.

When the electric currents are confined to a shell of which the thickness is infinitely small, and when they are continuously distributed through it, or distributed through it in such a manner as not to satisfy the condition by which a linear electromagnet is defined, the entire group of the complete circuits constitutes what is called a superficial electromagnet [or surface-electromagnet].

When electric currents are so arranged as to fill any solid portion of space, the group of the complete circuits constitutes a solid electromagnet.

It is clear that, in practice, electromagnets may be treated as linear, or superficial if the quantities which ought to be infinitely small, are merely very small compared with the dimensions of the magnets, and with the distances at which the electro-magnetic action are to be observed; and again, if wires, or linear currents of any kind, be disposed upon any surface or through any space, so that the distances between those which are adjacent are small compared with the dimensions of the circuits, or of the curves, or with the distances at which the magnetic actions are to be observed, the group may be considered as constituting practically a superficial electromagnet; and a solid electromagnet may be composed of a group of galvanic wires similarly arranged through a solid space.

536. *Linear Electromagnets.*—A linear electromagnet is completely specified when the form of the closed curve of the
current, and γ, the strength, are given.

Irrespectively of any theory, the term "electric current" will
often be made use of; but as the terms, literally interpreted,
imply a theory which, to say the least, is doubtful, it must be
borne in mind that they are not to be interpreted literally, and
that they are only used in this essay occasionally for convenience; and especially because of the almost universal use which
is made of them by writers on the same subject. The term
"galvanism" will often be used to denote the agency to which
the phenomena presented by continuous electric currents are
due, and *quantity of galvanism* in a linear conductor will be
measured according to the following standard :—

The strength of the current in a linear electromagnet into
the length of any part of the conductor in which it exists, is
the quantity of galvanism in that portion.

The term *intensity* will be used with reference to linear
currents, according to the following definition :—

The intensity of the galvanism in any part of a linear electro-
magnet is equal to the strength of the current, divided by the
area of the section of the conductor.

Hence in a linear conductor of which the section is not uniform
throughout, the intensity of the galvanism will vary inversely
as the section from one part to another of the conductor*.

537. *Superficial Electromagnets.*—*Def.* The quantity of galvanism on any small portion of the surface, divided by its area,
is the superficial intensity of the galvanism at that point.

If the superficial intensity† and the direction of the galvanism
is given at every point of a given surface, the specification of the
superficial electromagnet is complete. There are, however,
certain conditions to which such a specification is subject, and
an arbitrary specification, not satisfying them, will not correspond to any possible superficial electromagnet. The founda-

* [*Note, Oct.* 25, 1871.—I leave this section exactly as I find it in the old
manuscript, under protest that I do not now approve of the mode in which the
word "galvanism" is used in the terms which it proposes. Where these terms
occur henceforth it is because I have not invariably altered the manuscript to
substitute more convenient modes of expression.]

† [In 1871 we should rather say surface-intensity than superficial intensity.]

tion of these conditions is the fact that no incomplete circuit can exist permanently, from which it follows that all the currents, or the continuous superficial flux of electricity constituting a superficial electromagnet, must be resolvable into a group of closed galvanic currents.

This will lead to a condition which must be satisfied at every point of a purely superficial electromagnet, and again, a condition which must be satisfied at the boundary, if the surface be not closed. The mathematical expression of these conditions will be given later.

538. *Solid Electromagnets.—Def.* The intensity of the galvanism at any point within a solid electromagnet is the quantity of galvanism in a space of infinitely small dimensions round that point, divided by the volume of the space.

The complete specification of a solid electromagnet will be the expression of the intensity and direction of the galvanism at every point of it.

Here again there will be conditions to be satisfied by the specification, to express the fact that all the galvanism consists of a group of closed circuits.

539. After these preliminary explanations we may enter upon a regular analytical treatment of the subject; commencing with investigations of the conditions to which the distribution of galvanism in solid and in superficial electromagnets is subject.

Let u, v, w denote the components of the flux at any point (x, y, z) within a solid electromagnet; and, if there be besides a superficial distribution of galvanism on the bounding surface, let U, V, W be the components of the superficial flux at the point (x, y, z) when this point belongs to the surface. These quantities must satisfy the following conditions, in order that the galvanism expressed by u, v, w, U, V, W may consist of a group of closed circuits :—

$$\frac{du}{dx} + \frac{dv}{dy} + \frac{dw}{dz} = 0 \quad\ldots\ldots\ldots\ldots\ldots (1)$$

for every point (x, y, z) of the magnet, and

$$\frac{dU}{dx} + \frac{dV}{dy} + \frac{dW}{dz} + \left(\frac{dn}{dy} - \frac{dm}{dz}\right)(mW - nV)$$
$$+ \left(\frac{dl}{dz} - \frac{dn}{dx}\right)(nU - lW) + (lV - mU)\left(\frac{dm}{dx} - \frac{dl}{dy}\right) = lu + mv + nw \ldots(2)$$

for every point (x, y, z) of the surface of the magnet, the direction-cosines of a normal to the surface being denoted by l, m, n.

540. To demonstrate these conditions, let us consider an infinitesimal tubular portion of the magnet bounded by stream-lines *, and by the surface of the magnet, if any of these lines cut it. Let the stream-lines thus considered be infinitely near one another, so that the portion of the magnet contained by them may be a ring of infinitely small section, cut or not as the case may be, by the surface of the magnet. The conditions to be satisfied with reference to this portion of the magnet are, that the intensity of the galvanic stream at each point must be inversely proportional to the area of section perpendicular to the stream-lines of galvanism; and that if the ring be cut by the surface of the magnet, the incomplete galvanic arc thus existing within the magnet must be completed along the surface. Since the whole body may be divided into portions of this kind, we have a condition for every internal point, and by expressing that the superficial distribution U, V, W must be such as to complete circuits for the galvanic arcs, of which the ends are in the surface, the condition to which U, V, W are subject is obtained.

To investigate the condition for u, v, w, consider an infinitely small parallelepiped $\alpha\beta\gamma$, of which the centre is at (x, y, z), and the edges respectively parallel to OX, OY, OZ, and let the galvanic arcs into which the whole magnet is divided be supposed to be of sections so small that an infinite number of them will pass through this parallelepiped. The condition to be expressed will be that the sum of the products of the intensities into the sections at one set of the ends of these arcs shall be equal to the sum of the corresponding products at the other set of ends. The sums of these products for all the ends which lie on the two faces β, γ, of which the distances from YOZ are $x - \frac{1}{2}\alpha$, $x + \frac{1}{2}\alpha$ are respectively equal to

$$\left(u - \tfrac{1}{2}\alpha\frac{du}{dx}\right)\beta\gamma, \text{ and } \left(u + \tfrac{1}{2}\alpha\frac{du}{dx}\right)\beta\gamma$$

---

* [*Note, Oct.* 25, 1871.—This term (its introduction is I believe due to Rankine) is now much used in writings on hydrokinetics. It is substituted for "lines of galvanism," which I find in my old manuscript.]

and similarly, for the faces γ, α, we obtain the sums

$$\left(v - \tfrac{1}{2}\beta\frac{dv}{dy}\right)\gamma\alpha, \text{ and } \left(v + \tfrac{1}{2}\beta\frac{dv}{dy}\right)\gamma\alpha$$

and, for α, β, $\left(w - \tfrac{1}{2}\gamma\dfrac{dw}{dz}\right)\alpha\beta$, and $\left(w + \tfrac{1}{2}\gamma\dfrac{dw}{dz}\right)\alpha\beta$.

Now when u, v, and w are all positive, one set of the ends of the galvanic arcs will lie on the three faces of the parallele-piped of which the distances from the co-ordinate planes are respectively $x - \tfrac{1}{2}\alpha$, $y - \tfrac{1}{2}\beta$, $z - \tfrac{1}{2}\gamma$; and the other three faces will contain the other set of ends, and we must therefore have

$$\left(u - \tfrac{1}{2}\alpha\frac{du}{dx}\right)\beta\gamma + \left(v - \tfrac{1}{2}\beta\frac{dv}{dy}\right)\gamma z + \left(w - \tfrac{1}{2}\gamma\frac{dw}{dz}\right)\alpha\beta$$

$$= \left(u + \tfrac{1}{2}\alpha\frac{du}{dx}\right)\beta\gamma + \left(v + \tfrac{1}{2}\beta\frac{dv}{dy}\right)\gamma z + \left(w + \tfrac{1}{2}\gamma\frac{dw}{dz}\right)\alpha\beta,$$

whence* $$\frac{du}{dx} + \frac{dv}{dy} + \frac{dw}{dz} = 0. \qquad \text{[(1) of § 539].}$$

541.   To investigate the conditions for the surface of the body, it may be remarked that if there were no galvanic arcs from within, terminated at the surface, there might be no superficial galvanism, and that any superficial galvanism there could be must constitute a group of closed circuits; but that when there are interior galvanic arcs of which the ends lie on the surface, the superficial distribution must complete the circuits for them, besides containing any arbitrary distribution of closed circuits.   Hence, if P and P' be two points on a band of the surface between two lines of superficial galvanism infinitely near one another, β and β' the breadths of the band at these points, and I and I' the superficial intensities of the

---

* It is scarcely necessary to remark that this is the same as the "equation of continuity," for the motion of an incompressible fluid, of which the velocity at any point (x, y, z) is the resultant of u, v, w.   The condition that as much fluid leaves the parallelepiped $\alpha\beta\gamma$ as enters it, in a unit of time would lead to pre-cisely the same investigation as that of the text (*see* Duhamel's *Cours de Mécanique*, or *Cambridge and Dublin Mathematical Journal*, 1847, p. 282).   The electrical matter which may be imagined to be flowing through the body, must not become accumulated, nor leave a deficiency in any part.

[*Note, Jan.* 1872.   When this was written, upwards of twenty years ago, the investigation of the "equation of continuity" here referred to, adapted from Fourier, was but little known.]

galvanism; the values of the products $I.\beta$ and $I'.\beta'$ must differ by an amount equal to the sum of the strengths of the interior arcs of which the ends lie on the band between P and P'. Now if ds denote the length of an element of the band, the sum of the strengths of all the interior arcs having their ends on this part of the band will be

$$(lu + mv + nw).\beta.ds$$

and therefore if P and P' be situated at the two extremities of ds, we must have

$$I'\beta' - I\beta = (lu + mv + nw)\,\beta ds\,............(1);$$

or, if the symbol ð denote differentiation performed with reference to variations along the superficial stream-line through P,

$$ð\,(I\beta) = (lu + mv + nw)\,\beta ds\,...............(2).$$

Now let ϕ be such a function of x, y, z that the equation

$$\phi = k\,.............................(3),$$

with different constant values given to k, shall represent any set of surfaces cutting the surface of the magnet along the stream-lines; that is to say (as the direction cosines of the stream-line are proportional to U, V, W), let ϕ be any function satisfying the equation

$$U\frac{d\phi}{dx} + V\frac{d\phi}{dy} + W\frac{d\phi}{dz} = 0\,...............(4).$$

And, because the stream-line lies on the surface, we have

$$lU + mV + nW = 0\,....................(5).$$

If κ be the difference of the values of k for the two bounding stream-lines on the two sides of the band through P which we have been considering, we readily obtain, for the breadth of the band, the following expression:—

$$\beta = \frac{\kappa}{\left\{\left(m\frac{d\phi}{dz} - n\frac{d\phi}{dy}\right)^2 + \left(n\frac{d\phi}{dx} - l\frac{d\phi}{dz}\right)^2 + \left(l\frac{d\phi}{dy} - m\frac{d\phi}{dx}\right)^2\right\}^{\frac{1}{2}}}\,(6).$$

Equations (4) and (5) with

$$U^2 + V^2 + W^2 = I^2\,......................(7),$$

resolved for U, V, W, give -

$$U = \frac{\left(n\dfrac{d\phi}{dy} - m\dfrac{d\phi}{dz}\right)}{\Xi}$$

$$V = \frac{\left(l\dfrac{d\phi}{dz} - n\dfrac{d\phi}{dx}\right)}{\Xi} \quad \Bigg\} \quad \dots\dots\dots\dots(8),$$

$$W = \frac{\left(m\dfrac{d\phi}{dx} - l\dfrac{d\phi}{dy}\right)}{\Xi}$$

where, in virtue of (6) $\quad \Xi = \dfrac{\kappa}{I\beta}$ ....(9).

And from (8) we have—

$$\Xi(mW - nV) = (m^2 + n^2)\frac{d\phi}{dx} - l\left(m\frac{d\phi}{dy} + n\frac{d\phi}{dz}\right) = \frac{d\phi}{dx} - l\left(l\frac{d\phi}{dx} + m\frac{d\phi}{dy} + n\frac{d\phi}{dz}\right)$$

and therefore, if we put

$$l\frac{d\phi}{dx} + m\frac{d\phi}{dy} + n\frac{d\phi}{dz} = \Pi$$

we have

$$\frac{d\phi}{dx} = \Pi l + \Xi(mW - nV)$$

$$\frac{d\phi}{dy} = \Pi m + \Xi(nU - lW) \quad \Bigg\} \quad \dots\dots\dots\dots(10).$$

$$\frac{d\phi}{dz} = \Pi n + \Xi(lV - mU)$$

Differentiating (9) along the stream-line, we have

$$\frac{\eth(I\beta)}{ds} = -\frac{\kappa}{\Xi^2}\frac{\eth\Xi}{ds}.$$

Hence

$$\frac{\eth(I\beta)}{\beta ds} = \frac{-I}{\Xi}\frac{\eth\Xi}{ds} = -I\frac{1}{\Xi}\left(\frac{d\Xi}{dx}\frac{\eth x}{ds} + \frac{d\Xi}{dy}\frac{\eth y}{ds} + \frac{d\Xi}{dz}\frac{\eth z}{ds}\right)\dots(11).$$

Now

$$I\frac{\eth x}{ds} = U, \quad I\frac{\eth y}{ds} = V, \quad I\frac{\eth z}{ds} = W\dots\dots\dots\dots(12).$$

Using these in (11) and then putting for $U\dfrac{d\Xi}{dx}$ etc., the equivalent formulæ $\dfrac{d(U\Xi)}{dx} - \Xi\dfrac{dU}{dx}$, etc., and making use of equations

(8) we find

$$\frac{\eth\,(I\beta)}{\beta ds} = \frac{dU}{dx} + \frac{dV}{dy} + \frac{dW}{dz}$$

$$+ \frac{1}{\Xi}\left\{\frac{d\phi}{dx}\left(\frac{dn}{dy} - \frac{dm}{dz}\right) + \frac{d\phi}{dy}\left(\frac{dl}{dz} - \frac{dn}{dx}\right) + \frac{d\phi}{dz}\left(\frac{dm}{dx} - \frac{dl}{dy}\right)\right\}(13)$$

For $\dfrac{d\phi}{dx}$, $\dfrac{d\phi}{dy}$, $\dfrac{d\phi}{dz}$, substitute their values by (10): then, if we remark that

$$l\left(\frac{dn}{dy} - \frac{dm}{dz}\right) + m\left(\frac{dl}{dz} - \frac{dn}{dx}\right) + n\left(\frac{dm}{dx} - \frac{dl}{dy}\right) = 0\ldots(14),$$

since this is the condition that a factor, λ, may be found such that $\lambda\,(ldx + mdy + ndz)$ is a complete differential; we obtain

$$\frac{\eth\,(I\beta)}{\beta ds} = \frac{dU}{dx} + \frac{dV}{dy} + \frac{dW}{dz} + (mW - nV)\left(\frac{dn}{dy} - \frac{dm}{dz}\right)$$

$$+ (nU - lW)\left(\frac{dl}{dz} - \frac{dn}{dx}\right) + lV - mU\left(\frac{dm}{dx} - \frac{dl}{dy}\right)\ldots(15).$$

Hence equation (2) becomes

$$lu + mv + nw = \frac{dU}{dx} + \frac{dV}{dy} + \frac{dW}{dz} + (mW - nV)\left(\frac{dn}{dy} - \frac{dm}{dz}\right)$$

$$+ (nU - lW)\left(\frac{dl}{dz} - \frac{dn}{dx}\right) + (lV - mU)\left(\frac{dm}{dx} - \frac{dl}{dy}\right)[(2)\text{ of }\S\,539].$$

542. *Coroll.* The condition to be satisfied by the quantities U, V, W which express the distribution of galvanism in a superficial electromagnet is the following:—

$$0 = \frac{dU}{dx} + \frac{dV}{dy} + \frac{dW}{dz} + (mW - nV)\left(\frac{dn}{dy} - \frac{dm}{dz}\right)$$

$$- (nU - lW)\left(\frac{dl}{dz} - \frac{dn}{dx}\right) + (lV - mU)\left(\frac{dm}{dx} - \frac{dl}{dy}\right).$$

543. The second member of these equations is brought to another symmetrical form (simpler for some applications), by grouping in order of U, V, W, adding to it six balancing terms,

$$Ul\frac{dl}{dx} + Vm\frac{dm}{dy} + Wn\frac{dn}{dz} - Ul\frac{dl}{dx} - Vm\frac{dm}{dy} - Wn\frac{dn}{dz},$$

and observing that

$$l\frac{dl}{dx} + m\frac{dm}{dx} + n\frac{dn}{dx} = 0, \text{ etc.}$$

Thus for (2) of § 539 we have

$$lu + mv + nw = \frac{dU}{dx} + \frac{dV}{dy} + \frac{dW}{dz} + U\left(l\frac{dl}{dx} + m\frac{dl}{dy} + n\frac{dl}{dz}\right)$$
$$+ V\left(l\frac{dm}{dx} + m\frac{dm}{dy} + n\frac{dm}{dz}\right) + W\left(l\frac{dn}{dx} + m\frac{dn}{dy} + n\frac{dn}{dz}\right)\ldots(16).$$

544.   The mutual actions between electromagnets and common magnets, or between any part of an electromagnet and other partial or complete electromagnets or common magnets, may be determined by means of the expressions of §§ 525—529; and when the data are sufficient, the application of elementary statical principles leads to the solution of any problem that can be proposed.   The mode of specifying the distribution of galvanism in an electromagnet, explained in §§ 531—539, leads immediately, by means of Ampère's formula given above, §§ 525, 527, to proper expressions for the mutual action between any two solid electromagnets by means of four definite integrals representing the parts of that component due to the mutual actions of the solid and superficial parts of their distributions of electric current.

545.   A similar synthetical solution of the problem of determining the mutual action between an electromagnet and a common magnet, is obtained by first investigating a formula for the mutual action between an element of a galvanic circuit, and an infinitely small magnet, which may be done at once by means of the formulæ of §§ 526, 529, and the synthesis of a magnet explained in §§ 461, 462, and then applying statical principles to derive formulæ for the components (both of force and couple) of the mutual action.   It is sufficient here to indicate the method of proceeding, for such problems; and unnecessary to write down the formulæ, which, in fact, may always, when wanted, be written down at once from the formulæ of the preceding chapters, according to the principles which have been now explained.   Thus, write down the formulæ for the rectangular components of the force exerted by the electromagnet (u, v, w, U, V, W, § 539), upon a positive

unit pole, according to the formulæ of § 529; and for the components of couple which would be given by transferring the constituent forces from their supposed lines through the elements of electric current, to parallel lines through the magnetic pole. It will be found that in the integrals the components of couple disappear, and thus is proved Cor. 5 of § 549; that the resultant force is in a line through the pole. The expressions for the components of this force are, as mere inspection of the formulæ of § 529 proves, identical with those of § 517, (b). I proceed to propositions regarding electromagnetic force, the importance of which will appear from the application made of them in subsequent investigations.

546. *Proposition.*—The action of an infinitely small plane closed circuit on an element of another circuit, or on another complete electromagnet or magnet of any kind, is the same as would be produced by an infinitely small magnet, in the same position, with its axis perpendicular to the plane of the circuit*. [The proof is easily worked out from the formulæ of §§ 483, 485, 529.]

---

* [*Note added Jan.* 1872.]—Hence Ampère's theory of magnetism, according to which magnetization of steel or load-stone, or soft iron, or any other polar magnet (§ 549) consists of electric currents circulating round the molecules of the magnetized substance in planes perpendicular to the directions of magnetization. From twenty to five-and-twenty years ago, when the materials of the present compilation were worked out, I had no belief in the reality of this theory (compare § 602); but I did not then know that motion is the very essence of what has been hitherto called matter. At the 1847 meeting of the British Association in Oxford, I learned from Joule the dynamical theory of heat, and was forced to abandon at once many, and gradually from year to year all other, statical preconceptions regarding the ultimate causes of apparently statical phenomena. In a paper communicated to the Royal Society of London, 10th May 1856, under the title "Dynamical Illustrations of the Magnetic and the Heliçoidal Rotatory effects of Transparent Bodies on Polarized Light," [Art. xciii. of Reprint of Mathematical and Physical Papers (Vol. ii.)] after proving that the heliçoidal property shown by syrup, oil of turpentine, quartz crystals, etc., is due to a right or left-handed asymmetry in the constituent molecules, I made the following statement regarding the nature of magnetism :—

"The magnetic influence on light discovered by Faraday depends on the "direction of motion of moving particles. For instance, in a medium possess-"ing it, particles in a straight line parallel to the lines of magnetic force, dis-"placed to a helix round this line as axis, and then projected tangentially with "such velocities as to describe circles, will have different velocities according as "their motions are round in one direction (the same as the nominal direction of "the galvanic current in the magnetizing coil), or in the contrary direction. But "the elastic reaction of the medium must be the same for the same displace-"ments, whatever be the velocities and directions of the particles; that is to say, "the forces which are balanced by centrifugal force of the circular motions are "equal, while the luminiferous motions are unequal. The absolute circular "motions being therefore either equal or such as to transmit equal centrifugal

Cor. 1. The magnetic moment of the infinitely small magnet which produces the same magnetic effects as an infinitely small plane closed circuit is equal to the galvanic strength of the circuit, multiplied into the plane area which it encloses.

547. *Rule for Directions.*— The magnet must be so held relatively to the current which it represents, that if the circuit and it be placed at the centre of the earth, with its plane in the earth's equator, and with the current going round from east to west, the north polar side of the magnet shall be towards the earth's North Pole. [Mnemonic principle:—Remember that if terrestrial magnetism were due to currents in the earth's crust, their general direction would be "the way of the sun;" that is to say, from east to west.]

548. *Cor.* 2. The magnetic action of a linear electromagnet (§ 535) [that is to say, a galvanic circuit in an infinitely thin conducting ring] of any form is the same as that of a uniform magnetic shell (§ 506) of any shape having its edge coincident with the circuit, and having its magnetic strength numerically equal to the galvanic strength of the circuit. The rule for

---

"forces to the particles initially considered, it follows that the luminiferous
"motions are only components of the whole motion; and that a less lumi-
"niferous component in one direction, compounded with a motion existing in
"the medium when transmitting no light, gives an equal resultant to that of a
"greater luminiferous motion in the contrary direction compounded with the
"same non-luminous motion. I think it is not only impossible to conceive any
"other than this dynamical explanation of the fact that circularly polarized light
"transmitted through magnetized glass parallel to the lines of magnetizing
"force, with the same quality, right-handed always, or left-handed always, is
"propagated at different rates according as its course is in the direction or is
"contrary to the direction in which a north magnetic pole is drawn; but I
"believe it can be demonstrated that no other explanation of that fact is possible.
"Hence it appears that Faraday's optical discovery affords a demonstration of
"the reality of Ampère's explanation of the ultimate nature of magnetism; and
"gives a definition of magnetization in the dynamical theory of heat. The
"introduction of the principle of moments of momenta ('the conservation of
"areas') into the mechanical treatment of Mr Rankine's hypothesis of 'molecular
"vortices,' appears to indicate a line perpendicular to the plane of resultant
"rotatory momentum ('the invariable plane') of the thermal motions as the
"magnetic axis of a magnetized body, and suggests the resultant moment of
"momenta of these motions as the definite measure of the 'magnetic moment.'
"The explanation of all phenomena of electro-magnetic attraction or repulsion,
"and of electro-magnetic induction, is to be looked for simply in the inertia and
"pressure of the matter of which the motions constitute heat. Whether this
"matter is or is not electricity, whether it is a continuous fluid interpermeating
"the spaces between molecular nuclei, or is itself molecularly grouped; or
"whether all matter is continuous, and molecular heterogeneousness consists in
"finite vortical or other relative motions of contiguous parts of a body; it is
"impossible to decide, and perhaps in vain to speculate, in the present state of
"science."

directions is, that if the circuit be held so that in any part of it
the current is from east to west, then a point carried in a circle
round that part of the galvanic arc northwards above it and
southwards below it, will cut the shell through from its north
polar to its south polar side.

549. *Cor.* 3. A common magnet [or a *polar magnet* as I shall
henceforth call anything magnetized after the manner of a loadstone or a steel magnet] may be found which shall produce the
same action as any given complete electromagnet, upon other
magnets of either kind, or upon any portion of an electromagnet
[or arc of an electric circuit].

Cor. 4. The distribution of ordinary [or *polar*] magnetism
which produces the same force, according to the "electro-
magnetic definition" (§ 517), as a given electromagnet is
indeterminate.   [Because any lamellar distribution consisting of closed shells may (§ 512, Cor. 6) be superimposed on
a distribution of magnetism without altering the resultant
force electromagnetically defined in § 517. Compare below
§§ 584—588.]

Cor. 5. The mutual action between a magnetic point or pole,
that is, an end of an infinitely thin uniformly and longitudinally magnetized bar, and a complete electromagnet, is in a line
through that point.   [Compare §§ 526, 545.]

550. *Cor.* 6. The definition (1) of § 479 and the definition
of the potential with the propositions on which it is founded, as
set forth in §§ 481, 483 may be applied without alteration to an
electromagnet, as far as regards points external to the conducting matter through which the electric currents pass.

551. With regard to internal points, the definition given in
§ 517 for the resultant force requires no conventional understanding of an analogous character to that which was made in
the case of points in the substance of common magnets, and set
forth in the text and in the second foot-note of § 479. We cannot, as in the case of a common magnet, suppose a portion to be
cut from the substance of an electromagnet, without deranging
the magnetic condition of the remainder. If we imagine a
space hollowed out in the substance of an electromagnet,
we must suppose such arrangements made that the vacancy

shall only deflect, not interrupt the electric currents. If a small spherical portion, for example, be cut from an electromagnet, there may be either a gradual deflection of the current through some space round the part cut out; or the interrupted circuits may be completed by a condensation of electric current on the surface bounding the hollow. But it is satisfactory to know that the resultant magnetic force at any point within such a hollow space is infinitely little affected by the supposed deflection of the currents, when the space is infinitely small. This follows from the comparison of similar circumstances for similar hollows of different dimensions, which shows that the disturbing influence is in simple proportion to the linear dimensions of the hollow. Or, simply taking the triple integrals of §§ 545, 517 (*b*) or (*c*), and using them for a point, *P*, within the conducting substance, we see in a moment that the part of each integral belonging to any small space round *P* diminishes in proportion to the linear dimensions of this space when made infinitely small without change of shape or of position relatively to *P*. Hence there is no necessity for hollowing out a space in the electromagnet or of further considering the complicated circumstances referred to above, and the resultant force at any point within or without an electromagnet is the force which may be simply defined as the force expressed by the formulæ of § 528, according to the modes of specification and principles explained in §§ 536, 537, 538, 545; [that is to say, simply the formulæ (*b*) of § 517].

552. If an electromagnet consist of a number of conductors which when put together fit close to one another, without touching, or of a single wire of a rectangular or hexagonal section, rolled up with the different parts of the wire not touching one another, but lying close together so as to be separated by spaces infinitely small compared with the lateral dimensions of the wire; the preceding definition of the resultant force at any point of the magnet considered as a single solid electromagnet will give sensibly the same resultant force at neighbouring points whether in the substance of the conductor or in the interstitial space.

[*Addition and correction*, Oct. 27, 1871.—But even if the spaces between the different circuits, or the neighbouring por-

tions of one circuit constituting an ordinary artificial electro-
magnet, be not infinitely small or be infinitely great compared
with the sections of the conductors, the variation of force from
point to point between two neighbouring portions of circuit will
be small in comparison with the whole force generally, provided that the ratio of space occupied to whole space within the
bounds of the electromagnet be great in comparison with the
ratio of the diameter of the wire to the diameter of a section
of the electromagnet across all the circuits or wires. This is
easily proved from (c) of § 517. Consideration of the corresponding gravitational case is instructive. In the first place
for simplicity; consider a great spherical space, S, of radius, R,
with a great number, n, of equal homogeneous spheres of very
small radius, r, and density, ρ, distributed with average homo-
geneousness through it, so as to give an average density equal
to $\dfrac{nr^3\rho}{R^3}$. At the boundary of S the resultant force will be
approximately towards the centre and equal to

$$\frac{4\pi}{3}\frac{nr^3\rho}{R^3}R;$$

and at distance x from the centre, it will be approximately
towards the centre and equal to

$$\frac{4\pi}{3}\frac{nr^3\rho}{R^3}x.$$

The greatest deviation from these approximations would be
produced by taking one of the small constituent spheres from
a great distance, and bringing it into contact with the point
attracted, which would introduce a force amounting to

$$\frac{4\pi}{3}\rho r;$$

and therefore would produce but a small difference on either
the magnitude or the direction of the resultant force if $\dfrac{r}{x}$ is
small in comparison with $\dfrac{nr^3}{R^3}$. Generally, for any group of
molecules attracting according to the Newtonian law, if the product of the density into the diameter of a molecule be very small
in comparison with the product of mean density into diameter
of the whole; the masses of the molecules might be expanded

into the interstices so as to continuously occupy the whole
volume of the whole group, without producing anywhere more
than a very small change in the resultant force.]

553. A superficial distribution of electric currents gives the
same normal component, but different tangential components,
for the resultant magnetic force at points infinitely near it on
its two sides. The tangential component at one side is found
by compounding with a force equal and parallel to the tangential component force at the other side, a force perpendicular to
the stream-lines and equal to $4\pi I$, if I denote the surface
intensity of the electric stream. [These propositions are easily
proved from the surface term of the expression (*b*) of § 517,
applied to the present subject according to § 551. They are
in fact proved by equations (*o*) of § 517. Equations (*p*) of the
same section express in symbols the well-known corresponding
proposition in respect to a superficial distribution of matter
acting according to the inverse square of the distance, which in
words is,—that the tangential component is the same, for
points infinitely near one another on the two sides of the surface, but the normal components differ by $4\pi\rho$, if ρ denote the
surface density.]

554. *Original investigation of* § 517 (IX.) *referred to in* § 518.

"*Glasgow College,* 7*th November,* 1849.—Yesterday I fell upon
" a train of synthesis and analysis of galvanic distributions
" which I think will add much consistence and symmetry to
" the whole first part of my paper on magnetism (a portion of
" the first part was communicated on the 21st of June last, by
" Colonel Sabine, to the Royal Society), and it will help me in
" getting to work to write out the matter I have had so long
" in hand. It occurred to me to treat galvanic distributions
" according to the analogy of Chapter III., 'On the imaginary
" magnetic matter by which the polarity of a magnet may be
" represented [§§ 463—475 above];' thus, α, β, γ, being the
" components of the intensities of magnetization at (x, y, z),
" consider Ampère's imaginary currents round $dxdydz$. We
" have strength of current round OX, along faces $dxdy$, $dxdz$,
" $dxdy$, and $dxdz$, $= \alpha dx$.

" Consider all the partial currents parallel to OX. We have

" $-\beta dydx$, along one of the $dydz$ faces (that which corresponds
" to x, y, $z + dz$), and $\gamma dzdx$ along one of the $dzdx$ faces (that
" which corresponds to x, $y + dy$, z).

" The coincident face, $dydx$, of a contiguous elementary
" parallelepiped has

$$+ (\beta + \frac{d\beta}{dz} dz) \, dydz$$

" and the coincident face $dzdx$ of another contiguous parallel-
" epiped has

$$- (\gamma + \frac{d\gamma}{dy} dy) \, dzdx.$$

" Hence (as in Chapter III.) the share for the element
" $dxdydz$, of galvanism parallel to OX, is,

$$\left(\frac{d\beta}{dz} - \frac{d\gamma}{dy}\right) dxdydz.$$

" So for shares parallel to OY and OZ we find

$$\left(\frac{d\gamma}{dx} - \frac{d\alpha}{dz}\right) dxdydz$$

" and $$\left(\frac{d\alpha}{dy} - \frac{d\beta}{dx}\right) dxdydz.$$

" But at the surface of the magnet there is unneutralized
" galvanism. Hence, besides the internal distribution we have
" a superficial distribution ; and the share to a superficial ele-
" ment ds has, I find, for its components parallel to OX, OY,
" OZ, the following :—

$$- (\beta n - \gamma m) \, ds$$
$$- (\gamma l - \alpha n) \, ds$$
$$- (\alpha m - \beta l) \, ds$$

" and we verify that these are the components of a current in
" the surface by observing that

$$l (\beta n - \gamma m) + m (\gamma l - \alpha n) + n (\alpha m - \beta l) = 0;$$

" l, m, n being the direction cosines of a normal.

" This concludes the analogue of Chapter III.

" Let X, Y, Z be the components of the force at an external
" point P. We have", [formula IX. of § 517, which need not
be repeated here].

" Since the potential method cannot be applied where galvanic
" elements or incomplete circuits are considered, the following

" is the analogue of [§ 495] the section in Chapter IV., where a
" second or analytical (Poisson's original) demonstration is given
" of the equivalence of a certain determined distribution of ima-
" ginary magnetic matter to the given distribution of magnetism."
[Here follows in the manuscript memorandum, the investigation
(§ 518 above), which was communicated to the Royal Society,
June 20, 1850, and published in the *Transactions.*]

XXV. *On the Potential of a Closed Galvanic Circuit of any Form.*

[From the *Cambridge and Dublin Mathematical Journal*, 1850.]

The object of the following note is to point out an extremely interesting application of the principles explained by Professor De
Morgan in the preceding paper [of the *Cambridge and Dublin
Mathematical Journal*, 1850, on "Extension of the Word *Area*"],
which occurred to me in connexion with the determination of the
potential of an electromagnet in terms of the solid angle of a cone.

555. It has been shown by Ampère that a *closed galvanic
circuit in a re-entering curve of any form* produces the same
magnetic action as any infinitely thin sheet of steel, having
this curve for its edge, would produce if uniformly and normally magnetized. Now the resultant force of a magnet at any
point may be expressed, after the manner of Laplace, in terms
of the differential coefficients of a "potential function," and
therefore the same proposition is true for a closed galvanic
circuit *. When this is known to be true, for either a common or
an electro-magnet, the following definition may be laid down:—

---

* In other words, the quantity of *work* necessary to bring a magnetic pole
from any position in the neighbourhood of a closed galvanic circuit to any other
position does not *vary* with the form of the curve along which it is drawn from
one point to the other. There is however one remarkable difference between the
case of an electromagnet and that of any given steel magnet. In the case of an
electromagnet, although the quantity of work does not *vary* with the path, yet
it has determinately different values according as the path lies on one side, or on
another of any part of the galvanic wire circuit, or according to the convolutions
round any part of the wire which it may be arbitrarily chosen to make. Hence
arises the multiplicity of values of the potential at any point in the neighbourhood of an electromagnet noticed below. Yet for any one form of a magnetized
sheet of steel of the kind described in the text, agreeing, in the action which it
produces on all points not in its own substance, with the electromagnet, the
potential is perfectly determinate without a multiplicity of values; and the
difference in the two cases is accounted for when we consider that the magnetic
potentials at any two points infinitely near one another, on two sides of the
sheet of steel, differ by $4\pi\gamma$, where γ is a constant such that $\gamma\omega$ is the magnetic
moment of any infinitely small area ω of the sheet. The agreement in the
magnetic circumstances of the two cases fails for all points in the substance of
the magnetized steel. [Compare § 515, Cor. 2.]

556. *Def.* The potential at any point in the neighbourhood of a magnet is the quantity of work necessary to bring a unit north-pole (or the north-pole of an infinitely thin uniformly and longitudinally magnetized unit-bar) from an infinite distance to that point.

To determine the potential at any point due to a given closed galvanic circuit, let us imagine a magnetized sheet of steel (the form of the sheet is arbitrary, provided only that its edge coincide with the curve of the galvanic circuit), which according to Ampère produces the same magnetic action, and consequently the same potential, as a given closed circuit, to be divided into infinitely small areas. Then it is easily demonstrated, on the most elementary principles of the theory of magnetism, that the potentials at any point, P, produced by these areas, are proportional to the solid angles which they subtend at P; the true *sign* of the potential of any small area being obtained by considering the solid angle as positive, if the side of the area containing north poles, or negative, if the other side, be towards P. Hence the potential of the whole sheet of steel, at any point P, is proportional to the entire solid angle which it subtends at P; and consequently the potential of a closed galvanic circuit, at any point P, is equal to a constant (which may be taken as a measure of the strength of the gal-vanism, or as it is often termed, the "quantity" of the current) multiplied into the solid angle of the cone described by a straight line always passing through P, and carried round the circuit. In all cases, except those in which the galvanic circuit is contained in one plane, there will be positions of P for which this cone will be "autotomic"; and in many cases, especially the most common practical case of an electromagnet, in which the circuit consists of double or multiple concentric helices, with their ends connected, or of a single wire wrapped in a complex manner round a body of some irregular shape, so as to constitute most complicated curves of double curvature, there will be no position of the point P for which the cone is not excessively autotomic. The solid angle of such a cone, or the area enclosed by its intersection with a spherical surface of unit radius, having for centre its vertex, may be determined in a manner precisely similar to that which has been explained by Professor De Morgan for plane self-cutting curves, without any ambiguity as to the *circuit* by which the curve, when self-cutting*, is to be described, since the actual galvanic current is in a determinate circuit, and its projection, by the conical surface, on the surface of the sphere is to be described by the projection of a point moving along the electric conductor, either in the same direction as the current, or in the opposite, according to the convention we please to make. There is however a source of ambiguity which really affects the evaluation of the solid angle of a cone, or of the area of any given circuit described in a determinate manner on a spherical surface, and gives rise to a multiplicity of solutions of the problem, arising from the circumstance that of all the "primary parts" (only two in number if the circuit be not self-cutting) into which the spherical surface is divided by the curve, there is no reason for choosing one, more than another, as a zero space (or a space corresponding to the space exterior to a closed circuit in a plane)†.

557. When the value of the area, according to any one of these solutions, has been obtained, all the others may be deduced, by adding to it or subtracting from it any number of times the area of the whole spherical surface. Hence the most general expression for the solid angle of a cone described in a determinate manner, is

$$\sigma = \sigma_1 + 4i\pi,$$

where σ_1 denotes any one value and i any positive or negative integer. If too great a positive or too small a negative value be given to i, all the "primary spaces" of the spherical surface will be positive or all will be negative; and therefore if we wish to obtain only those solutions according to which some portion of the spherical surface is considered as *zero* or *external to the circuit*, a limited number only (not exceeding the number

---

* See note on the word "circuit" in the preceding paper [of the *Cambridge and Dublin Mathematical Journal*, year 1850, p. 140].

† Thus, if the given curve be a circle of the sphere, described in a given direction, and if θ denote the angular radius measured from that pole O, which would be *north* if the direction of describing the circle were from west to east; the area of the circuit is $+2\pi(1-\cos\theta)$ if the space on the other side of the circle from O be considered as the zero space, but it would be $-2\pi(1+\cos\theta)$ if the space in which O is situated were taken as zero, or external to the circuit. In general, the area of a circuit not self-cutting, on a spherical surface, will be either one of the two parts into which the spherical surface is divided, with the sign $+$, or the other part, with the sign $-$.

of primary parts into which the spherical surface is divided by the circuit) of values for i are to be admitted. The physical problem, however, requires no limitation to the range of values that may be given to i: for, if we take any two paths to the point P from an infinite distance, such that the space between them is *once* crossed by the galvanic circuit, the potential at P will differ by $4\pi\gamma$ according as it is estimated by one path or by the other; and therefore, by taking (for the sake of simplicity in the conception) different paths to the point P which go round a certain portion of the galvanic circuit once, twice, three times, four times, etc., in one direction, and again different paths which go round the same portion of the wire once, twice, three times, four times, etc., in the contrary direction, we obtain, according to the definition, an infinite number of values of the potential at the point P, which are successively expressed by the formula

$$v = v_1 + 4i\pi\gamma,$$

when we give i the values 1, 2, 3, 4, etc., and again the values $-1, -2, -3, -4$, etc.; v_1 being the potential estimated by a path, which makes none of those convolutions.

558. Hence we see that, to find the general expression for the potential at a point in the neighbourhood of an electromagnet, we may first choose some determinate path from an infinite distance to the point P, and investigate the value of the potential for it, which may be used as the value of v_1 in the preceding expression. If an infinite straight line in any direction, terminated at the point P, be the path chosen, the determinate potential will be found by considering, as the portion external to the circuit, the primary portion of the spherical surface described from P as centre, which is cut by this line. Hence, if we mark this primary portion with a zero, the number with which any other primary part is to be marked, according to Professor De Morgan's rule, will be got by drawing a line to any point within it, from any point O, in the external primary part, and counting the number of times it is cut by the curve; every time it is cut from right to left (with reference to a person walking from O, along it, on the convex surface of the sphere) being counted as $+1$, and every time it is cut in the other

direction, as −1; and the algebraical sum taken. When the number for each primary part has been thus determined, the sum of the areas of the different primary parts, each multiplied by its number (positive or negative, as the case may be), will be the required area of the circuit; and the potential at the centre of the sphere will be obtained by multiplying this by γ, the strength of the galvanic current. The absolute sign of the potential thus determined may be readily shown to be correct, if we agree to consider the potential due to terrestrial magnetism as on the whole positive for positions north, and negative for positions south of the magnetic equator; since, as is well known, currents round the earth, proceeding on the whole from east to west, would produce phenomena similar to the actual phenomena of terrestrial magnetism.

559. As an example, let us consider a conducting circuit which consists of twelve complete spires of a helix, and a line

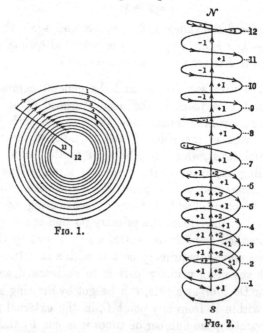

Fig. 1.

Fig. 2.

along the axis with two perpendicular portions connecting its extremities with those of the helix. The accompanying diagrams represent the projections, by radii, of the circuit, on a spherical surface in two different positions, viewed in each case from the interior of the sphere.

In the case illustrated by fig. (1), the centre of the sphere is nearly in a line with the axis of the helix, on the side towards the north pole* of the helix, and distant from it by about half the length of the axis. In the case illustrated by fig. (2), the centre of the sphere is in a perpendicular through a point of the axis, distant by about one-fourth of its length from the north pole of the helix, and is at about the same distance from the nearest part of the helix, as in the case of fig. (1); and the curve on the spherical surface is shown in the diagram, according to Mercator's projection with the great circle containing the axis of the helix as equator†. In each diagram the inner side of the spherical surface is shown.

560. The radii of the spheres being supposed to be equal in the two cases, if we denote their common value by r, and if A_1 and A_2 be the areas of the spherical curves represented in the diagrams, the zero or external portions on the spherical surfaces being taken as those which become infinite in the plane diagrams, the values of the potential at the centre of the sphere will be

$$\gamma \frac{A_1}{r^2}, \text{ and } \gamma \frac{A_2}{r^2},$$

respectively, for any paths from an infinite distance which do not lie round any portion of the galvanic wire, nor between any of the spires.

The area A_1 will be determined (in accordance with Professor de Morgan's rule‡) by finding the areas of the " primary

---

* The ends of the helix which would be repelled from the north and from the south respectively by the earth's magnetic action are, in the ordinary vague use of the term "pole," called the north and south poles of the electro-magnet.

† The diagram was actually drawn by tracing upon a cylindrical surface the shadow of a helix of twelve spires, ¾ in. in diameter and 4 in. in length, produced by a luminous point in the axis of the cylindrical surface; the axis of the helix being held in the plane through the luminous point perpendicular to the axis of the surface. On account of the narrowness of the band occupied by the diagram, the cylindrical surface very nearly coincided with the spherical surface, which in strictness ought to have received the shadow. After the shadow was thus traced, the cylindrical surface was unbent into a plane.

‡ In fig. (1), all the arrow-heads which are necessary for rendering determinate the "balances" for the primary parts are given; and the numbers expressing the balances are marked for the first six primary parts, commencing

parts," marked successively with the numbers 1, 2, ... up to 12, multiplying each area by the corresponding number, and taking the sum of the products. The area A_2 will be similarly determined by finding the areas of the primary parts in fig. (2), multiplying each by the positive or negative number with which it is marked, and taking the algebraic sum of the products.

---

with the outermost. In fig. (2), all the arrow-heads which are necessary to make the diagram represent determinately a closed circuit are indicated, except in a few places where the spaces are too confined for admitting of this being done in a clear manner; and the "balances" of all the primary parts are marked with numbers, except in the instance of a very small *triple* primary part, which is marked with three dots (...) instead of $+3$.

GLASGOW COLLEGE, *March* 25, 1850.

XXVI. [January, 1872.]

Chapter VII.—*On the Mechanical Values of Distributions of Matter*, and of Magnets.*

561. *Preliminary proposition.*—The work against mutual repulsions according to the inverse squares of the distances, required to produce any change in a distribution of matter, is equal to the augmentation which it produces in the value of the integral

$$\int_{-\infty}^{\infty}\int_{-\infty}^{\infty}\int_{-\infty}^{\infty} \frac{R^2}{8\pi}\, dx dy dz \dots\dots\dots\dots(1)$$

where R denotes the resultant force at x, y, z.

This is an obvious conclusion from the following investigation for the mutual potential energy (§ 503, Addition of date 11th December, 1871) of two distributions of matter; or, as for brevity we may call them, two bodies.

---

* "Matter" is here used conventionally and merely for brevity, to denote a substance fulfilling the conditions by which "imaginary magnetic matter" (§ 463) is defined; that is, substance of which any two small portions repel one another mutually with a force equal to the product of their quantities divided by the square of the distance between them. Either or both quantities may be negative, and the negative product of unlike masses indicates attraction. Not being in any way occupied with Kinetics at present, we suppose this imaginary matter to remain where it is placed until we please to move it; so that a "distribution" of it may be supposed to be either a rigid body or a flexible body, or a flexible and compressible body, held at rest by the necessary force, except when we suppose it to move; and then we perform work, positive or negative, upon it to whatever amount is necessary to produce, irrespectively of inertia, the supposed motion against or with the forces resulting from attraction or repulsion, which the portions of the matter moved experiences. All the formulæ and conclusions are applicable to real matter, gravitating according to the Newtonian law, if we substitute attraction for repulsion, that is to say, change the signs of each formula for force or work, and exclude negative matter. In applications of gravity, therefore, instead of the "mechanical value" or "potential energy" of a distribution of the imaginary magnetic matter, we have an "exhaustion of energy" (Thomson and Tait's *Natural Philosophy*, § 549) in a distribution of real matter.

Let ρ be the density at any point (x, y, z) of one of these bodies M; and let V' be the potential at the same point, due to the other body M'. Then denoting by Q the mutual potential energy of the two, we have

$$Q = \int_{-\infty}^{\infty} \int_{-\infty}^{\infty} \int_{-\infty}^{\infty} \rho V' dx dy dz \ldots\ldots\ldots(2).$$

We have by Poisson's theorem,

$$\rho = -\frac{1}{4\pi} \left(\frac{dX}{dx} + \frac{dY}{dy} + \frac{dZ}{dz} \right)$$

where X, Y, Z denote the components of the force at $(x, y, z,)$ due to the body M. This equation (as it also expresses Laplace's theorem for space containing none of the matter of M, since there $\rho = 0$;) holds throughout space. Hence for (2) we may write

$$Q = -\frac{1}{4\pi} \int_{-\infty}^{\infty} \int_{-\infty}^{\infty} \int_{-\infty}^{\infty} \left(\frac{dX}{dx} + \frac{dY}{dy} + \frac{dZ}{dz} \right) V' dx dy dz \ldots(3).$$

Hence by integration by parts

$$Q = \frac{1}{4\pi} \int_{-\infty}^{\infty} \int_{-\infty}^{\infty} \int_{-\infty}^{\infty} (XX' + YY' + ZZ') dx dy dz \ldots(4),$$

where $X' Y' Z'$ denote the components of the force at $(x, y, z,)$ due to M'.

Let now the second body consist of a distribution of matter coincident with the first and similar to it throughout, but let the whole quantity of matter in the second body be infinitely small and be denoted by dm, that of the first being denoted by m: we shall have

$$X' = \frac{dm}{m} X, \quad Y' = \frac{dm}{m} Y, \quad Z' = \frac{dm}{m} Z.$$

Instead of Q write now dE. We have

$$dE = \frac{1}{4\pi} \frac{dm}{m} \int_{-\infty}^{\infty} \int_{-\infty}^{\infty} \int_{-\infty}^{\infty} dx dy dz \, (X^2 + Y^2 + Z^2) \ldots(5).$$

This formula expresses the quantity of work required to add dm similarly distributed to a distribution m already made. Our supposed matter being not subject to the law of impenetrability, we might simply suppose the distribution of dm, precisely similar to that of m, to be given at an infinite distance and to be moved against the repulsion of m into coincidence: the work required is that which is denoted by dE. So far it is not necessary to suppose dm infinitely small. But if dm be infinitely small, the work required to bring it in infinitely smaller parts from infinite mutual distances into the supposed position of coincidence with the distribution of m, would involve only an infinitely small amount of the second degree of infinitesimals, on account of the mutual influences of the different parts of dm. Hence the formula (5) represents the work required to augment the supposed distribution from m to $m + dm$, by bringing altogether from a state of infinite diffusion the infinitesimal portion of matter dm; and therefore the integral of this formula from 0 to m is the whole work required to build up the distribution m from infinitely diffused matter. Now, with reference to the variation of m, each of X, Y, Z varies in simple proportion to m, and therefore the triple integral may be denoted by Cm^2, so that we have

$$dE = \frac{1}{4\pi}\, Cm\, dm,$$

which gives

$$E = \frac{1}{8\pi}\, Cm^2.$$

Finally eliminating C we have

$$E = \frac{1}{8\pi} \int_{-\infty}^{\infty} \int_{-\infty}^{\infty} \int_{-\infty}^{\infty} dx\,dy\,dz\, (X^2 + Y^2 + Z^2) \;\ldots\ldots(6).$$

The preceding deduction of the formula (4) from (2) *mutatis mutandis* allows us to come back to the following important alternative formula

$$E = \tfrac{1}{2} \int_{-\infty}^{\infty} \int_{-\infty}^{\infty} \int_{-\infty}^{\infty} \rho\, V dx\,dy\,dz \;\ldots\ldots\ldots\ldots(7).$$

The direct proof of this formula by integration with reference to m, commencing with an expression for dE derived from (2) is obvious.

562. The forces at points similarly situated relatively to similar bodies, are proportional to the linear dimensions of the bodies, and to their densities in corresponding places.

The values of (1) for similar bodies are therefore as the fifth powers of the linear dimensions, and as the squares of the densities. Hence if a homogeneous rectangular parallelepiped be divided into i^3 equal and similar parts, and these parts be separated to infinite distances from one another, the whole value of the integral (1) for the scattered parts is equal to $\frac{1}{i^2}$ of its value for the undivided body. It follows that if a finite body be divided into an infinite number of infinitely small parts, and these parts be separated to infinite distances from one another, the value of the integral (1) for all the parts becomes an infinitely small quantity of the same order as the square of the diameter of one of the parts. Hence the integral (1) *relatively to a finite body or distribution of matter*, composed of ultimately homogeneous continuous substance, expresses the work required to build it up out of infinitely small parts having the same density (or any other density not too infinitely great) and given at infinitely great distances from one another.

563. A complete analytical view of the circumstances contemplated in § 562 is, as is generally the case, easier than the quasi-elementary method, involving intricacies of language and perplexities of "compound proportion," to which, as the only alternative to utter vagueness, "popular" expositions are commonly restricted. At any point $(x, y, z,)$ let V be the potential and X, Y, Z the components of force due to a body M; and let m be its mass. Consider a similar distribution of matter of q-fold density at corresponding points, and of p-fold linear dimensions. The mass of this body will be p^3qm, and its potential and force-components at the point corresponding to $(x, y, z,)$ will be

$$p^2qV, \quad pqX, \quad pqY, \quad pqZ.$$

Hence if we put

$$E = \frac{1}{8\pi} \int_{-\infty}^{\infty} \int_{-\infty}^{\infty} \int_{-\infty}^{\infty} (X^2 + Y^2 + Z^2)\, dx\,dy\,dz,$$

that is to say, if E denote the mechanical value of the distribution M, the mechanical value of the supposed similar distribution of altered dimensions will be

$$p^5q^2E.$$

564. Considering now similar magnets of different dimensions, whether polar magnets or electro-magnets, we see from the fundamental formulæ (§§ 482, 483, 486, 544) that the forces at corresponding points are independent of the linear dimensions,

and are equal, with equal intensities of magnetization, when polar magnets are compared, or with intensities of electric currents inversely proportional to the linear dimensions of the bodies when electro-magnets are compared. Hence the values of the integral (1) of § 561 for similar magnets are simply proportional to their volumes; provided that, when polar magnets are compared their intensities of magnetization are equal, and when electro-magnets, the intensities of their electric currents are inversely proportional to their linear dimensions. Farther when polar magnets are compared, the proposition holds whether the polar or the electro-magnetic definition (§ 517) of resultant force through interiors is adopted. But an electro-magnet cannot be simply divided into parts infinitely small in all their dimensions each of which is an independent electro-magnet; and therefore the further consideration of electro-magnets must be deferred, while we use the divisibility of a polar magnet asserted in § 447, to investigate the mechanical value of a distribution of polar magnetism, after the manner of § 562.

565. At any point $(x, y, z,)$ let $\mathfrak{R}$ denote the resultant force due to a polar magnet; the definition of § 480 being adopted when $(x, y, z,)$ is in the substance of the magnet. The preliminary proposition (§ 561) is immediately applicable, and shows that the work required to produce any change in the relative position of a set of magnets is equal to the augmentation of

$$\int_{-\infty}^{\infty}\int_{-\infty}^{\infty}\int_{-\infty}^{\infty} \frac{\mathfrak{R}^2}{8\pi}\, dx\, dy\, dz \dots\dots\dots\dots(1).$$

Hence (§ 564) when a uniformly magnetized magnet is of such a shape that it can be divided into similar parts, the mechanical value of the whole is simply equal to the sum of the mechanical values of the parts; [a remarkable contrast to the corresponding proposition (§ 562) relative to a homogeneous distribution of matter]. In other words, the work required to separate to infinitely great mutual distances any number of parts, each similar to the whole, of a uniformly magnetized magnet, is zero. It follows that if an infinite number of infinitely small magnets, each distributed through a finite volume of space, with their magnetic axes parallel and with equal sums of magnetic moments in equal finite portions of that space, no work will be required to condense or rarefy the distribution without altering the proportions of mutual distances, or the direction of the magnetic axes relatively to the lines of these distances; provided that the condensation is never pushed so far as to bring the constituents within distances not infinitely great in comparison with the linear dimensions of the constituent magnets. This last proviso is unnecessary when the constituents are uniformly magnetized, all with the same intensity of magnetization, and are so shaped that when brought into contact in the supposed condensation they fit together and form a whole, similar in shape to each part.

566.   Consider now a bar or cylinder of uniformly and longitudinally magnetized substance, terminated by planes perpendicular to its length; and let i denote the intensity of the magnetization. This limit is approximately reached when the length of the bar is very great in comparison with its greatest transverse diameter. The corresponding distribution of imaginary magnetic matter consists (§ 473) of distributions of positive and negative matter, of surface density i on the two terminal planes. The resultant force at points infinitely near the edge of either of these planes is infinite; but notwithstanding this, it is easily proved that the value of the integral (1) is finite. If we suppose the bar to be at first infinitely short and to be gradually increased in length, the value of the integral (1), expressing the work required to draw the two terminal planes asunder against their mutual attraction, increases continuously from zero to a limiting value equal to twice the value of the corresponding integral for either of the terminal planes alone. Hence, because for similar bars the values of the integral are (§ 565) as the volumes of the bars, it follows that for bars of similar cross sections the integral has values proportional to the cubes of transverse dimensions and independent of the lengths, provided only that the length of each bar considered is very great in comparison with its greatest transverse diameter. Hence, if any polar magnet be divided into infinitely thin bars* along its lines of magnetiza-

---

* By an infinitely thin bar, I mean a bar of which the transverse diameters are all infinitely small in comparison with the length.

tion, and if these bars be separated to infinite distances from one another, the whole value of the integral (1) becomes infinitely small*.

567.   Hence if magnetized substance given in infinitely thin bars at infinitely great distances from one another be put together so as to form a polar magnet, the value of integral (1) for this magnet expresses the amount of work which was spent in thus building it up.   Neglecting then the (unknown) mechanical value of the material, supposed given in infinitely thin permanently magnetized bars at infinitely great distances from one another, and defining the mechanical value of a magnet as the amount of work required to build it up of such materials, we see that this is expressed by the integral (1) of § 565.

568.   The value of the integral (1) (§ 565) is zero, when the magnet consists of closed solenoids; because, in this case (§ 510 Cors. 2 and 3) $\mathfrak{R} = 0$ for every point.   This result might at first sight appear erroneous, because a finite positive amount of work is required to cut up a finite closed solenoid into bars and separate them to infinite distances from one another.   But it is verified by remarking that if each such bar, being of finite transverse dimensions, is split up into infinitely thin bars, work is gained by allowing these infinitely thin bars to repel one another to infinite mutual distances; and that the whole amount of work thus gained is exactly equal to what was spent in reducing the solenoid to separate finite bars.   Or vary the process by supposing a finite solenoid to be first split up into an infinite number of infinitely thin solenoids; then the sum of the infinitely great number of infinitely small amounts of work required to break these infinitely thin solenoids into bars and separate the bars to infinite mutual distances, is infinitely small.   In short the explanation of the apparent difficulty is contained in § 566.

569.   It is only for a magnet consisting of closed solenoids that $\mathfrak{R}$ is everywhere zero.   For every other magnet, the

---

* But if each of these bars be divided into lengths comparable with its transverse dimensions, and if these parts be separated to distances from one another infinitely great in comparison with their dimensions, the integral (1) acquires a finite value which is equal to the amount of work necessary to produce this separation.

integral (1) of § 565 has consequently a finite positive value.
This I shall now prove to be always less than

$$2\pi \int_{-\infty}^{\infty} \int_{-\infty}^{\infty} \int_{-\infty}^{\infty} i^2\, dx\, dy\, dz$$

(where i denotes the intensity of magnetization), except in the
extreme case of a magnet consisting of closed shells, when the
limiting value is reached.

As in the postscript to § 517, let, for any point (x, y, z), R
denote the resultant force according to the electro-magnetic
definition, and X, Y, Z its components; α, β, γ the component intensities of magnetization; $\mathfrak{R}$ (still as in § 565) the
resultant force according to the polar definition; and $\mathfrak{X}, \mathfrak{Y}, \mathfrak{Z}$,
$\mathcal{V}$ its components and its potential, so that

$$\mathfrak{X} = -\frac{d\mathcal{V}}{dx}, \quad \mathfrak{Y} = -\frac{d\mathcal{V}}{dy}, \quad \mathfrak{Z} = -\frac{d\mathcal{V}}{dz} \dots\dots\dots(2).$$

Let $\mathfrak{E}$ denote the value of the integral (1) of § 565; and E
the corresponding integral of the electro-magnetic resultant
force; that is to say, let

$$\mathfrak{E} = \frac{1}{8\pi} \int_{-\infty}^{\infty} \int_{-\infty}^{\infty} \int_{-\infty}^{\infty} \mathfrak{R}^2 dx\, dy\, dz \dots\dots(3),$$

$$E = \frac{1}{8\pi} \int_{-\infty}^{\infty} \int_{-\infty}^{\infty} \int_{-\infty}^{\infty} R^2 dx\, dy\, dz \dots\dots(4).$$

The formulæ (r) of the postscript to § 517, with (2) of the
present section give

$$R^2 = \mathfrak{R}^2 - 8\pi \left(\alpha \frac{d\mathcal{V}}{dx} + \beta \frac{d\mathcal{V}}{dy} + \gamma \frac{d\mathcal{V}}{dz} \right) + 16\pi^2 i^2.$$

Use this in (4); follow the usual process of integration by
parts, which gives

$$\int_{-\infty}^{\infty} \int_{-\infty}^{\infty} \int_{-\infty}^{\infty} \left(\alpha \frac{d\mathcal{V}}{dx} + \beta \frac{d\mathcal{V}}{dy} + \gamma \frac{d\mathcal{V}}{dz} \right) dx\, dy\, dz$$

$$= -\int_{-\infty}^{\infty} \int_{-\infty}^{\infty} \int_{-\infty}^{\infty} \mathcal{V} \left(\frac{d\alpha}{dx} + \frac{d\beta}{dy} + \frac{d\gamma}{dz} \right) dx\, dy\, dz \dots(5);$$

remark that [§ 473 (2)] $\dfrac{d\alpha}{dx} + \dfrac{d\beta}{dy} + \dfrac{d\gamma}{dz} = -\rho \dots\dots\dots(6)$,

where ρ denotes the density of the imaginary magnetic matter
which we substitute for the given magnet (when the polar defi-

nition is used for the force through the space occupied by it); and remark that according to the alternative formula (7) of § 561,

$$\mathfrak{E} = \tfrac{1}{2} \int_{-\infty}^{\infty} \int_{-\infty}^{\infty} \int_{-\infty}^{\infty} \rho \mathcal{V} dx dy dz \dots\dots(7).$$

So we find $E = \mathfrak{E} - 2\mathfrak{E} + 2\pi \int_{-\infty}^{\infty} \int_{-\infty}^{\infty} \int_{-\infty}^{\infty} i^2 dx dy dz$;

and therefore $\mathfrak{E} + E = 2\pi \int_{-\infty}^{\infty} \int_{-\infty}^{\infty} \int_{-\infty}^{\infty} i^2 dx dy dz \dots\dots(8).$

Now E has always a positive finite value except for the extreme case of a magnet consisting of closed shells, when it is zero, because (§ 512 cor. 6), $R = 0$ in this case for every point whether in the substance of the magnet or not. Hence the proposition is proved.

570. For $X^2 + Y^2 + Z^2$ take, in virtue of (c), § 517,

$$X \left(\frac{dN}{dz} - \frac{dM}{dy} \right) + Y \left(\frac{dL}{dz} - \frac{dN}{dx} \right) + Z \left(\frac{dM}{dx} - \frac{dL}{dy} \right),$$

and integrate by parts after the manner of § 518, but with infinities for limits. We thus find

$$E = \frac{1}{8\pi} \int_{-\infty}^{\infty} \int_{-\infty}^{\infty} \int_{-\infty}^{\infty} dx dy dz \left[L \left(\frac{dZ}{dy} - \frac{dY}{dz} \right) + M \left(\frac{dX}{dz} - \frac{dZ}{dx} \right) + N \left(\frac{dY}{dx} - \frac{dX}{dy} \right) \right] \quad (9),$$

or by § 517 (l)

$$E = \tfrac{1}{2} \int_{-\infty}^{\infty} \int_{-\infty}^{\infty} \int_{-\infty}^{\infty} dx dy dz (Lu + Mv + Nw) \dots (10).$$

This, which is the analogue to (7) of § 569, was discovered for fluid motion by Helmholtz, and given in his paper on Vortex Motion (Crelle's *Journal*, 1858, or, translation by Tait, *Philosophical Magazine*, 1867, second half year). Lastly, substituting for u, v, w their values by (a) of § 517, and integrating again by parts as before, we find

$$E = \tfrac{1}{2} \int_{-\infty}^{\infty} \int_{-\infty}^{\infty} \int_{-\infty}^{\infty} dx dy dz (\alpha X + \beta Y + \gamma Z) \dots (11).$$

The analogue to this is [compare § 503 (2)],

$$\mathfrak{E} = -\tfrac{1}{2} \int_{-\infty}^{\infty} \int_{-\infty}^{\infty} \int_{-\infty}^{\infty} dx dy dz (\alpha \mathfrak{X} + \beta \mathfrak{Y} + \gamma \mathfrak{Z}) \dots (12).$$

The addition of these two formulæ verifies (8) of § 569.

571. In a memorandum-book under date Oct. 16th, 1851, I find the following statement:—"I concluded that the value of

" a current in a closed conductor, left without electromotive
" force, is the quantity of work that would be got by letting
" all the infinitely small currents into which it may be divided
" along the lines of motion of the electricity come together
" from an infinite distance, and make it up.   Each of these
" 'infinitely small currents' is of course in a circuit which is
" generally of finite length.   It is the section of each partial
" conductor and the strength of the current in it that must be
" infinitely small."   A memorandum of principles and formulæ
proving this statement had been written a few days previously
(Oct. 13th, 1851).   A somewhat amplified statement of the
principle was first published, but without the formulæ, in 1860,
in the second edition of Nichol's *Cyclopœdia* (Article " Magnetism, Dynamical Relations of").   Though the subject does not
belong properly to the present volume, I append in 'footnotes the original memorandum*, and an extract from Nichol's

---

* *Memorandum, Oct.* 13, 1851.—Refers first to an erroneous temporary
conclusion which led me to think "that the value of a current in a closed
"conductor will be effected by steel magnets in its neighbourhood." "From
"this I was shaken a little by Faraday's finding (*Exp. Res.* § 1100) that steel
"does not do so well as soft iron," etc. [in respect to electro-magnetic induc-
tion], "and I soon saw that I must have fallen into some mistake. . . .
"I made out the true state of the case.   This is the explanation.   Let
"$\dfrac{dE}{ds}\dfrac{ds}{dt} . \gamma$ be the quantity of work done in time dt, by bringing a steel
"magnet towards a galvanic current, kept up, say, by a battery.   Then C,
"the electromotive force due to the chemical action, will be increased by
"$\dfrac{dE}{ds}\dfrac{ds}{dt}.$ Hence if k be the resistance in absolute measure

$$\gamma = \frac{C + \dfrac{dE}{ds}\dfrac{ds}{dt}}{k};$$

"so that if wdt denote the work

$$wdt = \frac{\dfrac{dE}{ds}\dfrac{ds}{dt}\left(C + \dfrac{dE}{ds}\dfrac{ds}{dt}\right)}{k}\, dt,$$

"and if Mdt be the mechanical equivalent of the chemical action (increased
"on account of the increased current), we have

$$Mdt = C\gamma dt = \frac{C\left(C + \dfrac{dE}{ds}\dfrac{ds}{dt}\right)}{k}$$

"Lastly, if Hdt be the heat developed, we have

$$JHdt = k\gamma^2 dt = \frac{\left(C + \dfrac{dE}{ds}\dfrac{ds}{dt}\right)^2}{k}\, dt;$$

"and therefore $\qquad\qquad JHdt = wdt + Mdt.$

*Cyclopædia**, containing the amplified statement. Defining then the dynamical value of an electro-magnet as the quantity of

---

"We conclude that the work actually spent, together with the mechanical
"equivalent of the chemical action, together produce exactly an equivalent
"of heat, and therefore no other effect. Hence the mechanical values of the
"current and of the magnet together are not altered. On the other hand,
"let two pure electro-magnets be brought towards one another. Adopting
"a notation corresponding to the former we have

$$w = \frac{dE}{ds}\frac{ds}{dt}\gamma\gamma' ; \quad \gamma = \frac{C + \frac{dE}{ds}\frac{ds}{dt}\gamma'}{k} , \quad \gamma' = \frac{C' + \frac{dE}{ds}\frac{ds}{dt}\gamma}{k'}$$

$$M = C\gamma = \frac{C\left(C + \frac{dE}{ds}\frac{ds}{dt}\gamma'\right)}{k} ; \quad M' = \frac{C'\left(C' + \frac{dE}{ds}\frac{ds}{dt}\gamma\right)}{k'}$$

$$JH = k\gamma^2 + k'\gamma'^2 = \frac{\left(C + \frac{dE}{ds}\frac{ds}{dt}\gamma\right)^2}{k} + \frac{\left(C' + \frac{dE}{ds}\frac{ds}{dt}\gamma\right)^2}{k'} = M + M' + 2w.$$

"Hence [J denoting Joule's equivalent] there is more heat evolved than
"$\frac{1}{J}(M+M'+w)$ by $\frac{1}{J}w$, and therefore the mechanical value of two cur-
"rents is diminished by $\frac{1}{J}wdt$ in the time dt."

* "*Electricity in motion.*—If an electric current be excited in a conductor,
"and then left without electro-motive force, it retains energy to produce heat,
"light, and other kinds of mechanical effect, and it gradually falls in strength
"until it becomes insensible, as is amply demonstrated by the initial experi-
"ments of Faraday and Henry, on the spark which takes place when a gal-
"vanic circuit is opened at any point, and by those of Weber, Helmholtz, and
"others on the electro-magnetic effects of varying currents. Professor W.
"Thomson has shown how the mechanical value of all the effects that a cur-
"rent in a closed circuit can produce after the electro-motive force ceases,
"may be ascertained by a determination, founded on the known laws of
"electro-dynamic induction, of the mechanical value of the energy of a cur-
"rent of given strength, circulating in a linear conductor (a bent wire, for
"instance) of any form. To do this, it may be remarked, in the first place,
"that a current, once instituted in a conductor, and circulating in it after
"the electro-motive force ceases, does so just as if the electricity had inertia,
"and will diminish in strength according to the same, or nearly the same,
"laws as a current of water or other fluid, once set in motion and left with-
"out moving force, in a pipe forming a closed circuit. But according to
"Faraday, who found that an electric circuit consisting of a wire doubled on
"itself, with the two parts close together, gives no sensible spark when
"suddenly broken, in comparison with that given by an equal length of wire
"bent into a coil, it appears that the effects of ordinary *inertia* either do not
"exist for electricity in motion, or are but small compared with those which,
"in a suitable arrangement, are produced by the 'induction of the current
"'upon itself.' In the present state of science it is only these effects that
"can be determined by a mathematical investigation ; but the effects of elec-
"trical inertia, should it be found to exist, will be taken into account by
"adding a term of determinate form to the fully determined result of the
"present investigation which expresses the mechanical value of a current in
"a linear conductor as far as it depends on the induction of the current on
"itself.
"The general principle of the investigation is this—If two conductors,
"with a current sustained in each by a constant electro-motive force, be

work specified in the statement quoted above in the text, we have in equation (5) a proof the first hitherto published, of thę assertion in the extract from Nichol's *Cyclopædia* quoted in the foot-note, that the dynamical value of a current in a closed circuit may be calculated by the formula (4). For let open magnetic shells (§§ 506, 548) be substituted for the "infinitely small currents" referred to in the preceding statement, supposed first to be in their actual positions in the electro-magnet composed of them; and let these shells be separated to infinite distances from one another. It is easily proved by considerations of infinitesimals analogous to those fully set forth in § 566, that when the shells are brought to infinite distances from one another, the value of E becomes zero; and, therefore, as the second member of (5) remains constant, the value of E before the circuits were separated, is equal to the addition of value which $\mathfrak{E}$ experiences during the process of separation,

---

"slowly moved towards one another, and there be a certain *gain of work* on "the whole, by electro-dynamic force operating during the motion, there "will be twice as much as this of work spent by the electro-motive forces "(for instance, twice the equivalent of chemical action in the batteries, should "the electro-motive forces be chemical) over and above that which they "would have had to spend in the same time, merely to keep up the currents, "if the conductors had been at rest, because the electro-dynamic induction "produced by the motion will augment the currents; while on the other "hand, if the motion be such as to require the *expenditure* of work against "electro-dynamic forces to produce it, there will be twice as much work "saved off the action of the electro-motive forces by the currents being dimin-"ished- during the motion. Hence the aggregate mechanical value of the "currents in the two conductors, when brought to rest, will be increased in "the one case by an amount equal to the work done by mutual electro-"dynamic forces in the motion, and will be diminished by the corresponding "amount in the other case. The same considerations are applicable to "relative motions of two portions of the same linear conductor (supposed "perfectly flexible). Hence it is concluded that the mechanical value of a "current of given strength in a linear conductor of any form, is determined "by calculating the amount of work against electro-dynamic forces, required "to double it upon itself, while a current of constant strength is sustained in "it. The mathematical problem thus presented leads to an expression for "the required mechanical value consisting of two factors, of which, one is "determined according to the form and dimensions of the line of the con-"ductor in any case, irrespectively of its section, and the other is the square "of the strength of the current. The mechanical value of a current in a "closed circuit, determined on these principles, may be calculated by means "of the following simple formula, not hitherto published :—

$$\frac{1}{8\pi} \int\!\int\!\int R^2 dx\,dy\,dz,$$

"where R denotes the resultant electro-magnetic force at any point (x, y, z). "This expression is very useful in the dynamical theory of magneto-electric "machines and electro-magnetic engines."—From Article "Magnetism, "Dynamical Relations of," Nichol's *Cyclopædia*, edit. 1860.

that is to say, is equal to the work spent in effecting this process.

572. Equation (5) expresses the following very remarkable proposition. The sum of the dynamical values of an electro-magnet and of any corresponding lamellar polar magnet is equal to 2π multiplied into the sum of the squares of the intensities of magnetization of all parts of the latter; the two species of dynamical value understood, being those defined in § 571 and § 567.

XXVII. [*Jan.* 1872.]

CHAPTER VIII.—*Hydro-kinetic Analogy.*

573. The hydro-kinetic analogy for the force of a polar magnet seems to have been first perceived by Euler. It requires the supposition of generation and annihilation of fluid in places of positive and negative magnetic polarity, if we adopt for "the resultant force" in the magnetic substance the definition proper for a polar magnet laid down in § 479; unless we limit the field of force considered, to places void of magnetized matter, whether external to the magnet or in hollows within it. Thus, if we consider all space as filled with an incompressible frictionless liquid initially at rest, and if at certain points, lines, surfaces, or volumes, we assume more liquid of the same density to be continuously generated, and at the same time in other places liquid in equal quantity to be continuously annihilated, the velocity of the resulting fluid motion would be the same in direction and magnitude as the resultant magnetic force due to a distribution of magnetism presenting unneutralized polarity, positive (or northern) in the places of the fluid analogue where there is generation, and negative (or southern) in the places where there is annihilation. There is, however, no interest in pursuing the consideration of this extension of the hydro-kinetic analogy through spaces occupied by magnetized matter, involving as it does the strained supposition of the generation and annihilation of matter in spaces through which the liquid is perfectly free to move.

574. On the other hand, the hydro-kinetic analogy limited to spaces unoccupied by magnetized matter is perfectly satisfactory,

as far as it goes. Let all these spaces be occupied by incompressible liquid, and let the magnetized matter be replaced by a rigid body perforated so as to constitute an infinitely numerous group of infinitely fine tubes fulfilling the following conditions :—Divide the whole surface of the magnet into infinitely small areas inversely proportional to the magnitudes of the normal component forces across them whether outwards or inwards. Because the surface integral of the normal component force for the whole surface of the magnet is zero, the number of these infinitesimal areas in that part of the surface where the normal component force is outwards must be equal to the number in the remainder of the surface. Now to pass to the fluid analogue; instead of the magnet substitute a rigid body perforated from each of the infinitesimal areas in the part of the surface where the normal component force is positive, by a single tunnel through to one of the areas in the other part of the surface. Let there be in the first place a piston in each of these tunnels or tubes, and apply force to it until it moves with such a velocity that the velocity of efflux at one end and influx at the other is numerically equal to the normal component of the magnetic force to be represented : and when this condition has been once reached let the pistons become dissolved into perfect liquid homogeneous with the rest. The solid with its perforations remaining a rigid tubular system, the liquid will continue for ever circulating through the tubes and the free external space : and its motion through all external space will be such that the velocity is everywhere of the same magnitude and in the same direction as the resultant magnetic force in the corresponding position relatively to the magnet. The proof of this proposition* is ;—that according to a well-known hydro-kinetic theorem, the motion of the liquid must be everywhere "irrotational" [Vortex Motion, § 59 (e)], and that if the normal component fluid velocity, or normal component force in the magnetic analogue, be given

---

* All the hydro-kinetic terminology and propositions used in the remainder of this volume are fully explained, with demonstrations when necessary, in the portion already published (in the *Transactions of the Royal Society of Edinburgh*, April 1867 and Dec. 1869) of a paper on "Vortex Motion," with the continuation of which I am at present occupied. References to it are given when necessary to justify any of the assertions in hydro-kinetic subjects made henceforward.

over the whole surface, the fluid motion or magnetic force is determinate through all external space (§ 591, Theorems 1 and 2). The permanence of the fluid motion fulfilling the same condition follows at once from the constancy of the circulation through each perforation [Vortex Motion, § 59 (*d*)], consequent upon the frictionless character which we assume the fluid to possess.

575. In the preceding statement no condition has been imposed as to the pairs of apertures in the surface of the rigid body substituted for a magnet, which are to be connected through the internal tubes; no such condition having been necessary, because we supposed the apertures over the whole surface to be inversely proportional to the magnitude of the normal component force. The statement may be varied thus:— take all that part of the surface for which the normal component force is outwards, and divide it in any manner into infinitesimal areas. From each point in the boundary of any one of these areas, draw a line through external space till it meets again, as it will meet again, the surface of the magnet. By doing this for every infinitesimal area of the boundary traversed outwards, a *corresponding area*, where the normal component force is inwards, is found, and the whole remainder of the surface is thus divided into areas corresponding to those chosen in the first part. Let the pairs of corresponding areas be connected by internal tubes. The remainder of the statement may be applied without alteration to this tubular arrangement. The fluid analogue thus constructed, will have the peculiarity, that each portion of fluid circulates for ever along one circuit (that is to say, closed curve).

576. The hydro-kinetic analogy is both more complete and more simple, it is in fact perfectly complete, and therefore perfectly simple, if instead of as in § 479 adopting the definition proper for a polar magnet (§ 549), we adopt the "electromagnetic definition" (§ 517 and postscript to § 517), for the resultant force at any point in the substance of the magnet, whether it be a polar magnet or an electro-magnet. *The resultant force defined "electromagnetically" for the space occupied by the magnet, and the resultant magnetic force according to the unambiguous definition for space not occupied by the*

magnet, agree everywhere in magnitude and direction with the velocity in a possible case of motion of an incompressible liquid filling all space. To prove this it is only necessary to remark that the sole condition that X, Y, Z, may be the velocity-components in a possible case of motion of an incompressible fluid, is that they fulfil the equation of continuity

$$\frac{dX}{dx} + \frac{dY}{dy} + \frac{dZ}{dz} = 0;$$

and we have seen (§ 517) that

$$\frac{dX}{dx} + \frac{dY}{dy} + \frac{dZ}{dz} = 0$$

throughout the substance of the magnet as well as through external space, if X, Y, Z denote components of the magnetic force. The component intensities of electric current in the electro-magnet producing this force are [§ 517 (a), (l)]

$$\frac{1}{4\pi}\left(\frac{dZ}{dy} - \frac{dY}{dz}\right), \quad \frac{1}{4\pi}\left(\frac{dX}{dz} - \frac{dZ}{dx}\right), \quad \frac{1}{4\pi}\left(\frac{dY}{dx} - \frac{dX}{dy}\right).$$

577. This proposition, which I found more than twenty years ago as an obvious deduction from my formulæ for electro-magnetic force, published in the *Transactions of the Royal Society* for June 1850 (§§ 515—518 above), is purely kinematical. Since that time it has acquired an interest which it did not then possess for me, in virtue of Helmholtz's splendid discovery of the dynamical laws of vortex motion*. I had not known more than that the distribution of "electro-magnetic" force through the substance of the magnet, as well as through external space, corresponded to a possible distribution of motion in a continuous incompressible fluid filling all space, and had no clue to the consequences of leaving a frictionless liquid to itself, with such a motion once established in it. By Helmholtz's theory, it is demonstrated that the fluid motion alters so as to always remain the representative of the electromotive force due to an electro-magnet continuously varied according to the following law. Lines of fluid matter which initially coincided with the lines of electric current in the electro-magnet initially

---

* Crelle's *Journal*, 1858, and (Tait's translation) *Philosophical Magazine*, July 1867.

replaced by the fluid, however they change in the subsequent motion, always mark the lines of electric current which must be constituted to produce the altered electro-magnet; and the whole amount of the intensity of the electric current crossing any area bounded by any closed curve passing always through the same fluid particles remains constant. It is unnecessary, however, to enter now on the wide hydro-kinetic subject thus indicated; although I cannot but refer to Helmholtz's theorem of vortex motion, not merely on account of its intrinsic beauty, but because I have found it of great value in assisting me to realize the purely kinematic representation of electro-magnetic force which fluid motion affords. The general hydro-kinematic analogy, and the dynamics of the irrotationally moving portions of the fluid, as they served me primarily twenty-four years ago in investigating the inverse problems, will be further considered in the following chapter.

578. The hydro-kinetic analogy is valuable in the mathematical theory of electro-magnetism as leading to a set of theorems respecting magnetic forces produced by electric currents, precisely analogous to those theorems of Green's respecting forces due to centres acting according to the Newtonian law, which I deduced in 1841 from an analogy with the "Uniform motion of heat in homogeneous bodies," by the investigation forming the first part of this volume (§§ 1—4 above). The following theorems I—III. are particular cases of the general proposition of § 576, and require no further demonstration.

579. *Theorem I.*—(Compare § 594 below.)—Considering all space as occupied by an incompressible frictionless liquid, let S be a closed surface, which (to facilitate conceptions) may be supposed to be constituted of a perfectly flexible and extensible membrane. At first let there be no motion of the liquid in any part of space, and then let any motion whatever be arbitrarily given to S, subject only to the condition of not altering the volume enclosed by it. The motion which is given to the liquid will be everywhere irrotational ("Vortex Motion," § 16 and § 60), and will therefore be continuously expressible throughout external space by a potential; and continuously expressible, likewise, through the internal space: but there

454 *A Mathematical Theory of Magnetism.* [XXVII.

will be a discontinuity at S; on the two sides of which the velocity-potential must differ by an amount equal to P, the impulsive pressure which would have to be applied to S to produce the actual motion instantaneously from rest. Divide S into infinitely narrow bands by lines corresponding to equal values of P, and in each of these bands let an electric current circulate of strength equal to $\dfrac{\delta P}{4\pi}$ where δP denotes the difference of the values of P at its two boundaries. The magnetic force produced by the distribution of electric currents thus constituted, will agree in magnitude and direction with the fluid velocity in the hydro-kinetic analogue. This proposition I used in a communication to the British Association at Oxford, in June 1847, "On the Electric Currents by which the Phenomena of Terrestrial Magnetism may be produced;" and it is referred to in the abstract of that communication (now reprinted in §§ 602, 603 below), which appeared in the yearly volume. It was probably one of five propositions which I wrote to Liouville in the September following (see § 589 below).

580. *Corollary.*—In the electro-magnetic analogue the direction of the electric current is perpendicular to the relative tangential motion of the liquid on the two sides of S, and the surface intensity of the electric current is equal to the relative tangential velocity divided by 4π.

581. *Example.*—Let S be kept of constant figure, and let the motion given to it be purely translatory. The liquid within it will move as if it were a rigid body. Hence the interior velocity-potential will be Ux, if U be the velocity, and if its direction be parallel to the axis of x. Hence if we consider a solid carried along through a frictionless liquid; determine the velocity and direction, relatively to the solid, of the liquid gliding along each part of its surface; and construct the analogous surface electro-magnet according to the rule of § 579; this distribution of electric currents will produce a uniform field of force, of intensity U throughout the space enclosed by the surface on which they are distributed, and will produce a resultant force at every external point, agreeing in magnitude and direction with the absolute velocity which the liquid is compelled to take in making way for the solid. The analytical

expression of this very interesting theorem is contained in (IX.) of § 517, applied to the case in which

$$\alpha = \frac{U}{4\pi}, \, \beta = 0, \, \gamma = 0.$$

582. *Theorem II.* (Includes the case § 581 of Theorem I.)— Let any motion of rotation be given to a rigid body in an infinite incompressible liquid. The magnetic analogue consists of a uniform current traversing the volume of the rigid body in lines parallel to the axis of rotation, and of intensity equal to twice the angular velocity; with the circuit completed superficially by the surface distribution constructed according to the rule of § 581. The resultant force of the completed solid and superficial electro-magnet (§ 535) thus formed will agree everywhere in magnitude and direction with the absolute velocity of the matter, whether solid or liquid, in the kinematic analogue. The analytical expression of this theorem (if we take the axis of the solid's rotation for the axis of x) is had by putting in (IX.) of § 517

$$\alpha = 0, \, \beta = -\zeta z, \, \gamma = \xi y.$$

583. *Theorem III.*—Consider a fixed rigid ring, having, for simplicity, but one perforation, and therefore giving duplex continuity to the space external to it. Let the whole of the external space be occupied by an incompressible frictionless liquid in a state of cyclic motion, with the ring for core. Take any surface S bounded by stream lines. This is necessarily a surface of duplex continuity enclosing the ring. On one of the stream lines forming a circuit of S, take i points corresponding to infinitely small differences of the velocity potential, each an exact sub-multiple $\frac{1}{i}$ of the "cyclic constant," or "whole circulation" (κ). Through these points draw equipotential lines on S, which therefore will each cut perpendicularly all the stream lines on S. In each of the infinitely narrow bands into which S is thus divided (constituting a geometrical circuit which crosses all the stream line circuits), let an electric current of strength $\frac{\kappa}{4\pi i}$ circulate. The resulting electro-magnetic force will be zero at every point within S, and will be equal to, and in the same direction as, the fluid velocity

in the space external to *S*. This interesting and important
proposition is perfectly analogous to that which is given by
Green for surface distribution of electricity and the resulting
electric force in Article 12 of his Essay (to which reference is
made in Thomson and Tait's *Natural Philosophy*, § 507, under
the designation "reducible case of Green's problem").

XXVIII. [*Nov.* 1871.]

CHAPTER IX.—*Inverse Problems.*

584. Inverse problems of magnetism are problems in which
the data are of magnetic force, and it is required to find distributions of magnetism or of electric currents by which the given
force can be produced.   They fall under two classes:—I. Those
in which the force is given for every point of space:—and II.
Those in which the force or some component of the force is
given through some portion of space, whether volume, surface,
or line;—and it is required, under certain limitations or conditions, to find distributions of magnetism or of electric currents
by which the given force can be produced. A complete and
unconditional solution of every problem of Class I. is, as we
shall immediately see, always easily found.

585. *Class I.*—First case, *polar definition* (§ 479 and Postscript to § 517) of *resultant force adopted*. In this case the
magnetic force is expressible by means of a potential, and
therefore the most general form of data is;—given the potential
at every point of space. Let *V* be its value at (*x, y, z*), so that
if $\mathfrak{X}$, $\mathfrak{Y}$, $\mathfrak{Z}$ denote the components of the magnetic force,

$$\mathfrak{X}=\frac{-dV}{dx}, \quad \mathfrak{Y}=\frac{-dV}{dy}, \quad \mathfrak{Z}=\frac{-dV}{dz}\dots\dots\dots\dots(1).$$

If *α, β, γ* denote the rectangular components of the required
magnetization, we have

$$\frac{d\alpha}{dx}+\frac{d\beta}{dy}+\frac{d\gamma}{dz}=\frac{1}{4\pi}\left(\frac{d^2V}{dx^2}+\frac{d^2V}{dy^2}+\frac{d^2V}{dz^2}\right) \quad [\S~517~(m)~\text{repeated}],$$

and *α, β, γ* may be any functions whatever which fulfil this
equation. Then as a particular solution we have

$$\alpha'=\frac{1}{4\pi}\frac{dV}{dx}, \quad \beta'=\frac{1}{4\pi}\frac{dV}{dy}, \quad \gamma'=\frac{1}{4\pi}\frac{dV}{dz}\dots\dots(2).$$

Let now α'', β'', γ'' denote any three functions whatever fulfilling the following equation :—

$$\frac{d\alpha''}{dx} + \frac{d\beta''}{dy} + \frac{d\gamma''}{dz} = 0 \quad\ldots\ldots\ldots\ldots\ldots(3).$$

The complete solution of the problem is,

$$\alpha = \alpha' + \alpha'', \quad \beta = \beta' + \beta'', \quad \gamma = \gamma' + \gamma''\ldots\ldots\ldots\ldots\ldots(4).$$

The arbitrary part α'', β'', γ'', of this solution consists of any distribution of magnetization agreeing everywhere in intensity and direction with the velocity and direction of a possible motion of an incompressible fluid through all space. When the given function V is such that its first and second differential coefficients

$$\frac{dV}{dx}, \quad \frac{d^2V}{dx^2}, \quad \frac{dV}{dy}, \quad \frac{d^2V}{dy^2}, \quad \frac{dV}{dz}, \quad \frac{d^2V}{dz^2}$$

are everywhere finite, there is nothing more to be said in respect to the preceding solution; but when the first differential co-efficients $\frac{dV}{dx}$, etc., though themselves everywhere finite, vary anywhere abruptly in their values, an interpretation of a sufficiently obvious character becomes necessary to deduce the solution from the preceding formulæ. Or the form of solution may be varied by introducing the proper formulæ [§ 473 (1)] for surface-distributions of the imaginary magnetic matter at the surfaces of discontinuity.

586. *Class I.*—Second case, *electro-magnetic definition adopted.* In this case the force, though expressible by means of a potential throughout every portion of space free from magnetized matter, is not so expressible through the substance of the magnet. Hence the data must be the intensity and direction of the resultant force at every point of space; but these data are not altogether arbitrary inasmuch as if X, Y, Z denote the three rectangular components of the force,

$$\frac{dX}{dx} + \frac{dY}{dy} + \frac{dZ}{dz} = 0 \qquad \text{[§ 517 (k) repeated]}.$$

Hence the problem is;—given X, Y, Z, each any function of (x, y, z), but subject to equation (k) of § 517; it is required to find three quantities α, β, γ such that

$$4\pi\left(\frac{d\gamma}{dy}-\frac{d\beta}{dz}\right)=\frac{dZ}{dy}-\frac{dY}{dz},\ 4\pi\left(\frac{d\alpha}{dz}-\frac{d\gamma}{dx}\right)=\frac{dX}{dz}-\frac{dZ}{dx},\ 4\pi\left(\frac{d\beta}{dx}-\frac{d\alpha}{dy}\right)=\frac{dY}{dx}-\frac{dX}{dy}$$

[§ 517 (l) repeated].

Of this problem the general solution is

$$\alpha=\frac{1}{4\pi}X+\frac{d\psi}{dx},\quad \beta=\frac{1}{4\pi}Y+\frac{d\psi}{dy},\quad \gamma=\frac{1}{4\pi}Z+\frac{d\psi}{dz}\ \ldots(5);$$

where ψ denotes any arbitrary function of (x, y, z). For simplicity we have supposed that there is no abrupt variation in the given values of X, Y, Z. The proper formulæ to suit the case of abrupt variations from one side to another of any surface, are easily found.

587. Remark that the arbitrary functions α'', β'', γ'', in the solution (4) of § 585 express any solenoidal distribution whatever with the solenoids all closed; and that the arbitrary part ψ in the solution (5) of § 586 expresses any lamellar distribution whatever with the shells all closed.

588. Remark also that the distribution of imaginary magnetic matter derivable (§ 473) from the solution of § 584, and of electric current derivable (§ 554) from the solution of § 585, are each determinate, and that it is only the distribution of magnetization which is affected by the arbitrary part of the solution in either case.

589. *Class II.*—For the present it is enough to consider the following typical problems of this class. Given the force through space external to a given closed surface S: required the distribution of imaginary magnetic matter, or of electric currents, or of magnetization; each distribution confined to an infinitely thin layer of matter coincident with this surface: and to investigate the determinacy of the solution in each case. With reference to these problems, I find a leaf of manuscript written in French, indorsed:—"Fragment of draft of letter "to M. Liouville, written on the Faulhorn, Sunday, September "12, 1847, and posted on the Monday or Tuesday week after, "at Maidstone. The letter has not been published yet, although "in Sept. 1848 I understood from M. Liouville in Paris, that he "had it for publication. Probably it has fallen aside and is "lost [? in consequence of the disturbed state of Paris at that

"time], which I should regret, as it contains my first ideas,
"and physical, especially hydro-dynamical, demonstrations of
"the theorems I am now about to write out for publication in
"my paper on magnetism for the Royal Society, from rough
"drafts written in August 1848.   W. T. *Oct.* 29*th*, 1849."
The "now" has been deferred until the present time,
November 20th, 1871. I am obliged to write from memory, as
I have not been able to recover any of those rough drafts. I
have added important details involving new ideas regarding
polycyclic* fluid motion, for much of which, as for the whole
terminology of multiple continuity, I am indebted to Helmholtz's paper on Vortex Motion.

590. First, with reference to the data, it must be remarked
that the force being by hypothesis due to polar magnets or
electro-magnets altogether within S, cannot be given arbitrarily
through the whole space external to that surface. It may
indeed be readily proved from a remarkable and important
proposition due to Gauss, to be found in Thomson and Tait's
Natural Philosophy, § 497, that if the potential were given for
any closed surface, lying altogether external to S, whether
enclosing S or not, and if not enclosing S, enclosing any portion
of external space however small, the force would be determinate
throughout the whole space external to S. The same may be
proved if (instead of the potential) the normal component force
were given over any surface whatever, external to S, and not
enclosing it, or over any simply continuous surface enclosing S.

At present, however, two cases only shall be considered:—
the potential given over the whole surface of S (Case 1), and
the normal force given over the whole of S (Case 2).

591. Preliminary Theorems 1—5.—*Theorem* 1 (Discovered
by Green). *The potential being given arbitrarily over* S, *the
resultant force is determinate through all external space, and a
determinate distribution of matter over* S, *acting according to the
inverse square of the distance, may be found which shall produce it.*

Theorem 2.—*The normal component force being given for* S,
the force is determinate through all external space, and a determi-

---

* "Vortex Motion," § 60 (z).

nate distribution of matter over S *acting according to the inverse square of the distance may be found which shall produce it,* provided that S is simply continuous.   [Compare § 207.]

Theorem 3.—*The potential being arbitrarily given for* S, *subject to the condition that its integral amount for the whole surface is zero; or the normal component force being arbitrarily given for* S, *subject to the condition that its integral amount for the whole surface is zero; the force in each case is determinate through all external space, and a determinate distribution of electric currents over* S *may be found which shall produce it,* provided that in the case in which the normal component force is given, the surface S is simply continuous.

Theorem 4.—If S be complexly continuous, let C_1, C_2, C_3, etc., be mutually irreconcilable closed curves encircling it, whether in contact with it, or in the space external to it.   If the continuity is *n*-fold, there are *n* such circuits.   *The normal component force being given arbitrarily for* S, *subject only to the condition that its integral amount for the whole surface is zero; and an arbitrary value* κ_1, κ_2, *etc., being given for the integral of the tangential component force round each of the circuits* C_1, C_2, *etc.: the resultant force is determinate through the whole space external to* S, *and a determinate distribution of electric currents over* S *may be found which shall produce it.*

Theorem 5.—When S is complexly continuous, *no distribution of matter over it can be found to produce force through external space fulfilling the conditions of Theorem* 4, *when the values of the cyclic constants* κ_1, κ_2, ... *are all finite; but if infinitely thin sheets of matter be introduced as barriers closing all the apertures of* S, *a determinate distribution of matter on these sheets and over* S *may be found which shall produce that force through all the space external to* S, *except the infinitely small parts of it occupied by the barriers.*

592.   *Demonstrations of Theorems* 1—4.—To prove Theorem 1, let the whole space within S and the whole space external to S, be occupied by homogeneous incompressible liquid, but let there be an infinitely thin vacuous space separating the external from the internal fluid.   Let equal impulsive pressures be ap-

plied in opposite directions, to the liquid surfaces on the two sides of this vacuous space, equal everywhere to the given value of the potential at the corresponding position in S, of the magnetic problem, the pressure being reckoned as positive when it is outwards from S on the external liquid, and inwards from S on the internal liquid. The motion will be irrotational throughout each portion of the fluid; and the initial velocity-potentials in portions of the fluid infinitely near one another on the two sides of S, will be equal to the given magnetic potential. Hence (§ 7) the given potential over S would be produced by a distribution of matter over S, having its surface density everywhere equal to the velocity of separation (reckoned negative when there is approach) of the two fluid surfaces divided by 4π*. By "velocity of separation" is meant the difference of the normal component velocities on the two sides of S.

593. *Demonstration of Theorem* 2.—With the same hydro-kinematic arrangement as in § 592, let the boundary of the fluid external to S be impulsively pressed so as to produce instantaneously a normal component velocity equal to the given normal component magnetic force. And let the bounding surface of the fluid within S be simultaneously acted on, with a pressure equal and opposite to that which produces the specified effect on the external fluid. The motion generated is irrotational through each portion of the fluid, and the potentials on the two sides of S, are each equal to the potential at S of the distribution of force through external space, which has for its normal component the given value for every point of S, the density of the determinate distribution of matter over S which would give that external distribution of force is, as in § 592, equal to the velocity of separation of the liquid surface, divided by 4π.

594. *Demonstration of Theorem* 3 (compare §§ 579, 580).— Let the whole of space be continuously occupied by homogeneous incompressible liquid, without any vacuous space at S;

---

* This is merely a hydro-dynamical proof of Green's celebrated theorem that a distribution of matter, acting according to the inverse square of the distance, over a surface S may be found determinately, which shall produce any arbitrarily given potential over the whole of S.

and, as immediate recipient for the action of force, imagine S to consist of a perfectly flexible and extensible membrane, separating the internal from the external fluid. Apply perpendicularly to this membrane an impulsive pressure which shall produce a normal component ·velocity equal to the external normal component force determinable from the given potential according to Theorem 1, when it is potential that is given, or equal to the given normal component force when it is force that is given. The motion is irrotational throughout each portion of the fluid; and the normal component velocities on the two sides of S, are everywhere equal to one another; but the tangential motions of the fluids, and therefore the velocity potentials, are unequal on the two sides. In the former case the velocity potential in the external fluid infinitely near S or in the latter case, the normal component velocity of the fluid on each side of S has specified values. In either case the determinate distribution of external force fulfilling the specified condition at S, whether as to potential or as to normal component, is produced (§§ 579, 580) by a determinate distribution of electric currents on S, fulfilling the following specification. The direction of the electric current is to be everywhere perpendicular to the direction of the slip in the fluid analogue; and the surface intensity of the current is to be equal to the velocity of the slip divided by 4π.

595. *Demonstration of Theorem* 4.—Let the same hydrokinematic arrangements as those in the demonstration of Theorem 3 be made, and in addition let each aperture of S be temporarily stopped by a perfectly flexible and extensible membrane, introduced merely as a recipient for the action of force. Let S be impulsively pressed so as to produce a normal component velocity equal to the given normal component force, and let uniform impulsive pressures equal respectively to κ_1, κ_2, κ_3, etc., be simultaneously applied to the barriers. The constancy ·of the difference, κ, of the potentials between contiguous portions of fluid on the two sides of each barrier, secures equality in the tangential component velocities, and therefore no "slip" between them. Suppose then the barriers annihilated. The determinate motion thus produced is irrotational throughout each portion of the fluid, and it fulfils

in the space external to S precisely the conditions which, when magnetic force is substituted for fluid velocity, are those specified in the enunciation of Theorem 4. Hence a determinate distribution of currents over S, answering to the same specification as that of Theorem 3, produces force in the space external to S which fulfils our present conditions, and thus Theorem 4 is demonstrated.

596. *Demonstration of Theorem 5.*—Let the apertures of S be stopped by material sheets of finite thickness. Imagine the matter of these sheets to be liquid, homogeneous with that occupying the rest of space, and continuous with the liquid supposed to occupy the interior of S. The boundary of the whole of this liquid is a simply continuous closed surface, consisting of the part of S not covered by the addition of the supposed barriers, and the two surfaces of each of these barriers. Let S' denote that part of the surface of S; and let B_1, B_1', B_2, B_2', etc., denote the surfaces of the barriers. As in the demonstrations of Theorems 1 and 2, let the external fluid be separated from the internal by an infinitely thin vacuous space over the whole bounding surface, and let pressure act so as to produce a given normal component in the external fluid next to S'; zero potential in the external fluid next to B_1, B_2, etc.; potentials 'equal to κ_1, κ_2, etc., in the external fluid next to B_1', B_2', etc.; and everywhere equal potentials in portions infinitely near one another, of the external and internal fluids. As in the demonstrations of Theorems 1 and 2, it is seen that there is a determinate distribution of matter over the whole bounding surface which shall produce the given normal component force over S', potential zero for B_1, B_2, etc., and potentials κ_1, κ_2, etc., for B_1', B_2', etc. If now the barriers be made infinitely thin, so that B_1 and B_1' shall be infinitely near one another, and B_2, B_2' infinitely near one another, and so on; the prescribed conditions are fulfilled by the distribution of matter determined for the limiting case thus reached. The distribution of imaginary magnetic matter on B_1, B_1' B_2, B_2', etc., may be explicitly determined by the following simple considerations. Consider an infinitely small column of the fluid between B_1 and B_1', bounded by any cylindrical or prismatic surface cutting the surfaces $B_1 B_1'$, at right angles, and enclos-

ing equal infinitely small areas on these surfaces. The density of the fluid being unity, the mass of this column will be At, if t denote the thickness of the space between B_1 and B_1', and A the area of either end of the column. This mass is acted on by an impulse $\kappa_1 A$, because by hypothesis one end of it experiences, during the initiating impulse, an impulsive pressure equal to κ_1 per unit area, and the other, zero pressure. Hence the velocity acquired by the infinitesimal column is $\frac{\kappa_1}{t}$. Let n denote the normal component velocity of the external fluid, which is equal for points infinitely near one another on the two sides of the barrier supposed infinitely thin. The velocity of separation of the fluid surfaces on each side of B_2, and the velocity of approach of the fluid surfaces on each side of B_1' will be each equal to $n + \frac{\kappa_1}{t}$. Hence the matter to be distributed over the two surfaces B_1, B_1' will be respectively, $\pm \frac{1}{4\pi}\left(n + \frac{\kappa_1}{t}\right)$. As $\frac{\kappa_1}{t}$ is infinitely great, the finite term n may be neglected, and therefore the densities on the two surfaces are $\pm \frac{\kappa_1}{4\pi t}$. These are (§ 472) precisely the densities of the positive and negative magnetic matter representing the free polarities on the two sides of a magnetic shell (§ 506) of strength $\frac{\kappa_1}{4\pi}$. The thickness t may, of course, be different in different parts of the shell, as is allowed in the general definition [§ 506 (1)] of a magnetic shell. The prescribed difference of potentials, κ_1, reckoned from B_1' through the external fluid to B_1, is verified by § 512, cor. 3.

597. Purely analytical proofs of theorems, including Theorem 1 and Theorem 2 above, are to be found in Thomson and Tait's *Natural Philosophy*, Appendix A. (*e*), and § 317, Example (3), and are included in §§ 206, 207 above [compare §§ 709—716 below]. These references supply also all that is necessary to eliminate all hydro-dynamical considerations from the preceding proofs of Theorems 3, 4, and 5. I therefore confine myself on the present occasion to the hydro-dynamical proofs now given; but remark that the analytical proofs are valuable in

respect to physical science as showing that in each case the integral

$$\iiint (X^2 + Y^2 + Z^2)\, dxdydz,$$

extended through external space is an absolute minimum [compare § 758 below] subject to the conditions prescribed in the enunciations of the several cases, and that the value of the same integral for the internal space is also a minimum subject to the conditions specified in the several demonstrations given above. From this, with §§ 567, 571 above, it follows that the dynamical value of the determinate distribution of imaginary magnetic matter on the surface S, which produces at that surface the prescribed potential of Theorem 1, or the given normal component force of Theorem 2, is less than that of any distribution of imaginary magnetic matter not confined to that surface, but still producing over it the same potential or the same normal component force; and that the electro-magnetic dynamical value of the determinate distribution of currents on S which produces at that surface the prescribed potential or the prescribed normal component force of Theorem 3, is less than that of any distribution of currents not confined to S, but still producing the same potential or the same normal component force over that surface.

598. To pass from a determinate distribution of imaginary magnetic matter, or a determinate distribution of electric currents, to a distribution of magnetization which shall produce the same resultant force, is as we have seen (§ 587) an indeterminate problem, even if the force is given throughout space. Still more is the problem indeterminate if the force be given in only one part of space, and it is required to find a distribution of magnetization in the remainder of space which shall produce that force. To find the complete solution of this problem with the proper arbitrary functions, we may proceed either from the determinate distribution of imaginary magnetic matter of § 591, Theorems 1 and 2, or from the determinate distribution of electric currents of § 591, Theorems 3 and 4, on the bounding surface. Our first step towards the complete solution shall be to find, from a determinate distribution of imaginary magnetic matter, or from a determinate distribu-

tion of electric currents, on a surface S, distributions of magnetization, confined to this surface, which shall produce the given external force.

599. Divide the whole superficial distribution of imaginary magnetic matter into an infinite number of equal parts, irrespectively of sign. As in § 523, join positive and negative parts in pairs chosen arbitrarily, by arbitrary curves all in the surface S, and lay solenoids of equal strengths along these curves. Thus on the surface S a distribution of tangential magnetization to a certain degree arbitrary is obtained, which shall produce through external space a determinate distribution of magnetic force fulfilling the prescribed surface condition. A complete representation of what is arbitrary in this solution consists of any distribution whatever of closed solenoids, each wholly coincident with S. Any such distribution of magnetization may (§ 510, Cor. 2) be superimposed on one fulfilling the prescribed condition without violating this fulfilment.

600. To proceed from surface distribution of currents to surface distribution of magnetism; (which if S is simply continuous can be done always, but if S is complexly continuous can only be done when every stream line bounds an area on S;) divide S by electric stream lines into an infinite number of bands of such breadths as to give equal strengths of current in them. This division must begin and end in points which for the present I call poles. There must therefore be at least two poles, and there may be any number, odd or even, greater than two. These poles I call north or positive when the electric currents in the bands encircling them are in the direction in which the hands of a watch, placed upon them facing outwards, would move. All the poles may be north poles or all south poles, or some may be north and some south. Commencing with any one of the poles, substitute a magnetic shell passing through it and lying altogether on S, for each band encircling it. If the whole surface can be thus exhausted the thing is done. If not, take next a pole on the unexhausted portion of surface and follow again the same rule; and so on until for each infinitely thin band of current, a magnetic shell has been substituted. Thus

we have (§ 508) a complex magnetic shell instead of the distribution of currents. Unlike the result of § 599, this result is
determinate, involving, however, one arbitrary constant. The
solutions thus obtained, differing according to the order in which
the two or more poles have been taken, are, each of them, fully
determinate. The difference between any two of them is
clearly a uniform magnetic shell of determinate strength coincident with the whole of *S*. The general solution comprehend-
ing them all, or any combination of them, is had by taking any
one of them and superimposing upon it a uniform magnetic
shell of arbitrary strength, coincident with the whole of *S*.
This arbitrary part of the general solution being a "closed
shell" (§ 512, Cor. 5) exercises no resultant force through
either external or internal space.

601. Consider lastly, the general problem of finding magnetization on and within any closed simply continuous surface *S*,
which shall produce the determinate external distribution of
force (§ 591, Theorems 1 and 2) due to any arbitrarily given potential or arbitrarily given normal component force, for every external point infinitely near *S*, with, of course, the condition that the
surface integral over the whole of *S* of the given potential or of
the given normal component force is zero. In Theorems 1, 2,
and 3 of § 591, proved in §§ 592, 593, and 594, we have seen
that a determinate distribution of imaginary magnetic matter,
or a determinate distribution of electric currents, over *S*, may
be found which shall produce the specified external distribution
of force. And in §§ 599 and 600 we have seen how in any
case when a surface distribution, either of magnetic matter or
of electric currents has been found, we can find synthetically a
surface distribution of magnetization which shall produce the
same external force ; this magnetization being purely tangential,
involving an arbitrary function when derived from imaginary
magnetic matter, and being purely normal, involving an arbitrary constant when derived from distribution of currents.

The complete solution of the present problem is obtained by
first assuming arbitrarily any distribution of magnetization
whatever within *S*, which may be altogether bodily magnetization spread through the interior, or altogether surface magnetization, whether tangential or normal or oblique, infinitely

close to the inside of *S*, or in part bodily magnetization, and
in part surface magnetization; then finding the external
potential or normal component force at points infinitely
near *S*, due to this magnetization, according as it is potential or normal component force that is given; then subtracting from the given potential or normal component force the
potential or normal component force due to the arbitrarily
assumed magnetization; and lastly, finding (at pleasure either)
a tangential or a normal distribution of magnetization on *S*
which shall produce potential or normal component force
equal to the difference. The surface-magnetization thus found,
compounded with the arbitrarily assumed magnetization, is
the most general distribution of magnetization within *S* which
can produce, at external points infinitely close to *S*, the given
potential or the given normal component force.

XXIX.—*On the Electric Currents by which the Phenomena of Terrestrial Magnetism may be produced.*

[From the *Report of the British Association for the Meeting of* 1867 in Oxford.]

602.   It is a well-known theorem, first demonstrated by
Green, that the action of a mass of any nature in attracting
an external point, may be represented by means of a distribution of matter of the same kind over the surface of the body;
that is to say, that a certain distribution of matter over the
surface of a body may be determined, which will produce
exactly the same force, whether of gravitation, of magnetism,
or of electricity as results from the body itself. Thus, by
applying this theorem to the case in which the force considered
is that of terrestrial magnetism, we see that a certain distribution of imaginary magnetic matter may be found which would
produce all the phenomena of terrestrial magnetism observed
at the surface of the earth or above it, except those which are
due to atmospheric or external sources of magnetism, if any
such exist. This proposition, although of great theoretical
interest, cannot be entertained as expressing a physical fact;
for there are only two ways in which we can conceive internal
sources of terrestrial magnetism to exist. We may either
imagine, as Gilbert did, the earth to be wholly or in part a

magnet, such as a magnet of steel, or we may conceive it to be an electro-magnet with or without a core susceptible of induced magnetism. In the present state of our knowledge this second hypothesis seems to be the more probable [? Feb. 4, 1872]; and indeed we have now many reasons for believing that the existence of terrestrial currents, producing wholly or in part the magnetic phenomena, is a physical fact. [The "earth currents" which render the localization of a fault in a submarine cable so difficult, certainly contribute to the resultant magnetic force observed at the earth's surface.] Connected with this it becomes an interesting question, whether mere electric currents could produce the actual phenomena observed. Ampère's electro-magnetic theory leads us to an affirmative answer, but an answer which must be regarded as merely theoretical; for it is absolutely impossible [compare § 546, foot-note] to conceive of the currents which he describes round the molecules of matter, as having a physical existence. The idea of an electro-magnet is what naturally presents itself when we endeavour to imagine a possible electrical theory of terrestrial magnetism; and the question which now occurs is this:—Can the magnetic phenomena at the earth's surface, and above it, be produced by an internal distribution of closed galvanic currents occupying a certain limited space below the surface? The answer is, that whatever be the form and magnetic contents of the earth, the same force as that which it exerts upon any exterior point may actually be produced by means of a distribution of closed electric currents on the surface. I have arrived at this result with the aid of Ampère's theory of the closed circuit, by means of the theorem of Green already mentioned, and by an analogous theorem of which a physical demonstration may be given by considerations connected with fluid motion. The steps in the analytical process of determining the required distribution of closed currents are as follows:—

603. Let V be the magnetic potential, according to Green's definition, at any exterior point P; $d\sigma$ an element of the surface; Δ the distance from $d\sigma$ to P; l, m, n the direction-cosines of the normal at $d\sigma$.

I. Find ρ, so that $\iint \dfrac{\rho d\sigma}{\Delta} = V$.

II. Find U^* so that $\dfrac{d^2U}{dx^2} + \dfrac{d^2U}{dy^2} + \dfrac{d^2U}{dz^2} = 0$ for internal points,

and $l\dfrac{dU}{dx} + m\dfrac{dU}{dy} + n\dfrac{dU}{dz} = \rho$ at the surface, or $\dfrac{dU}{d\nu} = \rho$.

III. Construct on the surface a " map of the values of U."

If wires be laid along the lines round the surface corresponding to sufficiently close equidifferent values of U, as indicated
by this map, and if currents of equal intensity be made to
circulate through them (each being a closed curve), the electro-
magnetic force that will result, upon external points, will be
the same as the force of terrestrial magnetism.

The explicit solution of this problem is very easy, when the
body considered is a sphere; as is actually the case, to a sufficient degree of approximation, with reference to the Earth.

Thus, if the potential at the surface be given by the equation

$$V = Y_1 + Y_2 + Y_3 + \text{etc.},$$

where Y_1, Y_2, etc., may be calculated for any latitude, by means
of the Gaussian constants [and a denote the radius of the
spherical surface], we readily find [Thomson and Tait's *Natural
Philosophy*, App. B. (52)]

$$\rho = \frac{1}{4\pi a}\{3Y_1 + 5Y_2 + 7Y_3 + \text{etc.}\},$$

$$U = -\frac{1}{4\pi}\{V + (Y_1 + \tfrac{1}{2}Y_2 + \tfrac{1}{3}Y_3 \text{ etc.})\}.$$

Hence we have the means of constructing an electro-magnetic
model of the earth, which would exhibit all the peculiarities
that can be expressed in a map constructed upon Gauss's
theory.

---

* [*Note*, Jan. 17, 1872.—This function is such that its surface value is
equal to the superficial function P of § 579, multiplied by 4π.]

CHAPTER X.   MAGNETIC INDUCTION.

On the Theory of Magnetic Induction in Crystalline and Non-Crystalline Substances.

XXX. [From the *Philosophical Magazine*, March 1851.]

604.   Poisson, in his mathematical theory of magnetic induction, founded on the hypothesis of "magnetic fluids" moveable within the infinitely small "magnetic elements" of which he assumes magnetizable matter to be constituted, does not overlook the possibility of these magnetic elements being non-spherical and symmetrically arranged in crystalline matter; and he remarks, that a finite spherical portion of such a substance would, when in the neighbourhood of a magnet, act differently according to the different positions into which it might be turned with its centre held fixed*. But "such a circumstance not having yet been observed†," he excludes the consideration of the structure which would lead to it from his researches, and confines himself in his theory of magnetic induction to the case of matter, consisting either of spherical magnetic elements, or of non-symmetrically disposed elements of any forms.   It is easy to conceive the modification which he would have introduced into his formulæ to make them applicable to a crystalline structure such as he describes; but, so far as I am aware, no writer has hitherto attempted to make this extension of Poisson's mathematical theory of magnetic induction.   Now, however, when a recent discovery of Plücker's has established the very circumstance, the observation of which was wanting to induce Poisson to enter upon a full treatment of the subject, the importance of working out a mathematical

---

* [" The substance of a homogeneous solid is called *isotropic* when a spheri-"cal portion of it tested by any physical agency exhibits no difference in "quality however it is turned.   Or, which amounts to the same, a cubical "portion cut from any position in an isotropic body exhibits the same qualities "relatively to each pair of parallel faces.   Or two equal and similar portions "cut from *any* positions in the body not subject to the condition of parallelism "(§ 675) are undistinguishable from one another.   A substance which is not "isotropic but exhibits differences of quality in different directions is called "æolotropic."—Thompson and Tait's *Natural Philosophy*, § 676.]

† " Mémoire sur le Magnétisme en Mouvement."   (*Mém. de l'Institut*, 1823, vol. vi. Paris, 1827.)   For quotations from this and the two preceding memoirs of Poisson, showing his theoretical anticipation of the discovery of magnecrystallic action, see the Appendix to this article.

theory of magnetic induction is obvious. On the other hand, in the present state of science, no theory founded on Poisson's hypothesis of "two magnetic fluids" moveable in the "magnetic elements" could be satisfactory, as it is generally admitted that the truth of any such hypothesis is extremely improbable. Hence it is at present desirable that a complete theory of magnetic induction in crystalline or non-crystalline matter should be established independently of any hypothesis of magnetic fluids, and, if possible, upon a purely experimental foundation. With this object, I have endeavoured to detach the hypothesis of magnetic fluids from Poisson's theory, and to substitute elementary principles deducible from it as the foundation of a mathematical theory identical with Poisson's in all substantial conclusions. In the present communication I shall state these principles, and point out what modifications of them may be required by a more complete experimental investigation of the subject than has yet been made; and, adopting them temporarily as axioms of magnetic induction, I shall give an account of some important practical conclusions deduced from them, by mathematical reasoning which I propose to publish on a future occasion.

Some explanations and definitions are prefixed to show the signification in which certain extremely convenient terms and expressions, occasionally employed by Faraday and other writers, will be used in what follows.

605. *Definition.*—The *force at any point due to a magnet* is the force which it would exert on the north pole of an infinitely thin, uniformly and longitudinally magnetized bar of unit strength placed at that point*, if it experienced no inductive action from the latter magnet.

Definition.—The *total magnetic force at any point* is the force

---

* "If two infinitely thin bars be equally, and each uniformly and longitudinally, magnetized, and if, when an end of one is placed at a unit of distance from an end of the other, the mutual force between these ends is unity, the magnetic strength of each is unity." (*Philosophical Magazine*, Oct. 1850, pp. 241, 242.) The definition of *magnetic force* in the text will agree precisely with the definition of "magnetic force in absolute measure" adopted by the Royal Society, in its "Instructions for making observations on terrestrial magnetism," if, in the definition of a unit bar, the unit of length understood be one foot, and the unit of force, a force which, if acting on a grain of matter, would in one second of time generate one foot per second of velocity. (See Admiralty Manual of Scientific Inquiry, pp. 16, 33, 37.)

which the north pole of a unit bar-magnet would experience from all magnets which exert any sensible action on it, if it produced no inductive action on any magnet or other body. Or,

The *total magnetic force at any point* is the quotient obtained by dividing the force experienced by either pole, placed at that point, of an infinitely thin bar, uniformly and longitudinally magnetized to a finite degree of intensity, by the infinitely small numerical measure of the magnetic strength of the bar; and its *direction* is that of the force experienced by the north pole of the bar.

Definition.—Any space at every point of which there is a finite magnetic force is called "a field of magnetic force;" or, *magnetic* being understood, simply "a field of force;" or, sometimes, "a magnetic field."

Definition.—A "line of force" is a line drawn through a magnetic field in the direction of the force at each point through which it passes; or a line touched at each point of itself by the direction of the magnetic force.

Definition.—A "uniform field of magnetic force" is a space throughout which the lines of force are parallel straight lines, and the intensity of the force is uniform.

Definition.—A substance magnetized so that the intensity and direction of magnetization at each point (§ 462) are represented by the diagonal of a parallelogram, of which the sides represent the intensities and directions at the same point in two other distributions, is said to possess a distribution of magnetism which is the resultant of these two superimposed, one on the other.

It is demonstrated by Poisson, that the force at any point due to a resultant distribution of magnetism is the resultant of

---

It may be remarked, that this unit of force will be the fraction $\frac{1}{g}$ of the weight, in any locality, of one grain of matter, if g denote the velocity acquired in one second by a falling body in that locality; and that it is therefore very nearly $\frac{1}{32 \cdot 2}$ of the weight, in any part of Great Britain or Ireland, of a grain. [*Addition, May* 30, 1872.—The units of mass and length now adopted are the gramme and the centimetre. As 32·2 feet is equal to 981·6 centimetres, we may take 982 as the number of absolute kinetic units of force, in the apparent force of gravity on one gramme of matter in these latitudes.]

the forces that would be produced at the same point if the
component distributions existed separately.

606. *Axioms of Magnetic Force.*

I. All mechanical action which a magnet experiences in
virtue of its magnetism is due to other magnets *.

II. The action between any two magnets is mutual.

III. The whole action experienced by any magnet is the
mechanical resultant of the actions which it would experience
from all the magnets in its neighbourhood, if each acted on it
as if the others were removed, the distributions of magnetism
in the two remaining unaltered.

607. *Laws of Magnetic Induction according to* Poisson's *Theory.*

I. When a given body, susceptible of inductive magnetization
(whether it be ferromagnetic or diamagnetic), is placed in the
neighbourhood of a magnet, it becomes magnetized in a manner
dependent solely on the field of force which it is made to occupy.

II. *Superposition of Magnetic Inductions.*—Different magnets
placed simultaneously in the neighbourhood of an inductively
magnetizable (ferromagnetic or diamagnetic) body induce in it
a distribution of magnetism which is the resultant of the
different distributions that would be induced by the separate
influences of the different magnets, each in its own position,
with the others removed.

608. The first of these two propositions merely implies that
any magnet, whether an electro-magnet, or a magnet consisting of
magnetized substance, which produces at each point of a certain
space the same "force" as another magnet of any kind, would
produce the same inductive effect on a magnetizable substance
occupying that space.   Everything that is known of inductive
action is consistent with it; and it is, I believe, universally
admitted as an axiomatic principle.

609. The second proposition, which asserts the mutual inde-
pendence of superimposed magnetic inductions, is equivalent to
an assertion that, if the force at every point of a magnetic field
be altered in a certain ratio, the magnetization of a substance
placed in it will be altered proportionately. This is undoubtedly

---

* This principle appears, from his *discovery* that the phænomena of terrestrial magnetism are produced by the earth acting as a great magnet, to have
been first recognised by Gilbert.

not a principle of universal application. It is not applicable
to steel, nor to the substances of which natural magnets are
composed; nor, in general, to substances possessing in any
degree that property of resisting magnetization or demagnetization, called by Poisson "coercive force," in virtue of which they
can permanently retain magnetism. Neither is it, as Joule's
experiments, and the more recent experiments of Gartenhauser
and Müller demonstrate, applicable to soft iron, except as an
approximate law of the magnetization when the magnetizing
force does not exceed certain limits of intensity. But, that it
is very approximately, if not rigorously, fulfilled in the magnetization of all homogeneous substances of very feeble inductive
capacity, and destitute of "coercive force" (as all known diamagnetics and all ferromagnetics which contain no iron or nickel,
or only very small proportions in chemical combination, appear
to be), is, I think, extremely probable   The foundation of a
complete theory of magnetic induction requires an experimental
investigation of the laws according to which the "coercive
force" acts in various substances, and of the variation of inductive capacity produced in soft iron, and it may be in other substances, by actual magnetization.   The following conclusions,
being mathematical deductions from the laws stated above, are
liable to modification, according to the deviations from those
laws which actual experiments may point out :—

610. 1. The determination of the conditions of magnetic
induction in a body of any kind in any circumstances may be
made to depend on a knowledge of the state of magnetization
induced in a homogeneous sphere of the same substance, placed
in a uniform field of magnetic force.

2. A homogeneous sphere of any substance placed in a
uniform field of force becomes uniformly magnetized in parallel
lines with an intensity which is independent of the radius of
the sphere.

[To prove this, imagine a uniformly magnetized sphere of
substance having infinite "coercive power." Let a spherical
portion be removed from its interior. The resultant force
at any point in the hollow will be (§§ 479, 473) that due
to "imaginary magnetic matter" or free polarity, as it may be
properly called, on the outer and inner spherical surface bound-

ing the magnetized matter which is left. The surface density of the polarity at any point of either surface will be equal to $i \cos \theta$, if i denote the intensity of the magnetization and θ the angle between the direction of magnetization and the radius through the point considered. The distribution on one alone of the spherical surfaces, according to a very elementary result of spherical analysis stated above in a foot-note on § 479 (and proved in the appended foot-note*), is parallel to the direction of mag-

---

* To find the resultant due to one such distribution of matter on a spherical surface, imagine first a solid material globe of uniform volume-density ρ throughout. By Newton's theorems for the attraction of a uniform spherical mass, acting according to his law of the inverse square of the distance, the resultant force at any point within the substance will be towards the centre, and equal to $\dfrac{4\pi\rho}{3}$ multiplied by the distance of the attracted point from the centre of the globe. Consider now two equal globes, one of uni-

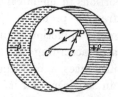

form positive matter and the other of uniform negative matter of the same density, the former repelling and the latter attracting a unit of positive matter (as in the electric and magnetic applications of the Newtonian law). Let them be placed with their centres C and C', at any distance apart less than the sum of their radii, and first imagine their materials to co-exist in the space common to the two spherical volumes, each acting as if the other were away. The resultant force at any point P within this space will be found by compounding a force equal to $\dfrac{4\pi\rho}{3} CP$ with a force $\dfrac{4\pi\rho}{3} C'P$, in the direction from P towards C', and therefore, according to the parallelogram of forces, will be in the direction PD parallel to CC', and will be equal to $\dfrac{4\pi\rho}{3} CC'$. This (as the positive and negative matters in the space common to the two spheres neutralize one another) is therefore the resultant force at P, due to uniform distribution of positive and negative matter in the two meniscuses formed by the non-coincident portions of the two spheres. Now let CC' become infinitely small, and ρ infinitely large, and denote by i the product $\rho CC'$, which we may suppose to have any value we please. The two meniscuses become a continuous superficial distribution of matter over a single spherical surface, having for surface-density $i \cos \theta$, at any point where the inclination of the normal to the diameter through CC' is θ. The resultant force is parallel to this diameter and of constant value equal to $\dfrac{4\pi i}{3}$ throughout the entire spherical space.

A similar investigation gives the resultant magnetic force at any point in the interior of a uniformly magnetized ellipsoid; but in this case it is convenient to consider components of magnetization and of force in the directions of the three principal axes. Thus if α, β, γ be the components of magnetization, and $\mathfrak{X}$, $\mathfrak{Y}$, $\mathfrak{Z}$ the components of the magnetic force according

netization, and equal to $\dfrac{4\pi i}{3}$; and therefore the two balance one another for every point within the supposed hollow space. The resultant force is therefore zero throughout this space. Replacing now the magnetized material in the hollow space, let the uniformly magnetized hollow sphere be placed in a uniform field of force, and instead of "coercive power," let its substance be endowed with such inductive susceptibility in each part of it, that by induction it shall remain uniformly magnetized. The magnetizing force actually experienced by any spherical portion of it is the same as if the surrounding substance were removed. Hence different equal spherical portions of the whole require equal inductive susceptibilities to keep them equally magnetized; and as we may suppose these spherical portions to be as small as we please, it follows that the inductive susceptibility must be equal throughout, and that if the substance be æolotropic its quality must be throughout similarly related to the force of the field. Conversely, the inductive magnetization experienced by a globe of homogeneous substance devoid of "coercive power" when placed in a uniform field of force, must be uniform and in parallel lines.]

3. If the sphere be of isotropic substance, the lines of its

to the polar definition, we find

$$\mathfrak{X}=\frac{4\pi\mathfrak{A}a}{3}, \quad \mathfrak{Y}=\frac{4\pi\mathfrak{B}\beta}{3}, \quad \mathfrak{Z}=\frac{4\pi\mathfrak{C}\gamma}{3},$$

where $\frac{1}{3}\mathfrak{A}$, $\frac{1}{3}\mathfrak{B}$, $\frac{1}{3}\mathfrak{C}$ denote the three elliptic integrals which appear in (6) of § 23, above, each with the factor $\sqrt{(1-e^2)}\sqrt{(1-e'^2)}$ retained. These expressions depend only on the proportions of the axes, and therefore the resultant force is zero in the hollow space left, when from a uniformly magnetized ellipsoid any similar ellipsoidal portion with principal axes in the same directions is removed. Hence the demonstration of the text proves that an ellipsoid of homogeneous substance, susceptible of magnetic induction, becomes uniformly magnetized when placed in a uniform field of force. An obvious extension of § 626, below, gives the following equations for determining a, β, γ, the components of the magnetization, in terms of F, G, H, the components of the force of the field, μ, μ', μ'' the principal susceptibilities, and (l, m, n), (l', m', n'), (l'', m'', n'') the three principal inductive axes, all specified with reference to the directions of the three principal axes of figure

$$\left(1+\frac{4\pi\mathfrak{A}}{3}\mu\right)la+\left(1+\frac{4\pi\mathfrak{B}}{3}\mu\right)m\beta+\left(1+\frac{4\pi\mathfrak{C}}{3}\mu\right)n\gamma=\mu\,(Fl\,+Gm\,+Hn)$$

$$\left(1+\frac{4\pi\mathfrak{A}}{3}\mu'\right)l'a+\left(1+\frac{4\pi\mathfrak{B}}{3}\mu'\right)m'\beta+\left(1+\frac{4\pi\mathfrak{C}}{3}\mu'\right)n'\gamma=\mu'\,(Fl'+Gm'+Hn')$$

$$\left(1+\frac{4\pi\mathfrak{A}}{3}\mu''\right)l''a+\left(1+\frac{4\pi\mathfrak{B}}{3}\mu''\right)m''\beta+\left(1+\frac{4\pi\mathfrak{C}}{3}\mu''\right)n''\gamma=\mu''\,(Fl''+Gm''+Hn'').$$

magnetization are in the same direction as the lines of force
in the field into which it is introduced, and the intensity of
magnetization is equal to the product of a constant (which may
be called *the inductive capacity of the substance*) into the intensity of the magnetizing force.

[For obvious reasons I now prefer a different definition of
inductive quality; and for the sake of brevity I prefer the one
word susceptibility to the two "inductive capacity." Instead
of the preceding definition, therefore, I shall henceforth adopt
the following:—

Definition 1.—*The magnetic susceptibility of an isotropic substance is the intensity of magnetization acquired by an infinitely
thin bar of it placed lengthwise in a uniform field of unit magnetic force.* And I add;—

Definition 2.—*The magnetic susceptibility, in any direction of
an æolotropic substance is the longitudinal component intensity of
magnetization experienced by an infinitely thin bar cut from the
substance in that direction, and placed lengthwise in a uniform
field of unit force.*]

4.   If the sphere be of crystalline substance, the lines of its
magnetization may not in general be in the same direction as
the lines of force of the field into which it is introduced; and
they are not so if the sphere, when free to turn round its centre,
is observed to be not in equilibrium.

611.   *Definition.*—*A principal axis of magnetic induction of a
substance* is a line in it, such that a spherical portion when introduced, with that line parallel to the lines of force, into a
uniform magnetic field, becomes magnetized in the direction of
those lines.

Definition.—*A principal inductive capacity of a substance,* or
*the inductive capacity of a substance in the direction of a principal
axis,* is the coefficient by which the intensity of the magnetizing force must be multiplied to obtain the intensity of magnetization when a spherical portion is introduced into a uniform
magnetic field, with a principal axis parallel to the lines of
force.

612.   5.   Any substance has through every point of it, three
principal axes at right angles to one another; and if the induc-

tive capacities with reference to three such axes be different, no other line through the same point is a principal axis*.

6. If the inductive capacities with reference to two principal axes through any point of a homogeneous substance be equal, every line in the plane of these two, or parallel to it, is a principal axis, and the inductive capacities with reference to all these principal axes are equal.

7. If the inductive capacities with reference to three principal axes through any point of a substance be equal, every line through the substance is a principal axis, and the inductive capacities with reference to all directions are equal; or the substance is destitute of magnecrystallic properties.

613. 8. A spherical portion of any homogeneous substance, supported in a uniform magnetic field in such a manner that it can turn freely in any manner round its centre which is immoveable, cannot be in equilibrium unless a principal axis be in the direction of the lines of force. If the three principal inductive capacities be unequal, the body will be in stable †¹ equilibrium with the principal axis of greatest inductive capacity, or in unstable equilibrium with either of the two other principal axes, in the direction of the lines of force. If the two less principal inductive capacities be equal to one another, the body will be in stable †² equilibrium with the principal axis of greatest inductive capacity in the direction of the lines of force, or in unstable equilibrium with the same axis perpendicular to the lines of force. If the two greater principal inductive capacities be equal to one another, the body will be in stable ‡ equilibrium with the plane of the corresponding principal axes parallel to the lines of force, or in unstable equilibrium with that plane perpendicular to the lines of force.

---

* Such, it may be expected, will be the magnetic circumstances in the case of any transparent substance which belongs to the optical class of "biaxal crystals;" and its three principal axes of magnetic induction will be the three rectangular axes deduced by Sir David Brewster from the "optic axes," and known in the undulatory theory as the principal axes of elasticity of the medium in which the undulations are propagated.

†¹,² In one respect the equilibrium might be said to be neutral rather than stable, since every position into which the body may be turned round the stable axis is a position of equilibrium.

‡ In two respects the equilibrium might be said to be neutral; since every position into which the body may be turned round the direction of the lines of force is a position of equilibrium, and every position into which it may be turned in the plane of the stable principal axes is a position of equilibrium.

614. 9. If a spherical portion, of volume σ, of a substance of which the three principal inductive capacities are A, B, and C, be held in a uniform magnetic field where the intensity of the force in absolute measure is R, with the three principal axes of induction inclined to the direction of the force at angles of which the cosines are respectively l, m, n, it will receive a state of magnetization which is the resultant of three states of uniform magnetization; one of intensity $A . Rl$, in the direction of the first principal axis; a second of intensity $B . Rm$, in the direction of the second principal axis; and a third, of intensity $C . Rn$, in the direction of the third principal axis; and it will experience a turning action, of which the mechanical definition is a couple, of moment

$$\sigma . R^2 . \{m^2n^2 (B - C)^2 + n^2l^2 (C - A)^2 + l^2m^2 (A - B)^2\}^{\frac{1}{2}}...(1),$$

in a plane of which the direction cosines* with reference to the three principal axes are respectively

$$\frac{mn (B - C)}{D}, \quad \frac{nl (C - A)}{D}, \quad \frac{lm (A - B)}{D} \quad........(2),$$

where D denotes the square root of the sum of the squares of the numerators of these three fractions, or the third factor of the preceding expression.

615. 10. If the sphere be infinitely small, and if it be put into a uniform or non-uniform field of force, the entire action which it experiences, whether directive tendency or tendency to move from one part of the field to another, is defined by the following proposition:—

The quantity of mechanical work which is required to bring the body from a position where the intensity of the force is R, and its direction cosines with reference to the three principal inductive axes l, m, n, to a position where the intensity of the force is R', and its direction cosines with reference to the three principal inductive axes in their new positions l', m', n', is equal to

$$\tfrac{1}{2}\sigma \{(Al'^2 + Bm'^2 + Cn'^2) R'^2 - (Al^2 + Bm^2 + Cn^2) R^2\}......(3).$$

11. If $A = B = C$, this expression becomes simply $\tfrac{1}{2}\sigma A$ $(R'^2 - R^2)$, and the proposition is equivalent to the mathe-

---

* Or the cosines of the inclinations of a perpendicular to the plane, to the three axes.

matical expression of Faraday's law regarding the tendency to places of stronger or of weaker force, of ferromagnetic or diamagnetic non-crystalline substances, on which some remarks [reprinted, §§ 647—668 below] are published in the *Philosophical Magazine* for October 1850.

616. 12. If, without moving its centre the ball be turned so that its three principal axes shall successively be in the direction of the lines of force (the field being non-uniform, but the body infinitely small), it will in each position experience a force in the line of most rapid variation of the "force of the field;" but the magnitude of the force will in general differ in the three positions, being proportional to A, B, and C respectively*. If

---

* Thus a ball cut out of a crystal of pure calcareous spar, which tends to turn with its optic axis perpendicular to the lines of force, and which tends as a whole from places of stronger towards places of weaker force, would experience this latter tendency less strongly when the optic axis is perpendicular to the lines of force than when it is parallel to them ; since, according to § 612 of the text, the crystal must have greatest inductive capacity or (the language in the text being strictly algebraic when negative quantities are concerned) least capacity for diamagnetic induction perpendicular to the optic axis. I am not aware that this particular conclusion has been verified by any experimenter ; but I am informed (Oct. 25, 1850) by Mr Faraday, that he finds a piece of crystalline bismuth to experience a different " repulsion" according as it is held with its magnecrystallic axis along or perpendicular to the lines of force in a non-uniform field ; the repulsion being less in the former case than in the latter, which agrees perfectly with the conclusions of the text, since, as a ball of bismuth would tend to place its magnecrystallic axis along the lines of force, that axis must, according to § 612, be the principal axis of greatest inductive capacity, or, bismuth being diamagnetic, the axis of least diamagnetic capacity.

It is right to add, that what, according to the theory explained in the text, must be the correct explanation of the peculiar phænomena of magnetic induction depending on magnecrystallic properties, was clearly stated in the form of a conjecture by Faraday in his 22d Series (2588) in the following terms:—" Or we might suppose that the crystal is a little more apt for mag-"netic induction, or a little less apt for diamagnetic induction, in the direc-"tion of the magnecrystallic axis than in other directions. But, if so, it "should surely show * * * in the case of diamagnetic bodies, as bismuth, a "difference in the degree of repulsion when presented with the magne-"crystallic axis parallel and perpendicular to the lines of magnetic force "(2552) ; which it does not do." (Read before the Royal Society, December 7, 1848.) The failure of the first experiment (2552) to detect this difference of action need not be wondered at, when we consider how minute it must probably be ; and the conjecture, apparently abandoned at the time by the author for want of experimental support, may be considered as fully established by his own subsequent experimental researches.

[The following appeared in the *Philosophical Magazine* for 1851, second half-year, under the title " Magnecrystallic Property of Calcareous Spar":—]

Extract from letter to the Editors.

Glasgow College, Nov. 7, 1851.— * * * * In the passage, as originally published (line 4 from beginning of foot-note), the word "more" occurred in the place of " less." The mistake was pointed out to me last April by Professor

each of these quantities be positive, the force on the ball in each position will be in the direction in which the force of the field increases; if any one of these quantities be negative, the force on the ball when the corresponding principal axis is in the direction of the lines of force, will be in the contrary direction, or that in which the force of the field decreases most rapidly.

617. 13. If A, B, and C be all positive, the body is called ferromagnetic; if they be all negative, it is called diamagnetic. No substance has as yet been found to have some of the quantities A, B, C positive, and others negative.

618. 14. If the inductive capacities be very small, all the preceding conclusions will be applicable to the actions experienced by bodies in air (ferromagnetic), or in any magnetizable fluid of either ferromagnetic or diamagnetic inductive capacity, provided, instead of the absolute inductive capacities of the substance in each case, we use for A, B, and C, or for the "principal inductive capacities" in the verbal enunciations, the excesses of the absolute principal inductive capacities of the substance, above the inductive capacity of the fluid.

619. Curious experiments might be made by means of a varying field of force occupied by a magnetizable fluid, and a ball of crystalline substance allowed to move freely in the line of most rapid variation of the force. If the inductive capacity (whether positive or negative) of the fluid be intermediate between the

---

Stokes, and I immediately requested you to correct it, which you accordingly did by an intimation in the "Errata." When the perplexity occasioned by the mistake is removed, it is obvious to any one reading the passage carefully, that the mistake itself was only a slip of the pen, as at the conclusion of the sentence it is asserted that a crystal of pure calcareous spar must have the "*least* capacity for diamagnetic induction, perpendicular to the optic axis."

This conclusion is verified by Dr Tyndall, who describes experiments, in a paper published in your September Number, by which it appears that the diamagnetic inductive capacity of calcareous spar in a direction parallel to the optic axis is to its diamagnetic inductive capacity perpendicular to the optic axis as 57 to 51.—I remain, gentlemen, your obedient servant,

WILLIAM THOMSON.

[We have also received a communication on this subject from Mr Tyndall, who in reference to a note received by him from Prof. Thomson, writes as follows :—"I have only to say that the facts are precisely what they are here "stated to be. Previous to writing the remarks in question, I looked to the "Errata, but not it seems with sufficient attention, for Professor Thomson's cor-"rection escaped me. Not only do our results agree in principle, but the same "substance and form of substance which Professor Thomson had referred to in "illustration of his theory was unwittingly examined by me in Berlin, and the "exact result which he had theoretically predicted arrived at by way of experi-"ment."—EDIT.]

greatest and the least of the absolute principal inductive capacities of the substances, the ball will be urged from places of
weaker towards places of stronger force when its axis of
greatest inductive capacity is placed along the lines of force,
and in the contrary direction when the axis of least inductive
capacity is placed in the same direction.

It would be easy to adjust the strength of a solution of sulphate of iron so as to satisfy this condition for a ferromagnetic
crystalline substance; but there might be great difficulty in
demonstrating by experiment the existence of the forces, on
account of their feebleness.

APPENDIX.

Quotations from Poisson *regarding Magnecrystallic Action.*

620. " . . . . . . la forme des élémens pourra aussi influer
"sur cette intensité; et cette influence aura cela de particulier,
"qu'elle ne sera pas la même en des sens différens. Supposons,
"par exemple, que les élémens magnétiques sont des ellipsoïdes
"dont les axes ont la même direction dans toute l'étendue
"d'un même corps, et que ce corps est une sphère aimantée par
"influence, dans laquelle la force coercitive est nulle; les
"attractions ou répulsions qu'elle exercera au-dehors seront
"différentes dans le sens des axes de ses élémens et dans tout
"autre sens; en sorte que, si l'on fait tourner cette sphère sur
"elle-même, son action sur un même point changera, en général,
"en grandeur et en direction: mais, si les élémens magnétiques
"sont des sphères de diamètres égaux ou inégaux, ou bien s'ils
"s'écartent de la forme sphérique, mais qu'ils soient disposés
"sans aucune régularité dans l'intérieur d'un corps aimanté par
"influence, leurs formes n'influeront plus sur les résultats qui
"dépendront seulement de la somme de leurs volumes, comparée
"au volume entier de ce corps*, et qui seront alors les mêmes en
"tout sens. Ce dernier cas est celui du fer forgé, et sans doute
"aussi des autres corps non cristallisés dans lesquels on a
"observé le magnétisme: mais il serait curieux de chercher si
"le premier cas n'aurait pas lieu lorsque ces substances sont

---

* [This error was corrected by Poisson himself in a subsequent memoir.]

" cristallisées ; on pourrait s'en assurer par l'expérience, soit en
" approchant un cristal d'une aiguille aimantée, librement sus-
" pendue, soit en faisant osciller de petites aiguilles taillées
" dans des cristaux en toute sorte de sens et soumises à l'action
" d'un très fort aimant."—Pp. 258, 259, *Mémoire sur la Théorie
du Magnétisme*, par M. Poisson. Lu à l'Académie des Sciences
le 2 Février, 1824. *Mém. de l'Inst.* 1821-22. Paris, 1826.

" . . . . . la forme des élémens et leurs positions par rapport
" aux plans fixes des coordonnées x, y, z, peuvent influer sur
" l'état magnétique de A, et sur les attractions ou répulsions
" qu'il exerce au dehors. Il pourrait même arriver que cette
" influence ne fût pas la même en tout sens, en sorte que, si A
" était une sphère homogène, et qu'on fît tourner ce corps sans
" déplacer son centre et sans rien changer aux forces extérieures
" ou à la fonction V, les actions magnétiques de A changeraient
" néanmoins en grandeur et en direction. Ce cas singulier,
" que nous avons déjà indiqué dans le préambule de ce Mémoire,
" ne s'étant pas encore présenté à l'observation, nous l'exclurons
" de nos recherches, quant à présent, et nous allons, en consé-
" quence, déterminer les relations qui doivent exister entre α',
" β', γ' *, et les quantités $\alpha_,$, $\beta_,$, $\gamma_,$ †, pour qu'il n'ait pas lieu."
—*Ibid.* p. 278.

621. The following explanation may serve to give an idea of
Poisson's mode of treating the subject of the last quotation, and
to show the relation it bears to the theory of which an outline
has been given above.

A sphere of any homogeneous magnetizable substance being
placed in a uniform field of force, intensity R, let the direction
of the force make angles whose cosines are l, m, n with three
rectangular axes fixed relatively to the substance ; and let α,
β, γ be the components of the induced magnetization. Poisson
deduces, from his hypothesis of magnetic fluids, equations‡

---

* Component intensities of magnetization.
† Components of the magnetizing force.
‡ The products of the first members of Poisson's three equations in p. 278
of his first *Mémoire*, into k, the ratio of the sum of the volumes of the magnetic
elements to the whole volume of the body, are respectively equal to the three
components of the intensity of magnetization (α, β, γ) ; and if A, B, etc., be
taken to denote the values of the products of k into Poisson's coefficients P, Q,
etc., respectively, the equations in the text coincide with those of Poisson.

which are equivalent to the following:—

$$\alpha = (Al\ + B'm\ + C''n)R$$
$$\beta = (A''l + Bm\ + C'n)\ R \qquad \qquad \text{......} \text{(4)},$$
$$\gamma = (A'l\ + B''m + Cn)R$$

where A, B, etc., are coefficients depending solely on the nature of the substance.   These equations are deducible from the axioms and the hypothetical principle of the superposition of magnetic inductions, stated above, without the necessity of referring at all to the hypothesis of "fluids."   All that remains of Poisson's theory is confined to the case of non-crystalline matter, with reference to which it is proved that A, B, and C must be equal to one another, and that each of the other six coefficients must vanish; and there is nothing to indicate the possibility of establishing any relations among the nine coefficients which must hold for matter in general.   I have found that the following relations, reducing the number of independent coefficients from nine to six, must be fulfilled, whatever be the nature of the substance:—

$$B'' = C',\quad C'' = A',\quad A'' = B'\text{......}\text{(5)},$$

the demonstration [added below, § 622] being founded on no uncertain or special hypothesis, but on the principle that a sphere of matter of any kind, placed in a uniform field of force, and made to turn round an axis fixed perpendicular to the lines of force, cannot be an inexhaustible source of mechanical effect. All the conclusions with reference to magnecrystallic action enunciated in the preceding abstract are founded on these relations.

[622. *Demonstration: January* 1872.—Because the field of force is uniform the dynamical action experienced by the magnetized sphere if of unit volume consists simply of a couple (§ 499) whose components are

$$(\beta n - \gamma m)R,\ \ (\gamma l - \alpha n)R,\ \ (\alpha m - \beta l)R \text{......}\text{(6)},$$

expressions which show that the axis of the resultant couple is perpendicular to (l, m, n).   Now remembering that the axes of co-ordinates are fixed relatively to the substance, suppose it to be turned, carrying OY and OZ with it round the axis OX, through an infinitesimal angle $d\phi$; and let ϕ denote the angle between the plane YOX and the plane of OX and (l, m, n).

The work done by the magnetized substance during this motion will be $(m\gamma - n\beta)\,Rd\phi$.........................(7),

which, if we put $l = \cos\theta$, $m = \sin\theta\cos\phi$, $n = \sin\theta\sin\phi$, and use (4), becomes

$R\{\sin\theta\cos\theta\,(A'\cos\phi - A''\sin\phi) + \sin^2\theta\,[(C - B)\sin\phi\cos\phi + B''\cos^2\phi$

$\qquad\qquad\qquad\qquad\qquad\qquad\qquad - C'\sin^2\phi]\}\,d\phi...(8).$

Integrating this expression from $\phi = 0$ to $\phi = 2\pi$, we find

$$R\sin^2\theta\,(B'' - C')\,\pi,$$

for the integral amount of work done during a revolution round OX. But this must be zero, for avoidance of the "perpetual motion*," since the body is brought back to its primitive position and physical condition at the end of the motion; and therefore $B'' = C'$. Similarly, by turning the body once round the axis OY, we prove that $C'' = A'$, and by turning it round OZ we prove that $A'' = B'$. Thus are established the three relations between the co-efficients expressed by equations (5) above.

623. To find a symmetrical expression for the work done in any infinitesimal rotation, remark that when l is constant we have

$$d\phi = -\frac{dm}{n} = \frac{dn}{m}.$$

Hence $\qquad (m\gamma - n\beta)\,d\phi = \gamma dn + \beta dm.$

Hence by (7) and corresponding expressions for the work done in infinitesimal rotations, round OY and OZ, we find for the whole work, dQ, done by any infinitesimal rotation whatever

$$dQ = R\,(\alpha dl + \beta dn + \gamma dm)\;...............(9).$$

Using in this for α, β, γ, their expressions by (4), as linear functions of l, m, n, and looking to the relations (5) established between the coefficients, we see that dQ is a complete differential of a quadratic function of l, m, n, as if these were three independent variables; and therefore by integration

$$Q = \tfrac{1}{2}\,(Al^2 + Bm^2 + Cn^2 + 2amn + 2bnl + 2clm)\,R^2...(10),$$

where a, b, c denote respectively the value of either members of the three equations (5). Hence by differentiation and comparison with (4),

$$\alpha = \frac{1}{R}\frac{dQ}{dl},\quad \beta = \frac{1}{R}\frac{dQ}{dm},\quad \gamma = \frac{1}{R}\frac{dQ}{dn}\;.........(11),$$

and $\qquad\qquad Q = \tfrac{1}{2}\,(\alpha l + \beta m + \gamma n)\,R\;...............(12).$

---

* See below, § 670, footnote.

This is necessarily equal to the exhaustion of energy (Thomson and Tait's *Natural Philosophy*, § 549) in letting the globule come from any place of zero magnetic force, into its actual position in the supposed magnetic field. Compare § 732, and § 722 (70) bis, and § 503 (2).

624. The elementary theory of the transformation of quadratic functions shows how, when A, B, C, a, b, c are known for any one set of three rectangular axes in the substance, we can find determinately by aid of the solution of a cubic equation, a set of three rectangular axes such that if we take them for axes of X, Y, Z, the coefficients of mn, nl, lm will vanish in the transformed quadratic function, and we should have simply

$$Q = \tfrac{1}{2} \left(A l^2 + B m^2 + C n^2 \right) R^2 \quad \dots\dots\dots (13),$$
and
$$\alpha = A l R, \quad \beta = B m R, \quad \gamma = C n R \dots\dots\dots (14).$$

Hence the propositions of §§ 612, 613, 614.]

XXXI. *Magnetic Permeability, and Analogues in Electro-static Induction, Conduction of Heat, and Fluid Motion.*

March 1872.

625. Supposing the coefficients A, B, C, and a, b, c of §§ 621—624, (5) and (10), to be known for a particular set of axes in a substance susceptible of magnetic induction, let it be required to find its susceptibility for magnetization in any given direction. Let a sphere of the substance be placed in a uniform field of force having components F, G, H parallel to the axes of co-ordinates. By § 623 (11) we have for the components of magnetization

$$\left. \begin{aligned} \alpha &= AF + cG + bH \\ \beta &= cF + BG + aH \\ \gamma &= bF + aG + CH \end{aligned} \right\} \quad \dots\dots\dots\dots\dots(1);$$

and denoting by i the intensity of the resultant magnetization, and l, m, n its direction-cosines,

$$i = \sqrt{(\alpha^2 + \beta^2 + \gamma^2)} \quad \dots\dots\dots\dots\dots(2),$$

$$l = \frac{\alpha}{i}, \quad m = \frac{\beta}{i}, \quad n = \frac{\gamma}{i} \quad \dots\dots\dots\dots\dots(3).$$

Conceive now an infinitely thin bar of the substance, of any length along the lines of magnetization, to be removed. The magnetic force in the hollow space will be compounded of the force of the field (F, G, H) and the force due to the free surface-polarity of the sphere; and therefore (§ 610, 2, foot-note) if we denote by X, Y, Z its components, we have

$$X = F - \frac{4\pi\alpha}{3}, \quad Y = G - \frac{4\pi\beta}{3}, \quad Z = H - \frac{4\pi\gamma}{3} \ \ldots(4).$$

It is this which is the magnetizing force actually experienced by the bar in its position as part of the sphere. The magnetization induced by it is of intensity $\sqrt{(\alpha^2 + \beta^2 + \gamma^2)}$, and is in the direction of the bar's length. Hence the magnetic susceptibility of the substance in the direction (3) of this bar is

$$\frac{\sqrt{(\alpha^2 + \beta^2 + \gamma^2)}}{\sqrt{(X^2 + Y^2 + Z^2)}} \ \ldots\ldots\ldots\ldots\ldots\ldots(5).$$

To find the magnetic susceptibility in any direction (l, m, n) explicitly in terms of l, m, n and the co-efficients A, B, C, a, b, c, all that is necessary is to eliminate $\alpha, \beta, \gamma, X, Y, Z, F, G, H$ from (5) by means of the nine equations (1), (3), (4). The algebraic process required involves only the solution of the three linear equations (1) for F, G, H. The simplified solution given in the following section may be regarded as algebraically equivalent to an expression of the preceding direct solution in terms of symmetrical functions of the roots of a cubic equation.

626.   To simplify let the axes of co-ordinates be chosen in the direction of the three principal axes (§ 611) of magnetic susceptibility. This makes $a = 0$, $b = 0$, $c = 0$, and we have

$$\alpha = AF, \quad \beta = BG, \quad \gamma = CH\ldots\ldots\ldots\ldots(6),$$

$$X = \left(1 - \frac{4\pi A}{3}\right) F, \ Y = \left(1 - \frac{4\pi B}{3}\right) G, \ Z = \left(1 - \frac{4\pi C}{3}\right) H \ (7).$$

Hence by (5), (3), and (2) we have, for the magnetic susceptibility in the direction l, m, n,

$$\frac{1}{\sqrt{\left(\dfrac{l^2}{\lambda^2} + \dfrac{m^2}{\mu^2} + \dfrac{n^2}{\nu^2}\right)}} \ \ldots\ldots\ldots\ldots\ldots\ldots(8),$$

where $\lambda = \dfrac{A}{1 - \dfrac{4\pi A}{3}}, \quad \mu = \dfrac{B}{1 - \dfrac{4\pi B}{3}}, \quad \nu = \dfrac{C}{1 - \dfrac{4\pi C}{3}} \ \ldots(9).$

627.   The coefficients denoted in (9) by λ, μ, ν are the three principal magnetic susceptibilities, as we see by considering the cases in which (l, m, n) coincides with the axes of co-

ordinates. By equations (9), conversely, for the inductive magnetization of a sphere when its principal susceptibilities λ, μ, ν
are given, we find

$$A = \frac{\lambda}{1 + \frac{4\pi}{3}\lambda}, \qquad B = \frac{\mu}{1 + \frac{4\pi}{3}\mu}, \qquad C = \frac{\nu}{1 + \frac{4\pi}{3}\nu} \dots (10).$$

628. In the exposition of Faraday's great electro-static
discovery, given above (§§ 36—50), I pointed out a perfectly
close analogy between the mathematical theories of the electro-
polar induction which he found to be experienced by insulators
in a field of electric force, of the inductive magnetization of
ferromagnetics, air, and diamagnetics, and of the conduction
of heat through a heterogeneous solid. This volume will end
with a fourth analogy (§§ 751—763, below), in which it will
be shown that precisely the same laws and mathematical expressions are applicable to the flow of a frictionless incompressible liquid, through a porous solid of infinitely fine texture,
when the motion of the liquid is throughout irrotational (or
such as may be produced from rest by any motion given to the
boundary of the liquid). The singular combination of mathematical acuteness, with experimental research and profound
physical speculation, which Faraday, though not a "mathematician," presented, is remarkably illustrated by his use of
the expression, *conducting power of a magnetic medium for lines
of force*, referred to in the foot-note to § 44, above. The analogue corresponding to conducting power of a solid for heat, or,
as it is shortly called, "thermal conductivity," is, in electrostatic induction, the "specific inductive capacity" of the
di-electric; in magnetism it is not what has hitherto been
called magnetic inductive capacity,—a quality which is negative
in diamagnetics, but it is Faraday's "conducting power for
lines of force;" and in hydrokinetics it is (§ 753, below) flux
per unit area, per unit intensity of energy. The common word
"permeability" seems well adapted to express the specific
quality in each of the four analogous subjects. Adopting it
we have thermal permeability, a synonym for thermal conductivity; permeability for lines of electric force, a synonym
for the electro-static inductive capacity of an insulator; magnetic permeability, a synonym for conducting power for lines

of magnetic force; and hydrokinetic permeability, a name for the specific quality of a porous solid, according to which, when placed in a moving frictionless liquid, it modifies the flow.

629. To find the relation between what has been called above magnetic susceptibility and magnetic permeability, consider a body with no intrinsic magnetization (§ 698, below) surrounded by air in a magnetic field. Let A be any infinitesimal area of its surface cutting perpendicularly one of the three principal inductive axes of the substance in its neighbourhood. Let ϑ be the normal component of the magnetization induced in the substance infinitely near A; and let N, N' be the values of the normal component force at external and internal points infinitely near A, the latter according to the polar definition (§ 517, *Postscript*). We have [§ 473 (1), and § 7]

$$N' = N - 4\pi\vartheta \dots\dots\dots\dots\dots (11).$$

Let now μ be the magnetic susceptibility in the direction of the normal, so that (§ 610, 3, definition 2) we have

$$\vartheta = \mu N' \dots\dots\dots\dots\dots\dots (12).$$

Eliminating ϑ from this by (11), we have $N' = N - 4\pi\mu N'$, and therefore

$$\frac{N}{N'} = 1 + 4\pi\mu \dots\dots\dots\dots\dots (13).$$

Hence (compare § 44, above) $1 + 4\pi\mu$ is the magnetic permeability of the substance in the direction of its principal axis perpendicular to A. Thus we see that if μ, μ', μ'' denote the three principal magnetic susceptibilities of a substance, and ϖ, ϖ', ϖ'' its principal magnetic permeabilities, we have

$$\varpi = 1 + 4\pi\mu, \quad \varpi' = 1 + 4\pi\mu', \quad \varpi'' = 1 + 4\pi\mu'' \dots\dots(14).$$

630. Experiment has hitherto given but little accurate knowledge of the magnetic susceptibilities of different substances. Comparisons of the susceptibilities of diamagnetics and feeble ferromagnetics with one another and with that of iron have been attempted; but the only determination in absolute measure hitherto made or even attempted is that of Thalén[*] for iron. He found the magnetic susceptibilities of different specimens to be very different. The greatest susceptibility which he found

---

[*] "Recherches sur les propriétés magnétiques du fer." Par T. R. Thalén. Extrait des actes de la Société Royale des Sciences d'Upsal. Série iiie. T. iv. Upsal, 1861.

was in some specimens of the best soft iron, and amounted to about 45. " Coercive force," the laws of which are at present wholly unknown, exists to a great degree in all varieties of iron and steel, including the softest iron ; and varies very much in the same specimen with its state of temper. It complicates excessively every investigation regarding the inductive qualities of iron and steel. On the other hand (and particularly now that the British Association has given to experimenters standards of electric resistance in absolute electro-magnetic measure*, and important contributions towards the general practice of the absolute system) it is a very easy thing to measure, with some degree of accuracy, the absolute value of the inductive quality of substances destitute of coercive force. (All fluids are necessarily so ; and, as stated in § 609, it is probable that all diamagnetics, and all homogeneous substances of feeble ferromagnetic quality, are nearly so.) As yet no such measurement has been made, but it is to be hoped that before long some experimenter will take up the subject.

631. Thalén's number, 45, gives, according to $(14), 1+4\pi\times45$, or about 566 for the permeability of the best soft iron. It has been stated that the inductive susceptibility of cobalt is greater than that of soft iron, but this seems to be by no means certain; and I believe it is certain that all other substances hitherto experimented on are less susceptible than iron. The permeabilities of all ferromagnetics exceed unity, but only by very small fractions, except the few so-called magnetic metals, or substances containing them in large proportion. It is also remarkable that no substance has been discovered for which the permeability falls short of unity by more than a very minute fraction, as is shown by the extreme feebleness of the forces due to diamagnetic induction in all cases which have been hitherto observed. If we knew something instead of nothing of the molecular theory of magnetic induction, we should probably see that the permeability of every substance must be positive.

---

* British Association Committee on Electric Measurement, appointed first in the year 1860, and reappointed after that from year to year. A reprint of its successive Reports collected is being made by the Committee, with permission of the Council of the British Association, and will soon be ready for publication in a separate form. [Published in 1873 by E, and F. N. Spon, London, under the title of " Reports of Electrical Standards," edited by Prof. F, Jenkin, F.R.S., LL.D.]

XXXII.   *Diagrams of Lines of Force; to illustrate Magnetic Permeability.*   [*May* 29, 1872.]

632.   The differential equation for lines of force in void space resulting from the Newtonian law is always integrable when the distribution is symmetrical round an axis, as was first shown in an article " On the Equations of Motion of Heat referred to Curvilinear Co-ordinates " in the *Cambridge Mathematical Journal*, Nov. 1843 [Art. IX. of my " Reprint of Mathematical and Physical Papers," Vol. I. University Press, Cambridge, 1882.] thus :—In the case of symmetry round an axis, take for co-ordinates x along the axis of symmetry, and y perpendicular to it in any plane through it.   Laplace and Poisson's equation becomes

$$\frac{d^2 V}{dx^2} + \frac{d^2 V}{dy^2} + \frac{1}{y}\frac{dV}{dy} = -4\pi\rho.$$

Therefore through void space,

$$\frac{d^2 V}{dx^2} + \frac{d^2 V}{dy^2} + \frac{1}{y}\frac{dV}{dy} = 0 \quad\ldots\ldots\ldots\ldots\ldots\ldots(1).$$

The differential equation of the lines of force is

$$\frac{dV}{dy}dx - \frac{dV}{dx}dy = 0.$$

This, in virtue of (1), is rendered integrable by the factor y, and therefore the integral equation of the lines of force is

$$\left.\begin{array}{c} \psi = \text{const.,} \\[2mm] \text{where} \quad \int\left(y\frac{dV}{dy}dx - y\frac{dV}{dx}dy\right) = \psi \end{array}\right\} \quad\ldots\ldots\ldots\ldots(2).$$

For example let $V = \dfrac{\mu x}{(x^2 + y^2)^{\frac{3}{2}}} - Fx \quad\ldots\ldots\ldots\ldots\ldots(3),$

so that the distribution of force is that of a uniform field, of intensity F, disturbed by the presence of an infinitesimal magnet, of magnetic moment μ, placed with its magnetic axis parallel to the lines of the undisturbed force.   We find

$$\psi = \frac{\mu y^2}{(x^2 + y^2)^{\frac{3}{2}}} + \tfrac{1}{2}Fy^2 \quad\ldots\ldots\ldots\ldots(4);$$

which, if we put $\dfrac{2\mu}{F} = a^3, \text{ and } \dfrac{2\psi}{F} = b^2 \quad\ldots\ldots\ldots\ldots(5),$

gives $y^2 = b^2 - \dfrac{a^3 y^2}{(x^2 + y^2)^{\frac{3}{2}}} \quad\ldots\ldots\ldots\ldots\ldots(6);$

or, resolved for x, $x = \left\{\left(\dfrac{a^3 y^2}{b^2 - y^2}\right)^{\frac{2}{3}} - y^2\right\}^{\frac{1}{2}} \quad\ldots\ldots\ldots\ldots(7).$

On account of the double sign of the radical in (6) we may, without loss of generality, suppose a always positive; and the branches of the curves corresponding to negative values of the radical will then correspond to the case in which the magnet is placed in the position in which, if it were rigidly magnetized, and free to turn, its equilibrium would be unstable. In these branches, which for brevity will be called *exflected*, y^2 is everywhere greater than b^2; while in the branches corresponding to a magnet placed in position of stable equilibrium, which will be called *inflected*, y^2 is everywhere less than b^2 Of the

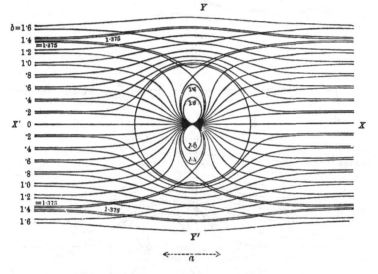

Radius of Circle $= a$.

Fig. 1.

annexed woodcuts*, fig. 1 represents the entire series of both sets of branches for all positive values of b^2; fig. 2 the whole series of inflected branches; fig. 3 the whole series of exflected branches; and figs. 4, 5, 6, 7 selections from the two sets to illustrate inductive influences of spherical bodies of various qualities, placed in a uniform current of incompressible frictionless liquid, or in uniform fields of electric or magnetic force.

---

* From photographs of large-scale diagrams calculated from equation (7), and drawn for the Natural Philosophy Class in the University of Glasgow about twenty-three years ago by Mr. D. Macfarlane, to illustrate fluid motion and the allied subjects of physical mathematics.

The two double points shown in figs. 1, 2, and 4 correspond to

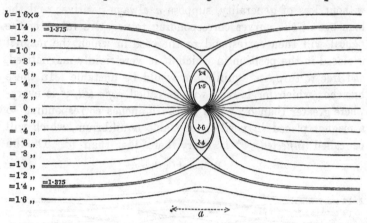

FIG. 2.

the pairs of equal roots $y = \dfrac{a}{\sqrt[3]{2}}$, $y = -\dfrac{a}{\sqrt[3]{2}}$, which the two

quintics $\qquad y^2(y^3 - b^2 y \pm a^3) = 0$

Radius of Circle $= a$.

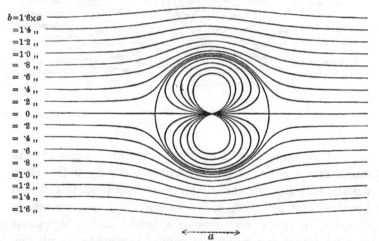

FIG. 3.

have when $b = \dfrac{\sqrt{3}}{\sqrt[3]{2}} = 1\cdot375$. A circle (fig. 4) described from the origin as centre through these double points, and therefore having $\dfrac{a}{\sqrt[3]{2}}$ for radius, cuts perpendicularly each of the inflected

$$\varpi = \infty.$$

$$r = \frac{a}{\sqrt[3]{2}} = \cdot794 \times a.$$

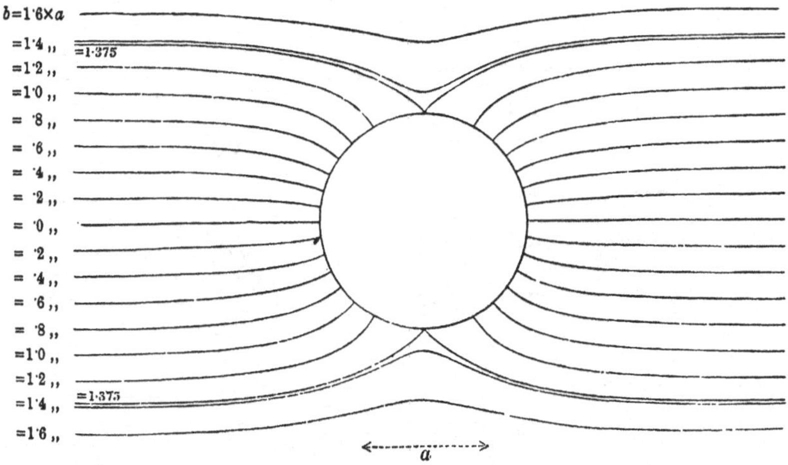

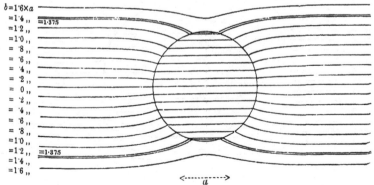

Fig. 4.

curves, except the one given by $b = \dfrac{\sqrt{3}}{\sqrt[3]{2}}$, which it cuts through

the double points at angles of $\pm \tan^{-1} \dfrac{1}{\sqrt{2}}$.

$$\varpi = 2\cdot8$$
$$r = 1\cdot1 \times a.$$

Fig. 5.

Of the exflected curves (fig. 3), that given by $b = 0$ consists of a circle of radius a, having its centre at the origin, together

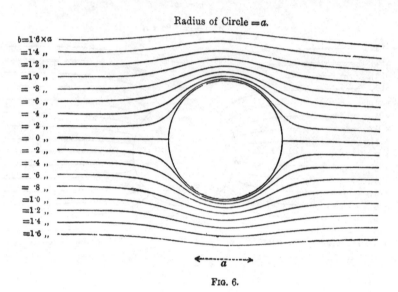

Radius of Circle $= a$.

$b = 1·6 \times a$
$= 1·4 ,,$
$= 1·2 ,,$
$= 1·0 ,,$
$= ·8 ,,$
$= ·6 ,,$
$= ·4 ,,$
$= ·2 ,,$
$= 0 ,,$
$= ·2 ,,$
$= ·4 ,,$
$= ·6 ,,$
$= ·8 ,,$
$= 1·0 ,,$
$= 1·2 ,,$
$= 1·4 ,,$
$= 1·6 ,,$

a

Fig. 6.

with the parts of the axis of x external to that circle, each doubled; or the same circle, together with the part of the axis of x within it, doubled.

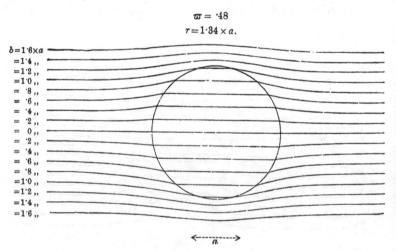

$\varpi = ·48$
$r = 1·34 \times a.$

$b = 1·6 \times a$
$= 1·4 ,,$
$= 1·2 ,,$
$= 1·0 ,,$
$= ·8 ,,$
$= ·6 ,,$
$= ·4 ,,$
$= ·2 ,,$
$= 0 ,,$
$= ·2 ,,$
$= ·4 ,,$
$= ·6 ,,$
$= ·8 ,,$
$= 1·0 ,,$
$= 1·2 ,,$
$= 1·4 ,,$
$= 1·6 ,,$

a

Fig. 7.

Fig. 4 represents the lines of electric force in the neighbourhood of an uncharged insulated metal globe placed in a uniform field of electric force. It also represents (§ 631) without sensible distinction the lines of magnetic force in the neighbourhood of a globe of soft iron in a uniform magnetic field. Fig. 6 represents the stream lines of a frictionless incompressible liquid passing a fixed spherical obstacle.

633. To investigate the relation of the lines of force in the neighbourhood of a solid globe of any ferromagnetic or diamagnetic homogeneous material destitute of intrinsic magnetism, put into a uniform magnetic field, with one of the three principal axes (§ 611) if the substance be not isotropic, placed parallel to the lines of force :—Let ϖ be the permeability of the substance (§ 629) and r the radius of the globe. The induced magnetization being (§ 610) uniform, and parallel to the lines of force of the field, its action through external space will (§ 610, foot-note) be the same as that of an infinitely small magnet at its centre. Hence, using the notation of (5) in (3), and instead of admitting the negative sign for the radical, taking the proper diamagnetic formula by itself, we have

$$(\text{external}) \quad \left\{ \begin{aligned} V &= \tfrac{1}{2}F\left\{\frac{a^3x}{(x^2+y^2)^{\frac{3}{2}}} - 2x\right\}\ \ldots\ldots(\text{ferromagnetic}) \\ V &= \tfrac{1}{2}F\left\{\frac{-a^3x}{(x^2+y^2)^{\frac{3}{2}}} - 2x\right\}\ \ldots\ldots(\text{diamagnetic}) \end{aligned} \right\} \ldots(8),$$

for the potential in external space due to the magnetism of the globe and the uniform force of the field. Throughout the internal space the force is (§ 610, foot-note) uniform, and its potential must be of the form Cx. Choosing C so that at the surface of the sphere (radius r) the external and internal potentials shall be equal, we find

$$(\text{internal}) \quad \left\{ \begin{aligned} V &= \tfrac{1}{2}F\left(\frac{a^3}{r^3} - 2\right)x \ \ldots\ldots(\text{ferromagnetic}) \\ V &= -\tfrac{1}{2}F\left(\frac{a^3}{r^3} + 2\right)x \ \ldots(\text{diamagnetic}) \end{aligned} \right\} \ldots\ldots(9).$$

From this and (8) we find, for the force at any point in the axis of x,

(external) $\begin{cases} X = F\left(1 + \dfrac{a^3}{x^3}\right) \quad \text{......(ferromagnetic)} \\[2mm] X = F\left(1 - \dfrac{a^3}{x^3}\right) \quad \text{......(diamagnetic)} \end{cases}$ ...(10);

and

(internal) $\begin{cases} X = F\left(1 - \tfrac{1}{2}\dfrac{a^3}{r^3}\right) \quad \text{......(ferromagnetic)} \\[2mm] X = F\left(1 + \tfrac{1}{2}\dfrac{a^3}{r^3}\right) \quad \text{......(diamagnetic)} \end{cases}$ ...(11).

For points in the axis of x infinitely near one another $x = r$, and (§ 629) we have

$$\frac{X \text{ (external)}}{X \text{ (internal)}} = \varpi.$$

Hence, by (10) and (11),

$$\left. \begin{aligned} \varpi &= \frac{2\left(1 + \dfrac{a^3}{r^3}\right)}{2 - \dfrac{a^3}{r^3}} \quad \text{..... (ferromagnetic)} \\[4mm] \varpi &= \frac{2\left(1 - \dfrac{a^3}{r^3}\right)}{2 + \dfrac{a^3}{r^3}} \quad \text{......(diamagnetic)} \end{aligned} \right\} \quad \text{.........(12)};$$

or (resolving for r)

$$\left. \begin{aligned} r &= a\sqrt[3]{\frac{\varpi + 2}{2(\varpi - 1)}} \quad \text{......(ferromagnetic)} \\[3mm] r &= a\sqrt[3]{\frac{2 + \varpi}{2(1 - \varpi)}} \quad \text{......(diamagnetic)} \end{aligned} \right\} \quad \text{......(13)}.$$

For great values of ϖ we have

$$r = \frac{a}{\sqrt[3]{2}}\left(1 + \frac{1}{\varpi}\right) \text{ approximately ....(14)}.$$

Hence for such values of ϖ as those discovered in soft iron by Thalén (§ 631 above) the value of r would be only greater by about $\frac{1}{500}$ part than that shown in fig. 4. The circles shown in figs. 5 and 7 were described with radii chosen at random. By measuring them in proportion to a in each case, I find the permeabilities of the inductively magnetized globes, whose influence on the lines of magnetic force is represented in those diagrams, to be respectively 2·8 and ·48.

XXXIII. *On the Forces experienced by Small Spheres under Magnetic Influence; and on some of the Phenomena presented by Diamagnetic Substances.*

[From the *Cambridge and Dublin Mathematical Journal*, May 1847.]

634. THE circumstance that a magnet* attracts small pieces of iron, is the phenomenon of magnetism which was first observed; and an analogous action, presented by rubbed amber, first drew attention to the phenomena of electricity. Now it has since been discovered that no mutual attraction or repulsion between two bodies can result from magnetism in one, unless the other be also magnetized, and that no electric force can exist unless each body be electrically excited. Hence it appears that the forces originally observed are the consequences of a temporary magnetic or electric state induced in a neutral body, when placed in the neighbourhood of a magnet or of an electrified body.

In the following paper the law of such phenomena with reference to magnetism† is considered. It is easily shown however that, by taking $i = 1$ in the formulæ obtained below, the corresponding results for small insulated conductors, electrified by influence, may be obtained, although the physical problems are entirely distinct.

635. We may commence by considering the case of a small sphere of soft iron, or of any other substance susceptible of magnetic induction; and it is easily shown that the formulæ expressing the results may be applied to the case of a small cube by merely altering the value of a certain coefficient; and in general to the case of a small portion of matter of any form, such that in whatever way it be turned, the resultant axis of magnetization, for the whole mass, shall coincide with the direction of the magnetizing force.

---

* Originally a piece of magnetic iron-ore or loadstone. The term may now be applied to any mass possessing permanent magnetism, and may even be extended to a galvanic wire of any form.

† This has not been made the subject of a special investigation by any writer, so far as I am aware, although the nature of the result, in the case of magnetism, appears to be entirely understood by Mr Faraday. Thus, from § 2418 of his *Experimental Researches* [quoted below, in the text (§ 646)] we might infer that a small sphere or cube of soft iron would in some cases be "urged along, and in others obliquely or directly across the lines of magnetic force;" and that all the phenomena would resolve themselves into this, that such a portion of matter, when under magnetic action, tends to move from places of weaker to places of stronger force.

636. It is well known [and proved in § 609 above] that if a small homogeneous sphere of soft iron, or of any other substance susceptible of magnetic induction, be placed in the neighbourhood of a magnet, it will become uniformly magnetized, throughout its mass, with an intensity numerically expressed by multiplying the magnetizing force, by a coefficient independent of the dimensions of the sphere. Thus if R denote the resultant force of the magnet, or the force that it would exert upon an imaginary unit of magnetism, at the position occupied by the sphere, of which we suppose the dimensions to be so small that R has sensibly the same value and direction throughout; and if κ be the intensity of the induced magnetism; we have

$$\kappa = \frac{3i}{4\pi} R \dots\dots\dots\dots\dots\dots\dots(1),$$

where i is a proper fraction (nearly equal to unity for soft iron) depending on the capacity of the substance for magnetic induction.

637. If the force R were rigorously constant in magnitude and direction throughout the whole space S occupied by the sphere, then there would be no resulting force tending to move the sphere ; as, for example, we may conceive it to be, without committing an appreciable error, in the case of a ball of iron of any ordinary dimensions magnetized by the terrestrial force. In the investigation which follows we shall therefore have to consider the small variation of R through the space S, but although considering the effect of this small variation in causing a moving force upon the magnetized sphere, we may neglect the deviation from rigorous uniformity of magnetization which it will produce.

638. Let X, Y, Z be the components of R at the point (x, y, z), which may be taken as the centre of the small sphere. At any point $(x + f)$, $(y + g)$, $(z + h)$, in the sphere, we shall have, for the components of the resultant force due to the magnet,

$$X + \frac{dX}{dx} f + \frac{dX}{dy} g + \frac{dX}{dz} h,$$

$$Y + \frac{dY}{dx} f + \frac{dY}{dy} g + \frac{dY}{dz} h,$$

$$Z + \frac{dZ}{dx} f + \frac{dZ}{dy} g + \frac{dZ}{dz} h.$$

By considering the effects of these forces upon the elements (as for instance thin bars, in the direction of magnetization) into which the magnetized sphere may be supposed to be divided, it is easily shown [§ 500 above], as has also been done by Poisson, that the components of the resulting force on the sphere are given by the equations

$$F = \frac{dX}{dx} \cdot \kappa\sigma \cdot l + \frac{dX}{dy} \cdot \kappa\sigma \cdot m + \frac{dX}{dz} \cdot \kappa\sigma \cdot n,$$

$$G = \frac{dY}{dx} \cdot \kappa\sigma \cdot l + \frac{dY}{dy} \cdot \kappa\sigma \cdot m + \frac{dY}{dz} \cdot \kappa\sigma \cdot n,$$

$$H = \frac{dZ}{dx} \cdot \kappa\sigma \cdot l + \frac{dZ}{dy} \cdot \kappa\sigma \cdot m + \frac{dZ}{dz} \cdot \kappa\sigma \cdot n,$$

where σ is the volume of the sphere, and l, m, n the cosines of the angles made by the direction of magnetization with the axes. Now since this direction is that of the force R, we have

$$l = \frac{X}{R}, \quad m = \frac{Y}{R}, \quad n = \frac{Z}{R}.$$

Hence, since $\kappa = \frac{3}{4\pi} i \cdot R$, we have

$$\left.\begin{aligned}
F &= \frac{3i}{4\pi} \sigma \left(X\frac{dX}{dx} + Y\frac{dX}{dy} + Z\frac{dX}{dz} \right) \\
G &= \frac{3i}{4\pi} \sigma \left(X\frac{dY}{dx} + Y\frac{dY}{dy} + Z\frac{dY}{dz} \right) \\
H &= \frac{3i}{4\pi} \sigma \left(X\frac{dZ}{dx} + Y\frac{dZ}{dy} + Z\frac{dZ}{dz} \right)
\end{aligned}\right\} \quad \dots\dots\dots\dots(2).$$

639. Now if R be due to any magnet, or to a closed galvanic current, $Xdx + Ydy + Zdz$ is necessarily a complete differential, and therefore we have

$$\frac{dY}{dz} = \frac{dZ}{dy}, \quad \frac{dZ}{dx} = \frac{dX}{dz}, \quad \frac{dX}{dy} = \frac{dY}{dx} \quad \dots\dots\dots(3).$$

Modifying the second members of (2) by means of these equations, we find

$$\left.\begin{aligned}
F &= \frac{3i}{4\pi}\sigma \left(X\frac{dX}{dx} + Y\frac{dY}{dx} + Z\frac{dZ}{dx} \right) = \frac{3i}{4\pi} \sigma \cdot R \frac{dR}{dx} \\
G &= \frac{3i}{4\pi}\sigma \left(X\frac{dX}{dy} + Y\frac{dY}{dy} + Z\frac{dZ}{dy} \right) = \frac{3i}{4\pi} \sigma \cdot R \frac{dR}{dy} \\
H &= \frac{3i}{4\pi}\sigma \left(X\frac{dX}{dz} + Y\frac{dY}{dz} + Z\frac{dZ}{dz} \right) = \frac{3i}{4\pi} \sigma \cdot R \frac{dR}{dz}
\end{aligned}\right\} \quad \dots(4).$$

From these we deduce

$$Fdx + Gdy + Hdz = \frac{3i}{4\pi} \sigma . RdR = d\left(\frac{3i}{8\pi} \sigma . R^2\right) \dots (5),$$

which expresses fully the result of equations (4).

640.   The interpretation of this result shows that a sphere of soft iron is urged in the direction in which the magnetizing force increases most rapidly; the components of the force in different directions being expressible by the differential coefficients of the function $\frac{3}{8\pi} \sigma R^2$. Thus in some cases it may actually be urged across the direction of the magnetizing force. For instance, if a ball of soft iron be placed symmetrically with respect to the two poles of a horse-shoe magnet, and at some distance from the line joining them, it will be urged towards this line in a direction perpendicular to it, although the magnetizing force is parallel to it; or if the magnetizing force be due to a straight galvanic wire, a ball of soft iron will be *attracted* towards the wire, although the force on an imaginary "magnetic point" is perpendicular to a plane through it and the wire.

641.   The positions of equilibrium of a small sphere acted upon by the magnetic forces alone, will be points in the neighbourhood of which R^2 is stationary in value, or points where $d(R^2) = 0$. This condition is satisfied by either $R = 0$, or $dR = 0$. Hence the sphere will be in equilibrium at points where the resultant magnetizing force vanishes; where it is a maximum or minimum; or where it is stationary in value.

642.   A position of stable equilibrium will be such that R^2 diminishes in every direction from it; and hence, if there be any point, external to the magnet, at which the resultant force has a maximum value, it would be a position of stable equilibrium for a small ball of soft iron, and any other position of equilibrium is essentially unstable.

643.   According to Mr Faraday's recent researches, it appears that there are a great many substances susceptible of magnetic induction, of such a kind that for them the value of the coefficient i is negative. These he calls diamagnetic substances, and, in describing the remarkable results to which his experiments conducted him with reference to induction in diamagnetic matter, he says: "all the phenomena resolve

themselves into this, that a portion of such matter, when under magnetic action, tends to move from stronger to weaker places or points of force*." This is entirely in accordance with the result obtained above; and it appears that the law of all the phenomena of induction discovered by Faraday with reference to diamagnetics may be expressed in the same terms as in the case of ordinary magnetic induction, by merely supposing the coefficient i to have a negative value†.

644. In the case of a diamagnetic sphere, the consideration of the stability or instability of equilibrium in different positions, is extremely interesting. Thus, at a point where R^2 is a minimum, a small sphere of diamagnetic matter will be in stable equilibrium; and this is actually the case at any point for which the force vanishes; even if we take into account the weight of the sphere, it is readily shown that stable positions of equilibrium may exist. Thus a hollow cylindrical bar-magnet (if sufficiently powerful), held with its axis vertical, would support a small diamagnetic sphere in a position of stable equilibrium at a point in the axis, a little below the lower end of the magnet. For, considering different points in the axis, we perceive that there is one below the lower end (at a distance $= \dfrac{a}{\sqrt{2}}$, if a, the radius of the cylinder, be very great compared with its thickness, and very small compared with its length, and if the distribution of magnetism be uniform) at which the resultant force is a maximum. If, on moving a small diamagnetic sphere upwards from this position, we arrive at a point where the force urging it upwards is greater than the weight, and then let it move freely from rest, it will oscillate about a position of stable equilibrium. It will probably be impossible ever to observe this phenomenon, on account of the difficulty of getting a magnet strong enough, and a diamagnetic substance sufficiently light, as the forces manifested in all cases of diamagnetic induction hitherto examined are excessively feeble.

---

* *Experimental Researches*, § 2418.

† The law of induction in a mass of any form, whether of magnetic or diamagnetic matter, may be stated as follows:—Let R be the magnetic force upon a point within an infinitely small spherical surface, described round a point P in the mass, resulting from the magnetism of all the matter external to this surface. The intensity of the magnetism at P is equal to $\frac{1}{3}\pi i R$, and its direction is that of the resultant force R.

645. A very curious phenomenon might readily be observed, according to the results given above, by placing two bar-magnets, with similar poles, in the neighbourhood of a ball of soft iron allowed to move in a horizontal straight line (or suspended in such a manner that any motion which can take place is in a circle of considerable radius). Thus if a pole, *S*, of a bar-magnet which we may regard for simplicity as very long and thin, be held in the neighbourhood, the ball will be drawn towards the point *A*, in which a perpendicular from *S* meets the line of motion, and *A* will therefore be a position of stable equilibrium. If now a pole *S'*, of an equally powerful magnet, be presented and held at an equal distance in *SA* produced, *A* will become an unstable position; and if the ball be placed in its line of motion, at any distance from *A* less than $\frac{SA}{\sqrt{2}}$, it will be *repelled* from *A*, although either magnet alone would cause it to move towards this point.

646. The result obtained above affords the true explanation of the phenomenon observed by Faraday, that a thin bar or needle of a diamagnetic substance, when suspended between the poles of a magnet, assumes a position across the line joining them. For such a needle has no tendency to arrange itself across the lines of magnetic force; but, as will be shown [§ 684, below] in a future paper, if it be very small compared with the dimensions and distance of the magnet (as is the case, for instance, with a bar of any ordinary dimensions, subject only to the earth's influence), the direction it will assume, when allowed to turn freely about its centre of gravity, will be that of the lines of force, whether the material of which it consists be diamagnetic, or magnetic matter such as soft iron: but Faraday's result is due to the rapid decrease of magnetic intensity round the poles of the magnet, and to the length of the needle, which is considerable compared with the distance between the poles of the magnet; and is thus explained by the discoverer himself. (§ 2269 of his *Experimental Researches.*) "The cause of the "pointing of the bar, or any oblong arrangement of the heavy "glass is now evident. It is merely a result of the tendency of "the particles to move outwards, or into the positions of weakest

" magnetic action*. The joint exertion of the action of all the " particles brings the mass into the position which, by experiment, " is found to belong to it."

St Peter's College, *May* 13, 1847.

XXXIV. Remarks on the Forces experienced by Inductively Magnetized Ferromagnetic or Diamagnetic Non-Crystalline Substances.

[From the *Philosophical Magazine*, October 1850.]

THE remarkable law laid down by Faraday in [§ 2418 of his *Experimental Researches*] his Memoir on the Magnetic Condition of all Matter [*Transactions Royal Society*, 1846, p. 21, or *Phil. Mag.* Vol. XXVIII., 1846], *that a small portion of diamagnetic matter placed in the neighbourhood of a magnet experiences a pressure urging it from places of stronger towards places of weaker force*, is a simple conclusion, derived from the mathematical solution of the problem of determining the action experienced by a small sphere of matter magnetized inductively, and acted upon in virtue of its induced magnetism. Without entering upon the analytical investigation, which will be found in [§§ 634 —646 above] a paper " On the Forces experienced by small Spheres under Magnetic Influence ; and on some of the Phenomena presented by Diamagnetic Substances†," I shall, in the present communication, state and explain briefly the result, and point out some remarkable inferences which may be drawn from it.

647. Let P be any point in the neighbourhood of a magnet, and let P' be a point at an infinitely small distance, which may be denoted by a, from P. Let R denote the force which a "unit north pole‡" if placed at P would experience, or, as it is called, " the resultant magnetic force at P;" and let R'

---

* The extreme feebleness of the diamagnetic action on account of which any small sphere or cube of the matter will experience very nearly the same force as if all the rest were removed, seems fully to justify this explanation.

† *Cambridge and Dublin Mathematical Journal*, May 1847.

‡ That is, the end of an infinitely thin uniformly and longitudinally magnetized bar of "unit strength" which is repelled *on the whole from the north* by the magnetism of the earth; "unit strength" being defined by the following statement :—

If two infinitely thin bars be equally, and each uniformly and longitudinally, magnetized, and if, when an end of one is placed at a unit (an inch, for example) of distance from an end of the other, the mutual force between these ends is unity; the magnetic strength of each is unity. The force R, defined in the text, is of course equal and opposite to the force that a "unit south pole" would experience if placed at P.

denote the same with reference to P'. Then, if a small sphere of any kind of non-crystalline homogeneous matter, naturally unmagnetic, but susceptible of magnetization by influence, be placed at P, it will experience a force of which the component along PP' is

$$A\sigma \cdot \frac{1}{2}\frac{R'^2 - R^2}{a},$$

where σ denotes the volume of the sphere, and A a coefficient depending on the nature of the substance. This coefficient, A [agreeing with the A, B, or C of §§ 614—618 applied to an isotropic substance] has a value a little less than $\frac{3}{4\pi}$ for soft iron, and it has very small positive values for all ferromagnetic substances containing little or no iron.

648. If it be true, as I think it must be, that the forces experienced by diamagnetic substances are occasioned by the influencing magnet magnetizing them inductively[*], and acting upon them when so magnetized, according to the established laws of the mutual action of two magnets, the preceding result will hold for all non-crystalline matter; and to apply A to a diamagnetic substance it will be only necessary to give it a negative value. [From § 630, § 628 (14), and § 627 we see that the extreme negative value conceivably admissible is $-\frac{3}{8\pi}$. Thus for every substance, whether ferromagnetic or diamagnetic, A is between $+\frac{3}{4\pi}$ and $-\frac{3}{8\pi}$.]

649. To interpret the result of § 647, we may remark, that by the elementary principles of the differential calculus as applied to the variation of a quantity depending on the position of a point

---

[*] The most natural explanation of the phenomena which he had discovered is suggested by Faraday in his original paper on the subject, and it is confirmed by the researches of subseqent experimenters, especially those of Reich and Weber, who have made experiments to show that a diamagnetic substance, under the influence of two magnets, will act upon one in virtue of the magnetization which it experiences from the other. The extreme feebleness of the polarity induced in diamagnetic substances is proved by Faraday in a series of experiments forming the subject of his last communication to the Royal Society; in which an attempt is made, by very delicate means, to test the induced current in a helix due to magnetization or demagnetization of a diamagnetic substance which it surrounds, but only negative results are obtained.

in space, it may be shown that the fraction $\dfrac{R'^2 - R^2}{a}$ is greater
when the point P' is chosen in a certain determinate direction
from P than in any other; that it is of equal absolute value,
but negative, if P' be chosen in the opposite direction; and
that it vanishes if P' be in a plane through P at right angles
to the line of those two directions. Hence it follows that the
resultant force upon the small sphere is along that line, in one
direction or the other, according as A is positive or negative,
and accordingly we draw the following conclusions :—

(1) A small ferromagnetic sphere in the neighbourhood of
a magnet, will experience a force urging it *in that direction in
which the "magnetic force" increases most rapidly.*

(2) A small diamagnetic sphere, in the neighbourhood of a
magnet, will experience a force urging it *in that direction in
which the magnetic force decreases most rapidly.*

(3) The absolute magnitude of the force in any case in
which the distribution of magnetic force in the neighbourhood
of the magnet is known, is the value which the expression in
§ 647 obtains when we give $\dfrac{R'^2 - R^2}{a}$ the value found by means
of the differential calculus, for a point P' at an infinitely small
distance PP' in the direction of the most rapid variation of the
magnetic force from P, the actual position of the ball.

650. It is deserving of special remark, that the direction of
the force experienced by the ball has no relation to the direction
of the lines of magnetic force through the position in which it
is placed. The mathematical investigation thus affords full confirmation and explanation of the very remarkable observation
made by Faraday (§ 2418 of his *Experimental Researches*), that
a small sphere or cube of inductively magnetized substance is in
some cases "urged along, and in others obliquely or directly
across the lines of magnetic force." It is in fact very easy to
imagine, or actually to construct, arrangements in which the resultant force experienced by a ball of soft iron, or of some
diamagnetic substance, is perpendicular to the lines of the
magnetizing force. For instance, if a ball of soft iron be placed
symmetrically with respect to the two poles of a horse-shoe
magnet, and at some distance from the line joining them, it will

be urged towards this line, in a direction perpendicular to it,
and consequently perpendicular to the lines of magnetizing
force in the space in which it is situated; and a ball of bismuth, or of any other diamagnetic substance, similarly situated,
would experience a force in the contrary direction. Or again,
if a ball of any substance be placed in the neighbourhood of
a long straight galvanic wire, it will be urged towards or from
the wire (according as the substance is ferromagnetic or diamagnetic) in a line at right angles to it, and consequently
cutting perpendicularly the lines of force, which are circles
with their centres in the wire and in planes perpendicular
to it.

651. The preceding conclusions enable us to define clearly
the sense in which the terms "attraction" and "repulsion"
may be applied to the action exerted by a magnet on a ferromagnetic and a diamagnetic body respectively. A small sphere
of ferromagnetic substance, placed in the neighbourhood of
a magnet, experiences in general, a force; but the term *attraction*, according to its derivation, means a *force towards;* and
if we apply it in any case, we must be able to supply an object for the preposition. Now, in this case the force is towards
places of stronger "magnetic force;" and hence the action
experienced by a ferromagnetic ball may be called an *attraction* if we understand *towards places of stronger force.* Places
of stronger force are generally nearer the magnet than places
of weaker force, and hence small pieces of soft iron are
generally urged, on the whole, towards a magnet (in consequence of which no doubt the term "attraction" came originally
to be applied): but, as will be seen below, this is by no means
universally the case; balls of soft iron being, in some cases,
actually repelled from the influencing magnet; and the term
"attraction" can only be universally used with reference to
ferromagnetic substances, on the understanding that it is
towards places of stronger force. The term "repulsion," the
reverse of "attraction," may, according to the same principles,
be applied universally to indicate the force with which a small
diamagnetic sphere is urged towards places of weaker force, or
repelled from places of stronger force.

652. The following passage, containing a statement of prin-

ciples on some of which Faraday himself lays much stress, but which have not, I think, been sufficiently attended to by subsequent experimenters, is quoted from the article in the *Mathematical Journal* already referred to. [Here comes quotation of § 646 above.]

653. It may be added to this, that the tendency of a bar, whether of ferromagnetic or of diamagnetic substance, in a uniform field of magnetic force, to take the direction of the lines of force, depends on the effect of the mutual action of the parts in altering the general magnetization of the bar, and is consequently so excessively feeble for any known diamagnetic substance that the most delicate experiments would in all probability fail to render it sensible*.

654. Faraday's law, stated at the commencement of these remarks, may be illustrated by some very curious although extremely simple experiments, which I shall now describe briefly†.

655. The special apparatus required is merely a long light arm (I have used one about four feet in length; but a much shorter rod, if suspended by a finer or by a longer torsion-thread, would have answered equally well) suspended from a "torsion-head" by means of a very fine wire, or thread of unspun silk fibres attached to it near its middle; and a case round it adapted to prevent currents of air from disturbing its equilibrium, but allowing it sufficient angular motion in a horizontal plane. A small ball of soft iron is attached to one end of the arm (or hung from it by a fine thread, which, for the sake of stability in many of the experiments, as for instance, experiments 2 and 3 described below, must not be too long), and a counterbalance is adjusted near the other end so as to make the arm horizontal. If only a small angular motion be allowed to the arm, the path of the ball will be sensibly straight, and we may consider that, by the arrangement which

---

* A very brief communication on this subject was laid before the British Association at the meeting of 1848, and is published in the Report for that year, under the title "On the Equilibrium of Magnetic or Diamagnetic Bodies of any form, under the Influence of Terrestrial Magnetic Force." [Art. XXXIV. Vol. I. of my Reprint of Mathematical and Physical Papers, University Press. 1882.]

† These experiments were shown, in illustration of lectures on magnetism in the Natural Philosophy Class in the University of Glasgow, during the Session 1848-49.

has been described, the ball is allowed to move with great
freedom in a straight line, but prevented from all other motion.

656. In making the experiments described below, it is convenient to have two stops so arranged that the motion of the
arm may be kept within any desired limits, and manageable
in such a way, that by means of them the arm may be rapidly
brought to rest in any position. In general, before commencing an experiment, the arm ought to be brought to rest
near one end of its course, and kept pressing very slightly
upon one of the stops by the torsion of the wire, which may
be suitably adjusted by the torsion-head, and the other stop
ought to be pushed away, so as to leave the arm free to move
in one direction.

657. *Experiment* 1.—Place a common bar-magnet with either
pole, the south, for instance, near the ball of soft iron in its
line of motion, but on that side towards which it is prevented
from moving by the stop. Taking another bar-magnet of
considerably greater strength than the former, bring its north
pole gradually near the fixed south pole of the other, in the
continuation of the line of motion of the iron ball. When
this north pole reaches a certain position, the arm will cease
to press on the stop, and if we push the north pole a little
nearer still, the arm will altogether leave the stop and take a
position of equilibrium, in which, after it is steadied (as may
easily be done by means of the stops), it will remain stable,
although the stops be removed entirely. If, by means of one
of the stops, the ball be pushed to any distance farther from
the magnets than this position of stable equilibrium, it will
return towards it when left free. If it be drawn a little nearer
by means of the other stop, and, when left for a few seconds,
it be found to continue pressing upon the stop, then, when
the stop is removed, the ball will return to that position of
stable equilibrium. If, however, it be very slowly drawn still
nearer the magnets, when it reaches a certain position it will
cease to press on the stop; and if after this it experience the
slightest agitation, or if it be drawn any nearer, it will leave
the stop and move up till it strikes the nearer magnet, in contact with which it will almost immediately come to rest. It
thus appears that there is a position of unstable equilibrium

for the ball between the former stable position and the nearer magnet. It is easy to arrange the torsion-head so that the torsion of the suspending-thread or wire may have as little effect as we please, by finding, by successive trials, either of these positions of equilibrium, subject to the condition that, when the magnets are removed, the torsion would not sensibly disturb the arm from the position so found.

658. After the explanations which have been given above, it is scarcely necessary to point out that the position of unstable equilibrium, determined in this experiment, is a point where the magnetizing force due to the south pole is destroyed by that of the more distant but more powerful north pole; and that the position of stable equilibrium is one where the excess of the magnetizing force due to the north pole, above that which is due to the less powerful south pole, has a maximum value with reference to points in the continuation, through the less powerful pole, of the line joining the two poles. If the poles were mathematical points, and the bars so long that their remote ends could produce no sensible action on the ball, the position of unstable equilibrium would of course be such that *its distances from the two poles would be directly as the square roots of the strengths of the magnets;* and, by the solution of a most simple "maximum problem," it may be shown that the stable position would be such that *its distances from the poles would be directly as the cube roots of the strengths.*

659. *Experiment 2.*—Place two equal bar-magnets symmetrically with reference to the line of motion, with similar poles at equal distances on two sides, in a perpendicular to this line, and, to make the best arrangement, let the lengths of the magnets be in the continuations of the lines joining their poles. Operating by means of the stops, in a manner similar to that described for the preceding experiment, it is readily ascertained that there are two positions of stable equilibrium for the ball at equal distances on two sides of the line joining the poles, and that the middle point of this line is a position of unstable equilibrium.

660. Here, again, the explanation is obvious. The positions of stable equilibrium being such that, with reference to points in the line of motion of the ball, the magnetizing force due

to the two similar poles may be a maximum, are readily found to be at distances $\frac{a}{2\sqrt{2}}$ on the two sides of the line joining the poles (the length of this line being denoted by a), if these be mathematical points, and if the lengths of the bars be so great that the distant poles produce no sensible effects.

661. *Experiment* 3.—Hold a common horse-shoe magnet with the line joining its poles perpendicular to the line of motion of the ball, and, by a suitable management of the stops and of the torsion-head, the existence of a force urging the ball perpendicularly across the "lines of force" towards the middle point of the line joining the poles, may be easily made manifest.

662. *Experiments on diamagnetic substances, and on ferro-magnetic substances of feeble inductive capacity.*—The phenomena discovered by Faraday relative to the action of magnets on substances not previously known to be susceptible of magnetic influence may be exhibited with great ease by means of the apparatus described above. Small balls of the substances to be experimented upon may be hung from one end of the balance (the ball of soft iron being of course removed) by fine threads of sufficient length to allow the arm, which may be of any substance containing no iron, to be out of reach of any sensible influence from the magnet employed. There is in these cases no difficulty, regarding the length of the suspending-thread, of the kind noticed above [§ 655] with reference to soft iron, as the magnetic forces experienced are never strong enough to produce lateral instability (that is, a want of stability in the line of motion), even with the lightest of the substances experimented on, unless the suspending thread be far longer than is necessary. In the experiments I have made, the threads bearing the small balls have not been more than four or five inches long. The diameters of the balls have been from a quarter of an inch to an inch, or an inch and a half. Instead of simple bar-magnets of steel, which are not powerful enough to be convenient for these experiments, I have used a bar electro-magnet of very moderate power, consisting of a helix and soft iron core. This core is a cylinder of about an inch in diameter and a foot and a half long,

with round ends (nearly hemispherical), which, when the core is in its central position, extend about an inch beyond the helix on each side. By these means the repulsion of balls of diamagnetic substance, and the attraction of very feebly ferro-magnetic substances, may be shown with great facility.

663. For example, I may mention that I have hung a small apple, whole, by a thread three or four inches long, and putting it at first at rest, pressing slightly (in virtue of torsion produced by the torsion-head mentioned above, § 656) upon one end of the soft iron core previously to the excitement of the electro-magnet, I have found that as soon as the galvanic current is produced, the apple is repelled away; and, by pushing forward the soft iron core, I have chased it across the field through a space of four or five inches.

664. I have also used the same apparatus to show that a body which is feebly attracted in air is repelled when immersed below the surface of a sufficiently strong solution of sulphate of iron in a small trough, so arranged that when, by the force of torsion, the body immersed in the liquid is made to press on a side of the trough, the electro-magnet may be placed with one end of its core pressing on the outside of the trough, close to the point where it is pressed upon by the body within. Using small glass balls (which, when empty, exhibit no sensible effects of the influence of the magnet), the magnetic conditions of different liquids filling them may be easily tested. Faraday's beautiful experiments on the relative magnetic capacities of solutions of sulphate of iron of different strengths, or rather, other experiments to illustrate the same principles, may be performed in an extremely convenient manner, by filling a glass ball of this kind with a solution, hanging it from one end of the arm, and, by a suitable adjustment of the weight at the other, immersing it below the surface of another solution contained in the trough. I have found that whenever the difference of the strengths of the two solutions was considerable, the ball immersed was attracted or repelled by the external magnet, according as the solution contained in the ball was stronger or weaker than the solution surrounding it.

On the Stability of small Inductively Magnetized Bodies in Positions of Equilibrium.

665.   In the paper [§§ 634...646 above] published in the *Cambridge and Dublin Mathematical Journal* (referred to above), I pointed out that a small ball of either ferromagnetic or diamagnetic substance placed in the neighbourhood of a magnet, and not acted upon by any non-magnetic force, is in *equilibrium* if it be in a situation where the "resultant force" (that which was denoted by R) is either a maximum or minimum, or "stationary" in value; that a diamagnetic ball is in *stable equilibrium* if, and not in stable equilibrium unless, it be situated where the force R is a minimum in absolute value; and that "if there be any "point external to the magnet, at which the resultant force has "a maximum value, it would be a position of stable equilibrium "for a small bar of soft iron, and any other position is essen-"tially unstable."   Shortly after the publication of that paper, I succeeded in proving that the resultant force cannot be an absolute maximum at any point external to a magnet, and consequently that no position of stable equilibrium for a ferromagnetic ball, perfectly free from all constraint, can exist.   I have very recently found that there may be points where the resultant force is an absolute minimum without being zero; and therefore there may be positions of stable equilibrium for a diamagnetic ball not included in the case of the force vanishing, noticed in the previous paper.   That case, however, affords the simplest illustration that can be given of that most extraordinary fact, that a solid body may be repelled by a magnet, or magnets, into a position of stable equilibrium.   If, for instance, we take the arrangement (described for Exp. 2, § 659 above) of two bar-magnets, fixed with similar poles near one another, we have obviously between these poles a point where the resultant force vanishes, and towards which consequently a small diamagnetic ball placed anywhere sufficiently near it would be repelled.   It is easily shown that, actually under the action of gravity, a ball of diamagnetic substance would be in stable equilibrium a little below this position, without any external support or constraint whatever, if only the magnets were strong enough.   It is, however, extremely im-

probable that any attempt to realize this by experiment will succeed, since, even in the most favourable cases, no diamagnetic repulsion upon a solid has yet been obtained which at all approaches in magnitude to the weight of the body. Still we must consider that a true theoretical solution of the celebrated physical problem* suggested by "Mahomet's coffin" has been obtained, which is not the least curious among the remarkable consequences of Faraday's magnetic discoveries.

On the relations of Ferromagnetic and Diamagnetic Magnetization to the Magnetizing Force.

666. In the mathematical investigation by which the result stated above was obtained, it is assumed that the magnetization of the substance of the ball in each case is proportional to the magnetizing force (although this assumption may of course be avoided by merely supposing μ to have a value varying with the force, which will not affect either the investigation or the form of the result). It appears to me very probable that this assumption is correct *for all known diamagnetic substances,* and *for homogeneous feebly ferromagnetic substances;* since [§ 606, Axiom II.] it is equivalent to an assumption that inductive magnetization of a substance does not impair or in any way alter its susceptibility for fresh magnetization by means of another magnet brought into its neighbourhood. This opinion cannot, however, at present be regarded but as a mere conjecture, being as yet unsupported by experiment. It is indeed directly opposed to the following conclusion to which M. Plücker arrives, from some of his experimental researches:—"J'ai déduit de "là cette loi générale, savoir: que le diamagnétisme décroît "plus vite que le magnétisme quand la force de l'aimant dimi-"nue, ou quand la distance des pôles augmente†:" but many

---

* It is, I believe, often thought that this problem is solved in the experiment in which a needle is attracted into a galvanic helix held with its axis vertical; but I have convinced myself that the needle always touches somewhere on the sides of the tube (if there be one round it) or on the wire of the helix; and I have also ascertained that, when a powerful helix is used with, in place of the needle, a tin-plate [iron] cylinder, even if it be very little less in diameter than the inner cylindrical surface of the helix, there is never stable equilibrium without contact between them. The phænomenon of a solid body, hovering freely in the air, in stable equilibrium, without any external support or constraint, has never, I am convinced, been witnessed as the result of any electric or magnetic experiment.

† Quoted from a paper in the French *Annales de Chimie et de Physique,*

of the curious phænomena from which M. Plücker was led
to this conclusion, and which he adduces in confirmation of
it, do not appear to me to support it, but rather to be connected with the peculiar magneto-inductive properties of crystalline or quasi-crystalline structure which he discovered
subsequently*; and with respect to those which appear at
first sight really to support it, I have conjectured that they
may admit of explanation solely on the principle expressed in
Faraday's law, quoted at the commencement of these remarks.
Thus, the experiments upon a watch-glass containing mercury,
placed at different distances from a magnet, which show
that the resultant force experienced by the watch-glass, in
virtue of its own magnetization as a ferromagnetic substance,
and the contrary magnetization of the diamagnetic mercury,
is sometimes increased by removing the whole to a slightly
greater distance from the magnet, do not prove that when the
magnetizing force is diminished the induced magnetization of
the mercury is diminished by a greater fraction of its former
amount than that of the watch-glass, but are most probably
to be explained by the circumstance that the "field of force"
occupied by the mercury and watch-glass when removed a
very short distance, is such that the mean value of the differential coefficient of the square of the force, with reference to
co-ordinates parallel to the direction of motion of the watch-
glass, is greater than the mean value of the same function,
through the field occupied when the watch-glass is in contact
with the magnet.   It is of course impossible to give more than
a general explanation such as this without some specific knowledge of the distribution of magnetic force in the neighbourhood of the actual magnet employed; but the phænomena
described by M. Plücker in this case are undoubtedly of a
kind that might be anticipated if a vertical bar-magnet be

June 1850, bearing the title, "Sur le Magnétisme et le Diamagnétisme:
par M. Plücker." This paper appears to be a *résumé* of the author's experimental researches and discoveries regarding magnetic induction, of
which detailed accounts have been published in various communications to
Poggendorff's *Annalen* in the course of the last two years.

* This connexion is recognised by the discoverer himself, as is shown by
the statement he makes at the commencement of § 4 of the paper already
referred to. Yet he mentions his experiments on cylinders of charcoal as
the foundation on which he establishes, as a general law, the conclusion
quoted in the text.

used, especially if the upper pole, over which the watch-glass is suspended, be flat. An electro-magnet with, for core, a hollow cylinder of soft iron open at the ends, would even *repel* a small ferromagnetic body capable of moving along the axis, in some positions, and *attract* it a little further off, since there would be variations of force in this case precisely similar to those explained with reference to points in the line of motion of the ball in Experiment 2, § 659 above.

667. The most striking experiments adduced by M. Plücker to support his hypothesis, that "diamagnetism increases more rapidly than magnetism" when the magnetizing force is increased, are those in which the force experienced by a small inductively magnetized body in a constant position is tested for different strengths of the same electro-magnet, produced by using a greater or less number of cells in the exciting battery. At the recent Meeting of the British Association in Edinburgh, I ventured to suggest *that a change in the distribution of magnetic force in the neighbourhood of the magnet, accompanying an increase or diminution in the strength of the galvanic current, might have contributed to produce some of the singular phœnomena which had been observed; and that there is some considerable change in the distribution of force in the neighbourhood of an electro-magnet with a soft iron core in a state of intense magnetization when, for instance, the strength of the current is doubled, seems extremely probable when we consider that a piece of soft iron in a state of intense magnetization cannot be expected to be as open to fresh magnetization as it would be if not magnetized in the first instance*.* On the same occasion I remarked, that some experiments made by Mr Joule in connexion with his researches on changes of dimensions produced in iron bars by magnetic influence, appeared to indicate diminished inductive capacities in states of intense inductive magnetization†. At that time I was not aware of the recent experimental researches of Gartenhauser and Müller on the magnetization of soft iron; but I have since met with a number of Poggendorff's *Annalen* (1850,

---

* [Embodied in Art. xxx. (§§ 604—624) above.]
† *Phil. Mag.* 1847, vol. xxx. pp. 76, 225. Also Sturgeon's *Annals*, Aug. 1840.

No. 3, published last April) containing an account of these researches*, which completely confirms the second part of the conjecture I had thrown out. Whether or not, however, the change in the distribution of force is of such a kind as to account for the phænomena by which M. Plücker supports the conclusion which has been quoted, it is impossible to pronounce without a complete knowledge of the circumstances. An *experimentum crucis* might be made by means of an electro-magnet without a soft iron core.

668.   In one respect M. Plücker's views receive a remarkable confirmation by Joule and by Gartenhauser and Müller's experiments, if it be true that a homogeneous *diamagnetic* substance is inductively magnetizable to an extent precisely proportional to the magnetizing force, or deviating less from this proportionality than the magnetization of soft iron. For if a complex body were made up consisting of a diamagnetic substance (either solid or in powder) and an extremely small quantity of soft iron in very fine powder or filings, spread uniformly through it; a small ball of this body would, when acted upon by a feeble magnetizing force, become on the whole magnetized like a ferromagnetic, and would be urged from places of weaker towards places of stronger force. If now the magnetizing force were gradually increased, the "resultant magnetic moment" of the complex body would at first increase, then, after attaining a maximum value, decrease to zero, after which it would become "negative," or the ball would be on the whole magnetized like a diamagnetic, and would be urged from places of stronger towards places of weaker force. Such, if I mistake not, is the bearing which M. Plücker expects of any complex solid consisting of a suitable mixture of ferromagnetic and diamagnetic substances; but mere experiments on soft iron, such as those of Joule and of Gartenhauser and Müller, do not render it probable that a homogeneous feebly ferromagnetic substance, containing no iron, or only a very small quantity and that chemically combined, should have its capacity for fresh magnetization

---

* "Ueber die Magnetisirung von Eisenstäben durch den Galvanischen Strom; von J. Müller."

diminished by the slight magnetization which the strongest magnetizing force that could be applied would produce*.

Row, Gare Loch, *Aug.* 21, 1850.

XXXV. ABSTRACTS OF TWO COMMUNICATIONS

[From the *Report of British Association for Belfast,* 1852.]

On certain Magnetic Curves ; with applications to Problems in the Theories of Heat Electricity, and Fluid Motion.

669. A method [§ 632 above], which had been given by the author in the *Cambridge Mathematical Journal,* Vol. IV., Nov. 1843†, for integrating the differential equations of the lines of force in any case of symmetry about an axis, is applied in this communication to the case of an infinitely small magnet placed with its axis direct or reverse along the lines of force of a uniform magnetic field. Diagrams [§ 632 above] containing the curves drawn accurately, according to calculations founded on the result of this investigation (corresponding to series of ten or twelve different values given to the constant of integration), were exhibited to the Section. Certain parts of these curves were shown in a separate diagram [§ 632, fig. 4], as constituting precisely the series of lines of electric force about an insulated spherical conductor under the influence of a distant electrified body; and the other parts, in a separate diagram [fig. 6], as constituting the lines of motion of a fluid mass in the neighbourhood of a fixed spherical solid, at considerable distances from which the fluid is moving uniformly in parallel lines so slowly as to cause no eddies round the obstacle. The circle representing the section of the spherical conductor, in the former of these diagrams, cuts the entire series of curves at right angles, with the exception of one curve, which it cuts through a double point at an angle of 45° to each branch. The circle representing the section of the spherical obstacle in the latter diagram, along with two infinite double branches consisting of the axial diameter produced externally in each direction, constitutes the limiting curve of the series shown, and is not intersected by any of them. A series of diagrams (deduced from

---

* [The last sentence of this article is cancelled from the reprint (July 5, 1872).]
† [Note of Feb. 22, 1884. Now republished, constituting Art. ix. of my "Reprint of Mathematical and Physical Papers," Vol. i. 1882. W. T.]

the former of these by describing a circle of the same size as
that shown in it, and drawing, on a smaller scale, as much of the
curves as lies without this circle) was shown as representing the
disturbed lines of magnetic force about balls of ferromagnetic
substance of different inductive capacities, placed in a uniform
magnetic field [one of these is shown in fig. 5 of § 632]; and
another series, similarly derived from the latter (that is, the
one representing the lines of fluid motion about a spherical
obstacle), was shown as representing the disturbance caused
by the presence of diamagnetic balls of different inductive
capacities in a uniform magnetic field [one of these is shown in
fig. 7 of § 632]. These two series of diagrams are also accurate
representations of the lines of motion of heat in a large homogeneous solid having heat uniformly conducted across it, disturbed by spherical spaces occupied by solid matter of greater
or less conducting power than the matter round them; the
two principal diagrams from which they are derived being the
corresponding representations for the cases of spherical spaces
occupied respectively by matter of infinitely great and infinitely
small conductivity. The author called attention to the remarkable resemblance which these diagrams bore to those which
Mr Faraday had shown recently at the Royal Institution to
illustrate his views regarding the action of ferromagnetics and
diamagnetics in influencing the field of force in which they
are placed; and justified and illustrated the expression "conducting power for the lines of force," by referring to rigorous
mathematical analogies presented by the theory of heat.

*On the Equilibrium of elongated Masses of Ferromagnetic Sub-
stance in uniform and varied Fields of Force.*

The fact, first discovered experimentally by Gilbert, that a
bar of soft iron, held by its centre of gravity in a uniform
magnetic field, settles with its length parallel to the lines of
force, is not explained correctly when it is said to be merely due
to the property of magnetic induction in virtue of which the
bar of soft iron becomes temporarily a magnet like a permanent
magnet in its position of stable equilibrium. For exactly the
same statement would be applicable to a row of soft iron balls
rigidly connected by a non-magnetic frame; yet such an arrange-

ment would not experience any directional tendency (since no one of the balls in it would experience either a resultant force or a resultant couple from the force of the field), unless in virtue of changes in the states of magnetization of the balls induced by their mutual actions. Hence the mutual action of the parts of a row of balls, and as is easily shown, of a row of cubes, or of a bar of any kind, must be taken into account before a true theory of their directional tendencies can be obtained. The author of this communication, by elementary mechanical reasoning founded on what is known with certainty regarding magnetic induction and magnetic action generally, shows that an elongated mass, in a uniform magnetic field, tends to place its length parallel to the lines of force, whether its inductive capacity be ferromagnetic or diamagnetic, provided it be non-crystalline, because if ferromagnetic it becomes more, or if diamagnetic, less intensely magnetized, if placed in such a position, than if placed with its length across the lines of force. But for all substances, whether ferromagnetic or diamagnetic, possessing so little capacity for induction as any of the known diamagnetics, this tendency, depending as it does on the mutual action of the parts of the elongated mass, is, and probably will always remain, utterly imperceptible in experiment. All directional tendencies in bars of diamagnetic substance which have yet been, and probably all which can ever be discovered by experiment, are due either to some magne-crystallic property of their substances, or to the tendency of their ends or other moveable parts, *from places of stronger towards places of weaker force*, in varied magnetic fields, or to these two causes combined, and in no respect to the inductive effects of the mutual influence of their parts. To consider the effects of a want of uniformity of the force, in a varied field, on the equilibrium of a ferromagnetic bar, the author quoted Faraday's admirable statement of the law regarding the tendency of a ball or cube of diamagnetic substance*, and referred to former papers [Arts. XXXIII. and XXXIV. above (§§ 634—668)], in which he had proved that, when applied to non-crystalline substances generally, with the proper modifica-

---

* [See Faraday's "Memoir on the Magnetic Condition of all Matter," *Transactions of the Royal Society*, 1846, page 21; or, *Philosophical Magazine*, Vol. XXVIII. 1846.]

tion for the case of ferromagnetics, it expresses with admirable simplicity the result of a mathematical investigation involving some of the most remarkable principles in the theory of attraction. From this it was shown, that if we conceive a ferromagnetic mass to be divided into very small cubes, each of these parts would, of itself, tend towards places of stronger force, and therefore that the bearing of the whole mass in a varied field will be produced partly by this tendency and partly by the tendency depending on the mutual inductive influence which alone exists when the field is uniform. The author then proceeded to illustrate these theoretical views by a series of experiments. In some of them a steel bar-magnet was used, and small soft iron wires, fixed in various positions on light wooden arms, were shown to be sometimes urged on the whole from places of stronger to places of weaker force by their tendency to get into positions with their lengths along the lines of force. In others, a ring electro-magnet, consisting of insulated copper wire, rolled fifty times round as closely as possible to the circumference of a circle of about 25 centimetres diameter, fixed in a vertical plane at right angles to the magnetic meridian, was used, and a single cube of soft iron, placed in an excentric position on a long narrow pasteboard tray centrally suspended in the field of force by unspun silk, was attracted into the plane of the ring ; but a row of three or four cubes placed touching one another in a line through the axis of suspension, settled as far from the plane as possible, in virtue of the tendency of an elongated mass to get its length along the lines of force. Two cubes placed in contact are found to be in stable equilibrium in the plane of the ring, or in oblique positions, or as far from the ring as possible, according to the greater or less distances at which they are placed in the tray, from the point of suspension. A number of equal and similar bars of a composition of wax and soft iron filings of different ferromagnetic strengths, suspended successively with their middle points in the centre of the magnet, settled in various positions. Those of them which were of greatest ferromagnetic capacity settled perpendicular to the plane of the ring or along the lines of force ; others, with a smaller proportion of iron filings, had positions of stable equilibrium both *in* the plane of the ring and *perpendicular* to it ; and others, with a still smaller proportion of iron filings, had their sole positions of stable equilibrium in the plane of the ring. The last-mentioned experiments illustrated very curiously the diminished proportion borne by the effects of mutual influence of the parts to those of a non-uniformity in the field of force, in similar bodies of smaller ferromagnetic capacity. [Compare last two sentences of § 670 below.]

XXXVI. *Remarques sur les oscillations d'aiguilles non cristallisées de faible pouvoir inductif paramagnétiques ou diamagnétiques, et sur d'autres phénomènes magnétiques produits par des corps cristallisés ou non cristallisés.*

[From the ' *Comptes Rendus*' of the French Academy, 1854, first half-year.]

" GLASGOW, le 22 mars 1854.

670. " J'ai lu aujourd'hui, dans les *Comptes Rendus* du 25 avril de l'année dernière, un Extrait de trois Mémoires de M. Matteucci relatifs au magnétisme, qui renferment un grand nombre d'observations intéressantes. J'y trouve la remarque que des aiguilles prismatiques de bismuth non cristallisé oscillent entre les pôles d'un aimant dans des temps égaux, lors même que leurs poids sont différents, *quand leurs longueurs sont les mêmes.* J'ai eu la pensée que la proposition serait encore vraie, lors même que cette dernière condition ne serait point remplie, ou du moins en y substituant cette autre condition moins absolue : *les longueurs des différentes aiguilles ne doivent point dépasser une petite fraction de la distance comprise entre les deux pôles de l'aimant.*

"Il me suffit, pour prouver cette proposition, de remonter à la raison donnée dès l'origine par M. Faraday de l'action éprouvée par une aiguille de bismuth non cristallisé placée entre les deux pôles d'un aimant : savoir que cette action est la résultante des tendances qu'éprouvent toutes les particules de l'aiguille à se transporter des *points où la force magnétique est la plus intense vers ceux où elle est la plus faible;* j'applique ici la théorie mathématique, présentée pour la première fois dans le *Journal de Mathématiques de Cambridge et de Dublin**.

---

* Des forces qui agissent sur de petites sphères soumises à des influences magnétiques ; aperçu de quelques phénomènes présentés par les substances diamagnétiques.

Cambridge and Dublin Mathematical Journal; mai 1847 [§§ 634...646 above].

Voyez aussi un article du *Philosophical Magazine,* octobre 1850, intitulé : "Remarques sur les forces qui agissent sur les substances ferromagnétiques

"Il est en effet démontré, dans cette investigation mathématique, qu'en désignant par μ un coefficient exprimant le pouvoir inductif de la substance (ce coefficient, positif pour les substances ferromagnétiques ou paramagnétiques, et négatif pour les substances diamagnétiques, exprime parfaitement la différence de propriétés, découverte par M. Faraday, et qui a servi de base à la division de tous les corps en deux classes, corps paramagnétiques et corps diamagnétiques); par σ le volume d'une particule du corps; par R la résultante des forces magnétiques qui s'exercent au point (x, y, z) du champ magnétique dans lequel il est placé, c'est-à-dire la force qui agirait sur un pôle magnétique égal à l'unité, ou sur l'unité de *magnétisme boréal*, ou de *matière magnétique imaginaire*, ou de *fluide magnétique* qui se trouverait en ce point. La force à laquelle sera effectivement soumise cette particule magnétisée par induction sera la résultante des trois forces X, Y, Z données par les trois équations [§ 639 (5) above]

$$X = \tfrac{1}{2}\mu\sigma\frac{d(R^2)}{dx}, \qquad Y = \tfrac{1}{2}\mu\sigma\frac{d(R^2)}{dy}, \qquad Z = \tfrac{1}{2}\mu\sigma\frac{d(R^2)}{dz}.$$

"Supposons que l'origine O des coordonnées soit placée au centre de la ligne qui joint les deux pôles de l'aimant, et que l'axe des coordonnées $X'OX$ coïncide avec cet *axe du champ magnétique*: la valeur de R^2 sera un minimum au point O relativement aux divers points de la ligne $X'OX$, et un maximum relativement aux points d'un plan équatorial qui lui serait perpendiculaire. On a, d'après cela, pour des points placés à une distance infiniment petite du point O,

$$R^2 = R_0^2 + Ax^2 - By^2 - Cz^2;$$

R_0 représente la valeur de R au point O, et A, B, C sont trois quantités positives.

"Supposons maintenant qu'un petit corps (de volume σ, de masse m, de pouvoir inductif μ) soit fixé à l'extrémité d'un bras rectiligne infiniment léger OM (de longueur a), qui puisse se mouvoir librement et uniquement autour de l'axe OZ, c'est-à-dire dans le plan YOX, et constitue ainsi ce qu'on nomme *un pendule magnétique simple;* l'équation de son mouvement sera

$$m\frac{d^2(a\theta)}{dt^2} = Y\cos\theta - X\sin\theta,$$

ou diamagnétiques non cristallisées magnétisées par induction" [§§ 647...668 above].

θ représentant l'angle MOX. Les expressions précédentes nous donnent
$$X = \mu\sigma Ax \text{ et } Y = -\mu\sigma By,$$
et comme on a géométriquement
$$x = a \cos\theta, \ y = a \sin\theta,$$
l'équation du mouvement devient
$$\frac{d^2\theta}{dt^2} = -\frac{\mu\sigma}{m}(A+B)\sin\theta\cos\theta.$$

Comme l'équation est indépendante de a, nous en concluons que : *le mouvement angulaire est indépendant du rayon du cercle dans lequel il s'effectue,* ou que *les oscillations de différents pendules* (définis comme nous l'avons fait) *autour du centre du champ magnétique sont isochrones, bien que leurs longueurs soient différentes.*

"La demi-période d'une oscillation infiniment petite est
$$\pi\sqrt{\frac{m}{\mu\sigma(A+B)}},$$
ou, si ρ représente la densité du corps,
$$\pi\sqrt{\frac{\rho}{\mu(A+B)}}.$$

(Il est évident que les oscillations d'un pendule magnétique infiniment petit autour d'un point qui ne possède aucune propriété de maximum ou de minimum magnétique, se feront dans des temps proportionnels aux racines carrées des longueurs, et suivront ainsi les mêmes lois que le pendule ordinaire, simple ou composé.)

"Ces conclusions sont applicables aux oscillations d'un petit corps d'une nature quelconque non cristallisé. Si μ est positif, c'est-à-dire si le corps est *paramagnétique,* les positions d'équilibre stable correspondront à $\theta = 0$ ou $\theta = \pi$, c'est-à-dire se trouveront sur l'axe. Si au contraire, μ est négatif, c'est-à-dire si la matière est diamagnétique, les positions d'équilibre stable répondront à $\theta = \frac{1}{2}\pi$ et $\theta = \frac{3}{2}\pi$, et se trouveront dans le plan perpendiculaire à l'axe, dans le plan *équatorial* du champ magnétique.

"Si l'on assemble une série de particules le long de la ligne OM, et si le pouvoir inductif, paramagnétique ou diamagnétique, est assez faible pour qu'elles n'exercent point une influence sensible les unes sur les autres, chacune d'elles sera

influencée comme si elle était isolée. Mais il a été démontré
que si elles sont formées de la même substance, leur mouvement angulaire sera le même si on les dérange de leur position
d'équilibre de la même quantité angulaire, et qu'elles ne soient
pas unies l'une à l'autre par un lien rigide. Nous en concluons
que les oscillations d'une aiguille (c'est-à-dire d'une barre dont
la longueur est un multiple très-élevé des dimensions latérales)
d'une substance paramagnétique ou diamagnétique non cristallisée, autour d'un point fixe placé au centre du champ magnétique,
sont indépendantes de sa masse et de sa longueur, et que la

demi-période d'une petite oscillation est égale à $\pi \sqrt{\dfrac{\rho}{\mu(A+B)}}$.

"Il est clair que les oscillations d'une barre cristallisée
ou non, seront indépendantes des dimensions latérales, pourvu
que celles-ci soient très faibles comparativement à sa longueur,
et qu'il n'y ait point d'influence inductive sensible exercée entre
ses diverses parties; et, par conséquent, que diverses aiguilles
prismatiques de la même longueur (même si cette longueur est
assez grande pour que les considérations précédentes soient
inapplicables), et d'une substance semblable et disposée semblablement, soit qu'elle soit ou non cristallisée, oscilleront dans le
même temps, quel que soit leur poids. Ce n'est qu'à des différences dans l'arrangement cristallin semblables à celles sur
lesquelles M. Matteucci a porté l'attention, et non pas à des
différences de poids, qu'il faut attribuer les variations qu'il a
observées dans les périodes d'oscillations de diverses aiguilles
cristallisées de même longueur.

"Les limites de la longueur d'une aiguille non cristalline
oscillant autour du centre d'un champ magnétique en deçà desquelles on peut appliquer les résultats précédents avec une
suffisante approximation, dépendent des dimensions et de la
forme de l'aimant, et en particulier de la disposition de ses
pôles. On peut observer qu'une aiguille paramagnétique d'une
trop grande longueur oscillera certainement plus rapidement
que la theorie ne l'indique, et qu'une aiguille diamagnétique
oscillera probablement d'autant plus lentement que sa longueur
sera plus grande, si sa longueur est telle que les équations précédentes ne puissent représenter ses mouvements avec une
rigueur suffisante.

"La détermination des mouvements de barres cristallines ou de masses d'une forme quelconque, dans les circonstances indiquées par M. Matteucci, peut s'effectuer sans difficulté en appliquant la théorie de l'induction magnétique dans les corps cristallins, dont les développements mathématiques ont été soumis, en 1850, à l'Association britannique à Édimbourg, et qui a été publiée depuis dans le *Philosophical Magazine.* On trouvera dans ce Mémoire*, et dans ceux que j'ai cités plus haut, la preuve que les phénomènes de direction que présente le bismuth cristallisé placé entre les pôles d'un aimant, et observés par M. Matteucci, trouvent leur parfaite explication dans la tendance que possèdent les molécules à se porter des points ou l'intensité magnétique est la plus grande vers ceux où elle est la plus faible ; combinée avec la *tendance directrice* qui dépend de ce dernier élément, et qui, ainsi que l'indique la théorie, résulte d'une inégalité du pouvoir inductif dans les diverses directions d'un cristal.

"J'ai lieu d'espérer que les raisonnements et les développements contenus dans ces Mémoires paraîtront suffisants pour m'autoriser à exprimer une opinion contraire à celle que M. Matteucci a avancée relativement aux phénomènes remarquables qu'il a observés.

"Puisque j'ai occasion de parler du passage (*Comptes Rendus*, t. XXXVI. p. 743) où M. Matteucci attribue à M. Tyndall la découverte d'une inégalité dans la répulsion diamagnétique présentée par les cristaux, suivant la position de l'axe du cristal, je crois nécessaire de faire remarquer que cette importante découverte est due à M. Faraday. M. Tyndall en rendant compte de ses recherches sur ce sujet (*Philosophical Magazine*, septembre 1851), cite les travaux antérieurs de M. Faraday (*Royal Society*, novembre 1850). Dans le paragraphe 2839 de ce Mémoire, M. Faraday énonce cette loi comme une conjecture en l'année 1848 (§ 2588) ; mais, faute d'expériences suffisantes, il ne s'y appesantit point : il revient sur ce sujet, à propos du bismuth cristallisé, dans le paragraphe 2839 de ce Mémoire, et réussit ensuite à vérifier ses prévisions par l'expérience (§ 2841). Plus tard, au sujet du spath calcaire

---

* Sur la théorie de l'induction magnétique dans les substances cristallisées et non cristallisées. *Philosophical Magazine;* mars 1851 [§§ 647...668 above].

(§ 2842), il dit notamment que *si l'axe optique est d'abord placé parallèlement à l'axe magnétique, puis perpendiculairement à cet axe, le corps sera plus diamagnétique dans la première position que dans la seconde,* et indique les défauts de sa disposition par suite desquels il ne peut vérifier cette proposition. M. Tyndall, en disposant l'expérience avec plus de précautions, réussit à en donner la démonstration expérimentale. Dans la communication à l'Association britannique que j'ai citée plus haut, j'ai fait remarquer moi-même, dès le mois d'août 1850, qu'il *doit* exister des différences dans les pouvoirs inductifs des corps cristallins suivant les diverses directions, et que c'était là la *seule explication possible* des phénomènes de direction cristallomagnétique découverts par Plücker et Faraday, et dans cette occasion je donnai les résultats particuliers au bismuth et au spath calcaire que l'expérience a confirmés depuis. C'est Poisson, le premier, qui a prévu les phénomènes cristallomagnétiques, dus à une différence dans les pouvoirs inductifs dans les différentes directions d'un corps cristallisé; mais il ne chercha point à vérifier la théorie qu'il émit alors, parce qu'il ne connaissait point de corps auxquels elle pût être applicable. Les expériences actuelles de M. Plücker et de M. Faraday ont été suggérées par le Mémoire que lut Poisson, à l'Académie, le 2 février 1842.

"Quand le pouvoir inductif des substances est tel, que les diverses parties exercent une action magnétique mutuelle les unes sur les autres, on ne peut plus supposer, comme nous l'avons fait, que l'aimant agit sur chaque particule comme si elle était isolée. Le fer doux offre l'exemple d'une substance pareille (le coefficient μ n'est, pour ce corps, qu'un peu inférieur à $\dfrac{3}{4\pi}$); cette influence mutuelle est ici la cause de phénomènes très-remarquables, surtout quand on fait les observations sur des masses allongées. La Note ci-après se rapporte à cette partie du sujet et aux expériences dont elle a été l'objet. J'ajouterai ici la description d'une expérience analogue à celle que fit M. Matteucci avec des cubes de bismuth cristallisé, fixés au bout d'une aiguille de sulfate de chaux dont les clivages plans étaient perpendiculaires à la longueur: dans la position d'équilibre stable, ces cubes étaient *aussi rapprochés que possible* des pôles de l'aimant. Fixez deux fines aiguilles de fer doux aux deux bouts d'une tige droite en bois (ou toute autre substance non sensiblement magnétique) et perpendiculairement à cette tige, suspendue par un fil au centre du champ magnétique, entre les deux pôles, et équilibrée de manière à ce que le plan des aiguilles de fer soit horizontal. Si la tige en bois n'est pas trop longue, elle se placera *perpendiculairement* à la ligne des pôles, c'est-à-dire que les aiguilles de fer doux, pour être en équilibre stable, devront être *aussi loin que possible* des pôles de l'aimant. Cette expérience peut être faite avec facilité, au moyen d'un simple aimant d'acier en fer à cheval. Le résultat observé est dû à la tendance qu'a chacune des deux aiguilles de fer doux, en vertu des actions mutuelles de ses différentes parties, à se placer parallèlement à la direction des forces. Le résultat de M. Matteucci doit être attribué à la tendance que possède chaque cube de bismuth, en vertu de sa structure cristalline, à placer son plan de clivage perpendiculairement à la direction de la force."

Note.—*De l'équilibre des masses allongées de substances ferromagnétiques dans des champs de force magnétique constante et variable.*

Le fait, découvert d'abord expérimentalement par Gilbert, qu'une barre de fer doux, fixée à son centre de gravité dans un champ magnétique uniforme, se place parallèlement à la direction des forces, n'est pas suffisamment expliqué quand on l'attribue uniquement à la vertu inductive que possède le fer doux de se transformer momentanément en un aimant semblable à un aimant permanent dans sa position d'équilibre stable. Car la même explication devrait s'appliquer à une rangée de sphères de fer doux assemblées à l'aide de joints non magnétiques; cependant un tel assemblage ne présenterait point de phénomène de direction (puisqu'aucune des sphères partielles ne recevrait l'action d'une force ou d'un couple résultant magnétiques) à moins que les sphères n'agissent les unes sur les autres, et qu'il ne se produise ainsi des changements dans leur état magnétique. Il faut donc admettre qu'il s'opère des actions mutuelles dans les différentes parties d'une rangée de sphères ou de cubes, ou simplement dans une barre, si l'on veut arriver à la vraie théorie des phénomènes de direction.

L'auteur de cette communication, à l'aide de raisonnements de mécanique élémentaire fondés sur les principes les mieux établis de l'induction magnétique et de l'action magnétique en général, fait voir qu'une masse allongée, ferromagnétique ou diamagnétique, placée dans un champ magnétique uniforme, tend à se placer parallèlement à la direction des forces, pourvu qu'elle ne soit point cristallisée : en effet, quand elle est ferromagnétique, elle est moins facilement magnétisée, quand on la place dans la position ci-dessus, que dans la position perpendiculaire; le contraire a lieu quand elle est diamagnétique.

Mais pour toutes les substances, des deux classes, qui possèdent un aussi faible pouvoir inductif que certains corps diamagnétiques connus, cette tendance qui résulte d'actions mutuelles intérieures ne peut être vérifiée par l'éxperience. Toutes les tendances directrices des barres diamagnétiques qui ont été jusqu'ici, et sans doute toutes celles qui seront encore découvertes par expérience, sont dues soit à quelque propriété cristallomagnétique, soit à la tendance des extrémités ou des autres portions mobiles à changer de place, de manière à ce que les molécules occupent les positions d'intensité magnétique minimum, ou à ces deux causes réunies, plutôt qu'aux effets inductifs mutuels. En étudiant les effets d'une force magnétique variable sur les positions d'équilibre d'une barre ferromagnétique, l'auteur cite l'admirable explication donnée par Faraday, de la loi relative aux tendances directrices d'une sphère ou d'un cube diamagnétiques, et rappelle que précédemment il a fait voir, qu'appliquée aux substances non cristallisées en général, avec les modifications convenables dans le cas où elles sont ferromagnétiques, cette loi exprime avec une admirable simplicité les résultats d'un travail mathématique comprenant quelques-uns des principes les plus remarquables d'une théorie de l'attraction.

D'après cette loi, on voit qu'en supposant une masse ferromagnétique divisée en cubes très-petits, chacune de ces parties tendrait d'elle-même vers la position d'intensité maximum, et qu'ainsi la position de la masse entière, dans le cas d'une force magnétique variable, serait due en partie à cette tendance et en partie aux actions intérieures mutuelles qui agissent seules, quand la force est constante. L'auteur a cherché à vérifier, par l'expérience, ces vues théoriques. Il a employé un barreau d'acier formant aimant et des fils minces de fer doux, fixés dans diverses positions sur une tige en bois; la tige en bois se plaçait de façon que les fils de fer ayant leur direction parallèle à celle de la force, les molécules fussent dans les positions d'intensité minimum. Dans une autre expérience, un anneau électromagnétique, formé de fils de cuivre isolés, roulés cinquante fois autour d'un cercle d'un diamètre égal à 25 centimètres, était fixé dans un plan vertical perpendiculairement au méridien magnétique; un simple cube de fer doux, placé excentriquement sur un plateau de carton mince suspendu à son centre par un fil de soie naturel dans le plan de la force, était attiré dans le plan de l'anneau; mais une suite de trois à quatre cubes placés au contact à la suite les uns des autres en ligne droite le long de l'axe de suspension, se plaçait aussi loin du plan que possible en vertu de la tendance d'une masse allongée, à placer sa plus grande dimension parallèlement à la direction de la force. Deux cubes placés au contact étaient en équilibre stable dans le plan de l'anneau ou dans une position oblique, ou aussi loin que possible de l'anneau, suivant la distance variable à laquelle on les plaçait sur le plateau au point de suspension. Des barres égales et semblables, formées par un mélange de cire et de limaille de fer doux et de puissances diamagnétiques différentes, suspendues successivement par leur point milieu, se fixaient dans des positions diverses: celles qui possédaient le plus grand pouvoir ferromagnétique se plaçaient perpendiculairement au plan de l'anneau ou dans la direction des forces ; les autres, celles qui contenaient moins de fer, avaient leur position d'équilibre à la fois dans le plan de l'anneau et perpendiculairement à ce plan ; et celles qui en contenaient encore moins, étaient en équilibre uniquement dans le plan de l'anneau.

Ces dernières expériences font voir d'une façon très-remarquable la part qu'il faut faire, dans cet ordre de phénomènes, aux actions mutuelles intérieures, et en même temps à la variation de la force [compare original, being last sentence of § 669]. Des mélanges de sable et de limaille de fer doux, placés dans des tubes de verre, feraient le même effet que les barreaux dont nous venons de parler et vaudraient peut-être mieux dans certains cas.

XXXVII. *Elementary Demonstrations of Propositions in the Theory of Magnetic Force.*

[From the *Philosophical Magazine*, April 1855.]

671. *Def.* 1. The lines of force due to any magnet or electro-magnet, or combination of magnets of any kind, are the lines that would be traced by placing the centre of gravity of a very small steel needle, perfectly free to turn about this point, in any position in their neighbourhood, and then carrying it always in the direction pointed by the magnetic axis of the needle.

Remark. Except in the case of symmetrical magnets, the lines of force will generally be lines of double curvature.

Def. 2. The lines of component force in any plane are the lines traced by placing the centre of a steel needle anywhere in this plane, and carrying it always in this plane in the nearest direction to that pointed by its magnetic axis; that is, the direction of the orthogonal projection of the magnetic axis on the plane; or the direction that the steel needle would point with its magnetic axis if placed with it in the plane, and left free to turn about an axis through its centre of gravity perpendicular to the plane.

672. *Prop.* I. If the line of component magnetic force through any point in a plane be curved at this point, the force will vary in a line perpendicular to the line of force in its plane, *increasing* in the direction *towards* the centre of curvature.

Let *EABF* (Fig. 1) be a line of component force in the plane

34—2

of the diagram, and let $GCDH$ be another near it, each and all between them being curved in the same direction, the arrow head on each indicating the way a north pole would be urged. Let AC, BD be lines drawn perpendicular to all the lines of component force between these two. Because of the curvature of these lines, the lines AC and BD (whether straight or curved) must be so inclined to one another that the portion CD cut off from the last shall be less than the portion AB cut off from the first. Let a north pole of an infinitely thin, uniformly and longitudinally magnetized bar, of which the south pole is at a great distance from the magnets, be carried from D to C along the line of component force through these points, from C to A perpendicular to all the lines of force traversed, from A to B again along a line of force, and lastly, from B to D perpendicular to the lines of force. Work must be spent on it in carrying it from D to C, and work is gained in passing it from A to B. Then, because no work is either gained or spent in carrying it from C to A or from B to D, the work gained in moving along AB cannot exceed the work spent in the first part of the motion, or else we should have [compare § 622 above] a perpetual development of energy from no source*, by simply letting the cycle of motion be repeated over and over again : and the

---

* [Note added *March* 26, 1855.]—It might be objected, that perhaps the magnet, in the motion carried on as described, would absorb heat, and convert it into mechanical effect, and therefore that there would be no absurdity in admitting the hypothesis of a continued development of energy. This objection, which has occurred to me since the present paper was written, is perfectly valid against the reason assigned in the text for rejecting that hypothesis; but the second law of the dynamical theory of heat (the principle discovered by Carnot, and introduced by Clausius and myself into the dynamical theory, of which, after Joule's law, it completes the foundation) shows the true reason for rejecting it, and establishes the validity of the remainder of the reasoning in the text. In fact, the only absurdity that would be involved in admitting the hypothesis that there is either more or less work spent in one part of the motion than lost in the other, would be the supposition that a thermo-dynamic engine could absorb heat from matter in its neighbourhood, and either convert it wholly into mechanical effect, or convert a part into mechanical effect and emit the remainder into a body at a higher temperature than that from which the supply is drawn. The investigation of a new branch of thermo-dynamics, which I intend shortly to communicate to the Royal Society of Edinburgh, shows that the magnet (if of magnetized steel) does really experience a cooling effect when its pole is carried from A to B, and would experience a heating effect if carried in the reverse direction. [See Art. XLVIII., part VII. of my "Reprint of Mathematical and Physical Papers," (Vol. I., page 291)]. But the same investigation also shows that the magnet must absorb just as much heat to keep up its temperature during the motion of its pole *with* the force along AB, as it must emit to keep from rising in temperature when its pole is carried *against* the force, along DC.

work spent along DC cannot exceed that gained from A to B, or
else we might have a perpetual development of energy from no

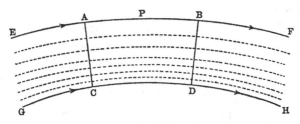

FIG. 1.

source, merely by reversing the motion described, and so repeat-
ing. The work spent and gained in the motions along DC and
AB respectively must therefore be exactly equal. Hence the
mean intensity of the force along CD, which is the shorter of the
two paths, must exceed the mean intensity of the force along
the other; and therefore the intensity of the force increases
from P in the perpendicular direction towards which the
concavity of the line through it is turned.

673. *Prop.* II. The augmentation of the component force in
any plane at an infinitely small distance from any point, towards
the centre of curvature of the line of the component force
through it, bears to the whole intensity at this point the ratio
of the infinitely small distance considered, to the radius of
curvature.

If, in the diagram for the preceding proposition, we suppose
AB and CD to be infinitely near one another, and each infinitely short, they will be infinitely nearly arcs of circles with
infinitely nearly equal radii. Hence the difference of their
lengths must bear to either of them the ratio of the distance
between them to the radius of curvature. But the mean
intensities along these lines must, according to the preceding
demonstration, be inversely as their lengths, and hence the
excess of the mean intensity in CD above the mean intensity
in AB must bear to the latter the ratio of the excess of the
length of AB above that of CD to the latter length; that is, as
has been shown, the ratio of the distance between AB and CD
to the radius of curvature.

674. *Prop.* III. The total intensity does not vary from any

point in a magnetic field to a point infinitely near it in a direction perpendicular to the plane of curvature of the line of force through it.

675.　*Prop.* IV.　The total intensity increases from any point to a point infinitely near it in a direction towards the centre of curvature of the line of force through it, by an amount which bears to the total intensity itself, the ratio of the distance between these two points to the radius of curvature.

These two propositions follow from the two that precede them by obvious geometrical considerations.

They are equivalent to asserting, that if X, Y, Z denote the components, parallel to fixed rectangular axes, of the force at any point whose co-ordinates are (x, y, z), the expression $Xdx + Ydy + Zdz$ must be the differential of a function of three independent variables.

Examination of the Action experienced by an infinitely thin uniformly and longitudinally Magnetized Bar, placed in a non-uniform Field of Force, with its length direct along a Line of Force.

676.　Let SN be the magnetized bar, and ST, NT' straight lines touching the line of force in which, by hypothesis, its extremities lie, and P a point on it, midway between them.　The resultant force on the bar will be the resultant of two forces pulling its ends in the lines ST, NT'.　If these two forces were equal (as they would be if the intensity of the field did not vary at all along a line of force, as for instance when the lines of force are concentric circles, as they are when simply due to a current of electricity passing along a straight conductor; or if P were in a situation between two dissimilar poles symmetrically placed on each side of it), the resultant force would clearly bisect the angle between the lines TS, $T'N$, and would therefore be perpendicular to the bar and to the lines of force in the direction towards which they are curved; that is (Prop. IV.), would be from places of weaker to places of stronger force, perpendicularly across the lines of force.　On the other hand, if the line of force through P has no curvature at this point, or no sensible curvature as far from it as N and S, the

lines NT and ST' will be in the same straight line, and the
resultant force on the bar will be simply the excess of the
force on one end above that on the other acting in the direction of the greater; and since in this case (Prop. IV.) there is

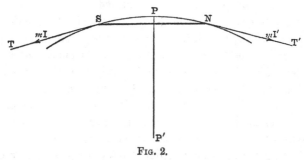

Fig. 2.

no variation of the intensity of the force in the field in a
direction perpendicular to the lines of force, the resultant force
experienced by the bar is still simply in the direction in which
the intensity of the field increases, this being now a direction coincident with a line of force.   Lastly, if the intensity
increases most rapidly in an oblique direction in the field, from
P in some direction between PS and PP', there must clearly
be an augmentation (a "component" augmentation) from P
towards P'; and therefore (Prop. IV.) the line through P must
be curved, with its concavity towards P', and also a " component" augmentation from N towards S, and therefore the
end S must experience a greater force than the end N.   It
follows that the magnet will experience a resultant force along
some line in the angle SNP', that is, on the whole from places
of weaker towards places of stronger force, obliquely across the
lines of force.

677. *Prop.* V. (*Mechanical Lemma.*)—Two forces infinitely
nearly equal to one another, acting tangentially in opposed directions on the extremities of an infinitely small chord of a circle,
are equivalent to two forces respectively along the chord and
perpendicular to it through its point of bisection, of which the
former is equal to the difference between the two given forces
and acts on the side of the greater; and the latter, acting
towards the centre of the circle, bears to either of the given
forces the ratio of the length of the arc to the radius.

The truth of this proposition is so obvious a consequence of "the parallelogram of forces," that it is not necessary to give a formal demonstration of it here.

678.   *Prop.* VI. A very short, infinitely thin, uniformly and longitudinally magnetized needle, placed with its two ends in one line of force in any part of a magnetic field, experiences a force which is the resultant of a longitudinal force equal to the difference of the forces experienced by its ends, and another force perpendicular to it through its middle point equal to the difference between the force actually experienced by either end, and that which it would experience if removed, in the plane of curvature of the line of force, to a distance equal to the length of the needle, on one side or the other of its given position.

NS being the bar as before, let I denote the intensity of the force in the field at the point occupied by N, I' the intensity at S, J the intensity at P on the line of force midway between S and N, and J' the intensity

at a point P', at a distance PP' equal to the length of the bar, in a direction perpendicular to the line of force. Then if m denote the strength of magnetism of the bar, mI and mI' will be the forces on its two

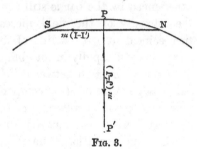

FIG. 3.

extremities respectively. Hence by the mechanical lemma, the resultant of these forces will be the same as the resultant of a force $m(I - I')$ acting along the bar in the direction SN, and a force perpendicular to it towards the centre of curvature, bearing the same ratio to either mI or mI', or to mJ (which is their mean, and is infinitely nearly equal to each of them), as NS to the radius of curvature, or (by Prop. II.) the ratio of the excess of the intensity at P' above that at P to the intensity at either, that is the ratio of $J' - J$ to J, and therefore itself equal to $m(J' - J)$. The bar therefore experiences a force the same as the resultant of $m(I - I')$ acting along it from S towards N, and $m(J' - J)$ perpendicularly across it towards P', through its middle point.

679.   *Cor.* The direction of the resultant force on the bar is

that in which the total intensity of the field increases most
rapidly; or, which is the same, it is perpendicular to the surface of no variation of the total intensity.

Prop. VII. The resultant force on an infinitely small magnet
of any kind placed in a magnetic field, with its magnetic axis
along the lines of force, is in the line of most rapid variation
of the total intensity of the field, and is equal to the magnetic
moment of the magnet multiplied by the rate of variation of
the total intensity per unit of distance; being in the direction
in which the force *increases* when the magnetic axis is "direct,"
(that is, in the position it would rest in if the magnet were free
to turn about its centre of gravity).

Cor. 1. The resultant force experienced by the magnet will
be in the contrary direction, that is, the direction in which the
total intensity of the field diminishes most rapidly, when it is
held with its magnetic axis reverse along the lines of force of
the field.

680. *Cor.* 2. A ball of soft iron, or of any non-crystalline
paramagnetic substance, held anyhow in a non-uniform magnetic
field, or a ball or small fragment of any shape, of any kind of
paramagnetic substance whether crystalline or not, left free to
turn about its centre of gravity, will experience a resultant force
in the direction in which the total intensity of the field increases
most rapidly, and in magnitude equal to the magnetic moment
of the magnetization induced in the mass multiplied by the
rate of variation of the total intensity per unit distance in the
line of greatest variation in the field. For such a body in such
a position is known to be a magnet by induction, with its
magnetic axis direct along the lines of force.

681. *Cor.* 3. A ball of non-crystalline diamagnetic substance
held anyhow in a magnetic field, or a small bar or fragment of
any shape of any kind of diamagnetic substance, crystalline or
non-crystalline, held by its centre of gravity, but left free to
turn about this point, experiences the same resultant force as a
small steel or other permanent magnet substituted for it, and
held with its magnetic axis reverse along the lines of force. For
Faraday has discovered, that a large class of natural substances
in the stated conditions experience no other action than *a*

tendency from places of stronger towards places of weaker force,
quite irrespective of the directions the lines of force may have,
and he has called such substances diamagnetics.

682. *Cor.* 4. A diamagnetic, held by its centre of gravity but
free to turn about this point, must react upon other magnets
with the same forces as a steel or other magnet substituted in
its place, and held with its magnetic axis reverse along the
lines of force due to all the magnets in its neighbourhood.

683. *Cor.* 5. Any one of a row of balls or cubes of diamagnetic substance held in a magnetic field with the line joining
their centres along a line of force, is in a locality of less intense
force than it would be if the others were removed; but any one
ball or cube of the row, if held with the line joining their centres
perpendicularly across the line of force, is in a locality of more
intense force than it would be if the others were removed.

684. *Cor.* 6. When a row of balls or cubes, or a bar, of perfectly non-crystalline diamagnetic substance, is held obliquely
across the lines of force in a magnetic field, the magnetic axis of
each ball or cube, or of every small part of the substance, is nearly
in the direction of the lines of force, but slightly inclined from
this direction towards the direction perpendicular to the length
of the row or bar. Hence, since the magnetic axis of every
part differs only a little from being exactly *reverse along the*
lines of force, the direction of the resultant of the couples with
which the magnets, to which the field is due, act on the parts
of the row or bar must be such as to turn its length along the
lines of force.

685. *Cor.* 7. The positions of equilibrium of a row of balls or
cubes rigidly connected, or of a bar of perfectly non-crystalline
diamagnetic substance, free to move about its centre of gravity
in a perfectly uniform field of force, are either with the length
along or with the length perpendicularly across the lines of
force: positions with the length along the lines of force are
stable; positions with the length perpendicularly across the
lines of force are unstable.

686. *Cor.* 8. The mutual influence and its effects, referred to
in Cors. 5, 6, 7, is so excessively minute, that it cannot possibly
have been sensibly concerned in any phænomena that have yet

been observed; and it is probable that it may always remain
insensible, even to experiments especially directed to test it.
For the influence of the most powerful electro-magnets induces
the peculiar magnetic condition of which diamagnetics are
capable, to so slight a degree as to give rise to only very feeble,
scarcely sensible, mutual force between the diamagnetic and the
magnet; and therefore the magnetizing influence of a neighbouring diamagnetic, which could scarcely, if at all, be observed
on a piece of soft iron, must be insensibly small on another
diamagnetic.

687.   *Cor.* 9.   All phænomena of motion that have been observed as produced in a diamagnetic body of any form or substance by the action of fixed magnets or electro-magnets, are
due to the resultant of forces urging all parts of it, and couples
tending to turn them; the force and couple acting on each
small part being sensibly the same as it would be if all the
other parts were removed.

688.   *Cor.* 10.   The deflecting power (observed and measured
by Weber) with which a bar of non-crystalline bismuth, placed
vertically as core in a cylinder electro-magnet (a helix conveying
an electric current), urges a magnetized needle on a level with
either of its ends, is the reaction of a tendency of all parts of the
bar itself from places of stronger towards places of weaker force
in its actual field.

The preceding investigation, leading to Props. VI. and VII.,
is the same (only expressed in non-analytical language) as one
which was first published in the *Cambridge and Dublin Mathematical Journal*, May 1846 [§§ 638—640 above].   The chief
conclusions now drawn from it, with particulars not repeated,
were stated in a paper entitled " Remarks on the Forces experienced by inductively magnetized Ferromagnetic or Diamagnetic
Substances," in the *Philosophical Magazine* for October 1850
[Article XXXIV. above].

GLASGOW COLLEGE, *March* 15, 1855.

XXXVIII. Correspondence with Professor Tyndall.

Letter to Professor Tyndall on the "Magnetic Medium," and on the Effects of Compression.

[From the *Philosophical Magazine*, April 1855.]

[*Editorial.*]—The following letter was received a few days ago. It was not written for publication, but the subject to which it refers being of general interest at present, I ventured to suggest to Professor Thomson the desirableness of having the letter printed. This he at once agreed to. With the exception of a paragraph relating to matters of a purely private nature, the letter appears as I received it.

JOHN TYNDALL.

March 24, 1855.

---

2 COLLEGE, GLASGOW, *March* 12, 1855.

689. MY DEAR SIR,—Allow me to thank you for the abstract of your letter on magnetism, and the copy of your letter to Mr Faraday, which I have recently received from you, and have read with much interest. I am still strongly disposed to believe in the magnetic character of the medium occupying space, and I am not sure but that your last argument in favour of the reverse bodily polarity of diamagnetics may be turned to support the theory of universally direct polarity. There is no doubt but that the medium occupying interplanetary space, and the best approximations to vacuum which we can make, have perfectly decided mechanical qualities, and among others, that of being able to transmit mechanical energy in enormous quantities (a platinum wire, for instance, kept incandescent by a galvanic current in the receiver of an air-pump, emits to the glass and external bodies the whole mechanical value of the energy of current spent in overcoming its galvanic resistance). Some of these properties differ but little from those of air or oxygen at an ordinary barometric pressure. Why not, then, the magnetic property? (of which we know so little that we have no right to pronounce a negative). Displace the interplanetary medium by oxygen, and you have a slight increase of magnetic polarity in the locality with a drawing in of the lines of force. Displace it with a piece of bismuth or a piece of wood, and a slight decrease of magnetic polarity through the

locality takes place, accompanied by a pushing out of the lines of force. A state of strain by compression may enhance, in the direction of the strain, that quality of the substance by which it lessens the magnetizability of the space from which it displaces air or "ether;" just as a similar state may enhance, in the direction of compression, the augmenting power of a paramagnetic substance.

690. By the bye, a long time ago (rather more than a year after the Edinburgh meeting of the British Association) I repeated with much pleasure some of your compression experiments, and found a piece of fresh bread instantly affected by pressure, so as always to turn the compressed line perpendicular to the lines of force, to whatever form the fragment was reduced. A very slight squeeze between the fingers was quite enough to produce this property, or again to alter it so as to make a new line of compression set equatorially. I repeated it a few days ago with the same results, and got a ball of bismuth, too, to act similarly. I remember formerly finding the bread *attracted* as a whole, instead of being repelled, as I expected from your results. I suppose, however, this must have resulted from some ferruginous impurities, which it may readily have got either in the course of the experiments with it, or in the baking. I mean to try this again*.

691. I do not quite admit the argument you draw from your compression experiments regarding the effect of contiguity of particles, because in fact we know nothing of the actual state of the molecules of a strained solid. You have made out a most interesting fact regarding their magnetic bearings; but experiments are neither wanted, nor can be made, to show any sensible effect whatever of the mutual influence of a row of small pieces of bismuth placed near one another, or touching one another. It is perfectly easy to demonstrate that it *must* be such as to impair the "diamagnetization" of each piece when the line of the row is parallel to the lines of force, and to enhance it when that line is perpendicular to the lines of force, but in each case to so infinitesimally minute a degree, as to be

---

* Prof. Thomson's supposition is correct; pure bread is *repelled* by a magnetic pole. I may remark that I am at present engaged in the further examination of the influence of compression, and have already obtained numerous instructive results.—J. T.

wholly inappreciable to the most refined tests that have ever
been applied. For let the lines of force be parallel to the line
shown in the figure, and act on a steel needle in the manner
there represented. Then, whatever hypothesis be true for

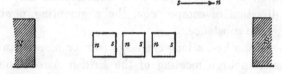

diamagnetism, there is not a doubt but that each piece is acted
on, and consequently reacts, precisely as a piece of steel very
feebly magnetized, with its magnetic axis reverse to that of a
steel needle free to turn, substituted for it, would do. Each
piece of bismuth therefore acts as a little magnet, having its
polarity as marked in the diagram, would do. Hence the
magnetizing force by which the middle fragment is influenced
is less than if the two others were away (this being such a
force as would be produced by a north pole on the left-hand
side of the diagram, and a south pole on the right). It is easily
seen, similarly, that if the line joining the centres be perpendicular to the lines of force, the magnetizing force on the space
occupied by the middle fragment is increased. Corresponding
assertions are true for the terminal fragments, although the
disturbing effect will be less on them in each case than in
the middle one. Hence the dia-
magnetization of each will be en-
feebled in the former case and
enhanced in the latter, by the presence of the others. It follows,
according to the principle of su-

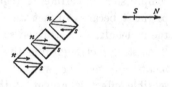

perposition of magnetizations, that if the line of the row be
placed obliquely across the lines of force, the magnetic axis of
each particle, instead of being exactly parallel to the lines of
force, will be a little inclined to them, in the angle between
their direction and the direction transverse to the bar. The
magnets causing the force of the field must act on the little dia-
magnets, each with its axis thus rendered somewhat oblique, so
as to produce on it a statical couple (as shown by the arrow-
heads), and the resultant of the couples thus acting on the frag-

ments will, when all these are placed on a frame, or rigidly connected, tend to turn the whole mass in such a direction as to place the length of the bar along the lines of force. Still, I repeat, this action, although demonstrated with as much certainty as the parallelogram of forces, is so excessively feeble as to be absolutely inappreciable. A fragment of bismuth, of any shape whatever, held in any position whatever in any kind of magnetic field, uniform or varying most intensely, only exhibits the resultant action of couples on all its small parts if crystalline, and of forces acting always according to Faraday's law on them if the field in which it is placed be non-uniform. Some phænomena that have been observed are to be explained by the resultant of forces from places of stronger to places of weaker intensity in the field, others by the resultant of couples depending on crystalline structure, and others by the resultant of such forces and couples co-existing; and none observed depend at all on any other cause.

692. I gave a very brief summary of these views (which I had explained somewhat fully and illustrated by experiments on paramagnetics of sufficient inductive capacity to manifest the effects of mutual influence, at the meeting at Belfast) as an abstract of my communication, for publication in the Report of the Belfast meeting of the British Association, where you may see them [§ 669 above] stated, I hope intelligibly. The experiments on the paramagnetics are very easy, and certainly exhibit some very curious phænomena, illustrative of the resultant effects due to the attractions experienced by the parts in virtue of a variation of the intensity of the field, and to the couples they experience when their axes are diverted from parallelism to the lines of force by mutual influence of the magnetized parts.

693. I had no intention of entering on this long disquisition when I commenced, but merely wished to try and briefly point out, that the assertions I have made regarding mutual influence are demonstrable in every case without special experiment, are confirmed amply by experiment for paramagnetics, and are absolutely incontrovertible, as well as incapable of verification by experiment or observation on diamagnetics.—Believe me, yours very truly,   WILLIAM THOMSON.

PROF. TYNDALL.

*On Reciprocal Molecular Induction : Letter from Professor
Tyndall to Professor W. Thomson, F.R.S.*

[From the *Philosophical Magazine*, December 1855.]

ROYAL INSTITUTION, *Nov.* 26, 1855.

694.　MY DEAR SIR, — The communication from Professor
Weber which appears in the present number of the *Philosophical Magazine*, has reminded me, almost too late, of your own
interesting letter on the same subject published in the April
number of this Journal.　A desire to finish all I have to say
upon this question at present induces me to make the following
remarks, which, had it not been for the circumstance just
alluded to, might have been indefinitely deferred.

With reference to the mutual action of a row of bismuth particles, you say that "it is perfectly easy to demonstrate that
"it *must* be such as to impair the 'diamagnetization' when the
"line of the row is parallel to the lines of force" (the "must,"
you will remember, is put in italics by yourself).　From this
you infer, that in a uniform field of force a bar of bismuth
would set its leng h along the lines of force.　Further on it is
stated that this action is " demonstrated with as much certainty
" as the parallelogram of forces ;" and you conclude your letter
by observing that "the assertions which I [yourself] have made
" are demonstrable in every case without special experiment, ...
" and are absolutely incontrovertible, as well as incapable of
" verification by experiment or observation on diamagnetics."

Most of what I have to say upon this subject condenses
itself into one question.

Supposing a cylinder of bismuth to be placed within a helix,
and surrounded by an electric current of sufficient intensity ;
can you say, *with certainty*, what the action of either end of
that cylinder would be on an external fragment of bismuth
presented to it ?

If you can, I, for my part, shall rejoice to learn the process
by which such certainty is attained : but if you cannot, it will,
I think, be evident to you that the verb " *must* " is logically
" defective."

We *know* that magnetized iron attracts iron : we *know* that

magnetized iron repels bismuth : this, so far as I can see, is
your only experimental ground for *assuming* that magnetized
bismuth repels bismuth, and yet you affirm that an action
deduced from this assumption " is demonstrated with as much
"certainty as the parallelogram of forces." Do I not state the
question fairly ? I can, at all events, answer for my earnest
wish to do so.

It is needless to remind one so well acquainted with the
mental experience of the scientific inquirer, that the very letters
which you attach to your sketch, page 291 [of *Philosophical
Magazine*, § 691 above], may tempt us to an act of abstraction
—a forgetfulness of a possible physical difference between the n
of iron and the n of bismuth—which may lead us very wide of
the truth. The very term " pole " often pledges us to a theoretic
conception without our being conscious of it. You are also well
aware of the danger of shutting the door against experimental
inquiry on an unpromising subject; and when you apparently
do this in your concluding paragraph, I simply accept it as a
strong way of expressing your personal conviction, that the action
referred to is too feeble to be rendered sensible by experiment.
Believe me, dear Sir, most truly yours,   JOHN TYNDALL.

*On the Reciprocal Action of Diamagnetic Particles : Letter from
Professor Thomson to Professor Tyndall.*

[From the *Philosophical Magazine*, January 1856.]

GLASGOW COLLEGE, *Dec.* 24, 1855.

695.   MY DEAR SIR,—I have been prevented until to day, by
a pressure of business, from replying to the letter you addressed
to me in the number of the *Philosophical Magazine* published at
the beginning of this month.

You ask me the question, " Supposing a cylinder of bismuth
"to be placed within a helix, and surrounded by an electric
"current of sufficient intensity; can you say, *with certainty*,
"what the action of either end of that cylinder would be on an
"external fragment of bismuth presented to it ?"

696.   In answer, I say that the fragment of bismuth will be *repelled* from either end of the bar provided the helix be infinitely

long, or long enough to exercise no sensible direct magnetic action in the locality of the bismuth fragment. I can only say this with the same kind of confidence that I can say the different parts of the earth's atmosphere attract one another. The confidence amounts in my own mind to a feeling of *certainty*. In every case in which the forces experienced by a little magnetized steel needle held with its axis reverse along the lines of force, and a fragment of bismuth substituted for it in the same locality of a magnetic field, have been compared, they have been found to agree. In a vast variety of cases, a fragment of bismuth has been found to experience the opposite force to that experienced by a little ball of iron, that is, the same force as a little steel magnet held with its axis reverse to the lines of force; and in no case has a discrepance, or have any indications of a discrepance, from this law been observed. I feel, therefore, in my own mind a certain conviction, that even when the action is so feeble that no force can be discovered at all on the bismuth by experimental tests, such in regard to sensibility as have been hitherto applied, the bismuth is really acted on by the same force as that which a little reverse magnet, if only feeble enough, would experience when substituted in its place. Now there is no doubt of the nature of the force experienced by the steel magnet, or by a little ball of soft iron, in the locality in which you put the fragment of bismuth. One end of a magnetized needle will be attracted, and the other end repelled by the neighbouring end of the bismuth bar; and the attraction or the repulsion will preponderate according as the attracted or the repelled part is nearer. There is then certainly repulsion when the steel magnet is held in the reverse direction to that in which it would settle if balanced on its centre of gravity. In every case in which any magnetic force at all can be observed on a fragment of bismuth, it is such as the steel magnet thus held experiences. Therefore I say it is in this case repulsion. But it will be as much smaller in proportion to the force experienced by the steel magnet, as it would be if an iron wire were substituted for the bismuth core. Yet in this case the repulsion on the bismuth is very slight, barely sensible, or perhaps not at all sensible when the needle exhibits most energetic signs of the forces it experiences. You know yourself, by your own experiments, how very small is even the *directive* agency experienced by a steel magnet placed across the lines of force due to the bismuth core. You may judge how much less sensible would be the attraction or repulsion it would experience as a whole, if held along the lines of force; and then think if the corresponding force experienced by a fragment of bismuth substituted for it, is likely to be verified by direct experiment or observation. I think you will admit that it is "incapable of verification," as well as "incontrovertible" by any collation of the results of experiments hitherto made on diamagnetics. As to the concluding paragraph of my letter which you quote, you do me justice when you say you accept it as an expression of my "personal conviction that the "action referred to is too feeble to be rendered sensible by "experiment." I will not maintain its unqualified application to all that can possibly be done in future in the way of experimental research to test the mutual action of diamagnetics under magnetic influence. On the contrary, I admit that no real physical agency can be rightly said to be "incapable of "verification by experiment or observation;" and I will ask you to limit that expression to experiments and observations hitherto made, and to substitute for the concluding paragraph of my letter the following statement [§ 686 above], written for publication three days later, and published in the same number of the Magazine as that to which you communicated my letter (*Phil. Mag.*, April 1855, p. 247). "The mutual influence" between rows of balls or cubes of bismuth in a magnetic field, "and its effects" in giving a tendency to a bar of the substance to assume a position along the lines of force, "are so excessively "minute, that they cannot possibly have been sensibly con-"cerned in any phænomena that have yet been observed ; and "it is probable that they may always remain insensible, even "to experiments especially directed to test them." I remain, my dear Sir, yours very truly, WILLIAM THOMSON.

DR TYNDALL.

XXXIX. *Inductive Susceptibility of a Polar Magnet.*

[*March* 1872. *Not hitherto published.*]

697. It is probable that every loadstone or steel magnet, or polar magnet of any kind, whatever degree of intrinsic magnetization it may possess, has also a susceptibility for magnetic induction, according to which, under the influence of other magnets brought into its neighbourhood, it will experience inductive magnetization temporarily superimposed upon its intrinsic magnetization. Hitherto experiment has given us little or no definite knowledge on this subject, or indeed generally on the relation between magnetic retentiveness and magnetic susceptibility. Waiting for more complete experimental investigation of the magnetic properties of matter, I shall assume as a typical magnetic solid, a rigid body possessing any degree of intrinsic magnetization in any direction, with perfect retentiveness; and having inductive quality defined by three principal magnetic susceptibilities along three principal rectangular axes of inductive capacity, in any given directions through it. The "rigid polar magnets" which we have hitherto considered are intrinsic magnets of zero susceptibility; and it now becomes necessary to define intrinsic magnetization for a substance of which the susceptibility is not zero.

698. *Def.* The intrinsic magnetization of a body is the resultant (§ 605) of the three intensities of magnetization found by cutting three infinitely thin bars from directions in it agreeing with its principal inductive axes, and testing them in a uniform magnetic field of air by measuring the couples which they experience when held at right angles to the lines of force. Before going on with the general problem of magnetic induction, we may consider the following particular case of it, merely as an illustration of this definition :—

699. *Problem.*—A solid sphere of uniform material, having μ, μ', μ'' for its three principal magnetic susceptibilities, and possessing intrinsic magnetization of intensity i in the directions specified with reference to the principal inductive axes by the direction-cosines, l, l', l'', is placed in air with no disturbing body in its neighbourhood : it is required to find its

actual magnetization. Let $-\xi, -\xi', -\xi''$, be the components of induced magnetization in the directions of the three principal axes; the required magnetization will be the resultant of

$$il - \xi, \quad il' - \xi', \quad il'' - \xi'' \dots\dots\dots\dots(1);$$

and therefore the problem is solved when ξ, ξ', ξ'' are determined. From the footnote to § 609, it follows immediately that the resultant force at any point within the sphere has for its components, in the directions of the principal axes,

$$-\frac{4\pi}{3}(il - \xi), \quad -\frac{4\pi}{3}(il' - \xi'), \quad -\frac{4\pi}{3}(il'' - \xi'') \dots\dots(2).$$

Now $-\xi, -\xi', -\xi''$ are the intensities of induced magnetization due separately to these three components of magnetizing force, and therefore (§ 610, *Def.* 2)

$$\xi = \mu \frac{4\pi}{3}(il - \xi), \quad \xi' = \mu' \frac{4\pi}{3}(il' - \xi'), \quad \xi'' = \mu'' \frac{4\pi}{3}(il'' - \xi'')\dots(3).$$

Solving these for ξ, ξ', ξ'', we have

$$\xi = \frac{\dfrac{4\pi\mu}{3} il}{1 + \dfrac{4\pi\mu}{3}}, \quad \xi' = \frac{\dfrac{4\pi\mu'}{3} il'}{1 + \dfrac{4\pi\mu'}{3}}, \quad \xi'' = \frac{\dfrac{4\pi\mu''}{3} il''}{1 + \dfrac{4\pi\mu''}{3}} \dots\dots(4),$$

and therefore (components of the whole magnetization)

$$il - \xi = \frac{il}{1 + \dfrac{4\pi\mu}{3}}, \quad il' - \xi' = \frac{il'}{1 + \dfrac{4\pi\mu'}{3}}, \quad il'' - \xi'' = \frac{il''}{1 + \dfrac{4\pi\mu''}{3}} \quad (5).$$

XL. *General Problem of Magnetic Induction.*

[*March* 1872. *Not hitherto published.*]

700. This problem is (§ 628) identical with the three general problems—electro-static induction through a heterogeneous insulating solid,—thermal or electric conduction through a heterogeneous conducting solid,—and (proved below, §§ 751—759) the flow of a frictionless incompressible liquid through a heterogeneous porous solid.

701. Let all space be occupied with matter of given permeabilities, $\varpi, \varpi', \varpi''$, along three principal inductive axes (l, m, n), (l', m', n'), (l'', m'', n''), (§ 611) through any point (x, y, z).

Let there be intrinsic magnetization (α, β, γ) at (x, y, z); and let constant electric currents be maintained having u, v, w for components of intensity at (x, y, z); subject to the condition (§ 540)

$$\frac{du}{dx} + \frac{dv}{dy} + \frac{dw}{dz} = 0 \dots\dots\dots\dots\dots(1).$$

Let ξ, η, ζ be the components of induced magnetization at (x, y, z). Then $\varpi, \varpi', \varpi''$, (l, m, n), (l', m', n'), (l'', m'', n''), $\alpha, \beta, \gamma, u, v, w$, being given for every point (x, y, z), it is required to find ξ, η, ζ. This is the general problem of magnetic induction. In it α, β, γ are absolutely arbitrary functions of (x, y, z); their values being zero in any part of space destitute of intrinsic magnetization: and u, v, w are arbitrary functions of (x, y, z), subject only to the condition (1); their values being zero throughout any portion of space through which there is no electric current.

702.   Let $\mathfrak{F}, \mathfrak{G}, \mathfrak{H}$ be the components of the resultant magnetic force according to the polar definition (§ 517, *Postscript*), calculated from the given intrinsic magnetization on the supposition of no induced magnetism; and F, G, H the components of the unambiguous resultant force (§ 551) calculated from the given electric currents. By § 545 and § 517 $(m), (n)$, and $(k), (l)$, we have

$$\left.\begin{array}{c} \dfrac{d\mathfrak{F}}{dx} + \dfrac{d\mathfrak{G}}{dy} + \dfrac{d\mathfrak{H}}{dz} = 4\pi\rho \\[2mm] \dfrac{d\mathfrak{H}}{dy} - \dfrac{d\mathfrak{G}}{dz} = 0, \quad \dfrac{d\mathfrak{F}}{dz} - \dfrac{d\mathfrak{H}}{dx} = 0, \quad \dfrac{d\mathfrak{G}}{dx} - \dfrac{d\mathfrak{F}}{dy} = 0 \end{array}\right\} \dots\dots(2),$$

where

$$\rho = -\left(\frac{d\alpha}{dx} + \frac{d\beta}{dy} + \frac{d\gamma}{dz}\right)$$

$$\left.\begin{array}{c} \dfrac{dF}{dx} + \dfrac{dG}{dy} + \dfrac{dH}{dz} = 0 \\[2mm] \dfrac{dH}{dy} - \dfrac{dG}{dz} = 4\pi u, \quad \dfrac{dH}{dx} - \dfrac{dF}{dz} = 4\pi v, \quad \dfrac{dF}{dy} - \dfrac{dG}{dx} = 4\pi w \end{array}\right\} \dots\dots(3).$$

Equations (2) suffice to determine $\mathfrak{F}, \mathfrak{G}, \mathfrak{H}$ from the data α, β, γ, by expressing that they are the differential coefficients of a function, and that that function is the potential of a distribution of imaginary magnetic matter having $-\left(\dfrac{d\alpha}{dx} + \dfrac{d\beta}{dy} + \dfrac{d\gamma}{dz}\right)$ for its density at (x, y, z), which we denote by ρ Similarly equations (3) determine F, G, H by virtually expressing that they are the components of the resultant magnetic force due to the given distribution of electric currents (u, v, w), and are therefore directly calculable from the data by the formulæ (b) of § 517 with F, G, H instead of X, Y, Z.

703. Let now
$$\underline{F} = \mathcal{F} + F, \quad \underline{G} = \mathcal{G} + G, \quad \underline{H} = \mathcal{H} + H \ldots\ldots\ldots(4).$$
The quantities $\underline{F}$, $\underline{G}$, $\underline{H}$ satisfy the equations
$$\frac{d\underline{F}}{dx} + \frac{d\underline{G}}{dy} + \frac{d\underline{H}}{dz} = 4\pi\rho,$$
$$\left. \frac{d\underline{H}}{dy} - \frac{d\underline{G}}{dz} = 4\pi u, \quad \frac{d\underline{F}}{dz} - \frac{d\underline{H}}{dx} = 4\pi v, \quad \frac{d\underline{G}}{dx} - \frac{d\underline{F}}{dy} = 4\pi w \right\} \ldots(5);$$
and these equations suffice to determine $\underline{F}$, $\underline{G}$, $\underline{H}$ fully, by virtually expressing that they are the sums of the two sets of components explicitly expressed in terms of the data, by the formulæ referred to in the preceding section. As we shall see immediately that we require from the data respecting intrinsic magnetization and electric currents nothing but the values of $\underline{F}$, $\underline{G}$, $\underline{H}$, we may simply regard these quantities as expressing the necessary data in this respect; and it is important to remark that they are unconditionally arbitrary for every point (x, y, z).

704. Let now the potential of the distribution of imaginary magnetic matter corresponding to the induced magnetism (ξ, η, ζ) be denoted by $\mathcal{V}$; that is to say, let $\mathcal{V}$ be the function of (x, y, z) which through all space satisfies the equation
$$\frac{d^2\mathcal{V}}{dx^2} + \frac{d^2\mathcal{V}}{dy^2} + \frac{d^2\mathcal{V}}{dz^2} = 4\pi\left(\frac{d\xi}{dx} + \frac{d\eta}{dy} + \frac{d\zeta}{dz}\right)\ldots\ldots(6);$$
and let
$$\mathcal{X} = -\frac{d\mathcal{V}}{dx}, \quad \mathcal{Y} = -\frac{d\mathcal{V}}{dy}, \quad \mathcal{Z} = -\frac{d\mathcal{V}}{dz}\ldots\ldots\ldots(7).$$
We shall see immediately that our problem is reduced to the determination of the single function $\mathcal{V}$; and we shall have simple equations [§ 705 (10)] giving explicitly the required components of induced magnetization ξ, η, ζ, in terms of the differential coefficients of this function.

705. Let I, I', I'' denote the components of the resultant of $\underline{F}$, $\underline{G}$, $\underline{H}$, and $\mathcal{S}$, $\mathcal{S}'$, $\mathcal{S}''$, the components of the resultant of $\mathcal{X}$, $\mathcal{Y}$, $\mathcal{Z}$, along the principal inductive axes. We have

$$\left.\begin{aligned}
&I=l\mathfrak{F}+m\mathfrak{G}+n\mathfrak{H}, &&I'=l'\mathfrak{F}+m'\mathfrak{G}+n'\mathfrak{H}, &&I''=l''\mathfrak{F}+m''\mathfrak{G}+n''\mathfrak{H}\\
&\mathfrak{F}=lI+l'I'+l''I'', &&\mathfrak{G}=mI+m'I'+m''I, &&\mathfrak{H}=nI+n'I'+n''I''\\
&\mathfrak{S}=l\mathfrak{X}+m\mathfrak{Y}+n\mathfrak{Z}, &&\mathfrak{S}'=l'\mathfrak{X}+m'\mathfrak{Y}+n'\mathfrak{Z}, &&\mathfrak{S}''=l''\mathfrak{X}+m''\mathfrak{Y}+n''\mathfrak{Z}\\
&\mathfrak{X}=l\mathfrak{S}+l'\mathfrak{S}'+l''\mathfrak{S}'', &&\mathfrak{Y}=m\mathfrak{S}+m'\mathfrak{S}'+m''\mathfrak{S}'', &&\mathfrak{Z}=n\mathfrak{S}+n'\mathfrak{S}'+n''\mathfrak{S}
\end{aligned}\right\}\dots(8).$$

The three principal magnetic susceptibilities (§ 629) being

$$\frac{\varpi-1}{4\pi},\ \frac{\varpi'-1}{4\pi},\ \frac{\varpi''-1}{4\pi};$$

the component intensities of induced magnetization along the principal inductive axes (to be denoted, § 712 below, by $\mathfrak{S},\ \mathfrak{S}',\ \mathfrak{S}''$) are

$$\mathfrak{S}=\frac{\varpi-1}{4\pi}(I+\mathfrak{S}),\ \ \mathfrak{S}'=\frac{\varpi'-1}{4\pi}(I'+\mathfrak{S}'),\ \ \mathfrak{S}''=\frac{\varpi''-1}{4\pi}(I''+\mathfrak{S}'')\dots(9).$$

Hence taking components along the axes of (x, y, z), and multiplying by 4π, we have

$$\left.\begin{aligned}
&4\pi\xi=\varpi(I+\mathfrak{S})l\ +\varpi'(I'+\mathfrak{S}')\,l'\ +\varpi''(I''+\mathfrak{S}'')l''\ -\mathfrak{F}-\mathfrak{X}\\
&4\pi\eta=\varpi(I+\mathfrak{S})m+\varpi'(I'+\mathfrak{S}')\,m'+\varpi''(I''+\mathfrak{S}'')\,m''-\mathfrak{G}-\mathfrak{Y}\\
&4\pi\zeta=\varpi(I+\mathfrak{S})n\ +\varpi'(I'+\mathfrak{S}')n'\ +\varpi''(I''+\mathfrak{S}'')\,n''\ -\mathfrak{H}-\mathfrak{Z}
\end{aligned}\right\}(10).$$

706. These three equations, together with the three equations by which $\mathfrak{X}$, $\mathfrak{Y}$, $\mathfrak{Z}$ might, according to §§ 518, 482, 483, be expressed in terms of ξ, η, ζ, suffice to determine the six unknown quantities ξ, η, ζ, $\mathfrak{X}$, $\mathfrak{Y}$, $\mathfrak{Z}$; but, by (7) and (6) introducing $\mathcal{V}$, we may eliminate those six unknown quantities, and obtain a single equation for the one unknown quantity $\mathcal{V}$, thus:

—Taking $\dfrac{d}{dx}$ of the first of the three equations (10), $\dfrac{d}{dy}$ of the second, and $\dfrac{d}{dz}$ of the third, adding and using (6) and (7), we find

$$\frac{d(\varpi\mathfrak{S}l+\varpi'\mathfrak{S}'l'+\varpi''\mathfrak{S}''l'')}{dx}+\frac{d(\varpi\mathfrak{S}m+\varpi'\mathfrak{S}'m'+\varpi''\mathfrak{S}''m'')}{dy}+\frac{d(\varpi\mathfrak{S}n+\varpi'\mathfrak{S}'n'+\varpi''\mathfrak{S}''n'')}{dz}$$
$$=\frac{d(\mathfrak{F}-\varpi Il-\varpi'I'l'-\varpi''I''l'')}{dx}+\frac{d(\mathfrak{G}-\varpi Im-\varpi'I'm'-\varpi''I''m'')}{dy}+\frac{d(\mathfrak{H}-\varpi In-\varpi'I'n'-\varpi''I''n'')}{dz}\ (11).$$

Substituting in this for $\mathfrak{S}$, $\mathfrak{S}'$, $\mathfrak{S}''$ their values by (8), then for $\mathfrak{X}$, $\mathfrak{Y}$, $\mathfrak{Z}$ by (7), and for I, I', I'' their values by (8), we have explicitly a linear differential equation of the second order with second member a known function of (x, y, z) to determine the unknown function $\mathcal{V}$.

707. The coefficients of $\dfrac{d\Psi}{dx}$, $\dfrac{d\Psi}{dy}$, $\dfrac{d\Psi}{dz}$ under the symbols $\dfrac{d}{dx}$, $\dfrac{d}{dy}$, $\dfrac{d}{dz}$ are related in the ordinary symmetrical manner to the coefficients which appear in the quadratic function

$$\frac{1}{8\pi}[\varpi(l\mathfrak{X}+m\mathfrak{Y}+n\mathfrak{Z})^2+\varpi'(l'\mathfrak{X}+m'\mathfrak{Y}+n'\mathfrak{Z})^2+\varpi''(l''\mathfrak{X}+m''\mathfrak{Y}+n''\mathfrak{Z})^2]\ldots\ldots\ldots(12)$$

when expanded; and it is unnecessary to write them out explicitly. A similar remark is applicable to the coefficients of $\underline{F}$, $\underline{G}$, $\underline{H}$ under differentiation in the second member. Denoting (12) by $\mathfrak{Q}$, and the same function of F, G, H by $\underline{Q}$, so that using again the notation of (8) for brevity, we have

$$\mathfrak{Q}=\frac{1}{8\pi}(\varpi\mathfrak{S}^2+\varpi'\mathfrak{S}'^2+\varpi''\mathfrak{S}''^2)\ldots\ldots(13)$$

and

$$\underline{Q}=\frac{1}{8\pi}(\varpi I^2+\varpi'I'^2+\varpi''I''^2)\ldots\ldots\ldots(14),$$

we see at once that the differential equation (11) may be written short, thus—

$$\left.\begin{array}{l}\dfrac{d}{dx}\dfrac{d\mathfrak{Q}}{d\mathfrak{X}}+\dfrac{d}{dy}\dfrac{d\mathfrak{Q}}{d\mathfrak{Y}}+\dfrac{d}{dz}\dfrac{d\mathfrak{Q}}{d\mathfrak{Z}}=\rho',\\[2mm]\rho'=\rho-\left(\dfrac{d}{dx}\dfrac{dQ}{d\underline{F}}+\dfrac{d}{dy}\dfrac{dQ}{d\underline{G}}+\dfrac{d}{dz}\dfrac{dQ}{d\underline{H}}\right)\end{array}\right\}\ldots\ldots\ldots(15).$$

where

Equations (10), similarly written short, are as follows:—

$$\left.\begin{array}{l}\xi=\dfrac{dQ}{d\underline{F}}+\dfrac{d\mathfrak{Q}}{d\mathfrak{X}}-\dfrac{1}{4\pi}(\underline{F}+\mathfrak{X})\\[2mm]\eta=\dfrac{dQ}{d\underline{G}}+\dfrac{d\mathfrak{Q}}{d\mathfrak{Y}}-\dfrac{1}{4\pi}(\underline{G}+\mathfrak{Y})\\[2mm]\zeta=\dfrac{dQ}{d\underline{H}}+\dfrac{d\mathfrak{Q}}{d\mathfrak{Z}}-\dfrac{1}{4\pi}(\underline{H}+\mathfrak{Z})\end{array}\right\}\ldots\ldots\ldots\ldots(16).$$

When, by the integration of (15), Ψ is determined, equations (16) give explicitly ξ, η, ζ, the components of the required magnetization.

708. I shall conclude with two slightly different demonstrations that, provided the permeabilities are everywhere positive, as (§ 631) we believe they must be for every substance in nature, there is one, and only one, value of Ψ for every point (x, y, z) if (15), with any given arbitrary function of (x, y, z) for its second member, be satisfied for every point of space. The first demonstration, to which I now proceed, is the more con-

venient for the magnetic or (§ 700) electric subject which we
have had hitherto under consideration; the second will be
added on account of convenience for the hydro-kinetic analogy.

709. *First demonstration of Determinacy and Singleness.*—Let
$\mathfrak{K}, \mathfrak{K}, \mathfrak{K}$ be any three real quantities, arbitrary functions of
(x, y, z). Consider the function

$$\mathfrak{P} = \frac{1}{8\pi}\left[\varpi\,(\mathfrak{S} - \mathfrak{K})^2 + \varpi'\,(\mathfrak{S}' - \mathfrak{K}')^2 + \varpi''\,(\mathfrak{S}'' - \mathfrak{K}'')^2\right]\dots(17),$$

and the triple integral

$$E = \int_{-\infty}^{\infty}\int_{-\infty}^{\infty}\int_{-\infty}^{\infty}\mathfrak{P}\,dx\,dy\,dz\dots\dots\dots(18).$$

(Compare §§ 503, 561, 206, 732, and 753—763). The function $\mathfrak{P}$
is necessarily positive, except in the particular case of $\mathfrak{S} = \mathfrak{K}$,
$\mathfrak{S}' = \mathfrak{K}', \mathfrak{S}'' = \mathfrak{K}''$, when it is zero. Remembering that $\mathfrak{S}, \mathfrak{S}'$,
$\mathfrak{S}''$ are linear functions of $\dfrac{d\mathfrak{V}}{dx}, \dfrac{d\mathfrak{V}}{dy}, \dfrac{d\mathfrak{V}}{dz}$, with given functions of (x, y, z) for their coefficients, apply the calculus of
variations to assign $\mathfrak{V}$, so that E may be a minimum. Using
for brevity the notation (7) of § 704, we have

$$\delta\mathfrak{P} = -\left(\frac{d\mathfrak{P}}{d\mathfrak{X}}\,\delta\frac{d\mathfrak{V}}{dx} + \frac{d\mathfrak{P}}{d\mathfrak{P}}\,\delta\frac{d\mathfrak{V}}{dy} + \frac{d\mathfrak{P}}{d\mathfrak{Z}}\,\delta\frac{d\mathfrak{V}}{dz}\right).$$

Hence, following the usual process of integration by parts, according to the calculus of variations, we find for the condition
that E may be a minimum,

$$\frac{d}{dx}\frac{d\mathfrak{P}}{d\mathfrak{X}} + \frac{d}{dy}\frac{d\mathfrak{P}}{d\mathfrak{P}} + \frac{d}{dz}\frac{d\mathfrak{P}}{d\mathfrak{Z}} = 0\dots\dots\dots\dots(19).$$

Now if we put

$$\mathfrak{L} = \mathfrak{K}l + \mathfrak{K}'l' + \mathfrak{K}''l'', \quad \mathfrak{M} = \mathfrak{K}m + \mathfrak{K}'m' + \mathfrak{K}''m'), \quad \mathfrak{N} = \mathfrak{K}n + \mathfrak{K}'n' + \mathfrak{K}''n''$$
which imply that $\Big\}\dots(20)$
$$\mathfrak{K} = \mathfrak{L}l + \mathfrak{M}m + \mathfrak{N}n, \quad \mathfrak{K}' = \mathfrak{L}l' + \mathfrak{M}m' + \mathfrak{N}n', \quad \mathfrak{K}'' = \mathfrak{L}l'' + \mathfrak{M}m'' + \mathfrak{N}n''$$

and look to equations (13) and (8) of § 707, we see that $\mathfrak{P}$ is the
same quadratic function of $\mathfrak{X} - \mathfrak{L}, \mathfrak{P} - \mathfrak{M}, \mathfrak{Z} - \mathfrak{N}$, that $\mathfrak{Q}$
is of $\mathfrak{X}, \mathfrak{P}, \mathfrak{Z}$. Hence $\dfrac{d\mathfrak{P}}{d\mathfrak{X}}, \dfrac{d\mathfrak{P}}{d\mathfrak{P}}, \dfrac{d\mathfrak{P}}{d\mathfrak{Z}}$ are linear functions of
$\mathfrak{X} - \mathfrak{L}, \mathfrak{P} - \mathfrak{M}, \mathfrak{Z} - \mathfrak{N}$; and if we denote by $\mathfrak{F}$ the same
quadratic function of $\mathfrak{L}, \mathfrak{M}, \mathfrak{N}$ that $\mathfrak{Q}$ is of $\mathfrak{X}, \mathfrak{P}, \mathfrak{Z}$, that is
to say, if we put

$$\mathfrak{F} = \frac{1}{8\pi}\left(\varpi\mathfrak{K}^2 + \varpi'\mathfrak{K}'^2 + \varpi''\mathfrak{K}''^2\right)\dots\dots\dots(21),$$

we have

$$\frac{d\mathfrak{P}}{d\mathfrak{X}} = \frac{d\mathfrak{Q}}{d\mathfrak{X}} - \frac{d\mathfrak{I}}{d\mathfrak{L}}, \quad \frac{d\mathfrak{P}}{d\mathfrak{Y}} = \frac{d\mathfrak{Q}}{d\mathfrak{Y}} - \frac{d\mathfrak{I}}{d\mathfrak{M}}, \quad \frac{d\mathfrak{P}}{d\mathfrak{Z}} = \frac{d\mathfrak{Q}}{d\mathfrak{Z}} - \frac{d\mathfrak{I}}{d\mathfrak{N}} \dots(22).$$

Hence (19) becomes

$$\frac{d}{dx}\frac{d\mathfrak{Q}}{d\mathfrak{X}} + \frac{d}{dy}\frac{d\mathfrak{Q}}{d\mathfrak{Y}} + \frac{d}{dz}\frac{d\mathfrak{Q}}{d\mathfrak{Z}} = \frac{d}{dx}\frac{d\mathfrak{I}}{d\mathfrak{L}} + \frac{d}{dy}\frac{d\mathfrak{I}}{d\mathfrak{M}} + \frac{d}{dz}\frac{d\mathfrak{I}}{d\mathfrak{N}} \dots.(23),$$

which, expanded in terms of $\mathcal{V}$, is a linear partial differential equation of the second order, with right-hand member a given function of (x, y, z). The fulfilment of this equation through all space is the sole condition which $\mathcal{V}$ must fulfil to make E a minimum. Now it is possible to assign $\mathcal{V}$ so as to make E a minimum, and therefore there exists a function $\mathcal{V}$ which satisfies equation (23) through all space. This is an obvious extension of Theorem 1, § 206. Demonstration 2 of § 206 extended in an obvious manner proves that no function differing at any point from one function which satisfies (23) through all space, can satisfy (23) through all space. Hence the solution of this equation is determinate and free from all ambiguity or multiplicity of values.

710. The extension of § 206, 2, gives the following useful theorems :—Let $\mathcal{V}$ be a function of (x, y, z) satisfying (23) through all space; let $\Delta\mathcal{V}$ be any function whatever of (x, y, z); let $\Delta\mathfrak{H}, \Delta\mathfrak{H}', \Delta\mathfrak{H}'', E(\Delta)$ be the values of $\mathfrak{H}, \mathfrak{H}', \mathfrak{H}'', E$, when $\Delta\mathcal{V}$ is substituted for V; and let $E + \Delta E$ be the value of E when $\mathcal{V} + \Delta\mathcal{V}$ is substituted for $\mathcal{V}$ Then—

Theorem I

$$\int_{-\infty}^{\infty}\int_{-\infty}^{\infty}\int_{-\infty}^{\infty} dxdydz\,(\varpi\mathfrak{H}\Delta\mathfrak{H} + \varpi'\mathfrak{H}'\Delta\mathfrak{H}' + \varpi''\mathfrak{H}''\Delta\mathfrak{H}'') = 0 \quad (24);$$

proved by the ordinary integration by parts of § 199, (*a*), (*b*), as extended in § 206, Demonstrations 1 and 2, and now further extended.

Theorem II. $\Delta E = E(\Delta) \dots\dots\dots\dots\dots\dots(25).$

This very important theorem is an instant consequence of Theorem I.

As $E(\Delta)$ is necessarily positive, a function $\mathcal{V}$, which satisfies (23), has the unique characteristic that every function differing from it gives a larger value to E.

711. The first member of (23) is identical with the first

member of (15). We may make the second member of (23) equal to the second member of (15), by taking

$$
\left.
\begin{aligned}
\mathfrak{K} &= -I + \frac{1}{\varpi}(I + \mathfrak{u}l + \mathfrak{v}m + \mathfrak{w}n), \quad \mathfrak{K}' = -I' + \frac{1}{\varpi'}(I' + \mathfrak{u}'l' + \mathfrak{v}'m' + \mathfrak{w}'n'), \\
\mathfrak{K}'' &= -I'' + \frac{1}{\varpi''}(I'' + \mathfrak{u}''l'' + \mathfrak{v}''m'' + \mathfrak{w}''n'')
\end{aligned}
\right\} \quad \ldots\ldots (26),
$$

where $\mathfrak{u}$, $\mathfrak{v}$, $\mathfrak{w}$ are any three quantities such that

$$
\frac{d\mathfrak{u}}{dx} + \frac{d\mathfrak{v}}{dy} + \frac{d\mathfrak{w}}{dz} = 0 \ldots\ldots\ldots\ldots\ldots\ldots (27).
$$

This we see at once by remarking that

$$
\left.
\begin{aligned}
4\pi \frac{d\mathfrak{H}}{d\mathfrak{L}} &= \varpi \mathfrak{K}l + \varpi' \mathfrak{K}'l' + \varpi'' \mathfrak{K}''l'', \quad \text{etc. etc.,} \\
\text{and} \qquad 4\pi \frac{dQ}{d\bar{F}} &= \varpi Il + \varpi' I'l' + \varpi'' I''l'', \quad \text{etc. etc.,}
\end{aligned}
\right\} \quad (28);
$$

and taking account of (8) and (5). Hence §§ 709, 710, with the values (26) for $\mathfrak{K}$, $\mathfrak{K}'$, $\mathfrak{K}''$, prove that there exists a function $\mathcal{V}$ satisfying the inductive equation (16) through all space; that this solution makes the triple integral E (18) a minimum; that if $\mathcal{V}$ be a function satisfying (15), and $\Delta\mathcal{V}$ any function whatever, $\mathcal{V} + \Delta\mathcal{V}$ substituted for $\mathcal{V}$ augments the value of E by the necessarily positive value of the triple integral found by substituting $\Delta\mathcal{V}$ for $\mathcal{V}$; and, therefore, that no function differing from one which satisfies (15) can also satisfy it.

712. *Preliminary to Second Demonstration of Determinacy and Singleness.*—First, it will be convenient to put the inductive equations (11) and (16) into a different form, a form suitable to the uniform reckoning of "resultant magnetic force," according to the "electro-magnetic definition" (§ 517, *Postscript*). Remembering (§§ 702, 704) that $\mathfrak{F}$, $\mathfrak{G}$, $\mathfrak{H}$ and $\mathfrak{X}$, $\mathfrak{Y}$, $\mathfrak{Z}$ are the components of the resultant forces calculated separately, according to the polar definition, from the intrinsic and induced magnetizations respectively, we see [§ 517 (r)] that

$$\mathfrak{F} + \mathfrak{X} + 4\pi (\alpha + \xi), \quad \mathfrak{G} + \mathfrak{Y} + 4\pi (\beta + \eta), \quad \mathfrak{H} + \mathfrak{Z} + 4\pi (\gamma + \zeta),$$

are the components of the resultant force of intrinsic and induced magnetizations together, according to the electro-magnetic definition. To these we must add F, G, H to find for the whole system (of inducing intrinsic magnetization and electric currents, and induced magnetization) the components of the resultant magnetic force, according to the electro-magnetic definition. Calling these X, Y, Z, and taking advantage of the short notation (4), we have

$$X=\underline{F}+\mathfrak{X}+4\pi(\alpha+\xi),\ Y=\underline{G}+\mathfrak{Y}+4\pi(\beta+\eta),\ Z=\underline{H}+\mathfrak{Z}+4\pi(\gamma+\zeta)\ (29).$$

Take now components of forces and of magnetizations along the principal inductive axes. Thus we have

$$S=I+\mathfrak{S}+4\pi(A+\mathfrak{I}),\ S'=I'+\mathfrak{S}'+4\pi(A'+\mathfrak{I}'),\ S''=I''+\mathfrak{S}''+4\pi(A''+\mathfrak{I}'')...(30),$$

where $S=Xl+Ym+Zn$, etc., implying $X=Sl+S'l'+S''l''$, etc. (31),

$$A=\alpha l+\beta m+\gamma n,\text{ etc., implying }\alpha=Al+A'l'+A''l'',\text{ etc. (32)},$$

and $\quad \mathfrak{I}=\xi l+\eta m+\zeta n$, etc., implying $\xi=\mathfrak{I}l+\mathfrak{I}'l'+\mathfrak{I}''l''$, etc. (33);

and I, I', I'', $\mathfrak{S}$, $\mathfrak{S}'$, $\mathfrak{S}''$ have still the same significance as that indicated in (8), §705, above. Now by (9) we have

$$4\pi\mathfrak{I}=(\varpi-1)(I+\mathfrak{S}),\ 4\pi\mathfrak{I}'=(\varpi'-1)(I'+\mathfrak{S}'),\ 4\pi\mathfrak{I}''=(\varpi''-1)(I''+\mathfrak{S}'')\ (34).$$

Hence eliminating $\mathfrak{I}$, $\mathfrak{I}'$ $\mathfrak{I}''$ from (30),

$$S=\varpi(I+\mathfrak{S})+4\pi A,\ S'=\varpi'(I'-\mathfrak{S}')+4\pi A',\ S''=\varpi''(I''+\mathfrak{S}'')+4\pi A''\ (35).$$

Put now

$$I+4\pi\frac{A}{\varpi}=C,\quad I'+4\pi\frac{A'}{\varpi'}=C',\quad I''+4\pi\frac{A''}{\varpi''}=C''...(36),$$

and let

$$\left.\begin{array}{l}Cl+C'l'+C''l''=\overline{F},\ \ Cm+C'm'+C''m''=\overline{G},\ \ Cn+C'n'+C''n''=\overline{H},\\ \text{implying}\\ C=l\overline{F}+m\overline{G}+n\overline{H},\ C'=l'\overline{F}+m'\overline{G}+n'\overline{H},\ C''=l''\overline{F}+m''\overline{G}+n''\overline{H}\end{array}\right\}\ (37).$$

By (35) we have

$$\mathfrak{S}=\frac{S}{\varpi}-C,\ \ \mathfrak{S}'=\frac{S'}{\varpi'}-C',\ \ \mathfrak{S}''=\frac{S''}{\varpi''}-C''\ ...\ (38).$$

Hence

$$\mathfrak{X}=\frac{S}{\varpi}l+\frac{S'}{\varpi'}l'+\frac{S''}{\varpi''}l''-\overline{F},\ \mathfrak{Y}=\frac{S}{\varpi}m+\frac{S'}{\varpi'}m'+\frac{S''}{\varpi''}m''-\overline{G},\ \mathfrak{Z}=\frac{S}{\varpi}n+\frac{S'}{\varpi'}n'+\frac{S''}{\varpi''}n''-\overline{H}\ (39).$$

713. Now let $\quad Q=\dfrac{1}{8\pi}\left(\dfrac{S^2}{\varpi}+\dfrac{S'^2}{\varpi'}+\dfrac{S''^2}{\varpi''}\right)$..............(40).

[Compare (13) of §707.] Substituting for S, S', S'' their values by (31), we have in Q a quadratic function of X, Y, Z (corresponding in the electro-magnetic formulæ to the function $\mathfrak{Q}$ of $\mathfrak{X}$, $\mathfrak{Y}$, $\mathfrak{Z}$ in the polar formulæ). Now (39) becomes

$$\mathfrak{X}=4\pi\frac{dQ}{dX}-\overline{F},\ \mathfrak{Y}=4\pi\frac{dQ}{dY}-\overline{G},\ \mathfrak{Z}=4\pi\frac{dQ}{dZ}-\overline{H}\ ...(41).$$

Eliminating $\mathfrak{X}$, $\mathfrak{Y}$, $\mathfrak{Z}$ by the condition that $\mathfrak{X}dx + \mathfrak{Y}dy + \mathfrak{Z}dz$ is a complete differential, we have

$$\frac{d}{dy}\frac{dQ}{dZ} - \frac{d}{dz}\frac{dQ}{dY} = \frac{1}{4\pi}\left(\frac{d\overline{H}}{dy} - \frac{d\overline{G}}{dz}\right), \quad \frac{d}{dz}\frac{dQ}{dX} - \frac{d}{dx}\frac{dQ}{dZ} = \frac{1}{4\pi}\left(\frac{d\overline{F}}{dz} - \frac{d\overline{H}}{dx}\right),$$
$$\frac{d}{dx}\frac{dQ}{dY} - \frac{d}{dy}\frac{dQ}{dZ} = \frac{1}{4\pi}\left(\frac{d\overline{G}}{dx} - \frac{d\overline{F}}{dy}\right) \quad\quad\quad \right\}\ ...(42),$$

three linear partial differential equations in X, Y, Z, equivalent to two independent equations, because $\frac{d}{dx}$ of the first added to $\frac{d}{dy}$ of the second and $\frac{d}{dz}$ of the third constitutes an equation in which each member is identically zero. Also, by (29), (5), (7), and (6), we have

$$\frac{dX}{dx} + \frac{dY}{dy} + \frac{dZ}{dz} = 0 \quad\quad\quad\quad\quad (43).$$

These four, (42) and (43,) equivalent to three independent equations, in which $\overline{F}$, $\overline{G}$, $\overline{H}$ are arbitrarily given functions of (x, y, z), determine fully and unambiguously the unknown X, Y, Z through all space, as will be proved immediately by the promised fresh demonstration. But first it may be remarked that one obvious way of dealing with them leads us back to our former analysis, thus:—The three equations (42) simply express that

$$\left(\frac{dQ}{dX} - \frac{1}{4\pi}\overline{F}\right)dx + \left(\frac{dQ}{dY} - \frac{1}{4\pi}\overline{G}\right)dy + \left(\frac{dQ}{dZ} - \frac{1}{4\pi}\overline{H}\right)dz$$

is a complete differential. Hence their most general integral is

$$4\pi\frac{dQ}{dX} = \overline{F} - \frac{d\mathcal{V}}{dx}, \quad 4\pi\frac{dQ}{dY} = \overline{G} - \frac{d\mathcal{V}}{dy}, \quad 4\pi\frac{dQ}{dZ} = \overline{H} - \frac{d\mathcal{V}}{dz} ...(44),$$

where $\mathcal{V}$ so far denotes an arbitrary function of (x, y, z). The first members here are merely short expressions for the linear functions of X, Y, Z which appear in (39) with S, S', S'' eliminated by (31). Solved for X, Y, Z, equations (44) give expressions which are the same as (29) with ξ, η, ζ eliminated by (10), and $\mathfrak{X}$, $\mathfrak{Y}$, $\mathfrak{Z}$ by (7); and eliminating by them X, Y, Z from (43) we have an equation for $\mathcal{V}$ identical with (11), which (§ 708) determines $\mathcal{V}$ unambiguously through all space.

714. *Second Proof of Determinateness and Singleness.*—Let K, K', K'' be any three arbitrarily given functions of (x, y, z); and put

$$\mathfrak{E}_1 = \int_{-\infty}^{\infty}\int_{-\infty}^{\infty}\int_{-\infty}^{\infty}\frac{1}{8\pi}\left[\frac{(S-K)^2}{\varpi} + \frac{(S'-K')^2}{\varpi'} + \frac{(S''-K'')^2}{\varpi''}\right]dxdydz \quad (45)$$

[where the suffix is appended to distinguish from the $\mathfrak{E}$ of §§ 729 ...731 below.]

Consider the problem of finding X, Y, Z so as to make $\mathfrak{E}_1$ a minimum, subject to (43). Denoting by λ an indeterminate multiplier, according to the ordinary method of the calculus of variations, make

$$\int_{-\infty}^{\infty}\int_{-\infty}^{\infty}\int_{-\infty}^{\infty} dxdydz\left[\frac{(S-K)^2}{\varpi}+\frac{(S'-K')^2}{\varpi'}+\frac{(S''-K'')^2}{\varpi''}+2\lambda\left(\frac{dX}{dx}+\frac{dY}{dy}+\frac{dZ}{dz}\right)\right] \tag{46}$$

unconditionally a minimum. The resultant equations are

$$\frac{dQ(S-K)}{dX}=\frac{1}{4\pi}\frac{d\lambda}{dx}, \quad \frac{dQ(S-K)}{dY}=\frac{1}{4\pi}\frac{d\lambda}{dy}, \quad \frac{dQ(S-K)}{dZ}=\frac{1}{4\pi}\frac{d\lambda}{dz} \tag{47},$$

where for a moment $Q(S-K)$ denotes the function integrated in (45). If we eliminate the unknown quantity λ from these by differentiation, we have three linear partial differential equations of the second order, equivalent to two, which with (43) determine the unknown functions X, Y, Z. Considerations corresponding perfectly to those of §§ 206, 709, 710, show that these equations can be satisfied through all space by real finite functions X, Y, Z, and that they cannot be satisfied by any functions differing in any part of space from one set of three functions which satisfy them. We have also, of course, theorems precisely corresponding to Theorems I. and II., (24) and (25) of § 710.

715. Now let $\quad K=\varpi C,\ K'=\varpi'C',\ K''=\varpi''C'' \ \ldots\ldots (48)$. This, as is easily seen from (37) and (40), gives

$$\frac{dQ(S-K)}{dX}=\frac{dQ}{dX}-\frac{1}{4\pi}\overline{F}, \quad \frac{dQ(S-K)}{dY}=\frac{dQ}{dY}-\frac{1}{4\pi}\overline{G}, \quad \frac{dQ(S-K)}{dZ}=\frac{dQ}{dZ}-\frac{1}{4\pi}\overline{H} \tag{49};$$

and the equations obtained by eliminating λ from (47) become identical with (42). It is thus proved that equations (42) and (43) determine X, Y, Z unambiguously through all space. With the particular values of K, K', K'' assumed in (48), we see by (38) that (45) becomes

$$\mathfrak{E}=\frac{1}{8\pi}\int_{-\infty}^{\infty}\int_{-\infty}^{\infty}\int_{-\infty}^{\infty} dxdydz\,(\varpi\mathfrak{S}^2+\varpi'\mathfrak{S}'^2+\varpi''\mathfrak{S}''^2)\ldots (50);$$

and therefore the problem of magnetic induction is reduced to making this configurational function a minimum, subject to the condition $\quad \dfrac{dX}{dx}+\dfrac{dY}{dy}+\dfrac{dZ}{dz}=0 \quad$ (43) of § 713 repeated.

716. Going back to the first proof of determinacy and singleness, and particularizing the values of $\mathfrak{K}$, $\mathfrak{K}'$, $\mathfrak{K}''$ of (26) by taking

$$\mathfrak{u}=-(4\pi\alpha+\underline{F}),\quad \mathfrak{v}=-(4\pi\beta+\underline{G}),\quad \mathfrak{w}=-(4\pi\gamma+\underline{H})\ldots(51)$$

which in virtue of (5) satisfies (27), the sole condition obligatory on $\mathfrak{u}$, $\mathfrak{v}$, $\mathfrak{w}$, we make the function $\mathfrak{P}$ of (17) equal to

$$\frac{1}{8\pi}\left(\frac{S^2}{\varpi}+\frac{S'^2}{\varpi'}+\frac{S''^2}{\varpi''}\right)\dots\dots\dots\dots(52),$$

easily proved from (8), (32), (36), and (38). Thus we have

$$E=\frac{1}{8\pi}\int_{-\infty}^{\infty}\int_{-\infty}^{\infty}\int_{-\infty}^{\infty}dxdydz\left(\frac{S^2}{\varpi}+\frac{S'^2}{\varpi'}+\frac{S''^2}{\varpi''}\right)\dots\dots(53),$$

and the problem of magnetic induction is reduced to making this configurational function a minimum, subject to the condition that $\mathfrak{X}dx+\mathfrak{Y}dy+\mathfrak{Z}dz$ is a complete differential, S, S', S'' being expressed by equations (38) and (8) in terms of $\mathfrak{X}, \mathfrak{Y}, \mathfrak{Z}$, the unknown quantities, and C, C', C'' three arbitrarily given functions of x, y, z.

717. A curious relation between the configurational functions (50) and (53) is proved thus:—Attending to (7) and remembering that $\mathfrak{Q}$ is a quadratic function of $\mathfrak{X}, \mathfrak{Y}, \mathfrak{Z}$, put

$$-\tfrac{1}{2}\left(\frac{d\mathfrak{V}}{dx}\frac{d\mathfrak{Q}}{d\mathfrak{X}}+\frac{d\mathfrak{V}}{dy}\frac{d\mathfrak{Q}}{d\mathfrak{Y}}+\frac{d\mathfrak{V}}{dz}\frac{d\mathfrak{Q}}{d\mathfrak{Z}}\right)$$

for it in (50) and perform integrations by parts. We thus find

$$\int_{-\infty}^{\infty}\int_{-\infty}^{\infty}\int_{-\infty}^{\infty}dxdydz\,\mathfrak{Q}=\tfrac{1}{2}\int_{-\infty}^{\infty}\int_{-\infty}^{\infty}\int_{-\infty}^{\infty}dxdydz\,\mathfrak{V}\left(\frac{d}{dx}\frac{d\mathfrak{Q}}{d\mathfrak{X}}+\frac{d}{dy}\frac{d\mathfrak{Q}}{d\mathfrak{Y}}+\frac{d}{dz}\frac{d\mathfrak{Q}}{d\mathfrak{Z}}\right)(54),$$

or by (13) and (15)

$$\frac{1}{8\pi}\int_{-\infty}^{\infty}\int_{-\infty}^{\infty}\int_{-\infty}^{\infty}dxdydz\,(\varpi\mathfrak{S}^2+\varpi'\mathfrak{S}'^2+\varpi''\mathfrak{S}''^2)=\tfrac{1}{2}\int_{-\infty}^{\infty}\int_{-\infty}^{\infty}\int_{-\infty}^{\infty}dxdydz\,\mathfrak{V}\rho'(55).$$

Now taking (53) substitute in it, for S, S', S'', their values by (38). We have immediately

$$E=\frac{1}{8\pi}\left\{\int_{-\infty}^{\infty}\int_{-\infty}^{\infty}\int_{-\infty}^{\infty}dxdydz\,[\varpi\,(\mathfrak{S}^2+C^2)+\varpi'(\mathfrak{S}'^2+C'^2)+\varpi''(\mathfrak{S}''^2+C''^2)]\right.$$

$$\left.+2\int_{-\infty}^{\infty}\int_{-\infty}^{\infty}\int_{-\infty}^{\infty}dxdydz\,(\varpi\mathfrak{S}C+\varpi'\mathfrak{S}'C'+\varpi''\mathfrak{S}''C''\right\}\dots\dots(56).$$

For C, C', C'', taking their values by (36), and attending to (8), (28), and (32), we have

$$\varpi\mathfrak{S}C+\varpi'\mathfrak{S}'C'+\varpi''\mathfrak{S}''C''=\varpi I\mathfrak{S}+\varpi'I'\mathfrak{S}'+\varpi''I''\mathfrak{S}''+4\pi\,(\mathfrak{S}A+\mathfrak{S}'A'+\mathfrak{S}''A'')$$

$$=\pi4\left(\mathfrak{X}\frac{dQ}{dF}+\mathfrak{Y}\frac{dQ}{dG}+\mathfrak{Z}\frac{dQ}{dH}\right)+4\pi\,(\mathfrak{X}a+\mathfrak{Y}\beta+\mathfrak{Z}\gamma).$$

Putting in the second member for $\mathfrak{X}, \mathfrak{Y}, \mathfrak{Z}$ their values,

$-\dfrac{d\mathcal{V}}{dx}$, etc., remembering that $\rho = -\left(\dfrac{d\alpha}{dx} + \dfrac{d\beta}{dy} + \dfrac{d\gamma}{dz}\right)$, and integrating by parts as usual, we find

$$\int_{-\infty}^{\infty}\int_{-\infty}^{\infty}\int_{-\infty}^{\infty} dxdydz\,(\varpi\mathcal{S}C + \varpi'\mathcal{S}'C' + \varpi''\mathcal{S}''C'')$$

$$= 4\pi\int_{-\infty}^{\infty}\int_{-\infty}^{\infty}\int_{-\infty}^{\infty} dxdydz\mathcal{V}\left(\frac{d}{dx}\frac{dQ}{d\overline{F}} + \frac{d}{dy}\frac{dQ}{d\overline{G}} + \frac{d}{dz}\frac{dQ}{d\overline{H}} - \rho\right) = -4\pi\int_{-\infty}^{\infty}\int_{-\infty}^{\infty}\int_{-\infty}^{\infty} \mathcal{V}\rho'\,(57),$$

the last step being simply an introduction of the notation of (15). Using this in (56), attending to (55) and (50), and transposing, we find

$$\mathfrak{E}_{1} + E = \frac{1}{8\pi}\int_{-\infty}^{\infty}\int_{-\infty}^{\infty}\int_{-\infty}^{\infty} dxdydz\,(\varpi C^{2} + \varpi'C'^{2} + \varpi''C''^{2})...(58).$$

Compare § 569 (7), (8) ; § 717 (55), (58); § 731 (99), (100).

718. The triple integral (53) denoted by E is of great importance, as being the expression for the whole kinetic energy in the hydro-kinetic analogue (Chapter XI. below). On account of the correspondence by opposites, which I perceived some years ago (§§ 733—739, below) between the forces experienced by solids held at rest in a moving liquid, and the forces experienced by magnetized matter in the corresponding cases of the magnetic analogue, I conclude that the diminution of the value of E produced by motion of any portion of matter, surrounded by space of uniform and isotropic permeability and not traversed by electric currents, is equal to the work required to effect the motion. Before proceeding to prove this proposition it is convenient to notice that the triple integral may be put into several other forms, each having a characteristic quality suitable for a class of applications.

719. These transformations will be simplified by, in the first place, substituting for electric currents, if there are any, distributions of intrinsic magnetization giving the same contributions to the values of S, S', S''; which may be done in an infinite variety of ways, as we see by the following considerations :—

For every closed circuit substitute (§ 548) an open magnetic shell producing the same potential as the circuit throughout space, except the portion occupied by the magnetized substance of the shell. The resultant force of the shell,

T. E.   36

reckoned in the magnetized substance according to the electro-magnetic definition (§ 517, *Postscript*), will throughout space be the same as that of the circuit. The values of $S, S'\ S''$, will be everywhere unchanged if the whole magnetized substance thus introduced be placed in space of zero susceptibility (or unit permeability), and be itself of zero susceptibility. But this cannot be if there are circuits completely imbedded in matter of other than zero susceptibility; if, for instance, part of the given system consists of an electric circuit through the aperture of a soft iron ring. Hence to avoid loss of gener-ality we must suppose some part, if not the whole, of the intrinsic magnetization, which we are now introducing, to be placed in portions of space having in the original data, sus-ceptibility different from zero. The magnetizing force in these portions of space will be altered by the substitution of mag-netization for electric current, but to make the whole external effect the same, we have only to add in them an intrinsic mag-netization equal to the inductive magnetization lost by the change.

720. As an illustration we may consider the familiar case of Ampere's electro-dynamic solenoid (§ 505, foot-note), with a soft iron core;—what is commonly called a bar electro-magnet. First, suppose there to be no soft iron core. We may do away with the current and substitute a uniformly and longitudinally magnetized bar of steel, with flat ends, occupying the whole internal space of the cylinder. This will, at every external point, give the same resultant force as the solenoid; and its resultant force, according to the electro-magnetic definition, will throughout its substance be the same as the resultant force of the solenoid throughout the cylindrical space between planes cutting it perpendicularly through its ends. In the substance of the steel magnet, the resultant force, according to the polar definition, will (§ 479) be merely the resultant of the force calculable from positive and negative planes of imaginary mag-netic matter coincident with its two ends; and this is what would be the magnetizing force due to the intrinsic magnetiza-tion of the steel if (§ 697) we attribute magnetic susceptibility to its substance, without depriving it of its intrinsic magnetiza-tion. It is of very small amount except very near the ends of

the bar, and is, throughout the interior, opposite in direction to the resultant force of the solenoid. To pass then from the case of a bar electro-magnet with core of soft iron or other substance susceptible of magnetic induction, to an arrangement producing the same external effects with intrinsic magnetization of the core instead of electric currents round it; we may first give to the core the intrinsic magnetization of the steel magnet we have just been considering, and superimpose upon this so much more of intrinsic magnetization as shall bring the whole magnetization of the core up to the resultant of the inductive magnetization which it has from the electric currents, and the uniform longitudinal magnetization which we attributed to the steel magnet. The core thus intrinsically magnetized and still retaining its magnetic susceptibility, will act the same upon all other magnets, and experience the same action from them, as the given electro-magnet. The same result may be also attained without attributing intrinsic magnetization to the core, in any case in which it is completely surrounded by matter of zero susceptibility; as is the case with an ordinary bar electro-magnet or horse-shoe electro-magnet, unless its ends be connected by an armature of soft iron or other susceptible substance (the substance of the electric conductor being supposed to be of zero-magnetic susceptibility). For in any such case the substance of the magnetic shells may be placed altogether outside the core of the electro-magnet, by hollowing them so that they may pass clear of the core round either end of it; or some of them round one end and some round the other so as to enclose the core among them. Then by supposing the substance of the shells to be of zero inductive susceptibility, we have a system in which the core is inductively magnetized in virtue of the intrinsic magnetization of the shells, to precisely the same degree as it was under the influence of the electric currents. The external resultant force is the same as that of the electro-magnet, being composed of a constituent due to the shells which is the same as that due to the electric currents, and a constituent due to the magnetization of the core, identical in the two cases.

721. Supposing then electric currents done away with by the process of § 719, we may simply take the data to be;—at any

point (x, y, z), intrinsic magnetization (α, β, γ), and inductive permeabilities $\varpi, \varpi', \varpi''$ along principal inductive axes (l, m, n), (l', m', n'), (l'', m'', n''). Thus (35) becomes

$$S = \varpi(\mathfrak{H} + \mathfrak{H}) + 4\pi A, \quad S' = \varpi'(\mathfrak{H}' + \mathfrak{H}') + 4\pi A', \quad S'' = \varpi''(\mathfrak{H}'' + \mathfrak{H}'') + 4\pi A'' \dots(59),$$

where $\mathfrak{H}$, $\mathfrak{H}'$, $\mathfrak{H}''$ denote the components along the principal inductive axes, of the resultant of $\mathfrak{F}, \mathfrak{G}, \mathfrak{H}$. Hence for $\dfrac{S^2}{\varpi}$ in (40) we may put $\left(\mathfrak{H} + \mathfrak{H} + 4\pi\dfrac{A}{\varpi}\right) S$, and so for the other terms. Now by the elementary formula for transformation of rectangular components, we have

$$(\mathfrak{H} + \mathfrak{H})S + (\mathfrak{H}' + \mathfrak{H}')S' + (\mathfrak{H}'' + \mathfrak{H}'')S'' = (\mathfrak{F} + \mathfrak{X})X + (\mathfrak{G} + \mathfrak{Y})Y + (\mathfrak{H} + \mathfrak{Z})Z \dots(60);$$

and because $(\mathfrak{F} + \mathfrak{X})\,dx\,(\mathfrak{G} + \mathfrak{Y})\,dy + (\mathfrak{H} + \mathfrak{Z})\,dz$ is a complete differential and $\dfrac{dX}{dx} + \dfrac{dY}{dy} + \dfrac{dZ}{dz} = 0$, we have

$$\int_{-\infty}^{\infty}\int_{-\infty}^{\infty}\int_{-\infty}^{\infty} dxdydz\,[(\mathfrak{F} + \mathfrak{X})X + (\mathfrak{G} + \mathfrak{Y})Y + (\mathfrak{H} + \mathfrak{Z})Z] = 0 \ (61).$$

Thus (53) becomes

$$E = \tfrac{1}{2}\int_{-\infty}^{\infty}\int_{-\infty}^{\infty}\int_{-\infty}^{\infty} dxdydz \left(\frac{AS}{\varpi} + \frac{A'S'}{\varpi'} + \frac{A''S'}{\varpi''}\right) \dots(62).$$

This is one of the transformed expressions promised in § 718.

722. To find the others, substitute for S, S', S'' their values by (59); and then remarking that, by the transformation of rectangular components,

$$(\mathfrak{H} + \mathfrak{H})A + (\mathfrak{H}' + \mathfrak{H}')A' + (\mathfrak{H}'' + \mathfrak{H}'')A'' = (\mathfrak{F} + \mathfrak{X})\alpha + (\mathfrak{G} + \mathfrak{Y})\beta + (\mathfrak{H} + \mathfrak{Z})\gamma \dots(63),$$

we find

$$E = \tfrac{1}{2}\int_{-\infty}^{\infty}\int_{-\infty}^{\infty}\int_{-\infty}^{\infty} dxdydz \left[(\mathfrak{F} + \mathfrak{X})\alpha + (\mathfrak{G} + \mathfrak{Y})\beta + (\mathfrak{H} + \mathfrak{Z})\gamma + \frac{4\pi A^2}{\varpi} + \frac{4\pi A'^2}{\varpi'} + \frac{4\pi A''^2}{\varpi''}\right] \dots(64).$$

Remarking that $(\mathfrak{F} + \mathfrak{X})\,dx + (\mathfrak{G} + \mathfrak{Y})\,dy + (\mathfrak{H} + \mathfrak{Z})\,dz$ is a complete differential, put

$$\mathfrak{F} + \mathfrak{X} = -\frac{dV}{dx}, \quad \mathfrak{G} + \mathfrak{Y} = -\frac{dV}{dy}, \quad \mathfrak{H} + \mathfrak{Z} = -\frac{dV}{dz} \dots(65).$$

Then integrating by parts in (64) as usual, we find

$$E = \tfrac{1}{2}\int_{-\infty}^{\infty}\int_{-\infty}^{\infty}\int_{-\infty}^{\infty} dxdydz \left[-V\rho + 4\pi\left(\frac{A^2}{\varpi} + \frac{A'^2}{\varpi'} + \frac{A''^2}{\varpi''}\right)\right] (66);$$

where [as in § 702 (2)]

$$\rho = -\left(\frac{d\alpha}{dx} + \frac{d\beta}{dy} + \frac{d\gamma}{dz}\right) = \frac{1}{4\pi}\left(\frac{d\mathfrak{F}}{dx} + \frac{d\mathfrak{G}}{dy} + \frac{d\mathfrak{H}}{dz}\right) \dots\dots(67).$$

Next, using in (66) the second of these expressions for ρ, and performing a set of integrations by parts: then putting

$$\mathfrak{F} = -\frac{d\mathfrak{A}}{dx}, \quad \mathfrak{G} = -\frac{d\mathfrak{A}}{dy}, \quad \mathfrak{H} = -\frac{d\mathfrak{A}}{dz} \dots\dots(68),$$

and performing another set of integrations by parts, we find the following two formulæ for E:—

$$E = \frac{1}{8\pi}\int_{-\infty}^{\infty}\int_{-\infty}^{\infty}\int_{-\infty}^{\infty} dxdydz\left[-(\mathfrak{F}+\mathfrak{X})\,\mathfrak{F} - (\mathfrak{G}+\mathfrak{H})\,\mathfrak{G} - (\mathfrak{H}+\mathfrak{Z})\,\mathfrak{H} + 16\pi^2\left(\frac{A^2}{\varpi} + \frac{A'^2}{\varpi'} + \frac{A''^2}{\varpi''}\right)\right](69);$$

$$E = \tfrac{1}{2}\int_{-\infty}^{\infty}\int_{-\infty}^{\infty}\int_{-\infty}^{\infty} dxdydz\left[-\mathfrak{A}(\rho+\sigma) + 4\pi\left(\frac{A^2}{\varpi} + \frac{A'^2}{\varpi'} + \frac{A''^2}{\varpi''}\right)\right]\dots\dots(70);$$

where

$$\sigma = \frac{1}{4\pi}\left(\frac{d\mathfrak{X}}{dx} + \frac{d\mathfrak{Y}}{dy} + \frac{d\mathfrak{Z}}{dz}\right) = -\left(\frac{d\xi}{dx} + \frac{d\eta}{dy} + \frac{d\zeta}{dz}\right) \dots(71).$$

Lastly, replacing in (70) ρ and σ by the first formula of (67) and the second of (71), integrating by parts, and using (68), we find

$$E = \tfrac{1}{2}\int_{-\infty}^{\infty}\int_{-\infty}^{\infty}\int_{-\infty}^{\infty} dxdydz\left[(\alpha+\xi)\,\mathfrak{F} + (\beta+\eta)\,\mathfrak{G} + (\gamma+\zeta)\,\mathfrak{H} + 4\pi\left(\frac{A^2}{\varpi} + \frac{A'^2}{\varpi'} + \frac{A''^2}{\varpi''}\right)\right] \;(70)\,bis;$$

which might have been had directly from (64) by taking the term $\mathfrak{X}\alpha + \mathfrak{Y}\beta + \mathfrak{Z}\gamma$ alone, and properly modifying the integral of it. Each of the three expressions (62), (64), (66), is remarkable as giving E by triple integration limited to space occupied by intrinsically magnetized matter: (although the integrations are marked as extending through all space, the evanescence of α, β, γ, A, A', A'', and ρ, wherever there is no intrinsic magnetization, limits the triple integrals to space where there is intrinsic magnetization). On the other hand, the expressions (70) and (70) *bis* are remarkable as giving E by triple integration through space occupied by matter possessing magnetization, whether intrinsically or by induction; that is to say, through those portions of space where there is intrinsic magnetization, and those portions where the permeability differs from unity. In (53) and (69) the integration extends generally through all space.

723. *Forces experienced by matter under magnetic influence.*— We shall still suppose, without loss of generality (§ 719), the electric currents in the given system to be done away with, and

a proper distribution of induced magnetization to be substituted for them. Let B be a portion of matter altogether surrounded by space of zero susceptibility or unit permeability. The force and couple experienced by B, regarded as a rigid body, is determinable by an application of § 500, when the whole magnetization (intrinsic and induced) of every part of B, and the resultant force at every point of its volume due to magnetization elsewhere, are known; or, *vice versâ*, when the magnetization of all other matter and the resultant force of B at every point of it, are known.

724. I shall conclude by adapting to our present case, in which part of the magnetization varies in virtue of magnetic induction, the method of § 502 for expressing the resultant of magnetic force on a rigid body, in terms of variations of a function of its co-ordinates, which in § 503 was worked out for the case of intrinsic (or rigid) magnetization alone. First, for a moment let the induced magnetization become rigid, and let all the given matter become unsusceptible of magnetic induction. Suppose the whole magnetized substance to be divided into infinitely small bars lying each in the direction of the magnetization, whether intrinsic or induced, or intrinsic and induced; and let W denote the amount of work which would be undone in separating these rigidly magnetized bars to infinite distances from one another. By (7) of § 569 we have

$$W = \tfrac{1}{2} \int_{-\infty}^{\infty} \int_{-\infty}^{\infty} \int_{-\infty}^{\infty} dx\,dy\,dz\, V(\rho + \sigma)\ldots\ldots\ldots(72).$$

725. Let now B denote any portion of the magnetized matter completely surrounded by space of zero susceptibility; and let Δ, prefixed to any function of (x, y, z), or to any configurational function of the system, denote augmentation produced by infinitesimal motion of B, the magnetization of B remaining unchanged (§ 72). The work required to produce this motion will be ΔW; and we have by (72)

$$\Delta W = \tfrac{1}{2} \int_{-\infty}^{\infty} \int_{-\infty}^{\infty} \int_{-\infty}^{\infty} dx\,dy\,dz\, [V\Delta(\rho + \sigma) + (\rho + \sigma)\Delta V]\ldots(73).$$

Now apply Poisson's equation

$$\frac{d^2 V}{dx^2} + \frac{d^2 V}{dy^2} + \frac{d^2 V}{dz^2} = -4\pi\,(\rho + \sigma)\ldots\ldots\ldots\ldots(74),$$

and we find, by two steps of integration by parts,

$$\int_{-\infty}^{\infty}\int_{-\infty}^{\infty}\int_{-\infty}^{\infty} dxdydz(\rho+\sigma)\Delta V=\int_{-\infty}^{\infty}\int_{-\infty}^{\infty}\int_{-\infty}^{\infty} dxdydz\, V\Delta(\rho+\sigma)..(75).$$

Hence instead of (73) we may write

$$\Delta W=\int_{-\infty}^{\infty}\int_{-\infty}^{\infty}\int_{-\infty}^{\infty} dxdydz\, V\Delta(\rho+\sigma)\ldots\ldots(76).$$

726. Consider now (part of the second member of this equation)

$$\int_{-\infty}^{\infty}\int_{-\infty}^{\infty}\int_{-\infty}^{\infty} dxdydz\, V\Delta\sigma.$$

Put in it [§ 722 (71)]

$$\Delta\sigma = -\left(\frac{d\Delta\xi}{dx}+\frac{d\Delta\eta}{dy}+\frac{d\Delta\zeta}{dz}\right)\ldots\ldots\ldots\ldots(77);$$

and perform integrations by parts as usual. We find

$$\int_{-\infty}^{\infty}\int_{-\infty}^{\infty}\int_{-\infty}^{\infty} dxdydz\, V\Delta\sigma=\int_{-\infty}^{\infty}\int_{-\infty}^{\infty}\int_{-\infty}^{\infty} dxdydz\left(\frac{dV}{dx}\Delta\xi+\frac{dV}{dy}\Delta\eta+\frac{dV}{dz}\Delta\zeta\right)$$
$$\ldots\ldots(78).$$

The second member of this expressed [§ 712 (33)], in terms of components along the principal axes of permeability, becomes

$$-\int_{-\infty}^{\infty}\int_{-\infty}^{\infty}\int_{-\infty}^{\infty} dxdydz\,(J\Delta\vartheta+J'\Delta\vartheta'+J''\Delta\vartheta'')\ldots\ldots(79),$$

where J, J', J'' denote the $\mathfrak{I}+\mathfrak{H}$, etc., of § 721, being the components of $-\dfrac{dV}{dx}$, $-\dfrac{dV}{dy}$, $-\dfrac{dV}{dz}$ along these axes. We have by (9)

$$\vartheta=\frac{\varpi-1}{4\pi}\,J,\ \vartheta'=\frac{\varpi'-1}{4\pi}\,J',\ \vartheta''=\frac{\varpi''-1}{4\pi}\,J''\ldots(80).$$

727. Remembering (§ 725) that Δ prefixed to any function of (x, y, z) denotes the augmentation which the function experiences when B is moved in any manner as a rigid body with its magnetization unchanged, while (80) expresses the actually varying inductive magnetization, we see that, throughout the volume of B,

$$\Delta\vartheta=\frac{J}{4\pi}\Delta\varpi+\frac{\varpi-1}{4\pi}\Delta_{\iota}J,\quad \Delta\vartheta'=\frac{J'}{4\pi}\Delta\varpi'+\frac{\varpi-1}{4\pi}\Delta_{\iota}J',$$

$$\Delta\vartheta''=\frac{J''}{4\pi}\Delta\varpi''+\frac{\varpi-1}{4\pi}\Delta_{\iota}J''\ldots(81),$$

where Δ_{ι} denotes augmentation produced by giving the actual motion to B, and ·moving all other magnetized matter as if rigidly connected with it, the axes of (x, y, z) being held fixed. Hence (79) is equal to

$$-\frac{1}{4\pi}\int_{-\infty}^{\infty}\int_{-\infty}^{\infty}\int_{-\infty}^{\infty} dx\,dy\,dz\,[(J^2\Delta\varpi + J'^2\Delta\varpi' + J''^2\Delta\varpi'')] -$$

$$\frac{1}{4\pi}\int_{-\infty}^{\infty}\int_{-\infty}^{\infty}\int_{-\infty}^{\infty} dx\,dy\,dz\,[(\varpi-1)J\Delta_1 J + (\varpi'-1)J'\Delta_1 J' + (\varpi''-1)J''\Delta_1 J'']\dots\dots(82),$$

where ϖ must be regarded as equal to unity through all space except that occupied by B. Now using the notation of § 730 (93), we have

$$\frac{1}{4\pi}(\varpi J\Delta_1 J + \varpi'J'\Delta_1 J' + \varpi''J''\Delta_1 J'') = \frac{dP}{d\frac{dV}{dx}}\Delta_1\frac{dV}{dx} + \frac{dP}{d\frac{dV}{dy}}\Delta_1\frac{dV}{dy} + \frac{dP}{d\frac{dV}{dz}}\Delta_1\frac{dV}{dz}\ (83);$$

and rectangular transformation gives

$$J\Delta_1 J + J'\Delta_1 J' + J''\Delta_1 J'' = \frac{dV}{dx}\Delta_1\frac{dV}{dx} + \frac{dV}{dy}\Delta_1\frac{dV}{dy} + \frac{dV}{dz}\Delta_1\frac{dV}{dz}\ (84).$$

Using these in the second term of (82) and performing integrations by parts, we reduce that term to

$$\int_{-\infty}^{\infty}\int_{-\infty}^{\infty}\int_{-\infty}^{\infty} dx\,dy\,dz\left[\frac{d}{dx}\frac{dP}{d\frac{dV}{dx}} + \frac{d}{dy}\frac{dP}{d\frac{dV}{dy}} + \frac{d}{dz}\frac{dP}{d\frac{dV}{dz}} - \frac{1}{4\pi}\left(\frac{d^2V}{dx^2}+\frac{d^2V}{dy^2}+\frac{d^2V}{dz^2}\right)\right]\Delta_1 V\,(85).$$

By (94) and (74) this becomes simply

$$\int_{-\infty}^{\infty}\int_{-\infty}^{\infty}\int_{-\infty}^{\infty} dx\,dy\,dz\,\sigma\Delta_1 V\dots\dots\dots\dots(86),$$

where σ must be regarded as zero through all space except that occupied by B.

728. Now from the definitions of Δ and Δ_1 it follows that $\Delta_1\sigma = \Delta\sigma$; and

$$\Delta_1\int_{-\infty}^{\infty}\int_{-\infty}^{\infty}\int_{-\infty}^{\infty} dx\,dy\,dz\,f(x, y, z) = 0\ \dots\dots(87),$$

where $f(x, y, z)$ denotes any function dependent on the configuration of the magnetized matter. Hence by taking $f(x, y, z) = \sigma V$ we see that

$$\int_{-\infty}^{\infty}\int_{-\infty}^{\infty}\int_{-\infty}^{\infty} dx\,dy\,dz\,\sigma\Delta_1 V = -\int_{-\infty}^{\infty}\int_{-\infty}^{\infty}\int_{-\infty}^{\infty} dx\,dy\,dz\,V\Delta\sigma\dots(88).$$

Substituting the second member of this equation for the second term of (82), and going back through (79) to (78): then transposing and halving, we find

$$\int_{-\infty}^{\infty}\int_{-\infty}^{\infty}\int_{-\infty}^{\infty} dx\,dy\,dz\,V\Delta\sigma = -\frac{1}{8\pi}\int_{-\infty}^{\infty}\int_{-\infty}^{\infty}\int_{-\infty}^{\infty} dx\,dy\,dz\,(J^2\Delta\varpi + J'^2\Delta\varpi' + J''^2\Delta\varpi'')\dots(89).$$

Finally, using this in (76), we find

$$\Delta W = \int_{-\infty}^{\infty}\int_{-\infty}^{\infty}\int_{-\infty}^{\infty} dx\,dy\,dz\,[\,V\Delta\rho - \frac{1}{8\pi}(J^2\Delta\varpi + J'^2\Delta\varpi' + J''^2\Delta\varpi'')\,]\ldots(90).$$

729. Now to prove § 718: let δ denote variation due to any motion of B as a rigid body, the magnetization of every portion of matter varying (according to its actual susceptibility) with the varying magnetizing force to which it is subjected. The work required to effect the motion of B, being infinitesimal, will be the same as if (according to the hypothesis of § 725) the actual magnetization were everywhere rigid. Hence if $\mathfrak{E} - c$ denote the work undone in removing B to an infinite distance from all other bodies possessing either intrinsic magnetization or magnetic susceptibility different from zero (that is to say, permeability differing from unity), and c a constant so far as the present variation is concerned [to be arbitrarily assigned later (731)], we have

$$\delta\mathfrak{E} = \Delta W\ldots\ldots(91).$$

730. Taking the variation of (66), § 722, we have

$$\delta E = -\tfrac{1}{2}\int_{-\infty}^{\infty}\int_{-\infty}^{\infty}\int_{-\infty}^{\infty} dx\,dy\,dz\,(V\delta\rho + \rho\delta V)\ldots(92),$$

as the term of the triple integral depending on $\dfrac{A^2}{\varpi} + \dfrac{A'^2}{\varpi'} + \dfrac{A''^2}{\varpi''}$ does not vary. Now putting

$$P = \frac{1}{8\pi}(\varpi J^2 + \varpi'J'^2 + \varpi''J''^2)\ldots\ldots(93),$$

we have, by (15),

$$\rho = -\left(\frac{d}{dx}\frac{dP}{d\frac{dV}{dx}} + \frac{d}{dy}\frac{dP}{d\frac{dV}{dy}} + \frac{d}{dz}\frac{dP}{d\frac{dV}{dz}}\right)\ldots\ldots(94).$$

Hence

$$\int_{-\infty}^{\infty}\int_{-\infty}^{\infty}\int_{-\infty}^{\infty} dx\,dy\,dz\,\rho\delta V = \int_{-\infty}^{\infty}\int_{-\infty}^{\infty}\int_{-\infty}^{\infty} dx\,dy\,dz\left(\frac{dP}{d\frac{dV}{dx}}\frac{d\delta V}{dx} + \frac{dP}{d\frac{dV}{dy}}\frac{d\delta V}{dy} + \frac{dP}{d\frac{dV}{dz}}\frac{d\delta V}{dz}\right)$$

As P is a quadratic function of $\dfrac{dV}{dx}, \dfrac{dV}{dy}, \dfrac{dV}{dz}$, the expression under the integral sign here is clearly a symmetrical function of $\dfrac{dV}{dx}, \dfrac{dV}{dy}, \dfrac{dV}{dz}$, and $\dfrac{d\delta V}{dx}, \dfrac{d\delta V}{dy}, \dfrac{d\delta V}{dz}$; and we may write it thus :—

$$\frac{dP}{d\frac{dV}{dx}}\frac{d\delta V}{dx}+\frac{dP}{d\frac{dV}{dy}}\frac{d\delta V}{dy}+\frac{dP}{d\frac{dV}{dz}}\frac{d\delta V}{dz}$$

$$=\frac{dV}{dx}\delta\frac{dP}{d\frac{dV}{dx}}+\frac{dV}{dy}\delta\frac{dP}{d\frac{dV}{dy}}+\frac{dV}{dz}\delta\frac{dP}{d\frac{dV}{dz}}-\frac{1}{4\pi}(J^2\delta\varpi+J'^2\delta\varpi'+J''^2\delta\varpi'')\ \ (95).$$

731.   Taking the first triple term alone and performing integrations by parts, we have

$$\int_{-\infty}^{\infty}\int_{-\infty}^{\infty}\int_{-\infty}^{\infty}dxdydz\left(\frac{dV}{dx}\delta\frac{dP}{d\frac{dV}{dx}}+\frac{dV}{dy}\delta\frac{dP}{d\frac{dV}{dy}}+\frac{dV}{dz}\delta\frac{dP}{d\frac{dV}{dz}}\right)$$

$$=-\int_{-\infty}^{\infty}\int_{-\infty}^{\infty}\int_{-\infty}^{\infty}dxdydz\,V\delta\left(\frac{d}{dx}\frac{dP}{d\frac{dV}{dx}}+\frac{d}{dy}\frac{dP}{d\frac{dV}{dy}}+\frac{d}{dz}\frac{dP}{d\frac{dV}{dz}}\right)=\int_{-\infty}^{\infty}\int_{-\infty}^{\infty}\int_{-\infty}^{\infty}dxdydz\,V\delta\rho\ (96).$$

Hence (92) becomes

$$\delta E=-\int_{-\infty}^{\infty}\int_{-\infty}^{\infty}\int_{-\infty}^{\infty}dxdydz\,V\delta\rho+\frac{1}{8\pi}\int_{-\infty}^{\infty}\int_{-\infty}^{\infty}\int_{-\infty}^{\infty}dxdydz\,(J^2\delta\varpi+J'^2\delta\varpi'+J''^2\delta\varpi'')\ (97).$$

Comparing this with (90), and remarking that, according to the definitions of Δ and δ (§§ 725, 729), we have $\Delta\rho=\delta\rho$, $\Delta\varpi=\delta\varpi$, $\Delta\varpi'=\delta\varpi'$, and $\Delta\varpi''=\delta\varpi''$, we see that

$$-\delta E=\Delta W\dots\dots\dots\dots\dots\dots(98),$$

which proves § 718.   In virtue of (98), (66), and (91) we may put

$$\mathfrak{E}=\tfrac{1}{2}\int_{-\infty}^{\infty}\int_{-\infty}^{\infty}\int_{-\infty}^{\infty}dxdydz\,V\rho\ \dots\dots\dots\dots(99).$$

By § 566 we see that this implies assigning to c of § 729 a value equal to the work which, *after B has been removed to an infinite distance*, must be undone to divide into infinitely thin bars every part of the system* possessing intrinsic magnetization and separate these bars to infinite distances from one another; their directions having been so chosen that when uninfluenced the magnetism of each is longitudinal. Thus we see that the function $\mathfrak{E}$ expressed by (99) is the "mechanical value" of the given magnetic system, according to the definition of § 567 extended to include material susceptible of magnetic induction along with intrinsically magnetized matter. It is essentially positive.   Were there no magnetic susceptibility in any of the

---

* Not omitting B though infinitely distant, if it has intrinsic magnetization.

material concerned, it would be identical with the $\mathfrak{E}$ of §§ 569, 570.

By (66) we have

$$\mathfrak{E} + E = 2\pi \int_{-\infty}^{\infty}\int_{-\infty}^{\infty}\int_{-\infty}^{\infty} dxdydz \left(\frac{A^2}{\varpi} + \frac{A'^2}{\varpi'} + \frac{A''^2}{\varpi''} \right) \ldots (100).$$

Compare § 717 (55), (58). For the particular case of zero susceptibility (or unit permeability) throughout the system, $\mathfrak{E}$ and E have the same significations as in § 569 above.

732. The expressions (62), (64), (66), (69), (70), (70) *bis*, for E, and (99) for $\mathfrak{E}$, depend on the exclusion of electric currents by which (§ 721) we simplified the formula for magnetic induction; but as (§ 719) this simplification did not involve any loss of generality, it is in reality proved that the configurational function E, expressed by the formula

$$E = \frac{1}{8\pi} \int_{-\infty}^{\infty}\int_{-\infty}^{\infty}\int_{-\infty}^{\infty} dxdydz \left(\frac{S^2}{\varpi} + \frac{S'^2}{\varpi'} + \frac{S''^2}{\varpi''} \right) \quad \text{[(53) of § 716 repeated]}$$

not involving the exclusion of electric currents, represents by its variations the forces experienced by detached portions of any system composed of intrinsically magnetized polar magnets, electromagnets, and inductively magnetized matter; thus:—The augmentation of this function produced by any motion of a rigid portion or portions of such a system, through space occupied by matter of zero susceptibility, is equal to the work gained by permitting the motion.

[Addition of date March 5th, 1884. The student is recommended to exercise himself by going through the whole investigation of §§ 700—732, for the simple case of equal permeability in all directions. It will then be seen that the seeming difficulties of the investigation as given above, are merely mathematical complexities essential to the expression of the formulae concerned, when the matter is aeolotropic.

In respect to "mechanical values" of magnetic and electromagnetic systems, the investigation for the case of isotropic matter is to be found in Article LXI. ("On the Mechanical Values of Distributions of Electricity, Magnetism, and Galvanism") of Vol. I., of my " Mathematical and Physical Papers." W. T.]

HYDROKINETIC ANALOGY FOR THE MAGNETIC
INFLUENCE OF AN IDEAL EXTREME DIAMAGNETIC.

*On the Forces experienced by Solids immersed in a Moving
Liquid.*

[From the *Proceedings of the Royal Society of Edinburgh* for Feb. 1870.]

733. Cyclic irrotational motion [*], [V. M. § 60 (z)] once established through an aperture or apertures, in a moveable solid
immersed in a liquid, continues for ever after with circulation
or circulations unchanged, [V. M. § 60 (a)] however the solid be
moved, or bent, and whatever influences there may be from other
bodies. The solid, if rigid and left at rest, must clearly continue
at rest relatively to the fluid surrounding it to an infinite distance, provided there be no other solid within an infinite distance
from it. But if there be any other solid or solids at rest within
any finite distance from the first, there will be mutual forces
between them, which, if not balanced by proper application of
force, will cause them to move. The theory of the equilibrium
of rigid bodies in these circumstances might be called Kinetico-
statics; but it is in reality a branch of physical statics simply.
For we know of no case of true statics in which some if not
all of the forces are not due to motion; whether, as in the case
of the hydrostatics of gases, thanks to Clausius and Maxwell,
we perfectly understand the character of the motion, or, as in
the statics of liquids and elastic solids, we only know that

---

[*] The references [V. M. §§] are to the author's paper on Vortex Motion,
recently published in the *Transactions* of the Royal Society of Edinburgh (1869),
which contains definitions of all the new terms used in the present article.
Proofs of such of the propositions now enunciated as require proof are to
be found in a continuation of that paper. [They are found in §§ 759—763, below.]

some kind of molecular motion is essentially concerned. The theorems which I now propose to bring before the Royal Society regarding the forces experienced by bodies mutually influencing one another through the mediation of a moving liquid, though they are but theorems of abstract hydrokinetics, are of some interest in physics as illustrating the great question of the 18th and 19th centuries:—Is action at a distance a reality, or is gravitation to be explained, as we now believe magnetic and electric forces must be, by action of intervening matter?

734. I. (Proposition.) Consider first a single fixed body with one or more apertures through it; as a particular example, a piece of straight tube open at each end. Let there be irrotational circulation of the fluid through one or more such apertures. It is readily proved [from V. M. § 63, *Exam.* (2.)]* that the velocity of the fluid at any point in the neighbourhood agrees in magnitude and direction with the resultant electro-magnetic force, at the corresponding point in the neighbourhood of an electro-magnet replacing the solid, constructed according to the following specification. The "core" on which the conductor is wound, is to be of any material having extreme diamagnetic inductive capacity †, and is to be of the same size and shape as the solid immersed in the fluid. The conductor is to form an infinitely thin layer or layers, with one circuit going round each aperture. The whole strength of current in each circuit reckoned in absolute electro-magnetic measure, is to be equal to the circulation of the fluid through that aperture divided by 4π. The resultant electro-magnetic force at any point will be numerically equal to the resultant fluid velocity at the corresponding point in the hydrokinetic system.

735. Thus, considering, for example, the particular case of a straight tube open at each end, let the diameter be infinitely small in comparison with the length. The "circulation" will exceed by but an infinitely small quantity the product of the velocity within the tube into the length. In the neighbour-

---

* Or from Helmholtz's original integration of the hydrokinetic equations.

† Real diamagnetic substances are, according to Faraday's very expressive language, relatively to lines of magnetic force, *worse conductors* than air. The ideal substance of extreme diamagnetic inductive capacity is a substance which completely *sheds off* lines of magnetic force, or which is perfectly impervious to *magnetic force* [or of zero "permeability," (§ 629)].

hood of each end, at distances from it great in comparison with the diameter of the tube and short in comparison with the length, the stream lines will be straight lines radiating from the end. The velocity, outwards from one end and inwards towards the other, will therefore be inversely as the square of the distance from the end. Generally at all considerable distances from the ends, the distribution of fluid velocity will be the same as that of the magnetic force in the neighbourhood of an infinitely thin bar longitudinally magnetized uniformly from end to end.

736. Merely as regards the comparison between fluid velocity and resultant magnetic forces, Euler's fanciful theory of magnetism (§ 573) is thus curiously illustrated. This comparison, which has been long known as part of the correlation between the mathematical theories of electricity, magnetism, conduction of heat, and hydrokinetics, is merely kinematical, not dynamical. When we pass, as we presently shall, to a strictly dynamical comparison relatively to the mutual force between two hard steel magnets, we shall find the same law of mutual action between two tubes, with liquid flowing through each, but with this remarkable difference, that the forces are opposite in the two cases; unlike poles attracting and like poles repelling in the magnetic system, while in the hydrokinetic analogue there is attraction between like ends and repulsion between unlike ends.

737. II. (Proposition.) Consider two or more fixed bodies, such as the one described in Prop. I. [§ 734]. The mutual actions of two of these bodies are equal, but in opposite direction, to those between the corresponding electro-magnets. The particular instance referred to above shows us the remarkable result, that through fluid pressure we can have a system of mutual action, in which like attracts like with force varying inversely as the square of the distance. Thus, considering tubes open at each end, with fluid flowing through them, if the exit ends be placed in the neighbourhood of one another, and the entering ends be at infinite distances, the mutual forces resulting will be simply attractions according to this law. The lengths of the tubes on this supposition are infinitely great, and therefore, as is easily proved from the conservation of energy, the quantities flowing out per unit of time are but infinitesimally affected by the mutual influence. [When any change is allowed in the relative

positions of two tubes by which work is done, a *diminution* of kinetic energy of the fluid is produced within the tubes, and at the same time an *augmentation* of its kinetic energy in the external space. The former is equal to double the work done; the latter is equal to the work done; and so the loss of kinetic energy from the whole liquid is simply equal to the work done.]

738. III. (Proposition.) Proposition II. holds, even if one of the bodies considered be merely a solid, with or without apertures; if with apertures, having no circulation through them. In such a case as this, the corresponding magnetic system consists of a magnet or electro-magnet, and a merely diamagnetic body, not itself a magnet, but disturbing the distribution of magnetic force around it by its diamagnetic influence. Thus, for example, a spherical solid at rest in the field of motion due to a fixed body through apertures in which there is cyclic irrotational motion, will experience from fluid pressure a resultant force through its centre equal and opposite to that experienced by a sphere of infinite diamagnetic capacity, similarly situated in the neighbourhood of the corresponding electro-magnet. Therefore, according to Faraday's law for the latter, and the comparison asserted in Prop. I. [§ 734], it would experience a force from places of less towards places of greater fluid velocity, irrespectively of the direction of the stream lines in its neighbourhood; a result easily deduced from the elementary formula for fluid pressure in hydrokinetics.

739. I have long ago shown [§ 646 above] that an elongated diamagnetic body in a uniform magnetic field tends, as tends an elongated ferromagnetic body, to place its length along the lines of force. Hence a long solid, pivoted on a fixed axis through its middle in a uniform stream of liquid, tends to place its length perpendicularly across the direction of motion; a known result (Thomson and Tait's *Natural Philosophy*, § 335). Again, two globes held in a uniform stream with the line joining their centres perpendicular to the stream, require force to prevent them from mutually approaching one another. In the magnetic analogue, two spheres of diamagnetic or ferromagnetic inductive capacity repel one another when held in a line at right angles to the lines of force. A hydrokinetic result similar

to this applied to the case of two equal globes, is to be found in Thomson and Tait's *Natural Philosophy*, § 332.

740. IV. (Proposition.) If the body considered in III., § 738 [be an infinitely small globe*, and] be acted on by force applied so as always to balance the resultant of the fluid pressure, calculated for it according to II. and III. for whatever position it may come to at any time, and if it be influenced, besides, by any other system of applied forces, superimposed on the former, it will move just as it would move, under the influence of the latter system of forces alone, were the fluid at rest, except in so far as compelled to move by the body's own motion through it. A particular case of this proposition was first published many years ago, by Professor James Thomson, on account of which he gave the name of "vortex of free mobility" to the cyclic irrotational motion symmetrical round a straight axis. [*Additional, Sept.* 14, 1872.—The same proposition holds for a globe of any dimensions, in a field of fluid motion consisting of circulation or circulations with infinitely fine rigid endless curve or curves for core, and no other rigid body in the liquid. Demonstration to appear in the *Proceedings of the Royal Society of Edinburgh* for 1871-2. †]

Extracts from two Letters to Professor Frederick Guthrie.

[From the *Philosophical Magazine* for June 1871.]

GLASGOW, *Nov.* 14*th*, 1870.

I HAVE to-day received the *Proceedings of the Royal Society* containing your paper "On Approach caused by Vibration," which I have read with great interest. The experiments you describe constitute very beautiful illustrations of the known theorem for fluid pressure in abstract hydrokinetics, with which I have been much occupied in mathematical investigations connected with vortex-motion.

741. According to this theorem, the average pressure at any point of an incompressible frictionless fluid originally at rest,

---

* [The proposition as originally published without limitation is obviously false, although that it is so I have only perceived to-day.—Sept. 2, 1872.]
† *Proceedings of the Royal Society of Edinburgh*, March 4, 1872.

but set in motion and kept in motion by solids moving to and fro, or whirling round in any manner, through a finite space of it, is equal to a constant diminished by the product of the density into half the square of the velocity. This immediately explains the attractions demonstrated in your experiments; for in each case the average square of velocity is greater on the side of the card nearest the tuning-fork than on the remote side. Hence obviously the card must be attracted by the fork as you have found it to be; but it is not so easy at first sight to perceive that the square of the average velocity must be greater on the surfaces of the tuning-fork next to the card than on the remote portions of the vibrating surface. Your theoretical observation, however, that the attraction must be mutual, is beyond doubt valid, as we may convince ourselves by imagining the stand which bears the tuning-fork and the card to be perfectly free to move through the fluid. If the card were attracted towards the tuning-fork, and there were not an equal and opposite force on the remainder of the whole surface of the tuning-fork and support, the whole system would commence moving, and continue moving with an accelerated velocity in the direction of the force acting on the card—an impossible result. It might, indeed, be argued that this result is not impossible, as it might be said that the kinetic energy of the vibrations could gradually transform itself into kinetic energy of the solid mass moving through the fluid, and of the fluid escaping before and closing up behind the solid. But "common sense" almost suffices to put down such an argument, and elementary mathematical theory, especially the theory of momentum in hydrokinetics explained in my article on "Vortex-motion,"* negatives it.

742. The law of the attraction which you observed agrees perfectly with the law of magnetic attraction in a certain ideal case which may be fully specified by the application of a principle explained in a short article [§ 733...740] communicated to the Royal Society of Edinburgh in February last [1870], as an abstract of an intended continuation of my paper on "Vortex-motion." Thus, if we take as an ideal tuning-fork two globes or disks

---

* *Transactions of the Royal Society of Edinburgh*, read 29th April, 1867.

moving rapidly to-and-fro in the line joining their centres, the corresponding magnet will be a bar with poles of the same name as its two ends and a double opposite pole in its middle. Again, the analogue of your paper disk is an equal and similar diamagnetic of extreme diamagnetic inductive capacity [§ 734]. The mutual force between the magnetic and the diamagnetic will be equal and opposite to the corresponding hydrokinetic force at each instant. To apply the analogy, we must suppose the magnet to gradually vary from maximum magnetization to zero, then through an equal and opposite magnetization back through zero to the primitive magnetization, and so on periodically. The resultant of fluid pressure on the disk is not at each instant equal and opposite to the magnetic force at the corresponding instant, but the average resultant of the fluid pressure is equal to the average resultant of the magnetic force. Inasmuch as the force on the diamagnetic is generally repulsion from the magnet, however the magnet be held, and is unaltered in amount by the reversal of the magnetization, it follows that the average resultant of the fluid pressure is an attraction on the whole towards the tuning-fork, into whatever position the tuning-fork be turned relatively to it. . . .

Nov. 23, 1870.

743. . . . There are, no doubt, curiously close analogies between some of the circumstances of motion in contiguous fluids of different densities, and the distribution of magnetic force in a field occupied by substances of different inductive capacities. Thus, if in a great space occupied by frictionless incompressible liquid denser in some portions than in others, a solid be suddenly set in motion, the lines of the fluid motion first generated agree perfectly [compare §§ 751...763 below] with the permanent lines of magnetic force in a correspondingly heterogeneous medium under the influence of a bar-magnet, to be substituted for the moveable solid and placed with its magnetic axis in the line of the solid's motion. As to amounts, the fluid velocity multiplied into the density is simply equal to the resultant magnetic force at each point, if the particular definition [the "electromagnetic definition" (§ 517, *Postscript*)] of the resultant magnetic force in a medium of

heterogeneous inductive capacity, given in the foot-note to [§ 516 above] § 48 of my paper on the "Mathematical Theory of Magnetism,*" be adopted. But here the analogy ends; the rigidity in virtue of which a solid moveable in a fluid medium differing from it in magnetic inductive capacity keeps its form, does not exist [contrast § 751 below] in the hydro-kinetic analogue. . . .

Report of an Address on the Attractions and Repulsions due to Vibration, observed by Guthrie and Schellbach.

[From the *North British Daily Mail* for Dec. 15, 1870; and *Proceedings of the Philosophical Society of Glasgow* for Dec. 14, 1870.]

744.   The speaker began by stating that interesting papers had recently appeared in the *Proceedings of the Royal Society* and the *Philosophical Magazine*, by Professor Guthrie, in which some very curious hydrokinetic phenomena were described. From hints and suggestions in his paper, it seems that Prof. Guthrie connected in his own mind these phenomena with possibilities of explaining some of the more recondite actions in nature; and he (the speaker) believed that what gave the great charm to these investigations for Prof. Guthrie himself, and no doubt also for many of those who heard his expositions and saw his experiments, was, that the results belong to a class of phenomena to which we may hopefully look for discovering the mechanism of magnetic force, and possibly also the mechanism by which the forces of electricity and of gravity are transmitted. The speaker, however, did not lay any stress at present upon the possibility of applying these results directly to explain magnetism. He believed, on the contrary, that the true kinetic theory of magnetism (and the ultimate theory of magnetism is undoubtedly kinetic) [compare § 290 and § 546, foot-note above] involves quite a different class of motions from those to which the beautiful phenomena discovered by Prof. Guthrie are due. He rather wished to point out the close con-nexion that existed between the laws of some of these actions and the laws of magnetism, which, while involving some remark-

---

* *Philosophical Transactions*, June 21, 1849. Published in Part I. for 1851. [§§ 504—523 above.]

able coincidences, involves certain contrasts decisive against any hypothesis, such as the ingenious one [§ 573 above] of Euler, explaining magnetism by fluid motion directly comparable with that which forms the subject of the present communication.

745. One of the most brilliant steps made in philosophical exposition of which any instance existed in the history of science, was that [§ 634 foot-note, and § 643 above] in which Faraday stated, in three or four words, intensely full of meaning, the law of the magnetic attraction or repulsion experienced by inductively magnetized bodies. He pointed out that a small globe or cube of soft iron tended in a certain direction when free to move in the magnetic field; while small detached fragments of inductively magnetized substances of the kind which he called diamagnetic, tended in the contrary direction; and that the precise specification of the direction in which the diamagnetic tended "was from places of stronger to places of weaker force."

746. By means of diagrams, the speaker then showed the action of magnets upon small pieces of soft iron in various positions, in the several cases in which the magnetic force is due to a bar-magnet, a horse-shoe magnet, and two bar-magnets placed side by side with their similar poles in the same direction. A diagrammatic illustration of "the lines of magnetic force," in the case of a bar-magnet, was also given. In the case of the horse-shoe magnet, it was pointed out that the small globe of soft iron would have a position of stable equilibrium in the line joining the poles, if free to move in the horizontal line bisecting that line at right angles; this stable position being the point of greatest force. The attraction experienced would be towards this point; so that if the globe were placed inside this point—that is to say, nearer the bend of the magnet—it would seem to be repelled on the whole by the mass of steel while moving towards the place of strongest force. In the case of two bar-magnets placed side by side [§ 645 above] with their similar poles in the same direction, it was pointed out that, for each pair of similar poles, there is a zero, or place of no force, mid-way between the two bars, and nearly in the line joining the ends. A globe of soft iron moveable midway between the

two bars is repelled, as it were, from each of the points of zero force, and finds a position of maximum force, which is one of stable equilibrium, on either side of either of the zeros.   Faraday's law [§ 634, foot-note above] showed that the soft iron was attracted from places of weaker to places of stronger force, quite irrespectively of the directions of the lines of force and thus summed up a great variety of very curious and puzzling phenomena in one sentence.

747.   This expression is perfectly applicable to small bodies at rest in an irrotationally moving fluid; with the substitution of "stream lines," instead of Faraday's "lines of magnetic force," and "greater or smaller fluid velocity," instead of "stronger or weaker magnetic force."

748.   Mathematicians were content to investigate the general expression of the resultant force experienced by a globe of soft iron in all such cases; but Faraday, without mathematics, divined the result of the mathematical investigation [§§ 638, 639, and §§ 671...681 above]; and, what has proved of infinite value to the mathematicians themselves, he has given them an articulate language in which to express their results.   Indeed, the whole language of the magnetic field and "lines of force" is Faraday's.   It must be said for the mathematicians that they greedily accepted it, and have ever since been most zealous in using it to the best advantage.

749.   Suppose a tube sunk in a perfect fluid, and the fluid by some means set to enter the one end and flow out by the other, the particles of it would follow the lines of magnetic force.   The magnetic field of force in the neighbourhood of a bar-magnet corresponded exactly with the straight tube taking water in at one end and discharging it at the other.   If two such tubes were presented with like ends to each other, they attracted, but with unlike ends, they repelled,—thus acting differently from two magnets placed in similar relative positions. But, except in being precisely opposite in direction, the resultant action between the supposed tubes and that between two bar-magnets follows rigorously the same law, both as to magnitude and as to line of. action.   This conclusion, and some others, containing the explanation of most of the experiments now to be shown to the Society, had been worked out mathe-

matically by the speaker, and communicated by him to the Royal Society of Edinburgh*.

750. It had been found by Faraday that the lines of magnetic force were diverted outwards from itself by a diamagnetic body placed in the field. If a body existed of extreme diamagnetic inductive capacity, the lines of magnetic force would pass altogether round it, and none of them through it. This is precisely the phenomenon, with reference to stream lines, which is met with in the hydrokinetic analogue. The speaker then drew attention to some small egg-shells which were suspended so as to move freely, each in a horizontal circle. By slightly waving the hand in front of the egg-shells they were attracted, and the same phenomenon was produced by holding in their neighbourhood a vibrating tuning-fork. This corresponded to the behaviour of a diamagnetic in the magnetic field, only that the direction of the motion was opposite. By means of a very delicate anemometer it was shown that the phenomena were independent of currents of air. The speaker showed that in whatever position, with one exception, the fork was held, the attraction was produced. The magnetic analogue to this fork would be a non-magnetic frame substituted for the tuning-fork, and bearing two small magnets laid across the ends, with similar poles pointing towards each other. In this case there would be a zero point in the middle, between the near poles. The same is true of the fluid velocity in the case of the tuning-fork. It would repel the suspended egg-shells from the zero point; but the experiment was one of too great delicacy for a lecture-room. Some very interesting experiments upon flames had been made by Mr Tatlock, his assistant, which the speaker had much pleasure in showing to the Society. A vibrating fork was supported horizontally, and the flame of a candle brought near the vibrating ends. All that part of the flame on a level with the fork was repelled, and bent down in the opposite direction, as if by a current of air. On the vibration being stopped, the flame at once assumed its upright form. A tall flame, obtained from ordinary coal gas, was next brought into

---

* *Proceedings, Royal Society, Edinburgh*, February 1870 [§§ 733—740, above.]

proximity to the vibrating fork, when the middle part of the flame was drawn out towards the fork, the upper and lower parts being repelled. In concluding, the speaker remarked, that it would be very wrong if he were to say that these experiments on the hydrokinetic analogue contained a direct opening up of the question of the mechanism of magnetic forces. They did not go any way towards explaining magnetic forces; but it was impossible to look upon them without feeling that they suggested the possibility of some very simple dynamical explanation.

XLII. *General Hydrokinetic Analogy for Induced Magnetism.*

February 1872. [*Compare* § 743 *above.*]

751. Imagine an infinitely fine-grained porous solid permeated by a frictionless incompressible liquid. The constitution of the supposed porous material will, for brevity, be designated as molecular, and although we might suppose it to depend on perforations in all directions, and everywhere opening into one another all through a continuous rigid solid, it will generally be more convenient to imagine it as made up of two classes of constituents;—(1) small detached rigid particles or molecules, each somehow held absolutely at rest, unless we find it convenient to apply force to it and move it: (2) closed infinitely fine curves of solid matter. It will be convenient to suppose each molecule to be a ring (that is to say a solid with at least one perforation through it); or at all events to suppose a considerable proportion of the molecules through any finite portion of space to be annular. This supposition gives the foundation (§§ 573...583 above) for the hydrokinetic analogue to a permanent polar magnet. Thus (§ 574) cyclic irrotational motion ["Vortex Motion," § 59 (*f*) and § 60 (*z*)*] through an infinitesimal solid ring constitutes a perfect analogy for an infinitely small portion of a permanent polar magnet. Again, when the kinematic analogy for a linear closed current (§ 535 above) is desired, we shall suppose an infinitely fine closed curve, which to avoid circumlocution I shall call an ityoid (*Proceedings, Royal Society of Edinburgh*, Dec. 18, 1871), of solid material to be placed, threading through among the interstices of the molecules and everywhere infinitely near the line of the electric current, but not in any case passing through the perforation of an annular molecule. By using a temporary membrane drawn across such an ityoid ("Vortex Motion," § 62)

---

* *Transactions, Royal Society of Edinburgh*, April, 1867 and Dec. 1869.

to generate cyclic irrotational motion, with no circulation through any other aperture than that of the ityoid itself, a perfect hydrokinetic analogue to the electro-magnetic effect of a fixed linear current of constant strength is obtained. An infinite number of ityoids placed infinitely near one another, nowhere in contact, but everywhere leaving sufficient interstices for the liquid to flow among them, gives the foundation for the hydrokinetic analogue to a solid electro-magnet (§ 535 above).

752. Let any cylindrical or prismatic portion of the supposed porous solid, terminated by planes perpendicular to the cylindrical surface or sides, be fixed in a tube of impermeable material fitting close to it all round, but leaving its ends free. This porous plug will constitute an obstruction, but not an absolute barrier, against the flow of a liquid through the tube. Imagine now two perfectly fitting frictionless pistons to be placed on the tube at any distance on the two sides of the plug, and let the whole space bounded by the pistons, the tube, and the impermeable constituents of the porous solid, be occupied by frictionless incompressible liquid. Let the liquid be set in motion by force applied to either or both the pistons. The motion will be determinate in every part of the fluid according to the condition [Thomson and Tait's *Natural Philosophy*, § 317, *Example* (3)] that the kinetic energy is less than that of any other motion of the liquid consistent with the given motion of the pistons. If the lengths of the clear portions of tube between the pistons and the two ends of the obstructing plug be very great in comparison with the diameter of the tube, it is easily seen that however coarse or heterogeneous be the porous material, the motion of the liquid will be sensibly uniform and in parallel lines through all the distant parts of the tube. But if the porous material be infinitely fine-grained and homogeneous as to the average structure of all equal and similar finite portions, the motion of the liquid will be uniform and in parallel lines at all finite distances on each side of the plug. If, as an extreme case, the plug be a continuous solid, with an infinite number of infinitely fine cylindrical perforations parallel to its length, the velocity of the liquid through it would be uniform, and would be to the velocity through the clear portions of the tube, in the inverse ratio of the areas traversed, that is to say,

in the ratio of the sectional area of the clear tube to the sum of the sectional areas of the perforations. The mass of the fluid in the perforations at any instant, would be to the mass in an equal length of the clear tube, as the sectional area of the tube to the sum of the sectional areas of the perforations; and therefore the kinetic energy of the whole motion in the perforations would be to the kinetic energy in an equal length of the clear tube, in the inverse ratio of the areas, that is to say, in the ratio of the whole sectional area of the tube to the sum of the sectional areas of the perforations. Hence, generally the greater the obstruction offered by a plug consisting of any kind of porous material, the greater will be the ratio of the kinetic energy of the liquid permeating through it, to that of the liquid moving freely in an equal length of clear tube; and (borrowing the word "permeability" from Le Sage), we may say that the permeability of the plug is inversely as the kinetic energy of the liquid permeating through it, when the velocity of the fluid in the clear parts of the tube is given.

753. If we were only occupied with hydrokinetics it would be natural to call the permeability of the clear parts of the tube unity. This would make unity the measure of perfect permeability, and would give always a proper fraction for the measure of the permeability of a porous solid. But in view of the magnetic analogy it is more convenient to call the permeability of some particular porous material unity, and to define the permeability of any other material as the number by which we must multiply the kinetic energy of the fluid permeating through a plug of it, to find the kinetic energy in a plug of equal length of the standard material fixed in the same tube. And further, for the magnetic analogy (compare § 732 above) it is convenient to attribute to the supposed liquid such a density that 4π times the kinetic energy of liquid permeating a solid of unit permeability, reckoned per unit volume of the whole space occupied by porous solid and liquid shall be equal to half the square of the "flux;" the word flux being borrowed from Fourier's theory of the conduction of heat and adapted to the use we have to make of it by the following definition :—

754. The component flux in any direction is the whole volume of the liquid traversing a plane perpendicular to this direction

per unit of area per unit of time. In the complicated motion of
the liquid through the interstices of the porous solid, the component velocity perpendicular to any plane may be in contrary
directions at different points of the plane; but in reckoning the
flux we must take the excess (positive or negative) of the
quantity crossing in the direction called positive above that
which crosses in the direction called negative. By considering
a tetrahedral portion of space (whether clear or occupied by
porous solid) bounded by three mutually rectangular planes
and a fourth plane cutting them all, we see immediately that
the composition of fluxes follows the ordinary law of the composition of velocities or the composition of forces; an elementary proposition due to Fourier.

755. Let X, Y, Z denote, for any possible motion of the
liquid, the components of flux at any point (x, y, z) referred to
rectangular co-ordinates. X, Y, Z must (§ 540 above) fulfil the
equation $$\frac{dX}{dx} + \frac{dY}{dy} + \frac{dZ}{dz} = 0 \quad \dots\dots\dots\dots\dots(1),$$
called the "equation of continuity."

756. In general the permeability of a porous solid may be
supposed to be different in different directions. When it is so
the structure is of course to be called æolotropic (Thomson and
Tait's *Natural Philosophy*, § 676; quoted above, § 604, foot-
note). Still denoting by X, Y, Z the components of flux in
three directions at right angles to one another, denote by Q the
kinetic energy per unit of volume, which must be a quadratic
function of X, Y, Z. Hence, by the ordinary analysis of quadratic functions, we see that there are three determinate directions (l, m, n), (l', m', n'), (l'', m'', n''), at right angles to one
another, to be called (according to analogy of ordinary usage)
the principal axes of permeability, and three determinate con-
stants ϖ, ϖ', ϖ'' to be called the principal permeabilities, in
terms of which we have the following expression for Q:—

$$Q = \frac{1}{8\pi} \left\{ \frac{(lX + mY + nZ)^2}{\varpi} + \frac{(l'X + m'Y + n'Z)^2}{\varpi'} + \frac{(l''X + m''Y + n''Z)^2}{\varpi''} \right\} \dots (2).$$

757. Now let us suppose the whole of space to be occupied
by a rigid porous solid of infinitely fine-grained texture with
different degrees of permeability and æolotropic quality in

different parts; and let a frictionless incompressible liquid
initially at rest fill all the interstices. In a portion M of the
porous solid (to represent the "inducing magnet" in the magnetic analogue), let some of the constituent molecules be annular, and let the apertures of some of the rings be temporarily
closed by infinitely thin flexible and extensible membranes.
(It is a matter of indifference whether there be other rings or
not either in M or elsewhere.) Let impulsive pressure be
applied to these membranes, uniform on each, but not necessarily of equal values for the different membranes; and instantly let all the membranes be dissolved. The motion of the
fluid will be everywhere irrotational and determinate ["Vortex
Motion," § 62 and § 62 (c)*], and will be of the class called
polycyclic ["Vortex Motion," § 60 (x)*]. The kinetic energy of
the whole fluid motion produced will [Thomson and Tait's
Natural Philosophy, § 317 *Example* (3)] be less than that of any
other motion consistent with the incompressibility of the fluid,
having the same normal component velocity at each point of the
supposed membrane surfaces. A partial application of the same
theorem shows that if we leave out of account the fluid motion
within any surface S, completely enclosing M, and consider the
normal component velocity as given at each point of this surface, the kinetic energy of the fluid motion through the rest of
space will be less than that of any other motion with the same
normal component velocity at each point of S.

758. To find the analytical expression of this condition let
$\iiint dxdydz$ denote integration through all space except that
enclosed by S. Then X, Y, Z must, subject to equation (1), be
such functions of (x, y, z) as to make $\iiint Q dxdydz$ a minimum.
Hence, λ denoting an indeterminate multiplier, we have

$$\left[\iiint \delta Q dxdydz + \lambda \left(\frac{d\delta X}{dx} + \frac{d\delta Y}{dy} + \frac{d\delta Z}{dz} \right) \right] = 0 \ \ldots\ldots(3).$$

Applying the usual process of integration by parts to the terms
involving λ, we find

$$\delta \iiint dxdydz\lambda \left(\frac{d\delta X}{dx} + \frac{d\delta Y}{dy} + \frac{d\delta Z}{dz} \right) = \iint dS\lambda (l\delta X + m\delta Y + n\delta Z)$$

$$- \iiint dxdydz \left(\frac{d\lambda}{dx}\delta X + \frac{d\lambda}{dy}\delta Y + \frac{d\lambda}{dz}\delta Z \right),$$

---

* *Transactions, Royal Society of Edinburgh,* April 1867 and Dec. 1869.

where $\iint dS$ denotes integration over the whole bounding surface of the space included in the triple integral, and l, m, n are the direction-cosines of the normal. For the infinitely distant parts of the boundary the double integral vanishes, as by hypothesis there is no motion there; and for the boundary of M (which is the remainder of the boundary of the space included in the triple integral) the double integral vanishes, because the condition that the normal component velocity is given over the boundary of M, requires that

$$l\delta X + m\delta Y - n\delta Z = 0.$$

Hence as Q involves only X, Y, Z, and not their differential coefficients, the variational equation (3) gives

$$\frac{dQ}{dX} = \frac{d\lambda}{dx}, \quad \frac{dQ}{dY} = \frac{d\lambda}{dy}, \quad \frac{dQ}{dZ} = \frac{d\lambda}{dz} \quad \ldots\ldots\ldots\ldots\ldots(4).$$

These equations, with (1) and (2), §§ 755, 756, and

$$lX + mY + nZ = N \quad \ldots\ldots\ldots\ldots\ldots\ldots(5).$$

for every point of the boundary of M, where N denotes the given normal component velocity, suffice to determine X, Y, Z for every point of space external to M. Comparing them with equations (43), (42), and (40) of § 713 above, we see that they are simply the equations of the magnetic induction through space external to M, due to any distribution of magnetization or of electric currents within M; if ϖ, ϖ', ϖ'' be the three principal magnetic permeabilities, and (l, m, n), (l', m', n'), (l'', m'', n'') the principal axes at any point (x, y, z); X, Y, Z the components of the resultant force at the same point according to the electromagnetic definition; and N its normal component at any point of a surface M, which completely encloses the inducing magnet.

759. Considering next the fluid motion within the space M, and its electro-magnetic analogue, we see from equations (42) of § 713 above, that

$$\frac{d}{dy}\frac{dQ}{dZ} - \frac{d}{dz}\frac{dQ}{dY}, \quad \frac{d}{dz}\frac{dQ}{dX} - \frac{d}{dx}\frac{dQ}{dZ}, \quad \frac{d}{dx}\frac{dQ}{dY} - \frac{d}{dy}\frac{dQ}{dX} \quad (6),$$

where they are not zero are equal to the component intensities of the electric flow (§ 539 above), at (x, y, z), in a determinate distribution of electric currents, which, with the magnetism induced by it throughout space, produces resultant electro-magnetic force (X, Y, Z) at any point (x, y, z). Suppose now any motion to be given (§ 751 above) to solid material in space external to M, or any cyclic irrotational motion of the liquid to be generated by the aid of membranes temporarily stopping apertures of solids in the space external to M; this will alter the motion already existing by compounding with it the motion which the supposed actions external to M would produce of themselves in the liquid if given motionless. Now from (4) it follows that throughout M, the values of the functions (6) are zero for the second supposed component of the motion. Hence, throughout M the functions (6) being linear functions of the flux components, remain unchanged in the altered motion of the liquid. It follows that their values through any portion of space, throughout which the molecular constitution of the solid matter is completely given, are determinable from the cyclic constants of the fluid motion through all the rings in this part of space, independently of the molecular constitution, or of circulations through apertures in other parts of space. From this, lastly, we see that if M be moved in any manner, translationally or rotationally, with all its parts kept rigidly connected, and the axes of co-ordinates moving along with it, and if it be brought to rest in an altered position, the values of the functions (6) will be the same as they were before the motion. This motion of M as a rigid body implies, of course, motions and changes of molecular arrangement in the solid matter of surrounding space which are altogether arbitrary, subject only to the condition of making way for M.

760. The analogy may be further extended to include the resultant force experienced by the inducing magnet, or by any moveable solid portion of matter experiencing its inductive influence. To do this, consider the effect of any variation of the solid matter concerned in the hydrokinetic analogue. First, it must be remarked that the effect of the change in the molecular distribution of the solid matter in the space M upon the motion of the fluid, cannot be determined from mere knowledge of the change which it produces in that average quality of the material which I have defined above (§ 752) as its permeability. For without changing the permeability we may so alter the molecular arrangement within M as to change to any degree we please

the flux of the fluid in this space, and therefore also the fluid motion through space external to M. Conceive, for instance, an infinitesimal molecular change to be produced which, without altering the "permeability" of the group, shall very much contract infinitesimal apertures through which there is circulation. This may be done either by altering the shapes of infinitesimal molecular rings, or by bringing other molecules towards the apertures of rings so as to obstruct passage through them. The circulation through each aperture remains ("Vortex Motion," § 59*) constant, but it is clear that the whole kinetic energy may be diminished as much as we please by the supposed process.

761. Let now A denote the solid matter in any portion of space which may be either the whole of M or altogether external to M. Let the permeability outside of A be uniform through some finite space all round it. Keeping A rigid throughout, alter its position infinitesimally; keep the permeability unchanged in the space immediately contiguous with it, by forces applied to surrounding molecules obliged to give way to it during its motion; and keep all other portions of solid matter in external space rigidly connected with one another. The work done by forces applied to A and the surrounding molecules to produce their supposed motions must be equal to the augmentation experienced by the integral $\int_{-\infty}^{\infty}\int_{-\infty}^{\infty}\int_{-\infty}^{\infty} Q\,dx\,dy\,dz$. This is the same as the amount of work required to give the corresponding motion to the portion of matter corresponding to A in the magnetic analogue; a consequence of § 731 above, with the consideration that both in the hydrokinetic system and the magnetic analogue, the values of $\dfrac{d}{dz}\dfrac{dQ}{dY} - \dfrac{d}{dy}\dfrac{dQ}{dZ}$, etc., are (§ 759 above) not altered by the supposed change of A's position.

762. The necessarily complicated character of the dynamical action required to produce the supposed motion of A and rearrangement of the surrounding molecules disappears altogether in the case in which a finite shell of space contiguous with A all

---

* *Transactions, Royal Society of Edinburgh*, April 1867 and Dec. 1869.

round is free from solid molecules. In this case the (generalized) component forces required to give any infinitesimal motion whatever to A (compare § 502 above), will be simply the differential co-efficients of Q with reference to the corresponding co-ordinates; and the forces required to balance A in any position, will be equal and opposite to these forces. Hence the force required to balance A in this case of the hydrokinetic system will be equal and opposite to the force required to balance a rigid body corresponding to A in the magnetic analogue. In the latter, the analogue to the space round A, clear of solids, but traversed by liquid, may [notwithstanding the different convention (§ 753 above) more generally adopted] be air. This particular convention being adopted for an instant, the magnetic analogue for all portions of space occupied by the "porous solid," described in § 751 above, or by continuous finite solid substance, will be diamagnetic material of any permeability from unity (that of air) to zero (that of ideal substance of extreme diamagnetic quality). The analogue of M may be either a real ordinary electro-magnet consisting of an electric current, or distribution of currents through solid conductors of diamagnetic material; or an ideal polar-magnet (§ 697 above) of diamagnetic inductive quality. But it is to be remarked that by choosing air for the magnetic analogue of space unobstructed by solids in the hydrokinetic system, we exclude all ferro-magnetic induction from the analogy.

763. Using now the general proposition of § 761, and making the proper particular suppositions regarding the moveable body A, we not only prove Propositions II. and III. of §§ 737, 738 above, but extend their application to real bodies of any degrees of diamagnetic inductive capacity instead of the ideal bodies of "extreme" diamagnetic quality (zero magnetic permeability) imagined in those propositions.

INDEX.

ACCUMULATOR, uniform current, § 408–411

Action of a small plane closed circuit on an element of another complete electro-magnet or magnet, § 546

Æolotropic, § 604, foot-note

Analogy, Hydrokinetic, §§ 573–583, 733–763

Atmospheric Electricity, early observers of, § 267

—— method of observing, §§ 262–266

—— new apparatus for observing, § 391

—— Notes on, §§ 392–399

—— Observations on, §§ 296–300

—— on the necessity for incessant recording, and for simultaneous observations in different localities to investigate, § 295

Atoms, size of, § 400

Attractions and repulsions due to vibration observed by Guthrie and Schellbach, Report of an address on the, § 744

Attraction of a uniform spherical surface on an external point, § 87

—— propositions in the theory of, §§ 187–205

CAPACITY of conductors, §§ 51–56

Cavendish, § 34, foot-note

—— ratio of the capacity of a disc to that of a sphere of the same diameter, § 235, foot-note

Certain partial differential equations, theorems with reference to the solution of, § 206

Coercive force, §§ 609, 630

Collector, water dropping, §§ 262, 266, 287

—— burning match, §§ 261, 286

Condenser, sound produced by the discharge of a, § 302

Conducting and non-conducting electrified bodies, on the attractions of, §§ 144, 148

—— sphere, determination of distribution on a, § 77

Conducting surfaces external and internal, § 97

Conductors, insulated, § 71

—— of electricity, § 68

Conditions to which the distributions of galvanism in solid and superficial electromagnets is subject, investigation of, §§ 539–546

Cone, area of segment cut from a spherical surface by a small, § 86

—— orthogonal and oblique sections of a small, § 85

—— the solid angle of a, § 81

Cones, definitions regarding, § 80

Contact electricity, new proof of, § 400

Coulomb's experiments, § 25

Crystalline and non-crystalline bodies, theory of magnetic induction in, §§ 604–624

Cyclic irrotational motion, § 733

DENSITY, electric, § 330

Diamagnetics, repulsion of, §§ 643–646

Diamagnetic particles, reciprocal action of, §§ 695, 696

Dielectric, §§ 36, 447

Dip, line of, § 441

Distribution of electricity on a circular segment of a sphere, §§ 231–248

Distribution of electricity, mechanical value of, §§ 695, 696

—— of magnetic matter necessary to represent the polarity of a given magnet, §§ 473, 474

Distributions of magnetism, solenoidal and lamellar, §§ 504–523

—— of matter, mechanical value of, §§ 561–563

ELECTRICITY, atmospheric, §§ 249–301

—— on the elementary laws of statical, §§ 25–50

—— conductors of, § 68

—— non-conductors of, § 67

—— of a charged conductor rests entirely on its surface, § 68

—— two kinds of, § 58

Electric current, strength of, § 532
—— —— accumulator, on a uniform, §§ 408–411
—— equilibrium, § 66
—— machines founded on induction and convection, §§ 416–425
Electrical density at any point of a charged surface, §§ 69, 93, 138
—— forces, superposition of, § 63
—— influence on an internal spherical conducting surface, §§ 102–105
—— —— on a plane conducting surface of infinite extent, §§ 106–112
—— quantity, § 61
Electrification of the atmosphere, what is known regarding the, §§ 253, 296–301
—— —— how experiments may be made for ascertaining the, §§ 254–262
Electrified bodies, law of force between, § 64
—— surface, repulsion on an element of an, § 88
—— spherical conductors, mutual attraction or repulsion between two, §§ 128–142
Electrometers and electrostatical measurements, Report on, §§ 341–390
—— classification of, §§ 343–385
Electrometer, definition of, § 341
—— absolute, §§ 307–309, 339, 358, 363
—— new absolute, §§ 364–367
—— divided ring, §§ 263–270, 345–357
—— electroscopic, § 305, foot-note
—— long range, §§ 383, 384
—— standard, §§ 379–382
—— portable, §§ 263, 277, 368–378
Electromagnet, definition of, § 434
Electromagnets, §§ 524–554
—— linear, § 536
—— superficial, § 537
—— solid, § 538
Electromotive force required to produce a spark in air between parallel metal plates at different distances, measurement of, §§ 320–340
Electroscope, Bennet's gold-leaf, § 387
—— Bohnenberger's modification of, § 388
Electrostatic force and variations of electric potential, relations between § 337
—— —— produced by a Daniell's battery, measurement of the, §§ 305–319
Electrophorus, reciprocal, § 427
Elements, division of surfaces into, § 79
Equilibrium, electric, § 66
Ellipsoid, attraction of a homogeneous,

on a point within or without it, §§ 21–24
Ellipsoid, uniform motion of heat in an, §§ 11–20

FARADAY'S researches, § 27
—— —— on electrostatic induction, § 36 etc.
—— —— on specific inductive capacity, § 46, etc.
—— Law, experimental illustrations of, §§ 654–664
—— —— deduced from the law of energy, §§ 674–687, 745–750
Ferromagnetic and diamagnetic magnetization, relations of, to the magnetizing force, §§ 664–668
Ferromagnetics, attraction of, §§ 634–642
Field of magnetic force, or field of force, § 605
Force at a point due to a magnet, § 605
—— analogy of, §§ 760–763
Forces experienced by inductively magnetized ferromagnetic or diamagnetic non-crystalline substances, remarks on, §§ 647–653
—— —— by matter under magnetic influence, §§ 723–732
—— —— by solids immersed in a moving liquid, § 733, etc.
"Frequency" electric, § 294

GALVANOMETER, § 341
—— mirror, § 350
Gauss, §§ 187–481
Geometrical slide, § 346
Green, essay on the application of mathematical analysis to the theories of electricity and magnetism, §§ 25, 156, 163, 167, 481
—— potential at a point, § 37, foot-note
—— quotation from, on some experiments by Coulomb, § 234
Guthrie, Professor, extracts from letters to, §§ 741–743

HARRIS on the law of electric force, examination of, § 26
Heat, uniform motion of, §§ 1–24
Heterostatic electrometers, § 385
Holtz's electrical machine, § 429

IDIOSTATIC electrometers, § 385
Images, electric, §§ 127, 208–230
Imaginary electrical points, § 116
—— magnetic matter, §§ 463–475
· Induced magnetism in a plate, §§ 156–162
Induction, magnetic, §§ 604, 624
—— plate, § 357

Inductive action, curved lines of, § 39
—— capacity of a substance, principal, § 611
Inductively magnetized bodies in positions of equilibrium, on the stability of, § 665
—— —— ferromagnetic or diamagnetic non-crystalline substances, remarks on the forces experienced by, §§ 647–653
Insulated sphere subjected to the influence of an electrical point, §§ 89–95
Inverse problems of magnetism, §§ 584–601
Intensity of magnetization, §§ 461, 462
Isothermal surface, § 1
Isotropic, § 604, foot-note

Laplace, § 481
Lamellar distribution of magnetism, characteristic of, § 514
Laws of statical electricity, on the elementary, § 25
—— of magnetic forces, §§ 452–453
Lamé's Memoir on Isothermal Surfaces, § 20
Lettres de *M.* William Thomson, A.M., Liouville, éxtraits de, §§ 208–220
Lines of electric force, §§ 39, 251, 256
—— of magnetic force, § 605
—— of force, diagrams of, §§ 632, 633
Liouville, sur un propriété de la couche électrique en equilibre à la surface d'un corp conducteur, § 163
—— note on the subject of electric images, §§ 221, 230
Lightning, on some remarkable effects of, observed in a farmhouse near Monimail, § 301
Leyden phial, capacity of a, §§ 51, etc.

Magnet, definition of a, § 434
Magnetic agency of the earth on a magnet, § 488
—— axis, §§ 440, 494
—— centre, § 494
—— field, § 605
—— force at any point, total, § 605
—— —— the characteristic of magnetism, §§ 432, 433
—— —— axioms of, § 606
—— induction, determination of the conditions of, § 610
—— —— general problem of, §§ 700–732
—— —— laws of, § 607
Magnetic induction, a principal axis of, § 611
—— inductions, superposition of, § 607

Magnetic moment, §§ 458–460
—— polarity, §§ 443–447
—— shell, §§ 506–512
—— solenoid, §§ 505, 507, 509, 511
—— strength, §§ 454–456
—— susceptibility, § 610
—— permeability, § 628
—— —— analogues of, §§ 625–631, 751–756
Magnetism, mathematical theory of, §§ 430, etc.
Magnetization, direction of, § 462
—— intensity of, § 461
—— intrinsic, § 698
Magnetized matter, mutual actions between any given portions of, §§ 476–501
Mathematical theory of electricity, actual progress in the, § 74
—— —— of electricity, objects of the, § 73
Measurement by electrometer, interpretation of, § 336
Mechanical theory of electricity, demonstration of a fundamental proposition in the, §§ 149–155
—— value of a distribution of electricity on a group of insulated conductors, § 138
Mouse-mill replenisher, § 426
Mutual action between two magnets consists of a force and a couple, §§ 496–501
—— —— between two magnets expressed in terms of a function of their relative position, §§ 502–505

Nicholson's revolving doubler, § 429

Oersted, § 524

Plane conducting surface, electrical influence on a, §§ 106–112
Plücker's hypothesis, § 666
Polar magnet, § 549
—— inductive suceptibility of a, §§ 697–699
—— —— mechanical values of, §§ 564–572
Polarity, § 443
Poles of a magnet, §§ 443, 549
Poisson, Memoirs of, on the mathematical theory of electricity, § 25
—— theory of magnetic induction, § 604
—— quotations from, regarding magnecrystallic action, with explanations, §§ 620, 621
Potential at a point, § 37, foot-note
—— at any point in the neighbourhood of or within an electrified body, § 129, foot-note

Potential, electric, § 335
—— of a magnetic shell at any point, § 512
—— of a closed galvanic circuit of any form, §§ 555–560
Potentials, equality and difference of, § 249, foot-note
Potential-Equalizer, §§ 422–426
Proof plane, §§ 25, foot-note, 35, 330

QUANTITIES of electricity, measurement of, § 328

REPLENISHER, §§ 352, 418–421, 427–429
Resultant electric force at a point, definition of, § 65
—— —— —— due to a uniform spherical shell, vanishes for any interior point, § 78
—— —— —— at any point in an insulating fluid, § 331
—— magnetic force at any point, §§ 479–515

SIZE of Atoms, § 400
Specific inductive capacity, § 45, etc.
Spherical conductors, geometrical investigations with reference to the distribution of electricity on, §§ 75, etc.
—— ——. geometrical investigations regarding, §§ 113–127
—— conducting surface, electrical influence on an internal point of a, § 102
—— surfaces of which the density varies inversely as the cube of the distance from a given point, attraction of, § 90
Solenoidal distribution of magnetism, characteristic of, § 513

Statement of the principles on which the mathematical theory of electricity is founded, §§ 57, etc.
Stratum of air between two parallel or nearly parallel plane or curved metallic surfaces maintained at different potentials, § 338
Strength of electric current, § 532
Superficial density of magnetic matter, §§ 471, 472
"Surface of the earth," definition of, § 250; generally negatively electrified, § 252

TELEGRAPH wire insulated in the axis of a cylindrical conducting sheath, electrostatic capacity of, §§ 54, etc.
Terrestrial electrification, extremely rapid variations of, § 259
—— magnetism, on the electric currents by which the phenomena of, may be produced, §§ 602, 603
Thalén, magnetic susceptibility of iron, § 630
The earth, a great magnet, § 436
The earth's action on a magnet, sensibly a couple, §§ 439, 442
Theory of electricity, on certain definite integrals suggested by problems in the, §§ 166–185
—— of magnetic force, elementary demonstrations of propositions in the, § 669
Tyndall, Professor, correspondence with, §§ 694–696

UNIT strength, § 647, foot-note

VARLEY'S instrument for generating electricity, § 428
Volta connection by flame, §§ 412–415

CAMBRIDGE: PRINTED BY C. J. CLAY, M.A. AND SON, AT THE UNIVERSITY PRESS.

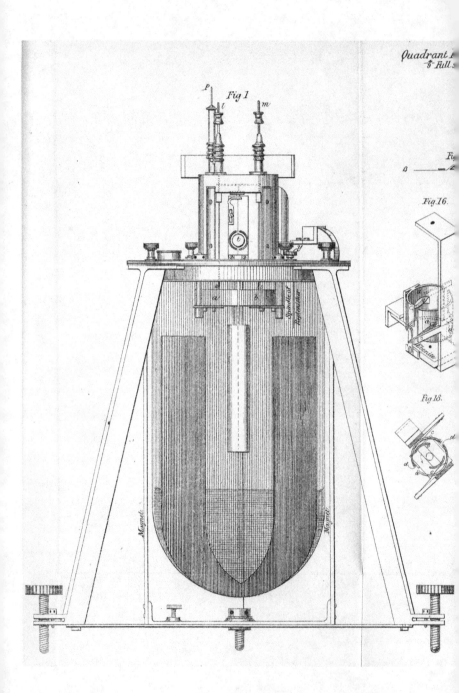

Fig 1

P   l   m

Fig
B

Fig.16.

Fig 18.

Spindle of Reproducer

Magnets.   Magnets.

Plate 1.

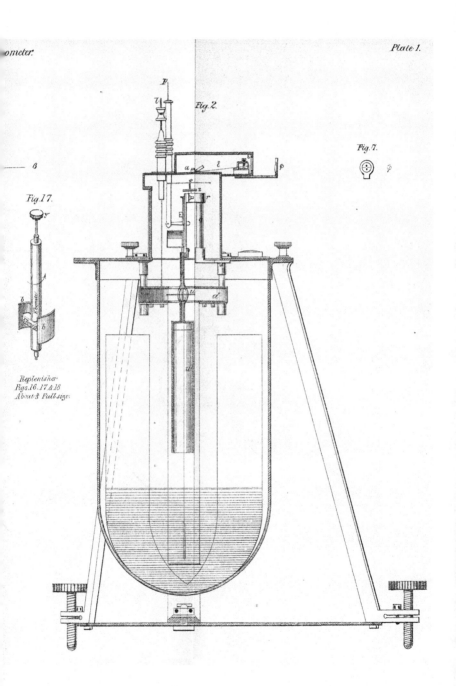

Fig.2.

Fig.7.

Fig.17.

Replenisher
Figs.16.17.&18
About ⅔ Full size.

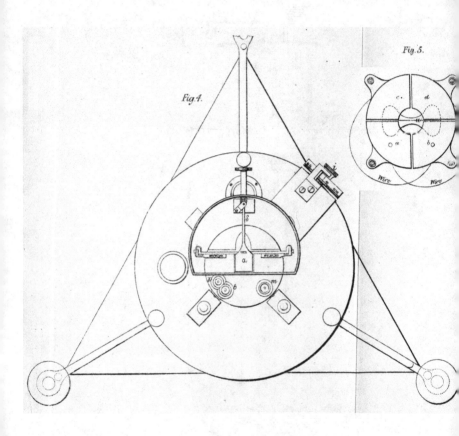

Fig. 4.

Fig. 5.

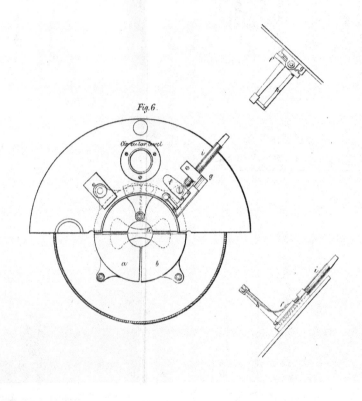

Fig. 6.

Circular level

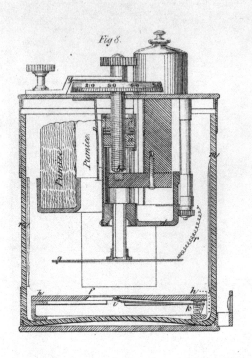

Fig 8.

2|0   5|0

Pumice.

Pumice.

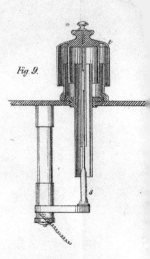

Fig.9.

Portable Electrometer
Figs 8. 9 & 10
⅔ Full size.

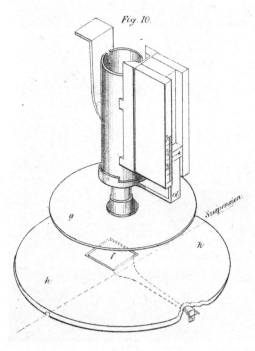

Fig. 10.

Suspension.

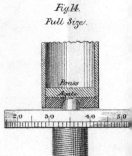

Fig.14.
Full Size.

2|0   3|0   4|0   5|0

Fig 12.

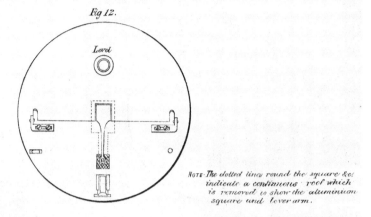

NOTE. The dotted lines round the square &c. indicate a continuous roof which is removed to show the aluminum square and lever arm.

Fig.13.

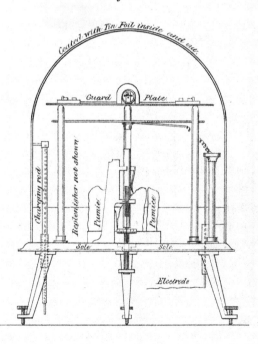

Fig 12.

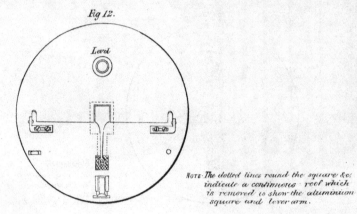

NOTE. The dotted lines round the square &c.
indicate a continuous roof which
is removed to show the aluminium
square and lever arm.

Fig. 13.

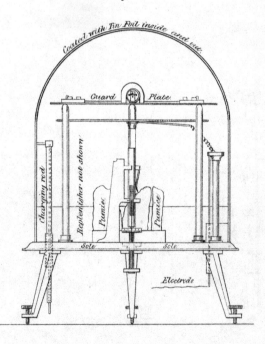

Plate 2.

Absolute Electrometer:

Fig. 2.

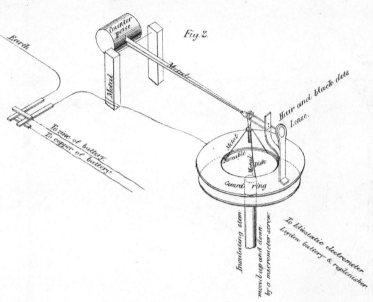

Earth

Counter poise

Metal

Metal

To zinc of battery.

To copper of battery.

Hair and black dots

Lense.

Metal

Moveable

Metal disk

Guard ring

Insulating stem

moves up and down
by a micrometer-screw.

To Idiostatic electrometer
Leyden battery & replenisher

Long Range Electrometer.
⅛ Full size.

Fig 15.

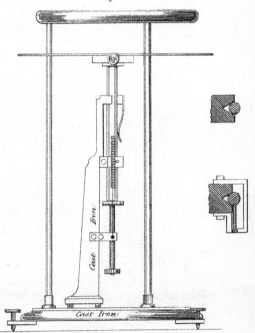

Cast Iron.

Cast Iron.

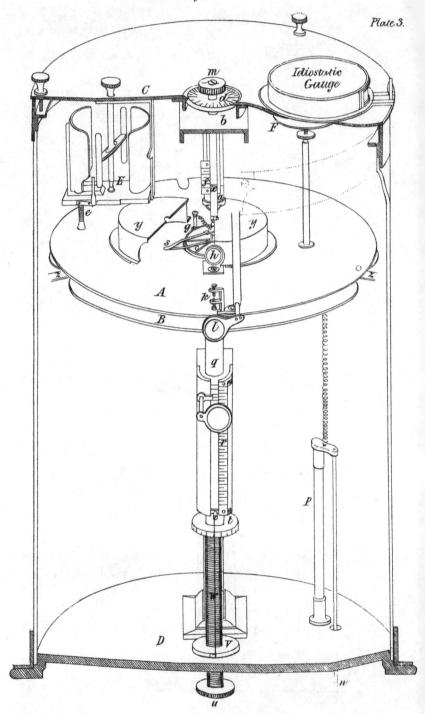